COMPUTATIONAL ELECTROMAGNETIC-AERODYNAMICS

IEEE Press
445 Hoes Lane
Piscataway, NJ 08854

COMPUTATIONAL ELECTROMAGNETIC-AERODYNAMICS

JOSEPH J.S. SHANG
Emeritus Research Professor of Wright State University
Emeritus Scientist of US Air Force Research Laboratory

IEEE Series on RF and Microwave Technology

IEEE
IEEE PRESS

WILEY

For general information on our other products and services or for technical support, please contact our Customer Care Department within the United States at (800) 762-2974, outside the United States at (317) 572-3993 or fax (317) 572-4002.

Wiley also publishes its books in a variety of electronic formats. Some content that appears in print may not be available in electronic formats. For more information about Wiley products, visit our web site at www.wiley.com.

Library of Congress Cataloging-in-Publication Data is available.

ISBN: 978-1-110-15592-8

10 9 8 7 6 5 4 3 2 1

CONTENTS

PREFACE

Tremendous amounts of research results have been accumulated to the information pool for plasmadynamics and plasma, which is the most common state of matter in the universe by either mass or volume. The underlying physics occurs mostly at the molecular and atomic scales and often under an extremely high enthalpy condition or in a strong electromagnetic field; these phenomena must be analyzed by physics-based modeling. The research progress faces formidable challenges because the compositions of non–equilibrium-ionized gas are always in transient quantum states where the fundamental chemical-physics processes are the least understood. Furthermore, the validating experimental data are also very sparse.

In spite of these limitations, thousands of conference papers and articles on this subject are still persistently releasing to the open literature annually. Meanwhile, the electromagnetics is increasingly recognized to broadening physical dimensions for interdisciplinary science and technology to meet engineering requirements. For the rapidly advancing scientific field, outstanding and fundamental technical references are published mostly in the period from the middle 1960s to later 1980s. Although all these illuminating accomplishments have withheld the test of time but have not been updated by converting information into knowledge for practical engineering applications. It is appropriate and timely that summarization of these impressive scientific accomplishments and a systematic review on the progress of this scientific discipline are needed for future development.

The inherent nature of interdisciplinary computational electromagnetic-aerodynamics links five major scientific disciplines from aerodynamics, electromagnetics, chemical-physics kinetics, plasmadynamics, to computational modeling and simulations. The knowledge of each individual discipline has been developed over centuries and numerous treatises on these subject areas have been published and firmly established for understanding. There are also some path-finding articles in classic magnetohydrodynamics that detail the magnetic field–dominated astrophysics around 50 years ago, but none have been documented for computational electromagnetic-aerodynamics to reach a comprehensive status in providing a practical working knowledge. For this reason, an attempt is made here to integrate these pertaining disciplines to supply a working knowledge for engineering and to suggest the best practice in applications. Hopefully this effort can be a viable reference for professional development and graduate studies.

The text is organized in 10 chapters, with the first three chapters consisting of introductions to fundamental knowledge of plasmadynamics, chemical-physics of ionization, formulations of classical magnetohydrodynamics, and their extension to low magnetic Reynolds number electromagnetic-aerodynamics conditions. In fact, most plasma-based flow control actuators, high-speed flows of interplanetary reentry, and ion thrusters in space exploration are operating under this environment.

For fundamental knowledge, a brief review of unique features of the plasma medium is presented in terms of the Debye shielding length, Coulomb and Lorentz forces, Joule heating, plasma frequency, and plasma waves. All these unique characteristics of ionized gas in an electromagnetic field are intermediately related to outstanding engineering applications of computational electromagnetic aerodynamics. Then the statistical thermodynamics that bridges the microscopic to macroscopic thermodynamics is introduced for the ionization processes. The mechanisms of photoionization, thermal excitation, and electron impact ionization are discussed on the basis of the most recent research findings.

The hierarchal relationship among fundamental formulations for plasmadynamics and their valid domains of approximation is outlined and reviewed next. The basic assumptions for the classic magnetohydrodynamics formulation are articulated, and detailed derivations from ideal to full magnetohydrodynamics equation, as well as the low magnetic Reynolds number equations system, are delineated. These governing equations, in fact, define the limitations and physical fidelity of all computational simulations. From the eigenvalues and eigenvectors analyses of the magnetohydrodynamics equations, correlations are revealed between propagating speeds of Alfven, entropy, contact surface, slow and fast plasma waves.

A peculiar property of plasma in an electromagnetic field is that it does not exist as an unrealistic one-dimensional phenomenon. Therefore, the shock jump condition must be formulated, at least by tangential and normal components across a discrete wave. The simplest shock in the electromagnetic field is essentially a two-dimensional oblique wave. When the macroscopic conservative equation is integrated across the shock, and it requires that the normal components of the magnetic flux density and the tangential component of the electric field intensity are continuously across any interface. As a consequence, the Rankine–Hugoniot relationship of gas dynamics across a shock is significantly modified in an ionized gas to become a unique feature of the plasma medium.

The next three chapters of the book provide in-depth and specific descriptions of numerical algorithms and procedures for solving Maxwell's equation in the time domain for computational electromagnetics, plasma wave propagation, and fluid dynamics. All these detailed numerical algorithms and computational procedures are the shared knowledge for the interdisciplinary computational modeling and simulation science and technology. The highlights are especially placed on overlapping areas of technical issues and realizable application opportunities.

Two major foci of the present objective are to integrate the interlinking computational modeling and simulation techniques of aerodynamics and electromagnetics. To make the procedure tractable, the computational approach is further restricted to the continuum regime so the approach is built on the frameworks of the time-dependent Maxwell equations and the compressible Navier–Stokes equations. It should be recognized that the Maxwell equations constitute the hyperbolic partial differential system and belong to a class of bondless initial-boundary-value problems. This unique feature poses a fundamental dilemma for all numerical analyses that must be carried out in a finite-size computational domain. However, the characteristic-based formulation can alleviate this constraint by achieving an invariant dependent variables grouping in the domain of dependence. This formulation permits all propagating electromagnetic waves exit through numerical boundary unimpeded. The computational electromagnetics technique is applying to the microwave propagation in waveguide, plasma wave diffraction and refraction, as well as the plasma diagnostic technique by microwave attenuation beyond Langmuir probes.

The last four chapters are focused on the modeling and simulation techniques for computational electromagnetic-aerodynamics. The most current progress in computational electromagnetic-aerodynamics is summarized for multifluid models, including chemical kinetics by nonequilibrium thermal excitations, chemical physics by electron impact ionization, transport property by gas kinetics, and numerical algorithms. The emphases are on the physical fidelity of computational models that describe the behavior of the direct current charge, dielectric barrier charges, microdischarge, and their interactions with aerodynamics through the mechanisms of Joule heating and Lorentz forces. Transport property by drift motions and diffusions of the charged particle is also described by the classical drift-diffusion theory. The interface boundary conditions between the chemically reacting media and solid surface are derived together with the requirements on electrodes in an externally applied electric field. The latter are described in detail in conjunction with the discharging electric circuit equation. On this frame of analysis, the secondary emission of electrons from the cathode, together for the self-limiting feature of dielectric barrier discharge by preventing transit to arc, is unambiguously illustrated. For the thermal ionizing model, the models of internal degrees of freedom for vibration relaxation and electron excitations are also described at practicing level. Some of the innovative numerical procedures are also included as bench marks for future progression.

It is important to acknowledge that the electromagnetic effect in most flow control applications appears only as a small perturbation. In general, the relative magnitude of aerodynamic inertia overwhelms the electromagnetic force by a thousand fold. In terms of magnetic Reynolds number for an electromagnetics–aerodynamics field that is measured by the ratio of inertia and the product of electric conductivity and magnetic permittivity $R_m = ul/(\sigma\mu_m)^{-1}$, its magnitude is much smaller than unity. The plasma-based actuators are therefore observed to be the mostly effectiveness for flow control at a critical point of a bistable aerodynamic state or to the high-frequency, low-amplitude oscillations. Thus, aerodynamic phenomena are singling out for possibly enhancements by infusing electromagnetic effects. For this purpose, the aerodynamic bifurcations of viscous–inviscid interaction, including flow separation, self-sustained oscillation, vortical dynamics, and hydrodynamic instability leading to turbulent flow, are specifically included. In this approach, the rudimentary numerical algorithms of computational electromagnetics and computational fluid dynamics are introduced and the best practices for computational electromagnetic aerodynamics are also encompassed.

In the last two chapters, the presentations of plasma-based actuators for flow control are divided into two major categories according to the mechanisms, whether via Joule heating or by Lorentz force. For the former, the experimentally validated applications in virtual leading-edge strake and a virtual variable inlet cowl are described. Meanwhile, viable amplifications to electromagnetic–aerodynamic interaction by viscous–inviscid interaction and externally applied magnetic field are demonstrated. The actuator by Lorentz force for stagnation point heat transfer mitigation, remote energy deposition, and flow control via dielectric barrier discharge for lifting surface enhancement are elaborated. A quantification for the flow control effectiveness by dielectric barrier charge is given, and the potential application to hydrodynamic stability is revisited. The successful applications of ion engine or ion thruster by electrostatic force and plasma micro jet for enhancing combustion stability are also included to illustrate the possible future explorations.

Here, I would like to express my personal gratitude for lifelong association in the vast arena of science with Professors Robert W. MacCormack of Stanford University, David I. Gottlieb (deceased) of Brown University, and Joseph L. Steger (deceased) of University of California at Davis. I am equally indebted to the interactions and collaborations with Professor Sergey T. Surzhikov of Russian Academy of Sciences and my colleagues; Drs. Wilbur L. Hankey, Miguel R. Visbal, and Datta V. Gaitonde of the Center of Excellence for Computational Simulation, United States Air Force Research Laboratory, as well as Professor George P. Huang of Wright State University and Professor Hong Yan of North Western Polytechnic University of China.

I have gratefully received editorial supports from Professor Frank K. Lu of University of Texas at Arlington and James A. Menart of Wright State University. Their generous act of friendship and meticulous efforts are truly a humbling experience to me. At last, but by no mean the least, I earnestly acknowledge the unfailing devotion from my wife Karen. I dedicate this book to her.

Joseph J.S. Shang

1

PLASMA FUNDAMENTALS

INTRODUCTION

Plasmas are the most common state of matter both by mass and by volume, although most of them are rarefied and intergalactic; nevertheless, plasma is one of the four fundamental states of matter. Langmuir (1928) first coined the word *plasma* in 1928. Plasma is an electrically conducting medium and has the basic tendency toward global electrical neutrality. The most fundamental definition of this state of matter is that the pairs of closely positioned positively and negatively charges in the plasma shield each other from an externally applied electric potential. This characteristic charge separation distance is the Debye shielding length and is the smallest dimension compared with all other macroscopic scales. Within this shielding sphere, the paired charged particles are never free from each other but interact permanently.

Plasma does exist not only in gas and liquid but also in a solid conductor. The electrons are bound but still can move within atomic or molecular structures between collisions. Although mass motion of the charges does not generally take place, when an external electromagnetic field is applied, the dynamic effect can always be observed in conduction, Hall Effect, and polarization.

Plasmas, like gases, have a negligible shear stress and therefore do not have a definite shape or volume. However, the electric-conducting feature makes it respond to electromagnetic fields, so plasma can be confined either by a solid container or by a magnetic field. Plasma also appears in a wide range of formations such as filaments, microdischarges, multiple layers in the presence of shock waves, and cellular structure

Computational Electromagnetic-Aerodynamics, First Edition. Joseph J.S. Shang.

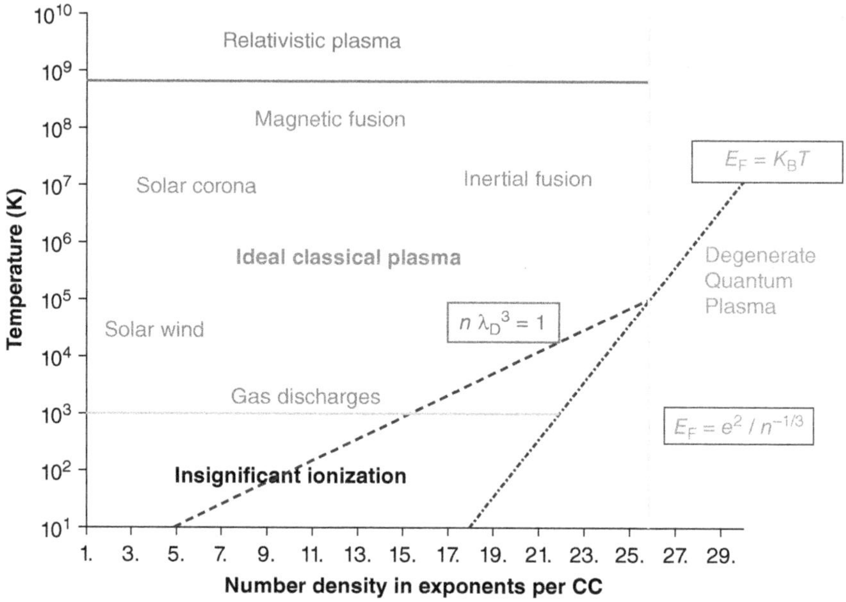

FIGURE 1.1 Classification for types of plasma.

in space. The spontaneous formations of unique spatial features on a wide range of length scales manifest the complexity of plasma. Because the electric conductivity is dependent on its charged particle number density and thermodynamic condition, the classification of plasma is usually by the electron number density and temperature in electron volts or the static temperature (Figure 1.1).

The topics of degenerated quantum plasma and relativistic plasma are beyond the scope of computational magneto-aerodynamics. For this reason, the present discussion is focused on the plasma domain that exists around atmospheric pressure, and on mixed thermal and nonthermal plasma. According to convention, the thermal condition of plasma is based on the relative temperatures of the electron, ion, and neutral components. By nonthermal plasma is meant that the ions and neutral particles have a much lower temperature than the electrons. For most gas discharges generated by electric currents, the electron temperature is at most 3 eV (around 10^5 K), while the ions and neutrals are at near room temperature. When ionization is achieved by a strong shock compression in high temperature or earth reentry, the temperatures of the plasma composition are on the same order, around 10,000 K or higher, but are not always equilibrated so as to remain in thermodynamic nonequilibrium (Jackson, 1975). Therefore, the plasma of interest is generally limited to an electron number density up to $10^{16}/cm^3$ and an overall temperature lower than 10^6 K.

From the discussion given earlier, plasma is an electric conducting medium but is globally neutral and characterized by a minimum macroscopic length scale between charges of opposite polarity known as the Debye shielding length. A strong electrostatic force exists between the paired charges; any small perturbation to the charges separation distance will trigger a high-frequency oscillation by the restoring force. This oscillatory motion is referred to as the plasma frequency that is distinguished from the lower frequency oscillation involving mass motion. In the presence of a sufficient number of charged particles, the oscillation also leads to a number of unique collective behaviors that are exemplified by the strong response to electromagnetic fields. Plasmas do not naturally conform to their surroundings but will respond to local and distant conditions by the Coulomb force and Lorentz acceleration. The other unique feature of quasi neutrality is that plasma stores the inductive energy and contributes to resistance and inductance when the current circuit forms a closed loop. Finally, the disturbance communication in plasma not only is through collision processes but also conveys by transverse wave with phase velocities greater than the speed of sound (Spitzer, 1956; Celexoix, 1960).

All physical phenomena of plasma are governing by the Maxwell and Navier-Stokes equations to become an interdisciplinary subject in macroscopic scales. Once the plasma is treated as a continuum, only the global thermodynamic and kinematic properties are needed to describe its behavior through a link between microscopic and macroscopic dynamics. This approach is accurate when the microscopic structure can be linked approximately to macroscopic motion by the Maxwell–Boltzmann distribution. However, the detailed wave–particle interaction will be unresolved.

1.1 ELECTROMAGNETIC FIELD

An electromagnetic field of plasma always consists of coexisted electric and magnetic fields. Charged particles interacting among themselves and with externally applied fields create induced components; the total field is a sum of applied and induced components. The magnitude of induced components can vary greatly in comparison with applied fields.

According to Coulomb's law, the electrostatic force between charged particles is a collinear force along the unit space vector between two charges separated by a distance r_{ij},

$$F_{ij} = \frac{1}{2\pi\varepsilon_o} \frac{q_i q_j}{r_{ij}^2} \hat{r}_{ij} \tag{1.1}$$

The magnitude of the force is directly proportional to the product of the two charges and inversely proportional to the square of the distance between them. In SI units, the electric permittivity ε_o has a value of 8.85×10^{-12} Farad/m (Krause, 1953). The charges q_i and q_j are measured in coulombs, which attract to each other of opposite polarity but repulse if the same. The resulting force per unit charge is defined as the electric field intensity E. Through this definition, the electrostatic force can be given as

$$F_i = Eq_i \tag{1.2}$$

The SI unit of electric field intensity is the newton per coulomb. However, in practical application, an equivalent unit is often given in volt per meter. In the case of multiple point charges, the principle of superposition determines the force on a particular charge. The entire electric field of a charged space is the vector sum of the field from all the individual source charges,

$$E_i = \frac{1}{4\pi\varepsilon_o}\Sigma\frac{q_i}{r_{ij}^2}\hat{r}_{ij} \tag{1.3}$$

In an electrically conducting medium, the charge particle movement induces current and exerts additional force on each other. In an isolator, the molecules of the dielectric material will polarize to reduce the net local field intensity. This effect is described by a dielectric constant κ and Coulomb's law becomes

$$F_{ij} = \frac{1}{4\pi\varepsilon_o\kappa}\frac{q_i q_j}{r_{ij}^2}\hat{r}_{ij} \tag{1.4}$$

Electric charge in motion constitutes a conductive electric current. In metallic conductors, the charge is carried by electrons with an elementary negative charge of 1.61×10^{-19} coulombs. In liquid conductors, such as electrolytes, the charge is carried by both positive and negative ions. The electric field compels the free charges into continuous motion and results in an electric current, which can be defined in terms of the electric flux vector per unit area.

The conductive current density considered in here is different from the convective and displacement current density. The convective current does not involve conductors and consequently does not satisfy Ohm's law. The current flows through an isolating medium such as the liquid, rarified gas, or vacuum, and the best example is an electron beam within a vacuum tube (Jahn, 1968). The displacement current arises from the time-varying electric field and is introduced by Maxwell to account for the generation of a magnetic field when the conductive current is zero. Without the concept of the displacement of current, the electromagnetic wave propagation would be impossible.

The conductive electric current density J has a physical dimension of coulomb per second or ampere,

$$J = \Sigma n_i q_i u_i = \Sigma \rho_i u_i \tag{1.5}$$

where n_i denotes the charge particle number density, u_i is the averaged velocity of the charged particles, and the current density ρ_i is simply the sum of the total electric charges per unit volume of species i.

A magnetic field is generated by an electric current and the orientation of this field is defined by the right-hand rule. The current may be due to electromagnetic fields by a permanent magnet, an electron beam, or a conductive current in current-carrying wire. The differential magnetic field intensity is governed by the Biot–Savart law for magnetostatics, which is the counter part of Coulomb's law for electrostatics. The field intensity is proportional to the product of a differential current element Idl and

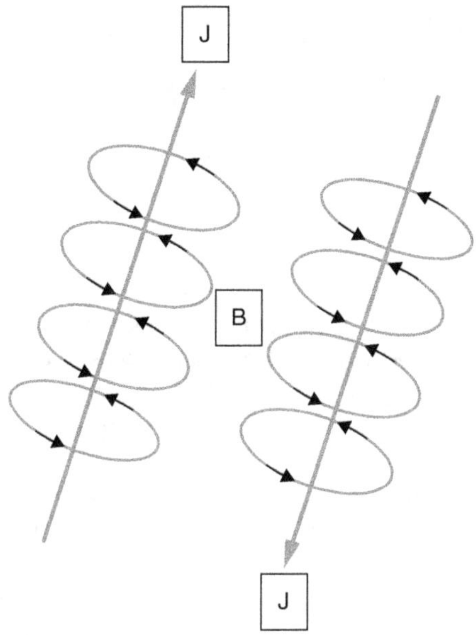

FIGURE 1.2 Induced magnetic field by electric current.

sine of the angle between the element and the line connecting the point of interest and is inversely proportional to the square of the mutual distance. The induced magnetic field line is continuous. In vector form, the Biot–Savart's law gives (Figure 1.2)

$$dH = \frac{Idl \times \hat{r}_{ij}}{4\pi r_{ij}^2} \tag{1.6a}$$

A current-carrying wire produces a magnetic field perpendicular to the electric current, and the magnetic flux density B has the physical unit of newton per amp-meter. In analogy to electrostatics, the intermediary B field is the magnetic flux density or is also called the magnetic induction. The connection between the magnetic field strength H and the magnetic flux density B in an isotropic medium is the constitute relation, $B = \mu B$, and μ is referred to as the magnetic permeability. In free space, it has the dimensional value of $\mu = 4\pi \times 10^{-1}$ hennery/meter.

If two current-carrying wires are brought into the vicinity of each other, each wire is surrounded by two individual magnetic fields leading to a force that acts on the wires. When the wires are carrying current in the same direction, the wires are attracted to each other, while wires carrying current move in opposite directions, the wires are repelled. The incremental force existing between the elements of the current path L_i and L_j is

$$dF_{ij} = \frac{\mu}{4\pi} \frac{J_i J_j}{r_{ij}^2} [dl_i \times (dl_j \times \hat{r}_{ij})] \tag{1.6b}$$

By definition of the magnetic induction is

$$dB_{ij} = (\mu J_j/4\pi)\left(dl_j \times \hat{r}_{ij}/r_{ij}^2\right) \tag{1.6c}$$

Analogous to that of electrostatics, the force on a current element becomes

$$dF_{ij} = J_i dl_i \times dB_{ij} \tag{1.7}$$

Basically, there are but two modes of the electromagnetic body force in an electrically charged medium that can exert on the charged particles. Figure 1.4 is the interaction of the electric field with the free charge density of the medium $F_e = \rho_e E$, which is known as the electrostatic force. The interaction of an externally applied magnetic field with the electric current that is driven by a force within the medium is $F_m = J \times B$ and it is the well-known Lorentz force.

As a consequence of a moving charge q in a steady and uniform magnetic field with an applied magnetic flux density B will be pushed by a force normal to B, namely,

$$F = qu \times B \tag{1.8}$$

Under the circumstance, the normal velocity component of the charge particles is always restricted in the plane perpendicular to a constant and steady magnetic B. The velocity component $u_{\|}$, which is parallel to B, will not be affected by the magnetic field but the velocity component u normal to B will be generated. Therefore, the charge particle motion in the normal plane to the magnetic field must follow a circular path and the charged particles and will execute a spiral motion. The gyro radius can be determined easily by the balance of the centrifugal acceleration and the electromagnetic force exerted on a singly charged particle, $eu_{\perp}B = mu_{\perp}^2/r_b$. The radius of the circular motion of a group of charged particles in the plane perpendicular the magnetic field is $r_b = mu_{\perp}/qB$, with an angular gyro velocity $\omega_b = qB/m$. The gyro radius is often referred to as the Larmor or cyclotron radius and can be evaluated on the basis of a single electron, $r_b = mu_{\perp}/eB$. Similarly, the angular gyro velocity has also been widely referred to as the Larmor or cyclotron frequency. The charge particle trajectory in three-dimensional electromagnetic field is then a helix with its axis parallel to B and a pitch of $2\pi u_{\|}/\omega B$.

The gyro frequency follows directly from its single-particle motion and can be given as

$$\omega_b = \frac{eB}{m} = 1.76 \times 10^{11} B\,(\text{rad/s}) \tag{1.9}$$

where the magnetic flux density B needs to be in the SI unit of webers/m^2.

The equation of motion for a charged particle in a uniform electromagnetic field is then

$$F = q(E + u \times B) = m\frac{du}{dt} \tag{1.10}$$

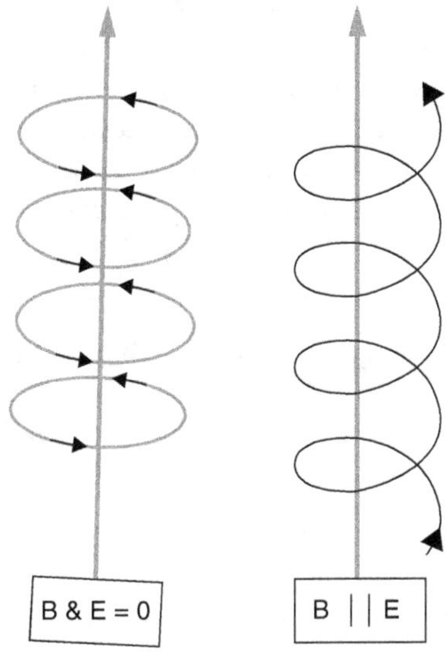

FIGURE 1.3 Trajectory of charge in electromagnetic field.

The electromagnetic field of moving charged particles therefore is always associated with a combination of electric and magnetic fields. A rectilinear acceleration is aligned with the externally applied electric field and a gyration intrinsically revolves around an applied magnetic field vector. In the absence of an electric field, the angular acceleration is restricted in the plane perpendicular to the magnetic field. When the electric field vector is applied in addition to the magnetic field, the charged particles will execute a persisted spiral trajectory as shown in Figure 1.3. The orientation of the resultant helical trajectory is dictated by the direction of the applied electric field intensity.

It becomes clear that the stationary charges generate an electrostatic field; in fact, the remotely acting Coulomb force is the genesis of the intrinsic characteristics of plasma in the Debye shielding length and the plasma frequency. While the direct current produces magnetostatic field, the dynamics of the electromagnetic field must more often be accompanied by a time-varying electric current density. In short, within a static electromagnetic field, the electric and magnetic fields are independent from each other whereas in dynamic electromagnetic fields the two fields are interdependent.

1.2 DEBYE LENGTH

To understand and to characterize the macroscopic behavior of a collection of charged particles, two fundamental parameters that associate with the electric properties of the ionized gas, namely the Debye length and plasma frequency, are paramount. The basic property of plasma is its tendency in maintaining electric neutrality (Langmiur, 1929).

This particular state requires an enormously large electrostatic force between an electron and an ion. A quantitative estimate of this dimension over which a deviation from charge neutrality may occur can be obtained by the Gauss law for electric field $\nabla \cdot D = \rho$. The electric displacement is related to the electric field intensity by a constitutive relationship $D = \varepsilon E$, where ε is the electric permittivity by the fact that the electric field intensity is inversely proportional to the gradient of the electric potential $E = -\nabla\varphi$. The resultant equation is known as the Poisson equation of plasmadynamics

$$\varepsilon\nabla^2\varphi(r) = -\rho_e = -e(n_i - n_e) \tag{1.11a}$$

In a thermodynamic equilibrium condition, the number densities of electrons and ions are given by the Boltzmann distributions,

$$n_e = ne^{e\varphi(r)/\kappa T} \quad \text{and} \quad n_i = ne^{-e\varphi(r)/\kappa T} \tag{1.11b}$$

Substituting these expressions into the Poisson equation and reducing the equation to one dimension in space yield

$$\varepsilon\frac{d^2\varphi}{dx^2} = en_\infty[\mathrm{e}^{(e\varphi/\kappa T_e)} - 1] \tag{1.12}$$

Expanding the exponential function in Taylor series in ascending order of $e\varphi/\kappa T$ yields

$$\varepsilon\frac{d^2\varphi}{dx^2} = en_\infty\left[\frac{e\varphi}{\kappa T_e} + \frac{1}{2}\left(\frac{e\varphi}{\kappa T_e}\right)^2 + \cdots\right] \tag{1.13}$$

Although the magnitude of $e\varphi/\kappa T$ may not be necessarily negligible over an entire domain between charged particles, the potential is known to drop rapidly with respect to the separation distance. Therefore, it is permissible to keep only the leading term, then

$$\varepsilon\frac{d^2\varphi}{dx^2} = \frac{e^2 n_\infty}{\kappa T_e}\varphi \tag{1.14}$$

This leads to the definition of the Debye shielding length

$$\lambda_d = \left[\frac{\varepsilon\kappa T_e}{e^2 n}\right]^{1/2} \tag{1.15}$$

The electric potential has been found to exhibit an exponentially decay behavior as a function of distance between charged particles,

$$\varphi = \varphi_o \exp(-|x|/\lambda_d) \tag{1.16a}$$

In practical applications, the Debye length can be calculated easily by

$$\lambda_d = 69(T/n)^{1/2}\text{m},\ T(\text{K});\quad \lambda_d = 7430(\kappa T/n)^{1/2}\text{m},\ T(\text{ev}) \tag{1.16b}$$

These simplified results for the Debye length are evaluated by the following constants: electron volt (ev) 1.1604×10^4 K, Boltzmann constant 1.3807×10^{-16} erg/K, elementary charge 4.8032×10^{-10} coulombs, and electric permittivity of free space 8.8542×10^{-12} farad/m.

The typical value of the Debye length in a magnetohydrodynamics generator is in the order of magnitude of the electron mean free path $O(10^{-6}\text{–}10^{-7}$ m) at a temperature of 2500 K and charge number density of $10^{20}/\text{m}^3$. In plasma generated by electronic impact at standard atmospheric conditions, the Debye length is 1.14×10^{-7} m.

Another way to estimate the separation distance between charged particles Δx is by recognizing that this position is the neutral state of the balanced average kinetic energy of the electron and the electric potential generated by the charged particles. The averaged separation distance can be obtained by integrating the one-dimensional Poisson equation of plasmadynamics

$$\frac{d^2\varphi}{dx^2} = -\frac{q}{\varepsilon} = \frac{n_e e}{\varepsilon} \tag{1.17}$$

Integrate consecutively with respect to x and set the farfield electric field to 0 ($E_o = 0$) to yield

$$\begin{aligned}\frac{d\varphi}{dx} &= \frac{n_e e}{\varepsilon}x + E_o \\ \Delta\varphi &= \frac{n_e e}{2\varepsilon}\Delta x^2\end{aligned} \tag{1.18}$$

The average energy needed to separate the charged particles by a distance Δx must overcome the total thermal potential. The mean separation distance between the charged particles is then

$$\begin{aligned}\frac{\kappa T_e}{2} &= \frac{n_e e}{2\varepsilon}\Delta x \\ \Delta x &= \left[\frac{\varepsilon\kappa T_e}{n_e e^2}\right]^{1/2}\end{aligned} \tag{1.19}$$

This charge separation distance is exactly the Debye length derived previously. The temperature of the free-electron component of the gas T_e is not necessarily equilibrated with the ions or neutrals. The Debye length has always been adopted as an index of the typical charge separation distance in plasmas that are sustained by the random thermal energy of the electrons. For this reason, the Debye length is often

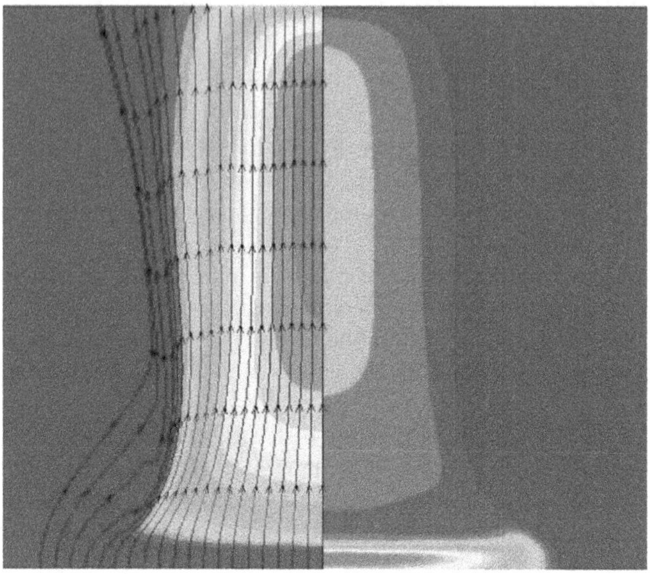

FIGURE 1.4 Cathode layer in a direct current discharge.

interpreted as a physical dimension for shielding the plasma from the effects of an externally applied electric field.

A most illustrative phenomenon of Debye shielding is the plasma sheath. The charge neutrality of plasma does not prevail in the immediate adjacent region to a solid surface, especially over electrodes of an applied electric field, and is the so-called plasma sheath. The Debye length gives the order of magnitude of the thickness of the actual sheath that separates the mass of the neutral plasma from the solid wall. This physical characteristic is easily identifiable in the cathode and anode layers of a glow discharge that is generated by a strong electric field (Figure 1.4). A major contributing factor to this behavior is the recombination mechanisms of chemical kinetics by depleting the charges of ionized gas when striking the surface. The electron absorption and emission properties of the solid surface also play an important role. This phenomenon is closely related to the catalyticity of the surface material.

1.3 PLASMA FREQUENCY

Since plasma has a tendency to be macroscopically neutral, it will always tend to return to its neural equilibrium state after a local perturbation. Unavoidably in this process, the inertia of electrons will overshoot and oscillate around the equilibrium condition with a characteristic rate known as the plasma frequency. The plasma frequency is a measure of the time scale of restoring forces that exerts on a displaced electron returning within Debye length to an ion and keep plasma to an electrically neutral state. In the absence of magnetic field and without externally applied thermal perturbation, the coupled ion is considered in a fixed location in uniform plasma that extends to infinite (Mitchner et al., 1973). The conservation equations of electron motion are

$$\begin{aligned} &\frac{\partial n_e}{\partial t} + \frac{d(n_e u_e)}{dx} = 0 \\ &mn_e \left[\frac{\partial u_e}{\partial t} + u_e \frac{\partial u_e}{\partial x} \right] = -en_e E \end{aligned} \tag{1.20}$$

The compatible electric field E can be determined by the classic Poisson equation of plasma dynamics

$$\varepsilon \frac{\partial E}{\partial x} = e(n_i - n_e) \tag{1.21}$$

The system of equations can be solved easily with a linearized approach by setting the known initial conditions and considering the displaced electron as a small perturbation,

$$n = n_o + \varepsilon n_1, \quad u_e = u_e + \varepsilon u_1, \quad E = E_o + \varepsilon E_1, \quad \varepsilon \ll 1 \tag{1.22a}$$

The resultant linearized governing equations to O(ε) become

$$\begin{aligned} &\frac{\partial n_1}{\partial t} + u_1 \frac{dn_1}{dx} = 0 \\ &m\frac{\partial u_1}{\partial t} + n_o \frac{\partial u_1}{\partial x} = 0 \\ &\varepsilon \frac{\partial E_1}{\partial x} = -en_1 \end{aligned} \tag{1.22b}$$

The perturbed oscillating variables are naturally definable by simple harmonic functions

$$u_1 = u_1 e^{i(\kappa x - \omega t)}, \qquad n_1 = n_1 e^{i(\kappa x - \omega t)}, \qquad E_1 = E_1 e^{i(\kappa x - \omega t)} \tag{1.22c}$$

Substituting these into the governing equations and eliminating n_1 and E_1 yields

$$-im\omega u_1 = -i\frac{n_o e^2}{\varepsilon\omega} \tag{1.23}$$

From which $\omega^2 = \frac{n_o e^2}{m\varepsilon}$. Thus the plasma frequency is

$$\omega_p = \left[\frac{n_o e^2}{m\varepsilon}\right]^{1/2} \quad \text{rad/s} \tag{1.24}$$

The practical approximate formula of the corresponding circular frequency in SI units is

$$\nu_p = \frac{\omega_p}{2\pi} \approx 8.97\sqrt{n_e}\ (\text{rad/s}) \tag{1.25}$$

The pertinent parameters for the plasma frequency calculation are the electric permittivity in free space of 8.8542×10^{-12} farad/m, the elementary charge $e = 1.6022 \times 10^{-19}$ coulombs, and the electron mass m of 9.1095×10^{-25} g (Raizer, 1991).

As an example, the electron number density in an electron-impact discharge has a value of around $10^{18}/\text{m}^3$ or $10^{12}/\text{cm}^3$. The plasma frequency yields a value of 8.7×10^9 Hz, which is within the H and X bands of the microwave frequency with the corresponding wavelength range between 2.4 and 4.25 cm. The frequency, however, is lower than the typical average electron collision frequency within a plasma generator of $2 \times 10^{11}/\text{s}$.

1.4 POISSON EQUATION OF PLASMADYNAMICS

The electrons are in thermodynamic equilibrium and, according to the statistic mechanics, have the Boltzmann distribution, $n_e = ne^{e\varphi(r)/\kappa T}$. The potential energy of an electron at a position r is $-e\varphi(r)$. Similarly, under known equilibrium condition, the ions retain the Boltzmann distribution, $n_i = ne^{-e\varphi(r)/\kappa T}$. The potential energy distribution for plasma is governed in general by the Poisson equation. Therefore, this equation of plasmadynamics is a fundamental formula in describing the collective behavior of plasma in a continuum. In this formulation, the individual charged particles are no longer recognized but to be described by the charge number density ρ_e to have a physical unit of coulomb per cubic meter. The sum of charges in an elementary volume can be given as

$$\rho_e = \lim_{\Delta v \to \varepsilon} \frac{\sum_i ze}{\Delta v} = e\sum_i n_i z_i \tag{1.26}$$

where n_i is the number density of the charge species and z_i denotes the number of elementary charges carried by the particles.

The total electric field generated by a group of charges can be evaluated by the principle of superposition. The integrated surface electric field over control surface must be balanced by the total number of charges within the control volume. Thus

$$\iint_i E(r)_i ds = \frac{1}{\varepsilon} \iiint \rho_e dv \tag{1.27}$$

By means of Stokes' divergence theorem, we have

$$\iiint \nabla \cdot E dv = \frac{1}{\varepsilon} \iiint \rho_e dv \tag{1.28}$$

This equality shall be held to an infinitesimal volume; in fact, it is one of the two Gauss laws of electromagnetics.

$$\nabla \cdot E = \frac{1}{\varepsilon} \rho_e \tag{1.29}$$

If the electric field intensity can be derived from the potential function $E = -\nabla\varphi$, which is the most prevailing condition in free space away from any solid surface, then the following equation becomes the well-known Poisson equation of plasma dynamics:

$$\nabla^2 \varphi = -\frac{\rho_e}{\varepsilon} \tag{1.30}$$

This equation can also be easily derived from Gauss' law, $\nabla \cdot D = \rho_e$ for the electric field of the Maxwell equations. The relationship between electric displacement D and electric field intensity E is given by a constitution coefficient ε known as the electric permittivity. The electric field intensity just equals to the negative gradient of electric potential φ.

1.5 ELECTRIC CONDUCTIVITY

The electric and magnetic fields in plasma always produce an electric current that is related to the diffusion velocities of the various charged particle species. Therefore, it is necessary to evaluate the electrical conductivity as microscopic properties together with the thermodynamic state of the plasma. Between collisions, the individual particles move in accordance with the $E \times B$ drift and diffusion velocities that are the driving mechanisms for the charged particle motions. Now the electrical resistivity is the resultant phenomenon of collision process. By introducing a mean or drift motion of the electrons in the gas, the electric current is

$$J = nqu = \frac{nq^2}{m\nu_e} E \tag{1.31}$$

In this equation, m is the mass of the electron and ν_c is the average electron collision frequency that is related to the atomic scale momentum transfer cross section and the random thermal motions of the individual particles. The current density is related empirically to the applied electric and magnetic fields by a bulk electric conductivity $J = \sigma(E + u \times B)$. The proportional constant is known as the electrical conductivity,

$$\sigma = \frac{nq^2}{m\nu_c} \tag{1.32}$$

In fact, it is the reciprocal of the electric resistance that has the SI unit of ohm-meter and is a function of the medium temperature. The electrical conductivity is expressed in reciprocal of ohm-meter or mho. For example, the fused quartz has the conductivity of 10^{-17}, the glass of 10^{-12}, sea water of 4, and silver of 6.1×10^7 mho/m. Note that Ohm's law at a point is $J = \sigma E$ and that J and E have the same direction in an isotropic medium and in the absence of a magnetic field. Under this special circumstance, electrical conductivity is a scalar quantity.

In the presence of a magnetic field, collisions play a similar role in determining the drift velocity of a charged particle. The magnetic field through the Lorentz acceleration generates a gyro motion of the charged particle with a gyro radius proportional to the velocity component perpendicular to and inversely proportional to the magnetic flux density B. The angular velocity of the gyro motion has been introduced, which has earlier been known as the gyro frequency. It may be clear the that helical trajectory of the charged particles has a wide and varying range of orientations in collision; the electrical conductivity is no longer a scalar quantity but is a tensor of rank 2.

The trajectory of the drift motion of a charged particle in an electromagnetic field is fascinating and complex. In the presence of both electric and magnetic fields, the charged particle in any inertial frame will execute a helical motion. The electric force usually is decomposed to components parallel and perpendicular to the magnetic field. The relationship between the transformed (reference) frame moving with the mean velocity of plasma and the laboratory (fixed) frame is $E_{\mathrm{tr}} = E + u \times B$. In the nonrelastivistic approximation, the transformed magnetic flux density is $B_{\mathrm{tr}} = B - u \times E/c^2$. For the case when coordinates system is moving with a velocity perpendicular to both the electric and magnetic field or on a laboratory frame, the helix trajectory will become a prolate cycloid with loops or the curtate cycloid (Jahn, 1968).

The ratio of the gyro frequency to the collision frequency of the charged particle is known as the Hall parameter ω_b/ν_c. This parameter characterizes the charged particle's motion as it responds to the applied E and B fields. The $E \times B$ component of the current is the so-called Hall current that defines the gyro motion. When $\omega_b/v_c \gg 1$, the charged particle rarely can execute one cycle of its motion before collision and hence cannot move in the helical trajectory. Thus, the primary component of the electric current is parallel to E. On the opposite limit, $\omega_b/v_c \ll 1$, the charged particles complete multiple gyro motions between collisions. As a consequence, the major component of the electric current will be in the direction of $E \times B$. Finally under the condition $\omega_b/v_c \sim 1$, the current will have comparable components parallel and perpendicular to E.

The Hall parameter depends on the charge-to-mass ratio and therefore has different value for various charged species. A singly charged ion and an electron have substantially different masses by a ratio of around 1836 to 1. This disparity complicates the formulation for the overall electrical conductivity. In practical applications in the presence of an applied magnetic field, the electrical conductivity has been formulated directly by using the plasma frequency, or by describing the modified electric field in parallel and normal component, equation (1.14) with respect to an applied magnetic flux density B_o. The derivation for the nonscalar electric conductivity is tedious. For purpose of illustration that the electric conductivity is really a tensor, an abbreviated derivation by Jahn is presented for the electric conductivity in a steady magnetic field (Jahn, 1968). Three distinct current components have been identified by the component, which lies along the electric field E and follows the drift $E \times B_o$, and the third component aligns with B_o but in proportional to the component of E parallel to B_o.

If B_o is aligned in the z coordinate of a Cartesian system, the conductivity appears as a tensor of rank 2,

$$\frac{\sigma}{\varepsilon} = \omega_p^2 \begin{bmatrix} b_{xx} & b_{xy} & 0 \\ b_{yx} & b_{yy} & 0 \\ 0 & 0 & b_{zz} \end{bmatrix} \tag{1.33}$$

In this matrix formulation,

$$\begin{aligned} b_{xx} = b_{yy} &\rightarrow \frac{v_c}{w_b^2 + v_c^2} \\ b_{zz} &\rightarrow \frac{1}{v_c} \\ b_{xy} = -b_{yx} &\rightarrow \frac{\omega_b}{\omega_b^2 + v_c^2} \end{aligned} \tag{1.34}$$

The gyro frequency is a rotational vector that requires a sign to distinguish the clockwise and counterclockwise rotations. Again, because of the substantial difference in the mass of an electron and an ion, the electrical conductivity tensors are qualitatively different from each other.

In partially ionized plasma, the electric conductivity is complicated by the huge disparity in mass of the electrons and ions. Therefore, a systematic approximation must be applied in the derivation. From the basic characterization of an electromagnetic field, it can be recognized that the B field plays no role in determining the component of the motion of particles parallel to it. Therefore, if one lets the magnetic field to be aligned with the Z coordinate, the particle motion can be decomposed into components parallel and perpendicular to the magnetic field. The effective electron conductivity may be written accordingly as

$$\sigma_{e,\perp} = \frac{\sigma_e}{1 + \beta_e^2}, \qquad \sigma_{e,\parallel} = \frac{\beta_e \sigma_e}{1 + \beta_e^2} \tag{1.35}$$

Introducing the slip factor $s = \beta_e \beta_i = \mu_e \mu_i B^2$ for weakly ionized plasma, the components of the electrical conductivity can be approximated as (Mitchner et al., 1973)

$$\sigma_\perp \approx \frac{\sigma(1 + s)}{(1 + s)^2 + \beta_e^2}, \qquad \sigma_\parallel \approx \frac{\beta_e \sigma}{(1 + s)^2 + \beta_e^2} \tag{1.36}$$

The electric conductivity in Cartesian tensor form appears as

$$\sigma = \begin{bmatrix} \sigma_\perp & -\sigma_\parallel & 0 \\ \sigma_\parallel & \sigma_\perp & 0 \\ 0 & 0 & \sigma_\perp \end{bmatrix} \tag{1.37}$$

Ohm's law subjected to constant driving fields E and B can be constructed by assigning a simplified scalar electrical conductivity as

$$E = \frac{1}{\sigma}[J + \beta_e (J \times B) + sB \times (J \times B)] \tag{1.38}$$

In this equation, the second term is immediately identifiable as the Hall current and the third term is often referred to as the slipping current unique to weakly ionized plasma.

1.6 GENERALIZED OHM'S LAW

A simplified relationship between electric current and electric field intensity that has been widely used in classic magnetohydrodynamics is the generalized Ohm's law. This approximation has the distinct advantage in compact formulation to bypass the complex and detailed description of electrical conductivity. The derivation and approximations of the law are based on the single-fluid, quasi-neutral plasma with single charged ions. First, the mass density, averaged mass velocity, and current density are defined as follows:

$$\begin{aligned} \rho &= Mn_i + mn_e \approx n_e(M = N) \\ u &= \frac{1}{\rho}(n_i M u_i + n_e m u_e) \approx \frac{Mu_i + mu_e}{M + m} \\ J &= e(n_i u_i + n_e u_e) \approx n_e(u_i - u_e) \end{aligned} \tag{1.39}$$

The equation of charged particles' motion in an applied steady electric and magnetic fields can be written as equation (1.39). Additional forces such as the gravitational or any nonelectromagnetic body force exerting on the plasma can also be included. But, first, the convective terms $(u \cdot \nabla)u$ or the divergence of the dyadic tensor $\nabla \cdot (\rho uu)$ for both electrons and ions is neglected, because the velocities of the mass-averaged organized motion for electrons and ions are assumed to be small. The shear stress is neglected for simplicity and the error is limited as long as the Larmor radius is much smaller than the other characteristic length scales of charged particles motion, the equations of charged particles motion can given as

$$\begin{aligned} Mn\frac{\partial u_i}{\partial t} &= en(E + u_i \times B) - \nabla p_i + Mng + P_{ie} \\ mn\frac{\partial u_e}{\partial t} &= -en(E + u_e \times B) - \nabla p_e + mng + P_{ei} \end{aligned} \tag{1.40}$$

The two equations describe the motion of the ions and electron, respectively. In equation (1.40), symbols M and m denote unit mass of ion and electron, respectively, and n is the number density of charged species. The sum of the above equations is

$$n\partial(Mu_i + mu_e)/\partial t = en(u_i - u_e) \times B - \nabla P + n(M + n)g \tag{1.41a}$$

In this equation, p_e and p_i are the partial pressures of the electrons and ions, respectively. The total momentum transferred to ions or electrons by the collision process are P_{ie} and P_{ei}, and they follow from Newton's third law, $P_{ei} = -P_{ie}$. The equation of motion of the single fluid in an electromagnetic field is then

$$\rho\partial u/\partial t = J \times B - \nabla p + \rho g. \tag{1.41b}$$

Multiply the equation of motion for ions by m, the equation of motion for electrons by M, and subtract the latter from the former to obtain

$$Mmn\frac{\partial(u_i - u_e)}{\partial t} = en(M + n)E + en(mu_i + Mu_e) \times B - m\nabla p_i + M\nabla p_e - (M + m)P_{ei} \tag{1.42}$$

By invoking the definitions of number density, mass-averaged velocity, and current density,

$P_{ei} = n^2e^2(u_i - u_e)/\sigma$, the above equation becomes

$$\frac{Mmn}{e}\frac{\partial}{\partial t}\left(\frac{J}{n}\right) = e\rho E - (M+m)ne\left(\frac{J}{\sigma}\right) - m\nabla p_i + M\nabla p_e + en(mu_i + Mu_e)\times B \tag{1.43a}$$

The last term can be rearranged as

$$\begin{aligned} mu_i + Mu_e &= Mu_i + mu_e + M(u_e - u_i) + m(u_i - u_e) \\ &= \frac{\rho}{n}u - (M-m)\frac{J}{en} \end{aligned} \tag{1.43b}$$

Dividing equation (1.43) by ρe yields

$$E + u\times B - J/\sigma = \frac{1}{\rho e}\left[\frac{Mmn}{e}\frac{\partial}{\partial t}\left(\frac{J}{n}\right) + (M-m)J\times B + m\nabla p_i - M\nabla p_e\right] \tag{1.44}$$

Further simplifications are possible by considering the temporal variation of the organized motion to be negligible and the relative lower mass of electrons to ions and neutrals. The resultant approximation is known as the generalized Ohm's Law, which describes the electric properties in an electrically conducting medium,

$$\sigma(E + u\times B) = J + \frac{1}{en}(J\times B - \nabla p_e) \tag{1.45}$$

In the classic formulation of the magnetohydrodynamics governing equation, the following expression is frequently adopted in magnetohydrodynamic formulation:

$$J = \sigma(E + u\times B) \tag{1.46a}$$

The physical interpretation of this equation is that the current density is driven by an effective electric field that is composed of the applied field E and an induced field $u\times B$.

The generalized Ohm's Law can be modified to include the Hall current as (Mitchner and Kruger, 1973)

$$J = \sigma[E + u\times B - \beta(J\times B)] \tag{1.46b}$$

where σ is the scalar electric conductivity, and hence the Hall current and frequency have been defined previously as $J\times B$ and $\beta = 1/en = B/\omega_c n$. Again the cyclotron or Larmor frequency is given as $\omega_c = eB/m$.

1.7 MAXWELL'S EQUATIONS

Maxwell's equations form the basic laws governing all phenomena of an electromagnetic field in free space and in media. These fundamental equations were established by James Maxwell in 1873 and verified experimentally by Heinrich Hertz in 1888. In 1905, Albert Einstein's special theory of relativity has further affirmed the rigorousness of the Maxwell's theory (Kong, 1986). This set of equations has been given in integral form, but for the present purpose, the differential form is presented as

$$\frac{\partial B}{\partial t} + \nabla \times E = 0 \tag{1.47a}$$

$$\frac{\partial D}{\partial t} - \nabla \times H + J = 0 \tag{1.47b}$$

$$\nabla \cdot B = 0 \tag{1.47c}$$

$$\nabla \cdot D = \rho_e \tag{1.47d}$$

The first equation is Farady's law, which relates the time rate of change of the magnetic flux density B (weber/m^2) to the curl of the electric field intensity E (volt/m). The second equation is the generalized Ampere's circuit law relating the time rate of the electric displacement D (coulomb/m^2) to the curl of the magnetic field strength H (ampere/m) and electric current density J (ampere/m^2). The last two equations are Gauss' laws for magnetic and electric fields, respectively. The first law simply requires that the divergence of B must vanish in any electromagnetic field to preclude the existent of a magnetic dipole. The second law defines the direct relationship between the divergence of the electric displacement D and the electric charge density ρ_e(coulomb/m^3).

From a pure mathematical point of view, Maxwell's equations have a closure issue, namely, the number of dependent variables is unbalanced by the number of unknowns. Meanwhile, for problem solving, the material media needs more characterization. This dilemma is removed by the constitutive relations. In an isotropic media, the required properties are

$$\begin{aligned} D &= \varepsilon E \\ B &= \mu H \end{aligned} \tag{1.48}$$

where ε denotes the electric permittivity in free space that it has a value of $4\pi \times 10^{-7}$ henry/m, and the symbol μ is the magnetic permeability in free space that has a value of approximately 8.85×10^{-12} farad/m.

It is learned that the two Gauss' laws, equations (1.14c) and (1.47d), are not independent equations. A conservation law for electric charge and current density has been regarded as a fundamental equation. This equation can be derived from the Faraday's and generalized Ampere's laws by taking the divergence of the latter and introducing the Gauss' divergence law for electric displacement, yielding

$$\frac{\partial \rho_e}{\partial t} + \nabla \cdot J = 0 \tag{1.49}$$

The charge conservation equation above has been demonstrated to be invaluable for verifying the global behavior for computational electro-aerodynamic simulations.

In summary, Maxwell's equations constitute a hyperbolic partial differential equation system and thus must be supplemented with boundary conditions and initial values wherever derivatives do not exist such as at the media interface. It is noted that in a single medium the electromagnetic field is continuous. However, at the boundary between two different media, the field may change abruptly in both magnitude and direction. It is convenient to describe the boundary condition by the tangential and normal components on the media interface. On a stationary boundary interface, the boundary conditions are

$$\begin{aligned} n \times (E_1 - E_2) &= 0 \\ n \cdot (B_1 - B_2) &= 0 \\ n \cdot (D_1 - D_2) &= \rho_s \\ n \times (H_1 - H_2) &= J_s \end{aligned} \tag{1.50a}$$

The first boundary condition indicates that the tangential electric field E is continuous across the interface boundary. Conversely, the normal component of the magnetic flux density B is also permitted to pass uninterrupted through the media interface. According to the third equation, the difference in normal component of the electric displacement D must be balanced by the surface charge density. It is important to note that the surface charge is the actual electric charge separated by a finite distance between charged particles but not the surface charge due to polarization. Finally, the discontinuity of the tangential components of the magnetic field strength H is equal to the surface current density J_s.

On a moving boundary, because the partial temporal derivatives are no longer commuted with the surface integral, the boundary conditions are required to formulate by the integral form of the Maxwell equations. Additional components via the normal velocity are required to append on the tangential formulation of the electric and magnetic field strength (Kong, 1986).

$$\begin{aligned} n \times (E_1 - E_2) - (n \cdot u_b)(B_1 - B_2) &= 0 \\ n \cdot (B_1 - B_2) &= 0 \\ n \cdot (D_1 - D_2) &= \rho_s \\ n \times (H_1 - H_2) + (n \cdot u_b)(D_1 - D_2) &= J_s \end{aligned} \tag{1.50b}$$

1.8 WAVES IN PLASMA

An ionized gas can generate a wide variety of oscillatory motions. Three types of waves in plasma have been classified as electromagnetic, hydromagnetic, and electrostatic (Spitzer, 1956; Celeroix, 1960). However, it is most unlikely that all of these idealized waves can exist in their pure form. The electromagnetic waves in the simplest cases are only transverse waves that are associated with the transverse motion of lines of magnetic induction. The tension in the lines of force tends to restore deformation back to the original line. These waves are exemplified by electromagnetic waves propagating in vacuum (Friedrichs et al., 1958). The phase velocity is on the order of the speed of light. Hydromagnetic or magnetohydrodynamic waves are similar to acoustic waves that are low-amplitude longitudinal, and compression waves motions are induced by the electromagnetic coupling of the electrons and ions. The wave propagation speed is governed by the speeds of sound. Finally, electrostatic waves are also a longitudinal wave and propagate because of thermal agitation (Landau et al., 1960).

Maxwell's equations govern the propagation of an electromagnetic wave. In order to describe wave motion for the purpose of better understanding, the plasma is considered as an unbound isotropic and stationary medium. The wave equation in plasma is derived by taking the temporal derivative of the generalized Ampere circuit law to get

$$\frac{\partial^2(\varepsilon E)}{\partial t^2} = \frac{1}{\mu}\nabla \times \frac{\partial B}{\partial t} + \frac{\partial J}{\partial t} \tag{1.51a}$$

then replaces the time derivative of the magnetic field strength H by Faraday's induction law

$$\frac{\partial^2(\varepsilon E)}{\partial t^2} = -\frac{1}{\mu}\nabla \times (\nabla \times E) + \frac{\partial J}{\partial t} \tag{1.51b}$$

and finally invokes a vector identity and substitutes into equation (1.51a).

$$\nabla \times (\nabla \times E) \equiv \nabla(\nabla \cdot E) - \nabla^2 E \tag{1.51c}$$

For an isotropic medium $D = \varepsilon E$ and $B = \mu H$ the basic electromagnetic wave equation appears as

$$\varepsilon\mu\frac{\partial^2 E}{\partial t^2} - \mu\frac{\partial J}{\partial t} = \nabla^2 E - \nabla(\nabla \cdot E) \tag{1.51d}$$

The electric permittivity and magnetic permeability are designated as ε and μ, respectively. The phase velocity is $(\varepsilon\mu)^{-1/2}$ and is the speed of light, which is $c = 2.998 \times 10^8$ m/s (De Hoffman et al., 1950). In discussing the wave motion, the orientation of the particle motion is relevant. From an earlier derivation for the generalized Ohm's law, the equation of charged particle motion has been described previously as equation (1.41) and is used as the reference for discussion

$$\rho\partial u/\partial t = J \times B - \nabla p + \rho g \tag{1.52}$$

If the electric intensity E and the current density J are parallel to the direction of wave propagation, only the electrostatic restoring force is presented. The longitudinal wave is the electrostatic wave. In this wave, the frequency is very high and the ions are unaffected by the electric field and the electrons retain the Boltzmann distribution.

The electric charge in the electrostatic wave is due to the divergence of the current density. There are two types of oscillations, namely the electron and ion oscillations. How the electron oscillations are excited in plasma is not clear, but these waves are damped as the wavelength approaches the Debye shielding length. The positively charged ion oscillations are waves of relatively low frequency. Electrical neutrality is preserved and the electrostatic wave usually behaves as an acoustic wave.

The hydromagnetic wave appears only in the presence of a magnetic field and at frequencies lower than the cyclotron frequency of the ions. The driving mechanism is the Lorentz acceleration $J \times B$. The oscillation is conserved to be waves in the line of force. The basic form is known as the Alfven (1950) wave, which is a purely magnetohydrodynamic phenomenon depending on the magnetic field and inertia of the medium. The wave velocity and the induced magnetic flux density are parallel to each other, while the mean magnetic field is normal to them. In the Alfven wave, the divergence of the medium velocity vanishes and the pressure gradient term becomes zero. For these reasons, the pressure and density do not vary in such transverse waves. Therefore, transverse wave motion is possible in an incompressible medium.

The phase velocity of these waves is the Alfven speed that is based on the component of the applied magnetic field in the direction of the wave propagation. From this view point, the Alfven wave is the vibration of a line of force. Its phase velocity is

$$u_p = B_o / \sqrt{\rho\mu} \tag{1.53a}$$

The normal stress component associated with lines of force is the well-known Maxwell tensile stress BB.

When the electric field intensity is perpendicular to the wave front, it will lead to an electromagnetic wave. The electrons will interfere with these transverse waves and increase the wave speed. When the electron number density exceeds a critical value with increasing plasma frequency, the electromagnetic wave cannot propagate through the plasma in the presence of an electromagnetic field. Another characteristic of wave motion in plasma is that the waves can propagate by different phase speeds. Since transverse waves can exist in both a compressible and incompressible medium, the electromagnetic waves in compressible medium have been referred to as magnetoacoustic waves (Sutton et al., 1965). Two phase velocities u_f and u_s, corresponding to the fast and slow waves, are

$$u_{f,s} = \frac{1}{2}\{\sqrt{c^2 + 2c[B/(\rho\mu)^{1/2}] + B^2/(\rho\mu)} \pm \sqrt{c^2 - 2c[B/(\rho\mu)^{1/2}] + B^2/(\rho\mu)}\} \tag{1.53b}$$

The phase velocities are dependent on different angles between the applied magnetic field and the direction of wave propagation. The velocity of the fast wave is always greater than the Alfven wave speed, and that of the slow wave is always less than it. The phase velocities of the electromagnetic waves degenerate to the speed of sound when the transverse magnetic field vanishes.

1.9 ELECTROMAGNETIC WAVES PROPAGATION

Both longitudinal compression and transverse waves can coexist in plasma. The transverse wave is associated with the magnetic field, and the mechanism of this wave is often interpreted as the tension of the magnetic lines of force that restores any disturbance to a straight line, thereby leading to a transverse oscillation. In this sense, the transverse wave always propagates in the perpendicular direction to a magnetic field and will cease when the magnetic field vanishes.

One of the most dramatic manifestations of plasma is the properties of an electromagnetic wave propagation. The intensity of an incident wave always attenuates in a weakly ionized gas, because it is a semiconductive medium or lossy material. The complete reflection of the electromagnetic wave at the interface of media also can occur when the frequency of the incident wave is lower than the plasma frequency and becomes as the cutoff frequency. Another unique feature of wave motion in plasma is that it is not solely depended on the collision mechanism as in acoustic wave. Multiple modes of the transverse wave also exist such as the Alfven wave associated with ions and the Whistler wave related to electrons. Both transverse waves are generated by an initial disturbance perpendicular to magnetic field. For this reason, these transverse waves are often regarded as the vibration of a line of magnetic force. All the direction-dependent disturbance propagations lead to a very complex wave system and introduce fascinating physics in resonance and energy damping. A systematic discussion of the plasma wave dynamics will be deferred until the governing equations of the classic magnetohydrodynamics are presented later.

The equation of electromagnetic wave propagation is

$$\varepsilon\mu\frac{\partial^2 E}{\partial t^2} - \mu\frac{\partial J}{\partial t} = \nabla^2 E - \nabla(\nabla \cdot E) \tag{1.54a}$$

The general solutions of a transverse plane wave are then

$$\begin{aligned} E &= E_o e^{i(\omega t \pm kr)} \\ J &= J_o e^{i(\omega t \pm kr)} \end{aligned} \tag{1.54b}$$

where $\kappa = (\varepsilon\mu\omega^2 - i\mu\sigma\omega)^{1/2}$ is a complex wave number where the electric conductivity was defined previously as $\sigma = n_e e^2 / m_e v_c$, and v_c characterizes the averaged electron collision frequency. In the absence of an external applied magnetic field, the magnitude of the induced magnetic flux density is on the order of $|E|/c$. If the effect of the magnetic flux density B is neglected in the generalized Ohm's law, the temporal behavior of the electric current density can be approximated as

$$\frac{1}{v_c}\frac{\partial J}{\partial t} \approx \sigma E - J \tag{1.55}$$

By satisfying the plasma wave equation and the above approximation to establish the relation between the electric field strength and current density, a dispersive relationship for a transverse plane wave was obtained by Mitchner and Kruger (1973),

$$\left(\frac{c\kappa}{\omega}\right)^2 = \left[1 + \frac{(\varepsilon - \varepsilon_o)}{\varepsilon} - \frac{(\omega_p/\omega)^2}{1 + (\nu_c/\omega)^2}\right] + i\left[\frac{(\omega_p/\omega)^2(\nu_c/\omega)}{1 + (\nu_c/\omega)^2}\right] \tag{1.56}$$

In this equation, $\omega_p = (n_e e^2/\varepsilon m_e)^{1/2}$ is the plasma frequency. The small difference between the electric permittivity in free space and plasma is ignored. The complex wave number can be expressed as $\kappa = \alpha + i\beta$ and then the electric field intensity becomes

$$E = E_o e^{-ax} e^{i(\beta x - \omega t)} \tag{1.57a}$$

The real number α is the attenuation constant, and β is the phase constant and related to the wavelength $\beta = 2\pi/\lambda$ and phase velocity of the wave $u_{\mathrm{ph}} = \omega/\beta$. It has also been pointed out by Mitchner et al. that $(\omega_p/\omega)^2/[1+(\nu_c/\omega)^2] \gg R$, where $R = (\varepsilon - \varepsilon_o)/\varepsilon_o$ is the so-called refractivity of the medium. In the limiting condition $\omega \gg \nu_c$, collisions are negligible and the dispersion relation becomes

$$\left(\frac{c\kappa}{\omega}\right)^2 = 1 - \left(\frac{\omega_p}{\omega}\right)^2 \tag{1.57b}$$

If $\omega > \omega_p$, κ is a real number and the wave propagates through the plasma without attenuation. However, if $\omega < \omega_p$ κ becomes an imaginary number and the solution $E = E_o \exp -(\alpha x + i\omega t)$ can no longer describe a propagating wave. It is referred to as an evanescent wave and on average it does not transport energy and will be reflected from the media interface. From the simplified analysis, for electromagnetic waves with frequencies lower than the plasma frequency, the incident wave will be reflected at the interface of plasma as exhibit in Figures 1.5 and 1.6.

The attenuation of an incident electromagnetic wave in plasma is depicted by computational results at a frequency lower than the plasma frequency $\omega_p = (n_e e^2/\varepsilon m_e)^{1/2}$. The numerical results describe the rapid dissipation of wave amplitude in terms of the penetration distance versus the ratio of frequencies in decibels and the phase shift.

In fact, the dissipation is a manifestation of the Debye shielding of charges within plasma. While collisions tend to modify the attenuation of the electromagnetic wave, but the general behavior of the cutoff frequency is still similar. By meaning of a critical electron number density, the wave propagation pattern is established.

$$n_{ec} = \frac{\varepsilon m_e \omega^2}{e^2} \tag{1.58a}$$

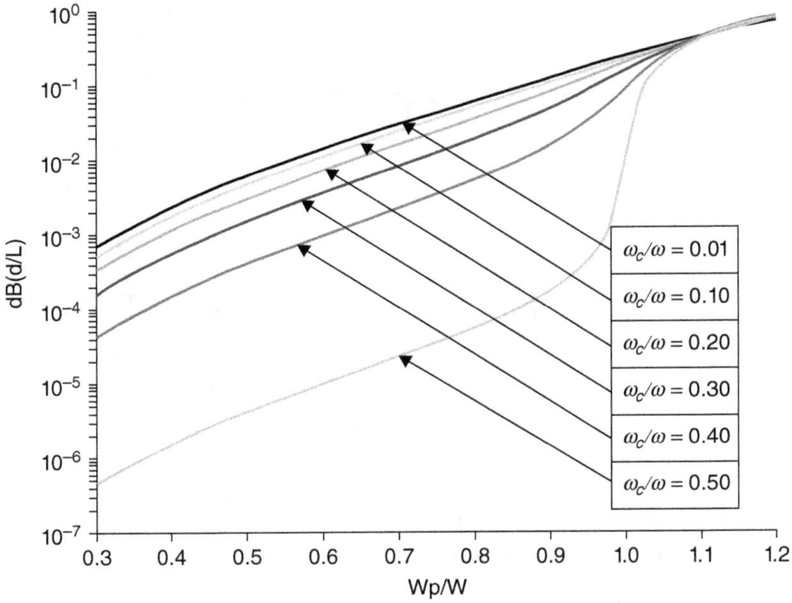

FIGURE 1.5 Incident wave attenuation.

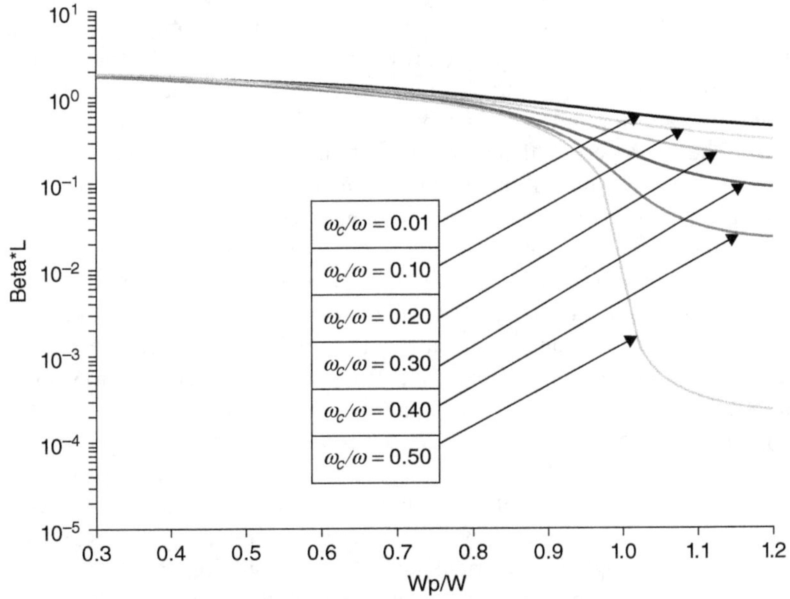

FIGURE 1.6 Incident wave phase relationship.

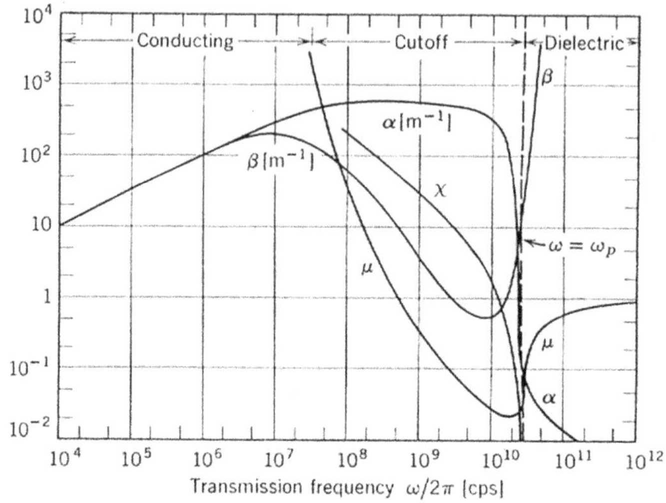

FIGURE 1.7 Amplitude-frequency relation of electromagnetic wave. From Head, M., and Wharton, C. *Plasma Diagnostics with Microwaves*. Used with permission from John Wiley & Sons, 1965.

If $n_e < n_{ec}$ ($\omega > \omega_p$), the electromagnetic waves can propagate through a collision-free plasma, but the wave amplitude will be attenuated by collisions. If $n_e > n_{ec}$ ($\omega < \omega_p$), the electromagnetic waves now may still be able to propagate through the plasma but are strongly attenuated. The $1/e$ depth of penetration is inversely proportional to the square root of product of the frequency of incident wave, the electric permeability and conductivity of the medium $1/\sqrt{\pi\omega\mu\sigma}$. Both the wavelength and the wave attenuation rate are implicit functions of the free electron density and the collision frequency. These properties determine the ability of an electromagnetic wave to penetrate the plasma by describing the phase change and reflection coefficients at the interface of the medium. Once the real and imaginary wave numbers are known, the electron number density and average plasma collision frequency can be found.

As an illustration of the peculiar behavior of electromagnetic wave attenuation in plasma is the gist of plasma diagnostics and the communication blackout in the earth reentry. The propagation constants as a function of frequency in typical laboratory hydrogen plasma are given by Heald and Wharton (Heald et al., 1965) as shown in Figure 1.7. The data is collected at an electron number density of $10^{13}/cm^3$. Three frequency domains are labeled on the top of the graph. In the intermediate frequency spectrum, $\nu < \omega < \omega_p$, an electromagnetic wave will not propagate through the plasma that is similar to that of the cutoff frequency of waveguides. When an incident microwave has a frequency lower the cutoff frequency, the wave will not able to propagate through the electrically conducting medium.

The refractive index and attenuation are related to the complex propagation constants via the attenuation and phase coefficients,

$$\alpha = \chi\omega/c \quad \text{and} \quad \beta = \mu\omega/c \tag{1.58b}$$

In the absence of a magnetic field, these indexes can be obtained by assuming the Lorentz conductivity that is accurate when the collision frequency is independent of electron velocity as the limiting condition.

The most interesting electromagnetic wave propagation phenomenon occurs in the intermediate frequency spectrum where the frequency of the incident wave lies in between the collision frequency of momentum transfer and the plasma frequencies, $\nu < \omega < \omega_p$. The real refractive index and the attenuation indexes of the incident wave can be approximated as

$$\mu \approx (\nu\omega_p / 2\omega^2)(1 - 5\nu^2 / 8\omega^2 + \omega^2 / 2\omega_p^2)\ \chi \approx (\omega_p / \omega)(1 - 3\nu^2 / 8\omega^2 - \omega^2 / 2\omega_p^2) \tag{1.58c}$$

And the plasma penetration depth can be given as

$$\delta \approx (c / \omega_p)(1 + 3\nu / 8\omega^2 + \omega^2 / 2\omega_p^2) \tag{1.58d}$$

From the above approximations, the penetration depth of the evanescent wave is practically constant in the interior of the frequency region. The physical dimension of the depth is comparable to the wavelength of the plasma frequency to have a value of a few millimeters.

At the high-frequency domain, the incident electromagnetic wave has a greater frequency than the plasma frequency, $\omega > \omega_p$. Under this circumstance, the plasma behaves as a dielectric in which the electromagnetic wave propagates through with a very low attenuation similar to that of a relatively low loss medium.

1.10 JOULE HEATING

The Joule heating is also known as Ohmic and resistive heating. It is the process by which the passage of an electric current through a medium releases heat. In other words, the electromagnetic field does work to overcome the resistive force that equals to the dissipated heat in response to resistance. The amount of energy released is proportional to the square of the current. This relationship is known as Joule's first law and the IS unit of the energy was subsequently named after Joule by the symbol J (J = 1 Newton-meter, J = 10^7 erg).

Joule heating is the consequence of the interaction of a charged particle in an electric circuit accelerated by an electric field but must give up some of their kinetic energy every time they collide with ions of the conductor. The increase in kinetic or vibrational energy of the ions manifests itself as heat and an increase in the temperature of the conductor.

The most general and fundamental formula for Joule heating is

$$q = E \cdot J = \frac{J^2}{\sigma} = \sigma E^2 \tag{1.59a}$$

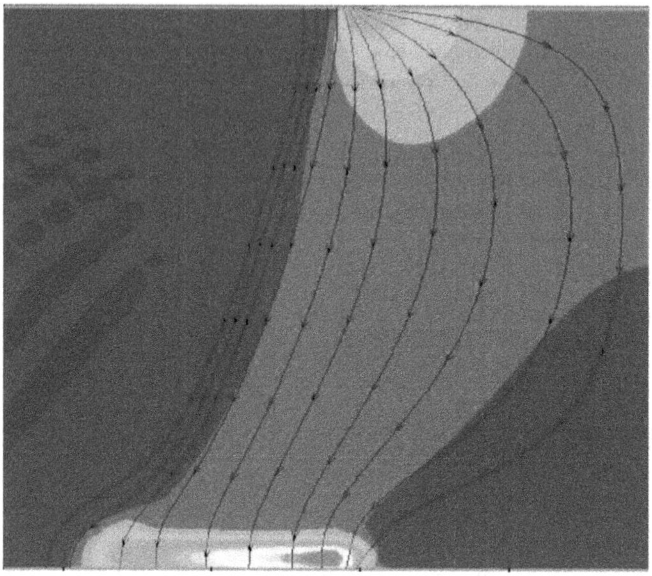

FIGURE 1.8 Concentrated regions of Joule heating in a direct current discharge.

The electric current in a circuit is the volumetric integral of the electric current density

$$I = \iiint J dv \tag{1.59b}$$

The total energy dissipated in the conductor in time is then $W = \int E \cdot I dt$ and known as Joule's law. An identical formula can also be used for AC power, except that the parameters are averaged over one or more cycles. In these cases of a phase difference between I and E, a multiplier $\cos \phi$ must be included.

The composite graph by the computational electric current density contours together with the electric field vector traces is given in Figure 1.8, and from these data the Joule heating can be evaluated. For the computational simulation of a direct current discharge, the anode is placed at the upper border of the figure and cathode is located in the opposite lower boundary. The gap distance between electrodes is 5 cm. As shown from the figure, the electric field intensifies at the sharp edge of the anode. The electron number density has a maximum value of $1.7 \times 10^{10}/\text{cm}^3$ locally. This value is 3.4 times greater than the infinite parallel electrode counterpart. At the connecting edge of the anode, the anode layer is completely distorted by the stronger local electric field. From the contour presentation, the cathode layer can be clearly discerned even through the discharge structure over the cathode and is no longer uniform. The current density is clustered locally. Thus, the Joule heating is concentrated along the outer edge of cathode closest to the anode, at an applied direct current electric field of 840 volt and the calculated total current of 4.88 mA. In the graphic example, the Joule heating of the glow discharge at the pressure of 5 Torr is 4.10 J (Shang et al., 2009; Shang et al., 2012).

1.11 TRANSPORT PROPERTIES

Diffusion is one of the major complications in the analysis of plasma dynamics. The diffusive mass flux is generated by three driving mechanisms due to the gradients of species concentration, temperature, and external forces exerted on the different species. From the kinetic theory of dilute gases, the description of diffusion does not extend to the presence of internal freedoms in molecules. In an electromagnetic field, additional random motions by the electric statistical force and by Lorentz acceleration are recognized as new mechanisms for the forced diffusion. In a diffusing mixture, the velocities of individual species can be significantly different from each. However, the peculiarity of the Debye length of plasma creates a unique constraint known as ambipolar diffusion (Howatson, 1975). The species diffusion velocity u_i is defined by a relative velocity component with respect to a local mass-averaged velocity of the gas mixture,

$$u = \frac{\sum_i \rho_i u_i}{\sum_i \rho_i}. \tag{1.60}$$

Here ρ_i is the density of species i. The diffusion velocity of species i is defined by the difference of the mass-averaged value and the species diffusion velocity on a stationary frame of reference. By definition, the sum of mass diffusion of all species is identical to zero. $\Sigma \rho_i u_i = 0$.

A landmark of the kinetic theory of a diluted gas mixture is its ability to describe the transport properties of any gaseous mixture. To be consistent with the kinetic model of the internal structure of a gas, the transport properties of the gas mixture for thermal diffusion, viscosity, and thermal conductivity shall be calculated from Boltzmann equation by using the Chapman–Enskog expansion (Chapman et al., 1950). The thermal diffusivity is

$$D_{i,j} = 1.858 \times 10^{-3} \sqrt{T^3 \frac{M_i + M_j}{M_i M_j}} / \sigma_{i,j}^2 \Omega^{(1,1)} \tag{1.61}$$

The molecular viscosity is given as

$$\mu = 2.67 \times 10^{-5} \sqrt{M_i T} / \sigma_i^2 \Omega^{(2,2)} \tag{1.62}$$

and the thermal conductivities are

$$\kappa_{i,m} = 1.989 \times 10^{-4} \sqrt{\frac{T}{M_i}} / \sigma_i^2 \Omega^{(2,2)} \tag{1.63}$$

$$\kappa_{i,p} = 2.519 \times 10^{-4} \sqrt{\frac{T}{M_i}} / \sigma_i^2 \Omega^{(2,2)} \tag{1.64}$$

for a monatomic and a polyatomic gas, respectively (Bird et al., 1960; Hirshfelder et al., 1954).

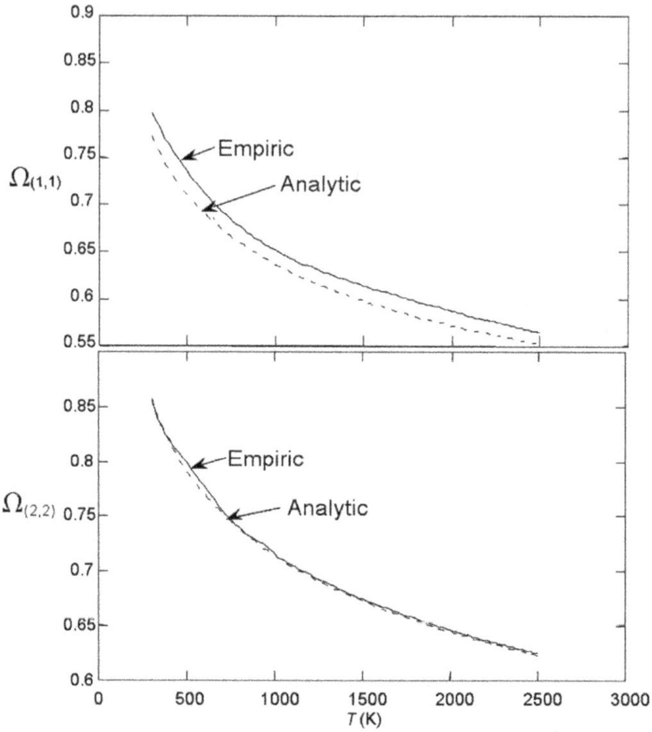

FIGURE 1.9 Comparison of collision integrals for transport properties.

All collision integrals and cross sections can be obtained from either the Lenard–Jones potential for nonpolar molecules or a polarizability model for ion-neutral nonresonant collisions by the more recent works by Capitelli et al. (2000) and Levin and Wright (2004). For the purpose of understanding, a very wide range of temperatures from 200 to 28,000 K by the 11 species of ionized air is calculated. According to the study of Capitelli et al., the calculated collision integrals are not dramatically dependent on the specific form of the potential models. The collision integral is a dimensionless parameter that is generated by a triple integral process whose value depends on the dynamics of binary collisions, and thus is controlled by the intermolecular force law.

The comparison of collision integrals by analytic results and an existent empirical database is presented in Figure 1.9. The graph depicts a comparative study of the integral $\Omega^{(1,1)}$ from two different approaches for molecular hydrogen over a temperature range up to 2500 K. The analytic result using the Lenard–Jones potential is consistently 3.1% lower than the polynomial fitted data; therefore, the diffusion coefficient of hydrogen is anticipated to be proportionally greater than data at an identical thermodynamic state. On the contrary, the reduced collision integral $\Omega^{(2,2)}$ has the nearly identical value to that of an empirical database. The depicted comparison, in a sense, is a study of the state-of-the-art status in transport properties calculations.

The agreement between analytic results and empirical database reflects an impressive accumulation of knowledge for transport properties, but the database does not include ionized gases. The analytic approach for transport properties evaluation must be used for electromagnetic–aerodynamics interaction.

The variation of these products of collision integral and cross section is less pronounced in the high-temperature region to reflect a linear dependence to the local temperature and molecular weight of the species. Although there is substantial progress in the kinetic theory of gases for calculating transport properties, the specific comparison of the calculated transport properties still yields a difference of 16.3%–24.2% in viscosity and thermal conductivity for high-temperature air using different sets of collision integrals in the temperature range from 300 to 30,000 K.

The Chapman–Enskog theory gives the viscosity and thermal conductivity for a gas mixture. These formulas are also known as Wilke's mixing rule

$$\begin{aligned} \mu_{i,j} &= \sum \frac{x_i \mu_i}{\sum x_i \phi_{i,j}} \\ \phi_{i,j} &= \frac{1}{\sqrt{8}} \left(1 + \frac{M_i}{M_j}\right)^{-1/2} \left[1 + \left(\frac{\mu_i}{\mu_j}\right)^{1/2} \left(\frac{M_j}{M_i}\right)^{1/4}\right]^2 \end{aligned} \tag{1.65}$$

and

$$\begin{aligned} \kappa_{i,j} &= \sum \frac{x_i \kappa_i}{\sum x_i \phi_i} \\ \phi_{i,j} &= \left[1 + \left(\frac{\kappa_i}{\kappa_j}\right) \left(\frac{M_j}{M_i}\right)^{1/4}\right]^2 \Big/ \left[8\left(1 + \frac{M_i}{M_j}\right)\right]^{1/2} \end{aligned} \tag{1.66}$$

In these equations, the symbol x_i denotes the molar fraction of the components for the gas mixture. The transport properties of the present analysis are derived from kinetic theory via the collision integrals and collision cross sections.

For the purpose of understanding, a very wide range of temperatures from 200 to 28,000 K by the 11 species of air mixture is calculated. According to the study of Capitelli et al. (2000), the calculated collision integrals are not dramatically dependent on the specific form of the potential models. The collision integral is a dimensionless parameter that is generated by a triple integral process whose value depends on the dynamics of binary collisions and thus is controlled by the intermolecular force law.

The variation of these products of collision integral and cross-section is less pronounced in the high-temperature region to reflect a linear dependence to the local temperature and molecular weight of the species. Although there is substantial progress in the kinetic theory of gases for calculating transport properties, the specific comparison of the calculated transport properties still yields a difference of 16.3%–24.2% in viscosity and thermal conductivity for high-temperature air using different sets of collision integrals in the temperature range from 300 to 30,000 K.

1.12 AMBIPOLAR DIFFUSION

In an ionized gas, the random or thermal motion of a charged particle is the result of mutual collisions plus forces acting on the particle by an electromagnetic field. The acceleration of an electron has a magnitude of eE/m in the negative direction of E. After numerous collisions, the average kinetic energy reaches a constant value and the force diffusion has directional average velocity. This motion is the well-known drift velocity. A similar drift velocity also is attained by the positively charged ion but in the positive E direction, and, due to the greater mass of the ions, its drift velocity is much lower than that of the electron.

The averaged energy gain between collisions is dependent on the ratio of E/p. The proportionality constant between the drift velocity and the electric field $u_e = -\mu_e E$ is called mobility,

$$\mu_e = \frac{u_e}{E} = \frac{e}{m_e \nu} \tag{1.67a}$$

In this equation, ν is the averaged collision frequency for momentum transfer, and the mobility of the drift velocity of ion can also be given as

$$\mu_i = e/m_i \nu \tag{1.67b}$$

The self-diffusion or ordinary diffusion is proportional to the concentration gradient of the number particle density ∇n. The diffusion coefficient determined by the elementary kinetic theory of gases is proportional to the mean random velocity and the mean-free path between collisions. The rate of flow particles per unit area is

$$\Gamma = -D\nabla n; \quad D \sim \lambda u \tag{1.68}$$

From the viewpoint of the conservation of number particle density, the mass flux density is the Fick's second law of diffusion. This relationship is valid for both electrons and ions. Since the mean-free path of electrons is greater than that of ions, the electron's random velocity is much greater; it follows that $D_e > D_i$.

The diffusion of charged particles is related to mobility; both arise from the random motion and unbalanced collision force. In one-dimensional motion of an ion,

$$u_i = \frac{\Gamma_i}{n_i} = \frac{1}{n_i} D_i \frac{\partial n_i}{\partial x} = \frac{1}{p_i} D_i \frac{\partial p_i}{\partial x} \tag{1.69}$$

The gradient of the partial pressure $\partial p_i/\partial x$ must be balanced by the total force acting on the ion. For the present consideration, the force is exerted by the electric field, thus $eEn_i = \partial p_i/\partial x$. By the definition of mobility, we have

$$\frac{u_i}{\mu_i} = E = \frac{1}{en_i} \frac{\partial p_i}{\partial x} = \frac{1}{en_i} \frac{u_i p_i}{D_i} \tag{1.70}$$

or

$$\frac{D_i}{\mu_i} = \frac{p_i}{en_i} = \frac{\kappa T}{e} \tag{1.71}$$

This relationship between mobility and diffusion is known as the Einstein relation (Surzhikov et al., 2004).

The often encountered charge separation is the result of the disparity in the diffusion of electrons and ions. The electromagnetic field augments the drift velocity of the ions and retards the electrons. When this process reaches a state of local equilibrium for the drift velocity, this resultant process is called ambipolar diffusion.

Again, by a simple one-dimensional analysis, the limiting and reasonable estimated value of ambipolar diffusion coefficient can be found. Since

$$-D_e\frac{\partial n_e}{\partial x} - n_e\mu_e E = -D_i\frac{\partial n_i}{\partial x} + n_i\mu_i E \tag{1.72a}$$

and $n_i = n_e$ for a globally neural plasma.

The ambipolar diffusion coefficient can be written as

$$D_a\frac{\partial n}{\partial x} = -D_e\frac{\partial n}{\partial x} - n\mu_e E = -D_i\frac{\partial n}{\partial x} + n\mu_i E \tag{1.72b}$$

Eliminating E to yield

$$D_a = \frac{D_e\mu_i + D_i\mu_e}{\mu_e + \mu_i} \tag{1.72c}$$

In general, $\mu_e \gg \mu_i$, and the ambipolar diffusion coefficient can be approximated as

$$D_a = \frac{D_e\mu_i + D_i\mu_e}{\mu_e} \tag{1.73}$$

There are two conditions commonly encountered in practical applications. First, at the same temperature for electron and ion by thermal ionization,

$$D_e\mu_e = D_i\mu_i \tag{1.74a}$$

We then have

$$D_a = 2D_i \tag{1.74b}$$

Second, in electron impact ionization, $T_e \gg T_i$ thus $D_e\mu_e \gg D_i\mu_i$. The ambipolar diffusion coefficient can then be approximated as

$$D_a = D_e(\mu_i/\mu_e) = (\kappa T_e/e)\mu_i \tag{1.74c}$$

The diffusion by a magnetic field can be deduced by the equation of electron motion,

$$n_e m v u = n_e e u \times B - \nabla p_e \tag{1.75}$$

The effects of the magnetic field on the diffusion are twofold by the peculiarity of the biased orientation of an externally applied magnetic field and the interaction of the Hall current $eu \times B$.

The gradient of the partial pressure gradient transverse to B corresponds to a diffusion coefficient perpendicular to the magnetic field

$$D_{e\perp} = \frac{D_e}{1 + (\omega_b / \nu_c)^2} \tag{1.76}$$

The diffusion components paralleled to B are altered by both the partial pressure gradients and magnetic field to yield,

$$D_{e\|} = D_e \frac{\omega_b / \nu_c}{1 + (\omega_b / \nu_c)^2} \tag{1.77}$$

Recall the Larmor frequency is $\omega_b = eB/m$ and ν is the averaged collision frequency for the momentum transfer.

REFERENCES

Alfven, H., *Cosmical Electrodynamics*, Clarendon, Press, Oxford, 1950.

Bird, R.B., Stewart, W.E., and Lightfoot, E.N., *Transport Phenomena*, John Wiley & Sons, New York, 1960.

Capitelli, M., Gorse, C., and Longo, S., Thermophysics and heat transfer, *J. Collision Integrals High-Temperature Air Species*, vol. 14 no. 2, 2000, pp. 259–268.

Celcroix, J.L., *Introduction to the Theory of Ionized Gases*, Interscience Publishers, New York, 1960.

Chapman, S., and Cowling, T.G., *The Mathematical Theory of Non-Uniform Gases*, 2nd ed., Cambridge University Press, 1964.

De Hoffmann, F., and Teller, E., Magneto-hydrodynamic shocks, *Phys. Rev.*, vol. 80, no. 4, pp. 692–703, 1950.

Friedrichs, K.O., and Kranzer, H., Nonlinear wave motion, Notes on magnetohydrodynamics, VIII, N.Y. University Rept. NYO 6486, July 31, 1958.

Heald, M.A., and Wharton, C.B., *Plasma Diagnostics with Microwaves*, John Wiley & Sons, New York, 1965.

Hirschfelder, J.O., Curtiss, C.F., and Bird, B.B., *Molecular Theory of Gases and Liquids*, 2nd ed., John Wiley & Sons, New York, 1954.

Howatson, A.M., *An Introduction to Gas Discharge*, Pergamon Press, Oxford, 1975.

Jackson, J.D., *Classical Electrodynamics*, 2nd ed., John Wiley & Sons, New York, 1975.

Jahn R.G., *Physics of Electric Propulsion*, McGraw-Hill, New York, 1968.

Kong, J.A., *Electromagnetic Wave Theory*, John Wiley & Sons, New York, 1986.

Krause, J.D., *Electromagnetics*, 1st ed., McGraw-Hill, New York, 1953.

Landau, L.D., and Lifshitz, E.M., *Electrodynamics of Continuous Media*, Pergamon Press, New York, 1960.

Langmuir, L. Oscillations in ionized gases, *Proc. Natl. Acad. Sci.*, vol. 14, no. 8, 1928.

Langmuir, L. The interaction of electron and positive ion space charge in cathode sheath, *Phys. Rev.*, vol. 33, no. 6, 1929, p. 954.

Levin, E., and Wright, M.J., Collision integrals for ion–neutral interactions of nitrogen and oxygen, *J. Thermophys. Heat Transfer*, vol. 18, no. 1, 2004, pp. 143–147.

Mitchner, M., and Kruger, C.H., *Partially Ionized Gases*, John Wiley & Sons, New York, 1973.

Raizer, Y.P., *Gas Discharge Physics*, Springer-Verlag, Berlin, 1991.

Shang, J.S., Huang, P.G., Yan, H., and Surzhikov, S.T., Computational simulation of direct current discharge, *J. Appl. Phys.*, vol. 105, 2009, pp. 023303-1-14.

Shang, J.S., Surzhikov, S.T., and Yan, H., Hypersonic nonequilibrium flow simulation based on kinetics models, *Frontiers Aerospace Eng.*, vol. 1, no. 1, November 2012, pp. 1–12.

Spitzer, L., *Physics of Fully Ionized Gases*, Interscience Publishers, New York, 1956.

Surzhikov, S.T., and Shang, J.S., Two-component plasma model for two-dimensional glow discharge in magnetic field, *J. Comput. Phys.*, vol. 199, no. 2, September 2004, pp. 437–464.

Sutton, G.W., and Sherman, A., *Engineering Magnetohydrodynamics*, McGraw-Hill, New York, 1965.

2

IONIZATION PROCESSES

INTRODUCTION

The ionization process of atoms or molecules is through energizing the electrons causing them to leave stable orbit revolving around the nucleus and becoming free-moving charges. The typical ionization potential of an atom or a molecule for most gas species is on the order of 10 eV (1 eV = 1.602×10^{-19} J), which is generally much greater than the energy levels of all other internal degrees of excitation. The excited state also decays spontaneously to a lower energy state by emitting photon in a time scale of 10^{-8} seconds (Raizer, 1966; 1991). Thus, the ionized state must be maintained by persistent energy sustainment. The main mechanisms for ionization consist of thermal, electron impact, chemical reactions, photoionization, and microwave or electron cyclotron resonance. In practical applications, the ionization process is traditionally classified by the pressure condition in which the plasma is generated and sustained (Table 2.1).

In a low-pressure environment up to a level of few hundred torr (mm Hg), the processes widely used in laboratories are discharges by glow, capacity and induction coupled, cascaded arc, and microwave. The last process for discharge generation is also called by other terms such as radio frequency (RF) or microwave, helicon, and electron cyclotron resonance discharge. The generated plasma commonly has a relatively low charged number density and low current density with a value around $10^9/cm^3$ and a few mA/cm^2, respectively. Most of these discharges are referred to as nonthermal plasma because the heavier ions and neutrals exist at room temperature

Computational Electromagnetic-Aerodynamics, First Edition. Joseph J.S. Shang.

TABLE 2.1 Typical ionization potentials of gases

Species	Ar	He	H	H_2	O	O_2	N	N_2	NO
I (eV)	15.8	24.6	13.60	15.40	13.62	12.2	14.55	15.58	9.25

and only the electrons attain the temperature of few electron volts (1.16×10^4 K). Thus, the nonthermal plasma is always in a nonequilibrium thermodynamic state (Clark et al., 1964).

Most plasmas studied by computational magneto-aerodynamics are generated at atmospheric pressure in the form of corona, dielectric barrier, capacitive (or RF by a frequency around 13.6 MHz), thermal, and arc discharge. Except for arc and thermal ionizations that can attain a temperature of 50,000 K or higher, the rest are referred to as cold or nonthermal plasma and exist in a nonequilibrium thermodynamic state. For thermal ionization, the high-temperature or high-enthalpy medium is the result of strong shock compression and finite-rate chemical reactions in hypersonic flows that stands out from ionization by an externally applied electric field. The greater mobility of the free electrons and their vastly different collision cross sections from ions for momentum and energy transfer always lead to the nonequilibrium thermodynamic condition for a weakly ionized gas. The charge number density of thermal ionization is slightly higher than that by electron impact to reach a value of 10^{14}/cc or more, but the electric conductivity is still very low (generally lower than 100 mho/m) with the possible exception of the arc discharge.

The plasma generated at high pressure has the advantage of an accentuated Joule heat effect and the additional radiative energy transfer mechanism. The high-pressure plasma generator is used and is often called as a plasmatron coupled with induction, resonance, microwave, and arc systems. The power output of this generator ranges from hundreds to thousands of kilowatts. The inductive coupled ionization at RF to energize a helical flow structure is most often used. On the contrary, the plasmas generated in the microwave frequency are sustained by dissipated energy from an induced alternative current (AC) in waveguide by a transverse electromagnetic wave. Similarly, plasma generations are accomplished by discharge in a resonator, and interacting with a plane electromagnetic wave. The optical or laser discharge is also possible, although the absorption coefficient of plasma in spectra of light falls off steeply with frequency but still can be made to be very efficient.

Any steady discharge can be developed by a plasmatron that is based on ionization by forcing a cold gas jet stream through an enclosed electric field. The energy transfer is provided by either an inductive current or resonance of a microwave that matches the plasma frequency. The microwave plasmatrons are not much different from the waveguide discharge scheme. It can have very high energy conversion efficiency up to 90%. The generated process attains a temperature ranging from 4000 to 6000 K, which is lower than RF plasmatrons with a range from 9000 to 10,000 K.

In this chapter, a more detailed description for the ionization is focused on the basic mechanisms of photoemission, thermal, and electron impact discharges. Since the ionization occurs at the atomic and molecular scales, the process will be studied by the models of quantum chemical-physics kinetics.

2.1 MICROSCOPIC DESCRIPTION OF GAS

In computational simulation of high-temperature gas dynamics, the overall behavior of gas motion is described by continuum theory. The macroscopic properties of the fluid such as pressure, density, mass-averaged velocities, and specific internal energy are treated by kinetic theory. However, the individual properties of the particle in a highly excited state must be evaluated by the law of quantum mechanics. Quantum theory recognizes that the energy transfer through absorption and emission is not continuous but through discrete quanta, and the size of quantum depends on the natural frequency of wave motions (Leighton, 1959).

Planck discarded the ancient maxim governs the energy distribution by considered the energy of an atom has no fixed value but depended on the frequency of oscillatory motion (Moore, 1963),

$$\varepsilon = h\nu \tag{2.1a}$$

where the Planck constant in SI unit is 6.6252×10^{-27} erg-s and ν denotes the fundamental frequency. From spectroscopic measurement of hydrogen atom, a regular relationship between the frequency and the atomic line of hydrogen in the visible region of the spectrum was discovered. This finding leads to discrete energy level between quantum states from the first Bohr orbit or the ground state is given by $n = 1$. The quantum transition of energy between $n > 1$ and ground state is (Leighton, 1959)

$$\varepsilon_n - \varepsilon_1 = \frac{2\pi^2 Z^2 e^4 m}{h^2}\left(1 - \frac{1}{n^2}\right) \tag{2.1b}$$

The coefficient $2\pi^2 Z^2 e^4 m / h^2$ is known as the Rydberg constant, which has a value of 109,737 cm^{-1} and the symbol Z is the number of elementary charges in an atom.

The governing wave mechanics of quantum mechanics is the Hamiltonian of a conservative system for kinetic and potential energy (Leighton, 1959). The second-order partial differential equation of quantum mechanics of a single particle of mass m and potential function V is the Hamiltonian operator that forms the Schroedinger wave equation,

$$\frac{h^2}{2m}\nabla^2\varphi - V(r)\varphi + ih\frac{\partial\varphi}{\partial t} = 0 \tag{2.2}$$

Quantum numbers are the indisputable conclusions based on many physical observations. From pure mathematics' view point, the quantum numbers are eigenvalues of the partial differential equation for each degree of freedom in space. The eigenfunction φ of the three-dimensional wave equation is sometimes called probability amplitude function that is required to be single-valued, finite, and continuous for all physically possible values in space. The solutions of the wave equation represent a set of discrete stationary energy state for an atom, characterized by quantum number n, l, and m_l for three-dimensional motions.

TABLE 2.2 Electron configurations of gaseous atoms

Shell	K	L	M	N
Subshell/element	1s	2s 2p	3s 3p 3d	4s 4p 4d 4f

The quantum numbers describe the energy state of an atom that can be briefly summarized by following a single or multiple electrons. The dynamic states of an atom under the influence of a magnetic field in rotational and spinning motions are designated by quantum numbers of n, l, m_i, and s. The principal quantum number n is a discrete integer number that is originated from Bohr orbits in radial coordinate, $n = 1$, 2, 3,.... The quantum number l denotes the azimuthal quantum number related to angular momentum, $l = 0, 1, 2,... (n - 1)$. The magnetic quantum number is related to orientations of electrons under influence of an external magnetic field. The particular behavior of electronic motion reveals that the reaction of angular momentum vector in a field of magnetic force is characterized as $m_i = \pm l, \pm(l - 1)$, ... Finally, the spin quantum number of atom is designated by s, and it has a fixed value for a single-electron atom of 1/2 , and the component of spin quantum number in an external magnetic field is identified as m_s (Clark et al., 1965).

In principle, one can construct on eigenfunction corresponding to the energy states in which a definite set of orbits is occupied, but the energy state cannot specify which electron occupies a particular orbit. Besides, there are no distinct orbits of electron, which perhaps shall be described as the shells as electron clouds about the nuclei (Vincenti et al., 1965).

Four different elements or subshells exist for each set of energy levels or shells K, L, M, and N: they are designated as S, P, D, and F (Hertzberg, 1944). The electron configurations of gaseous atoms can be given as (Moore, 1963) in Table 2.2.

In each shell, the energy levels are further distributed into subshells or elements. In general, the transition of energy level is from the higher s subshell to the lowest p subshell. Analogously, f subshell has a higher energy level than the d subshell. In most atoms, the energy of the various clouds or subshells lies in the order of *1S*, *2S*, *2P*, *3S*, *4S*, *4P*, *6S*, *4F*, *5D*, *7S*, *6D*, etc., where the *1s* orbit has the lowest energy. For example, *3p* means the quantum numbers $n = 3$ and $l = 1$. The notation of $1S^2$ denotes that two electrons are in the *1s* subshell. And $4P^3$ means that three electrons are in the 4p subshell. The general energy level specification of a configuration is *1S*, *2P* for the various subshells with the exponent indicating the number of electrons in the L shell, namely, $(1S)^2(2S)^2(2P)^6$. Therefore, the ground state of a nitrogen atom is described as S^2P^3 and for atomic oxygen is S^2P^4. The energy of next level of energy state is the energy necessary to remove an electron completely from the field of the nucleus and is the potential of ionization.

The slightly different energy levels in a quantum state are indicated by the upper left index *1*, *2*, or *3* on subshell or element; $^2S,^2P,^2D$... and the quantum number l for doublets terms are indicated as $^2S_{1/2}, ^2P_{1/2}, ^2P_{3/2}, ^2D_{5/2...}$. For example, the lowest orbits of one electron of hydrogen atom is in the 1S orbit and the ground state is $^2S_{1/2}$. The lowest electron configuration of nitrogen atom has three electrons $^2P^3$, which give the terms $^4S, ^2D, ^2P$, and the ground state is therefore $^4S_{3/2}$. The invert triplet ground states of oxygen are given as $3p_2, 3p_1, 3p_0$.

The Pauli Exclusion Principle governs the basic rule of quantum mechanics, namely that no two electrons of the same atom can have the same four quanta numbers of the same atom. For example, the unexcited nitrogen and oxygen atom have electron configuration of $(1s)^2(2s)^2(2p)^3$ and $(1s)^2(2s)^2(2p)^4$, respectively.

The molecular quantum number can be described similarly as that of the atom, except in molecular spectra, the electron spin is important. Both the spin and orbital magnetic moments of the electron and their interaction must be considered (Hertzberg, 1950). The molecular orientation is also analyzed by the quantum mechanics in which the rotational momentum has different directions. Thus, the difference in quantum number of molecular and atom appears as an increased distinguishable states or the degeneracy of the same energy level.

Five quantum numbers specify the spin–orbit interaction of molecules (Clark et al., 1965): Λ denotes the component of electron orbital momentum. The orbital momentum is designated as Σ state for $\Lambda = 0$, and the electron states are invariant about the symmetric axis. The electronic state is denoted by Π for $\Lambda = 1$, and the Δ state is characterized by $\Lambda = 2$. In addition, the $\pm$ sign designates whether a reflection in a plane the symmetric axis lies, and the subscripts g (gerde) and u (ungerde) tell whether the wave function remains invariant or not, upon an inversion at the center of symmetry; $\Sigma_g^+, \Sigma_u^+, \Sigma_g^-$ and Σ_u^-.

The total spin angular momenta of the electrons are designated as S. The commonly adopted notations v and J are the vibrational and rotational quantum numbers, respectively. Finally, the orbital angular momentum about the axes perpendicular to the internuclear axis is identified as K. For example, the most diatomic molecule in the electronic ground state is designated as $^1\Sigma$ for $\Lambda = s = 0$ and with a statistical weight of unity.

The number of energy levels and the number of possible quantum transitions between molecules are much greater than those for atoms. Therefore, molecular spectra are much more complex than those of atoms (Zeldovich et al., 2002). In addition, the individual spectral lines are closely packed in certain region that they are almost continuous. In order to describe the simultaneous changes in the vibrational and electronic energy states of a molecule, information from vibronic spectroscopy is often included such as the bandwidth of the electronic excited states or stable and unstable molecules.

Another designation of the quantum number for molecule is consistent with respect to that of the atom, except that the transition in a band system is important. For example, for an oxygen molecule, in the Schumann–Runge band (B→X) the energy states are designated as $B^3\Sigma_u^-$ to $X^3\Sigma_g^-$. For a nitrogen molecule transition of the second positive band, C→B are described as $C^3\pi_u$ to $B^3\pi_g$, the transition from the first positive band; B→A is given as $B^3\pi_g$ to $A^3\Sigma_u^+$. These processes are also the most encountered transitions in air ionization. Finally, in the Kaplan forbidden band A→X, the end state is presented as $X^1\Sigma_g^+$. The intensity of allowable vibronic transitions is governed by the Franck–Condon principle (Figure 2.1) (Zel'dovich et al., 2002).

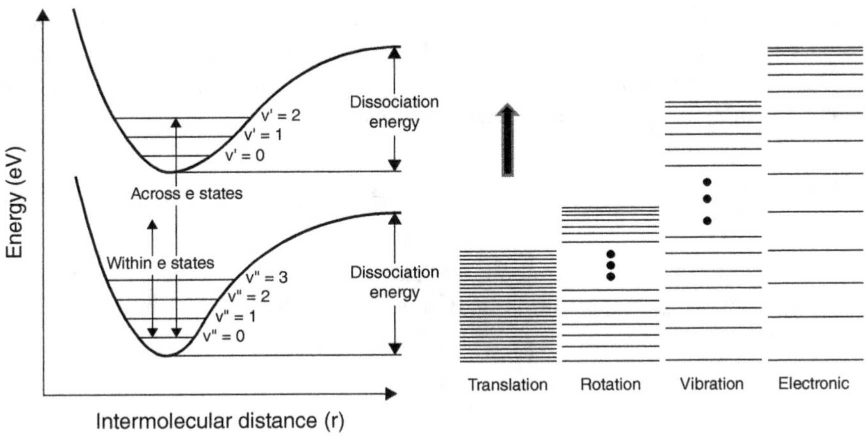

FIGURE 2.1 Quantum spectrum of internal degree of freedom of gas molecule.

The specific state and energy of molecules and atoms are treated by their internal degrees of freedom. The translational and rotational internal degrees of gas molecules are fully excited at the standard atmospheric condition. Far above that temperature or more precisely above a certain energy level, the vibrational modes of molecules begin to excite. As the energy of the gas is further elevated by any external sources, dissociation will take place and the atomic constituents will ionize. Meanwhile, free charges will generate a long-range electrostatic Coulomb field.

The energy of a molecule is measured above its zero-energy or ground state and is the sum of the energy of translational, rotational, vibrational, and electronic degrees of freedom.

$$e_i = e_t + e_r + e_v + e_e \tag{2.3a}$$

For atoms, the total energy consists of only the translational and electron energy,

$$e_i = e_t + e_e \tag{2.3b}$$

The inception of the quantum concept to gas dynamics by Planck in 1901 has a deep root in electromagnetics. Quantum mechanics dominated atomic and physics of elementary particles. The collision and the energy states are quantum restricted; in other words, the energy is transferred in discrete units and the transfer process is instantaneous because there is no continuity between the energy states.

According to de Broglie's hypothesis, the wavelength of a particle of mass m and the particle's velocity u are related by $mu = h/\lambda$, where the Planck constant has a value in Joules of $6.62 \times 10^{-34}\text{J} \times \text{s}$. In free space with a bounding dimension l, there is an integral number of half wavelengths, $l = n\lambda/2$. The allowed energy level for the translational degree of freedom is then

$$e_t = \frac{h^2}{8m}\left[\left(\frac{n_x}{l_x}\right)^2 + \left(\frac{n_y}{l_y}\right)^2 + \left(\frac{n_z}{l_z}\right)^2\right] \tag{2.4}$$

The vast majority of energy levels of translational motion are so closely packed relative to the datum κT that the energy levels are practically continuously distributed.

From the quantum solution of rotational motion of a diatomic molecule modeled by a rigid dumbbell, the acceptable energy levels are

$$e_r = \frac{J(J+1)h^2}{8\pi^2 I}, \quad J = 0, 1, 2, \ldots \tag{2.5a}$$

where I is the moment of inertia about two axes of rotation for a diatomic molecule, and the degeneracy of the partition function is $g_i = (2J + 1)$. The characteristic rotational temperature is given as

$$\Theta_r = h^2/8\pi^2 Ik \tag{2.5b}$$

The values have been found from spectroscopic analysis. For atomic nitrogen, oxygen, and nitric monoxide, the characteristic temperatures are 2.86, 2.07, and 2.42 K. Therefore, it may conclude that the rotational degree of freedom of air is fully exited at the standard atmospheric condition.

Quantum mechanics assigns the discrete vibrational energy levels for a simple harmonic vibrator as

$$e_v = (v + 1/2)hv, \quad v = 0, 1, 2, 3 \ldots \tag{2.6a}$$

The levels are not degenerate because the ground state has a nonzero value, and the frequency is

$$v_v = \sqrt{f/2\pi^2 m} \tag{2.6b}$$

for both atoms of a diatomic gas having the same mass. One can write that the lowest level of the vibrational excitation is $hv/2$, and the oscillator shall always have a residual energy greater than 0. It is known that the diatomic molecule does not behave like a simple harmonic vibrator but an anharmonic vibrator to allow dissociation. For a homopolar binding molecule such as hydrogen or oxygen, quantum mechanics permits the energy levels as (Clark et al., 1964)

$$e_v = (v + 1/2)hv_e - (h^2 v_e^2/4d_e)(v + 1/2)^2 \tag{2.6c}$$

where v_e is the frequency of the lowest vibrational state and d_e is the dissociation energy of the molecule measured from the zero energy reference. The characteristic temperatures for vibration and dissociation are $\Theta_v = hv_e/k$ and $\Theta_d = d_e/k$. For atomic oxygen, the respective characteristic temperatures are 2230 K and 59,000 K, while for hydrogen, the values are 6000 K and 50,000 K. From these values, the vibrational quantum levels for oxygen and hydrogen are estimated to be around 50 and 13.

The electronic or nuclei energy is from ionic or homopolar bonding that are intimately connected to the electronic structure of the atom. A third kind of attraction that is the van der Waals binding, but the quantum energy levels are degenerated. For the present purpose, the electronic contribution to the internal energy can be given as functionary,

$$e_e = e_e(g_i, \Theta_i, \Theta_i/T), \quad i = 0, 1, 2, \ldots \tag{2.7}$$

In this equation, g_i is the degeneracy and Θ_i denotes the characteristic temperatures for electronic excitation. In order to describe the electronic configuration of an atom, Pauli' exclusive principle must be held for that no two electrons in the same atom can have the same set of quantum numbers.

A thermodynamic system consisting of a group of molecules or atoms of the N_i number of population at different energy levels is defined by the accepted terminology as the population distribution or macrostate. At any instant the population distribution can be different, but the total energy of the system is

$$E = \sum_j N_j e_i \tag{2.8}$$

It is important to point out an essential idea of statistical mechanics that is accounting for all of the most probable distributions or macrostates. Thermodynamic equilibrium of a system contains the maximum number of macrostates.

2.2 MACROSCOPIC DESCRIPTION OF GAS

The bridge between the microscopic and macroscopic descriptions of gas is established by the distribution functions based on statistic mechanics and the concept of probability. There are two statistics for accounting the microstates according to even or odd number of elementary particles—Bose–Einstein and Fermi–Dirac—which lead to the thermodynamic probability

$$P = \prod_j \frac{(N_j + g_j - 1)!}{(g_j - 1)!N_j!} \text{ (Bose–Einstein, even)} \tag{2.9}$$

$$P = \prod_j \frac{g_j!}{(g_j - 1)!N_j!} \text{ (Fermi–Dirac, odd)} \tag{2.10}$$

In these equations, g_i is the degeneracy that indicates that an energy level can be constituted by a number of different states. The values of degeneracy are obtained from quantum mechanics from the eigenvalues analysis or spectroscopic measurements.

The most probably macrostate is found by using the Sterling formula for the factorial function and by assuming the probability function as a continuous function of N_j's. The maximum probability is solved by the Lagrange's method of undetermined multiplier. In spite of the difference between the two probabilities, the most probable macrostate of a system is nearly identical,

$$(N_j)_{\max} = g_j/(e^{\alpha+\beta e_j} - 1) \quad \text{or} \quad (N_j)_{\max} = g_j/(e^{\alpha+\beta e_j} + 1) \tag{2.11a}$$

Determination of the coefficients α and β must be achieved at the limiting condition. Under this situation, the value of unity of both most probably microstates is negligible in comparison with the exponential term and the microstate becomes

$$N_j^* = N\frac{g_i e^{-\beta\varepsilon_j}}{\Sigma g_i e^{-\beta\varepsilon_j}} \tag{2.11b}$$

where β is determined to be $\beta = 1/\kappa T$. For all practical purposes, β is the actual distribution over the energy group of particles in a system in equilibrium. This important and fundamental distribution is the Boltzmann distribution or frequently called the Maxwell–Boltzmann distribution. In fact, this distribution establishes the definable relationship between the properties of microscopic and macroscopic states. The sum in the dominator of equation (2.11b) is the molecular partition function Z, which distributes the energy of all particles among all the energy groups

$$Z(V,T) = \sum_i g_i \exp(-e_i/\kappa T) \tag{2.12}$$

From the classification of the internal degrees of excitation in translation, rotational, vibrational, and electronic modes, the partition function of a molecule is the product of all

$$\begin{aligned} Z(V,T) &= [\Sigma g_i \exp(-e_{i,t}/\kappa T)][\Sigma g_i \exp(-e_{i,r}/\kappa T)][\Sigma g_i \exp(-e_{i,v}/\kappa T)] \\ &\quad \times [\Sigma g_i \exp(-e_{i,e}/\kappa T)] \\ &= Z_t(V,T)Z_r(V,T)Z_v(V,T)Z_e(V,T) \end{aligned} \tag{2.13}$$

It is important to know that the factorization property of the partition functions is only valid so long as each energy mode can be assigned an energy level that is independent from the other energy modes. In most circumstances, the coupling between modes is ignored.

From the partition function, all thermodynamic properties of a system can be easily evaluated from statistics mechanics. Hence,

$$\begin{aligned} E &= N\kappa T^2 \left(\frac{\partial \ln Z}{\partial T}\right)_v \\ P &= N\kappa T \left(\frac{\partial \ln Z}{\partial V}\right)_t \\ S &= N\kappa \left[\left(\ln\frac{Z}{N} + 1\right) + T\left(\frac{\partial \ln Z}{\partial T}\right)_v\right] \end{aligned} \tag{2.14}$$

One needs to bear in mind that equations (2.14) have their theoretical limits. At the condition that involves strongly interacting particles, the results of statistic mechanics will fail. Thus, these relationships are applicable only to perfect gas in equilibrium condition.

The partition functions of internal degrees of freedom from quantum mechanics can be given as

$$Z_t = \left(\frac{2\pi m\kappa T}{h^2}\right)^{3/2} V$$

$$Z_r = \frac{8\pi^2 I\kappa T}{\sigma h^2}; \ \sigma \text{ symmetric factor}$$

$$Z_v = \frac{1}{2\sinh(h\nu/2\kappa T)} \text{ or } \frac{\exp(-h\nu/2kT)}{1-\exp(-h\nu/\kappa T)} \text{ Simple harmonic oscillator} \quad (2.15a)$$

$$Z_v = \exp\left(-\frac{\Theta_v}{2T}\right)\frac{[1-\exp[-v_{\max}+1)\Theta_v/T]}{1-\exp(-\Theta_v/T)} \text{ Anharmonic oscillator}$$

$$Z_e = g_o + g_1\exp(-e_1/\kappa T) + g_1\exp(-e_1/\kappa T) + \cdots$$

The vibrational excitations described by the simple harmonic and anharmonic oscillator models at high vibration quanta are significantly different. Figure 2.2

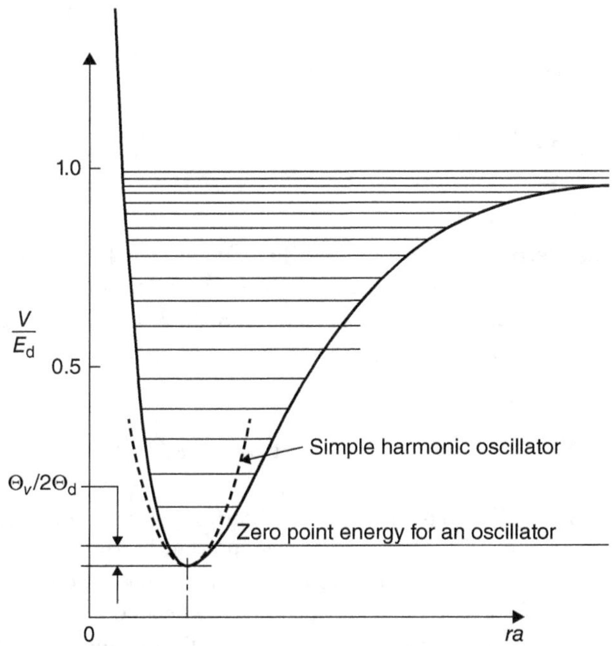

FIGURE 2.2 Vibration energy levels.

displays the needed improvement to physical fidelity for vibration degree of freedom by the simple harmonic oscillator mode based on Morse potential. The partition function is inaccurate at highly excited quantum states. However, there is no close form expression for describing the electronic excitation; thus, the partition function by anharmonic oscillator is often adopted.

Some typical partition functions of air components are included at here for better understanding. The ground state of molecular oxygen has a triplet state due to the effects of electron spin, which is designated as $^3\Sigma$ with two uncoupled electron spins; that is, $S = 1$ and $\Lambda = 0$. The partition function for molecular oxygen is

$$Z_e = 3\exp(-e_o/\kappa T)[1 + 2/3\exp(-11,300/T) + 1/3\exp(-18,900/T) + \cdots] \tag{2.15b}$$

For the atomic oxygen that has a triplet electronic ground state,

$$\begin{aligned} Z_e = 5\exp(-e_o/\kappa T)[1 + 3/2\exp(-228/T) + 1/5\exp(-327/T) \\ \exp(-22,700/T) + \exp(-48,400/T) + \cdots] \end{aligned} \tag{2.15c}$$

One notes immediately that the characteristic temperatures of 228 and 327 K are relatively low at the lowest ground states. Therefore, the electronic excitation partition function of atomic oxygen can be approximated as

$$Z_e \cong 9\exp(-e_o/\kappa T) + \cdots \tag{2.15d}$$

The electronic partition function for nitrogen atom is

$$Z_e = 4\exp(-e_o/\kappa T)[1 + 10/4\exp(-27,500/T) + 6/4\exp(-41,200/T) + \cdots] \tag{2.15e}$$

The final example of the electronic partition function is for the nitro monoxide. This chemical species is always presented in the ionization of air at atmospheric condition. Its ground state is characterized at $^2\pi$ with $S = 1/2$; thus, a multiplet of the state is splitting to states of $^2\pi_{1/2}$ and $^2\pi_{3/2}$. The electronic partition function becomes

$$Z_e = 2\exp(-e_o/\kappa T)[1 + \exp(-178/T) + \cdots] \tag{2.15f}$$

In the above expressions, equation (2.15f), partition function contains any additional terms with a magnitude less than a magnitude of $\exp(-50{,}000/T)$ has been truncated. The above electronic partition functions, in fact, reflect a huge quantum jump between the equilibrium temperature above the ground state.

It must also be noted that only the partition function of the translational energy mode contains dependent variables in volume. According to equations (2.14) and

(2.15a), the only internal mode of translation can contribute to pressure of a thermodynamic system. The rest of internal energy modes would not contribute to the pressure of the system.

2.3 CHEMICAL REACTIONS AND EQUILIBRIUM

Our understanding of chemical kinetics is based on the interatomic and intermolecular collision process. For finite rate chemical reactions, the concept of local equilibrium must be used in the computational procedure. The relaxation time scale in a nonequilibrium situation is thus not unique. For this reason, the designation of a chemical reacting flow field as equilibrium, frozen, and finite-rate by comparing the relative time scales between molecular reaction and organized gas motion lacks rigor. The relaxation time scale may be adopted only as a convenient way to simplify analysis.

A chemical reaction by collision processes always contains reactants, activated complex, and product. A transitory compound is formed by accumulating a sufficient amount of activation energy for reaction to take place. This amount of energy, required to move the reactant over the energy barrier in order for the reaction to begin, is usually greater than the standard heat of formation. Conditions of the chemical reacting system affecting the rate of chemical reactions in macroscopic states are numerous, such as concentrations of the chemical composition, temperature, pressure, and the presence of catalyst or inhibitor. However, the complexity of chemical reactions has not been completely resolved, because on microscopic scales, the energy of relative motion of colliding particles, and orientation of particles at collision, and the so-called steric factor are insoluble issues in quantum chemistry (Kuo, 2005).

A chemical reaction is an irreversible process. The chemical composition change in a system takes place toward its equilibrium value. Any chemical reaction can be expressed as a group of simple one-step reactions according to permissible chemical kinetics:

$$\Sigma v_i' C_i \rightleftarrows \Sigma v_i'' C_i \tag{2.16}$$

The symbols v' and v'' are referred to as the stoichiometric coefficients, and C_i is the species concentration in mole per unit volume. In some way, equation (2.16) may be viewed as a statement of conservation of the total number of atoms of species in the one-step reaction process. The original law of the mass action is an empirical relationship that has been confirmed by numerical experimental observations that is

$$dc_i/dt = \kappa \prod_{i=1}^{n} (C_i)^{v_i} \tag{2.17}$$

It states that the rate of reaction of a chemical species is proportional to the products of the concentrations of the reacting chemical species, each concentration being raised to a power equal to the corresponding stoichiometric coefficients. The coefficient κ is the specific reaction rate constant. It is independent of the species concentration and depends only on the temperature

$$\kappa = aT^b \exp(-E_a/R_u T) \tag{2.18}$$

This equation is referred as the Arrhenius law. The leading term aT^b represents the collision frequency of the reacting species. The exponential term is known as the Boltzmann factor that indicates the fraction of collisions that have energy levels greater than the activated energy to initiate the chemical reaction. All these parameters are unique for a specific chemical reaction.

The direction of chemical reaction proceeds from a state of higher to a lower state of potential

$$\sum_{i=1}^{n} \nu_i' \phi_i > \sum_{i=1}^{n} \nu_i'' \phi_i \tag{2.19}$$

In this equation, the chemical potential of species i per molecule is introduced as $\bar{\mu}_i$, and the first law of thermodynamics, including chemical reaction, becomes (Anderson, 1989)

$$dE = TdS - PdV + \sum_{i=1}^{n} \phi_i dN_i \tag{2.20}$$

The chemical reactions are still explicitly excluded from the system by the condition that N_i shall be varied independently of each other. For an irreversible change occurring in a constant mass mixture, the following generalization of thermodynamic quantities, the Gibbs thermodynamic potential with chemical reactions can be written as

$$dG = -SdT + Vdp + \sum_{i=1}^{n} \phi_i dN_i \tag{2.21a}$$

For a perfect gas, the internal energy is a function of temperature only,

$$Tds_i = dh_i - (1/\rho_i)dp_i \tag{2.21b}$$

Recall that the partial pressure of the species i is $p_i = \rho_i(R/w_i)T$. It follows that

$$T(\partial s_i/\partial p_i)_T = -1/\rho_i \quad \text{and} \quad T(\partial s_i/\partial p_i)_T = -R/p_i w_i \tag{2.21c}$$

Integrate the above equations at constant temperature to get a useful result.

$$s_i(p_i, T) - s_i(p^*, T) = -\left(\frac{R}{w_i}\right) \log\left(\frac{p_i}{p^*}\right) \tag{2.22}$$

This equation is derived for a gas in local thermodynamic equilibrium. Thus, a thermally perfect gas is equivalent by assuming that the individual molecules of the gas exert no long-range force on each other; thus, the potential energy of interaction is zero. The entropy S of the system will be a maximum once equilibrium has been reached and simultaneously the Gibbs potential has attained the minimum value for a fixed thermodynamic state.

Since the partial pressure of each chemical species is uniquely determined by the species concentration, the system is in chemical equilibrium if the partial pressures of the reactants satisfy the following condition:

$$\prod_{i=1}^{n}\left(\frac{p_i}{P*}\right)^{(v_i'-v_i'')} = \exp\left(-\frac{\Delta\phi}{RT}\right) \tag{2.23}$$

For convenience, an arbitrarily chosen value of the reference pressure can be set as one atmosphere. An equation for the equilibrium constant dependent on temperature can be then given as

$$\kappa(T) = \prod_{i=1}^{n} p_i^{(v_i'-v_i'')} \tag{2.24}$$

The rate of chemical reaction is usually studied as a function of the species concentration of the reactants and the temperature. The rates of forward and backward reactions are

$$\kappa_f(T) = \prod_{i=1}^{n} C_i^{v_i'} \quad \kappa_b(T) = \prod_{i=1}^{n} C_i^{v_i''} \tag{2.25}$$

And the net rate of progression of the nth reaction is the simply the difference of $k_f - k_b$. Arrhenius recognized that one of the most striking features of the chemical reaction rate is the strong temperature dependence. His observation led to the Arrhenius hypothesis. In fact, the Arrhenius equation is given as

$$\frac{d\ln(\kappa_f)}{dt} = \frac{E_a}{\kappa T^2} \tag{2.26}$$

For a nonequilibrium process, the total differential of the Gibbs free energy ($G = H - TS$) is

$$dG = \left(\frac{\partial G}{\partial T}\right) + \left(\frac{\partial G}{\partial p}\right) dp + Tds_{\text{irrev}} \tag{2.27a}$$

The relationship between the species concentration and the chemical reaction rate can be found from the first law of thermodynamics $dG = Vdp - SdT$. By a term-by-term comparison, it is shown that

$$Tds_{\text{irrev}} = \sum_i (\partial G/\partial N_i)dN_i \tag{2.27b}$$

At equilibrium, $dS_{\text{irrev}} = 0$, or the summation of the Gibbs free energy for all reacting species vanishes, which provides a mean for determining equilibrium chemical composition. The condition for equilibrium has an alternate, namely,

$$\Sigma \nu_i G_i = 0 \tag{2.27c}$$

Thus, the law of mass action becomes formally a relationship between the partial pressure of the reactants and the temperature of the reacting system

$$\kappa_p(T) = \Pi p_i^{\nu_i} = \exp\left(-\frac{\Delta G}{\kappa T}\right) \tag{2.28}$$

The values of $\kappa_p(T)$ can be obtained from the experimental observation or can always be calculated from statistic thermodynamics (Shang, 1991).

2.4 SAHA EQUATION OF IONIZATION

The equilibrium ionization of the mixture of electrons, ions, and neutral particles can be analyzed by the law of mass action and the electronic partition function. For a monatomic gas under the simplifying condition when the excitation above the ground state is ignored, the analytic formulation for ionization will reduce to the result by Saha (1920) that was derived from the classical thermodynamics.

The ionization process is rather complex; it includes inelastic collision, charge exchange, recombination of three-body collision, dissociation, and negative ions. In addition to the collision mechanisms for ionization, photoionization is also possible. Ionization of an atom is achieved by energizing the orbiting electrons, and ionization of molecules takes place by energy transfer through internal modes according to molecular structure. For a better understanding of the equilibrium ionization process, the equilibrium reactants of neutral particles, ions, and electrons can be analyzed by elementary one-step reactions. For a simple ionization reaction,

$$A^+ + e \rightleftarrows A \tag{2.29}$$

In this equation, A represents a neutral atom or molecule, A^+ is the corresponding singly charged ion, and e is an electron.

The law of mass action gives

$$\frac{N_a}{N_{a^+} N_e} = \frac{Z_a}{Z_{a^+} Z_e} \exp\left(\frac{I}{\kappa T}\right) \tag{2.30a}$$

where I is the ionization potential and Z_i's are the partition functions of the reactants. Since the net electric charge of the original neutral gas must always remain zero, we have $N_{a^+} = N_e$. By introducing the degree of ionization

$$\alpha = N_{a^+} / N_a^o \tag{2.30b}$$

and $N_a^o = \rho V/m_a$, we have then $N_{a^+} = \alpha N_a^o, N_e = \alpha N_a^o, N_a = (1-\alpha)N_a^o$. The law of mass action leads to

$$\frac{\alpha^2}{1-\alpha} = \frac{m_a}{\rho V}\left(\frac{Z_{a^+}Z_e}{Z_a}\right)\exp\left(-\frac{\Theta_i}{T}\right) \tag{2.31}$$

The symbol Θ_i is a characteristic temperature for ionization defined by $\Theta_i = I/\kappa$. The partition functions for each electronic species are the products of its translational and internal degrees of freedom. The partition functions of internal degrees of polyatomic molecular species shall include the rotational and vibration modes

$$Z_e = Z_{e,t}\Pi Z_{e,\text{int}}, Z_{a^+} = Z_{a^+,t}\Pi Z_{a^+,\text{int}}, Z_a = Z_{a,t}\Pi Z_{a,\text{int}} \tag{2.32a}$$

The only internal energy by an electron is associated with its spin, and only two permissible quantum energy states at the ground energy level. Thus $\Pi Z_{e,\text{int}} = Z_{e,\text{spin}} = 2$ and the translational partition function is

$$Z_{e,t} = V(2\pi me\kappa T/h^2)^{3/2} \tag{2.32b}$$

Finally, by recognizing the mass m_a can be determined from the equation of the state, $p/\rho = (1+\alpha)(\kappa/m_a)T$, the law of mass action equation becomes

$$\frac{\alpha^2}{1-\alpha^2} = \frac{1}{p}\left(\frac{2\pi m_e}{h^2}\right)^{3/2}\frac{2\Pi Z_{a^+,\text{int}}}{\Pi Z_{a,\text{int}}}\exp\left(-\frac{\Theta_i}{T}\right) \tag{2.33}$$

When A is a monatomic gas, the partition functions are those of the electronic excitation only. Since the masses of ions and molecules are nearly identical, the translational partition functions of these species are thus equal. If excitation above the ground state is ignored, the above equation is further simplified as

$$\frac{\alpha^2}{1-\alpha^2} = C\frac{T^{3/2}}{p}\exp\left(-\frac{\Theta_i}{T}\right) \tag{2.34}$$

This is actually the form that Saha (1920) first derived from classical thermodynamics. This simple equation provides invaluable insights. At low degrees of ionization or for a weakly ionized gas, the degree of ionization is proportional to the three-quarter power of the system temperature and inversely proportional to the square root of the pressure. However, it needs to be borne in mind that this observation is valid only under the simplified assumptions.

2.5 IONIZATION MECHANISMS

Ionization is the limiting case of electronic excitation when a bounded electron leaves the atom and passes into the continuous spectrum. The lowest ionization potential of the majority of gas atoms and molecules vary between 7 and 15 eV (I/k ~ 80,000–170,000 K). Therefore, the statistical weight of a freed electron is very high. The degree of ionization is increasing with temperature and the state of gas rarefaction.

The equilibrium ion or electron concentration satisfies a relationship identically to the law of mass action for chemical reaction and dissociation. Thus, the ionization or dissociation of atoms and molecules has been treated as a chemical reaction. The Gibbs free energy of a unit mass of ionized gas can be written as

$$G(V,T) = -\sum_i N_i \kappa T \ln \frac{Z_i e}{N_i} - N_e \kappa T \ln \frac{Z_e e}{N_e} \tag{2.35}$$

where Z_i and Z_e are the partition functions of an ion and an electron. In thermodynamic equilibrium and at the constant temperature and volume (T and V), the Gibbs energy is the minimum with respect to the number of particles.

Thermal excitation is the most natural process of ionization, because the energy source of the reacting mixture exceeds the ionization potential and the energy transfer between the heavy molecules is extremely efficient. The energized molecules distribute the gained energy to different internal degrees of freedom by the instantaneous quantum transitions and cascading processes. In a high enthalpy environment, the quantum jump may follow the climbing-the-ladder path via each adjacent quantum levels or the rapid multiquantum levels transition by the so-called big bang transition. Meanwhile, the chemical composition also undergoes a drastic change that leads to a chemical reaction dominant phenomenon. The vastly different characteristic temperatures of internal degrees of freedom require a corresponding different number of collisions to a new energy state. Thus, the relaxation process also becomes very pronounced. The system is no longer able to maintain an equilibrium state. When the degree of ionization is sufficiently high, the presence of a Coulomb force field introduces a strong forced diffusion mechanism for the transport properties. Nevertheless, the details of ionization mechanisms are identical for either thermal excitation or the electron impact process.

The ionization of an atom can be divided into mechanisms by the inelastic collisions between electrons and electrons and heavier atoms and molecule, as well as photoionization, electric impact, and charge exchange. The inelastic collision is a gas kinetic process or a purely thermal collision. However, even the high kinetic energy of slow particles is not effective in ionization because velocities of the heavy particles required the velocities to be comparable to the electron velocity in atoms that has a value of 10^8 cm/s.

Photoionization occurs when the electron acquires an energy exceeding its binding energy to an atom, molecule, or ion. Therefore, the ionization process is referred to as the bound-free transition. The excess of the photon energy over the binding energy is transferred into kinetic energy of the free electron. The cross section for the absorption of a photon $h\nu$ in the nth quantum state is

$$\sigma_{\nu n} = \frac{64\pi^4}{3\sqrt{3}} \frac{e^{10} m z^4}{h^6 \nu^3 n^5} = 7.9 \times 10^{-18} \frac{n}{z^2} \left(\frac{\nu_n}{\nu}\right)^3 \text{ cm}^2 \tag{2.36}$$

This equation is also known as Kramer's formula. Here z is the number of the elementary charges.

The rate equation for electric impact ionization from fundamental collision dynamics is

$$\frac{dn_e}{dt} = \alpha\, n_n n_e - \beta n_i n_e^2 \tag{2.37}$$

The value α and β are the coefficients of ionization and recombination by principle of detailed balancing,

$$\alpha = \kappa(T_e)\beta \tag{2.38a}$$

And the equilibrium constant can be determined by the law of mass action,

$$\kappa(T_e) = \frac{n_e n_i}{n_n} = \frac{g_i}{g_n}\frac{2(2\pi m_e \kappa T_e)^{3/2}}{h^3}\exp\left(-\frac{I}{\kappa T_e}\right) \tag{2.38b}$$

In this equation, g_i and g_n are the statistical weights of the energy level of the ion and the neutral molecule. Under gas discharge conditions, ionization by electron impact dominates over the photoionization because the cross sections of photoionization close to the ionizing threshold are rather high. However, photoionization always supplies the seeding electrons that start the electron avalanches in streamer propagation or microdischarge. It is important to note that the electron impact is proportional to the electron density while photoionization is proportional to the radiation frequency.

Finally, in certain weakly bonded atoms, an electron may be lost by a strong valence interaction with another atom or with an ion by a charge exchange. All these aforementioned mechanisms have a wide range of alternative mechanisms for ionization and the processes are very complex quantum physical–chemical interactions.

Ionization of a molecule has a significant and additional process by thermal excitation of internal degrees of freedom of molecule that can interact with the electronic structure. This process occurs in the high-temperature environment by exothermic chemical reactions and in hypersonic flows by strong shock wave compression. The ionization process is further complicated by the energy-cascading processes among the molecular internal excitations and enhanced interactions with media interface by the catalytic effect. On the contrary, the absorption, emission, and trapping of charged particles by a solid surface are closely tied to the largely unknown atomic and molecular binding energy of the surface material.

The free electrons are generated mostly by electronic collisions and, to some extent, by detachment of electrons from metastable molecules. Theoretical and numerical analysis for electron collisions has a widely used formula by Townsend

$$\alpha = Ap\exp[-B(p/E)] \tag{2.39}$$

For electrons, the average energy gained by a positive ion between collisions is proportional to *E/p* and the constants *A* and *B* are determined from experimental observation. The theoretical base of this formulation can be recognized immediately if $T_e \sim E/p$ in that the fraction of electrons in Maxwellian spectrum that is capable of ionizing an atom is known to be proportional to $\exp(-I/\kappa T_e) \sim \exp[-B(p/E)]$. The depletion mechanisms of electrons are the attachment and dissociation recombination. The depleting attachment like ionization is related to the drift motion of electrons. The dissociation recombination is the fastest mechanism of bulk recombination, which is temperature and pressure dependent to have a range from 10^{-6}–10^{-3} s, and the released energy is mostly transferred to the excitation energy of the atom.

Positively charged ion generation is also by electronic collision and the depletion mechanisms are dissociation and ion–ion recombination. Ion–ion recombination is the main mechanism for charge neutralization as that of electron attachment. At low pressure, the process is similar to charge transfer. At moderate pressure, the recombination is preceded through a triple collision. The maximum recombination coefficient is reached at the atmospheric pressure condition.

The negatively charged ion is generated by the attachment or dielectric recombination mechanism. This is an important, and sometimes, the main process of producing negative charges. However, the electron lifetime of the attachment is very short to be around 10^{-8} s. Electron attachment impedes discharge breakdown and makes it difficult to sustain the ionization state. The depletion mechanisms of negative ions are dissociative and detachment recombination. The recombination coefficient can be determined experimentally. At the very low negatively charge component concentration, a typical characteristic decay time is around 10^{-3} s.

The other ionization and neutralization mechanisms are also known as radiative recombination that is a two-way process of the electronic bound-free transition. When electron received energy from quanta of energy or photon to break up the electron bound is photoionization. The reverse process to photoionization is radiative recombination, where the potential energy and relative kinetic energy of the recombining electron–ion or ion–ion pair are released as a quantum of radiation. The reverse photoionization is also referred to as the Bremssstrahlung and to have a continuous emission and absorption spectrum. The probability of Bremsstrahlung associated with a neutral atom is much smaller than the probability with ion.

In summary, this area of scientific discipline is extremely challenged by virtue of deficient basic knowledge in quantum chemical physics. Additional issues are the initial values and boundary conditions for analyzing these quantum mechanics phenomena (Shang et al., 2014). In a high-temperature environment, the catalytic property of the media interface and electron losses to diffusion toward this boundary are very limited, and those losses are irreversible and beyond our ability by using even the *ab initio* approach in quantum chemical physics (Eyring et al., 1944).

2.6 PHOTOIONIZATION

In 1905, Einstein successfully applied Planck's quantum hypothesis to study photoionization (Pai, 1962). The kinetic energy of the electron is described by a simple equation

$$\frac{1}{2}mv^2 = h\nu - \phi \tag{2.40}$$

In this equation, ϕ is the energy needed from the photon or the electromagnetic wave to free the surface electron from the bonding energy. The supply energy seems to be continuous, but energy is actually transferred in discrete units by quantum jumps and energy transfer is instantaneous because there is no continuity between the state before and after energy transfer. The photoionization is basically a free-bound radiative transition associated with a two-body recombination reaction,

$$h\nu + A \rightleftarrows e + A^{+} \tag{2.41}$$

If the energy of a photon is greater than the ionization potential of the atom, it absorbs the radiant energy of the photon and ionization takes places. The excess photon energy over the bounding potential is transformed into kinetic energy of the free electron. The kinetic energy of the emitting electron is a function of the frequency of the incident light, $h(\nu - \nu_o)$, where ν_o is a minimum frequency below which no electron can be emitted. Varying the intensity of the incident light at a constant frequency does not affect the kinetic energy of the emitted electron but merely changes the number per unit time. The simplest estimate of the ionization rate is by the black body radiation in plasma as

$$\frac{dn_e}{dt} = \alpha \frac{2\pi\kappa^3}{h^2}\frac{T^3}{h^2} \exp\left[\left(-\frac{eI}{\kappa T}\right)^2 + \frac{2eI}{\kappa T} + 2\right] \tag{2.42}$$

where α is the coefficient for photon absorption and I is the first ionization potential of an atom.

Using the principle of detailed balancing, the photoionization cross section of a hydrogen-like atom is given by Kramers' formula. The basic characteristic of the capture cross section is inversely proportional to the cube power of the frequency and the cross section that has the maximum value at the absorption threshold $\nu = \nu_o$.

At the energy threshold for ionization excitation, the cross sections for inelastic collisions of heavy particles are several orders magnitude smaller than the cross sections of inelastic electron collisions. There is also a fundamental difference between electron impact and photoemission effects. Note that the number density of ionization generated by electron impact is proportional to the electron number density. On the contrary, the ionization by photoemission is proportional to the radiation density. Therefore, only in a rarefied gas environment, the photoionization can dominate over that of electron impact for ionization.

When rapid heating of gas such as by a strong shock compression, ionization by heavy molecule collisions is most important, and photoemission provides a small number of electrons playing a seeding process. Similarly, photoionization plays a crucial role in electric breakdown. Photoionization cannot compete with the electron impact ionization under discharge conditions, but it supplies seeding electrons to start the avalanche and streamer propagation. The collision cross sections for photoionization close to breakdown are high and the gas has few quanta above the ionization potential ($h\nu > I$) (Table 2.3) (Raizer, 1991).

TABLE 2.3 Collision cross sections for photoionization of gases

Species	Ar	N	O	O_2	N_2
$\sigma \times 10^{-18} cm^2$	35	9	2.6	1	26

2.7 THERMAL IONIZATION

Ionization by thermal collision takes place in high temperature by energized particles collisions. From the Maxwell velocity distribution, the rate of ionization is proportional to the total number of the participating neutral particle number and the ratio of ionizing collisions with sufficient energy. Another important parameter for the ionization rate is the mean free path of the gas medium and is ultimately related to the collision cross section. All these processes occur at the atomic and molecular scales. In linking the details of atomic and molecular structure to the global thermodynamic behavior of a gas mixture, statistically, the dynamics of a real gas is considered to consist of translational, rotational, vibrational, and electronic internal degrees of freedom. For translational motion, the majority of the quantum energy levels are closely spaced, relative to the datum level. Thus, the energy states are nearly continuously distributed. Since only a few intermolecular collisions are required to achieve equilibrium in the translational and rotational modes, these two internal energy states are generally considered to be fully excited and in thermodynamics equilibrium.

The thermodynamic properties of a gas at high temperature differ considerably from the standard atmospheric condition; hence, they affect the shock-layer thickness over blunt body as shown in Figure 2.3. In hypersonic flow, the thermal state by a

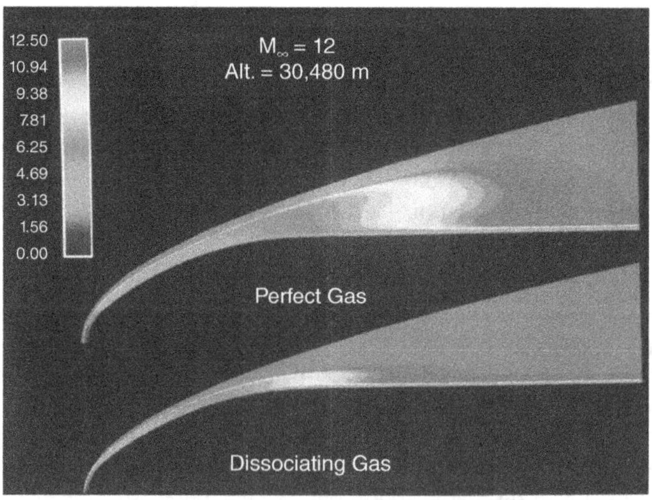

FIGURE 2.3 Shock layer structures of perfect and real gases.

strong shock compression is proportional to the free stream Mach number. The Mach number at a moderate hypersonic speed is sufficient to the chemical composition change of the atmospheric air by shock compression in the shock layer. For example, when the air temperature is higher than 1500 K, the excitation of the vibrational degree of freedom of the diatomic molecules, oxygen and nitrogen, becomes significant. As the temperature is increased, nitric oxide is formed by chemical reaction followed by the dissociation of oxygen, nitric oxide, and nitrogen. Only when the temperature is elevated beyond 9000 K does the ionization of air components occur. In addition to the effect of changing chemical composition with respect to temperature, the transport processes are affected by the individual species concentration, gas density, and the electrostatic force by the electrically charged species. The energy transfer among different internal degrees of freedom is controlled by the collisions process. Hence, the energy above the ground state of quantum transitions from each internal degrees of excitation is drastically different. Thermodynamic state is rarely in an equilibrium condition.

The separation of the internal structure beyond translational excitation becomes increasingly uncertain. For example, the centrifugal field of rotation affects vibration, and the vibration motion changes the moment of inertia of gyration. Vibrational excitation and dissociation are also closely coupled. This is because energy is stored in the vibrations and the rate of dissociation is dependent on the vibrational state of the molecules. There are possible mechanisms to transit a molecule from the vibrational states into the dissociation of free atom. One is the climbing-the-vibrational-ladder theory; the other is the theory of direct transition by violent collision with heavy particles—big bang. Regardless of the uncertainty in these processes and known discrepancy is existed from experimental evidences, the vibrational relaxation precedes the onset of dissociation, and the rate of dissociation is strongly dependent on the vibrational state of molecules.

The additional complication of ionization is introduced by the long-range electrostatic Coulomb field leads to forced diffusion and radiation. The avalanche of ionization has been described by Park (1989). In spite of the complex interconnected internal excitations of molecule, ionization has been analyzed as a chemical reaction. Our understanding of chemical kinetic is based on the intermolecular collision process. Therefore, the reaction induced by radiation will be excluded, but the radiative energy transfer is retained in the global energy conservation equation. For finite rate chemical reactions, the concept of local equilibrium must be used and the relaxation time scale in the nonequilibrium condition is not unique and is included by chemical-physics kinetic models.

At the microscopic scale, the interaction occurs at the atomic and molecular structures that are beyond the realm of gas kinetics theory. Therefore, the chemical reaction must be modeled by chemical kinetics or quantum chemical-physical kinetics. All chemical reactions can be expressed as a collection of elementary reactions,

$$\sum_i v'_{ii} X_i \Leftrightarrow \sum_i v''_i X_i \tag{2.43}$$

here ν_i' and ν_i'' are the stoichiometric coefficients for reactants and products, respectively. These coefficients are integers that denote the number of molecules partaking in the reaction, and X_i is the mole concentration of the reacting species.

The law of mass action is actually an empirical formulation confirmed by experimental observations. The law states that the rate of generation and depletion of a chemical species is proportional to the products of its concentration to a power corresponding to the stoichiometric coefficient ν_i'. The forward and backward reactions are

$$R_{f,i} = \kappa_{f,i}(T)\Pi(X_i)^{\nu_i'} \quad R_{b,i} = \kappa_{b,i}(T)\Pi(X_i)^{\nu_i''} \tag{2.44a}$$

where $\kappa_{f,i}(T)$ and $\kappa_{b,i}(T)$ are the reaction rate constants that are functions of temperature only and independent of the concentrations of the reacting species. The net rate of change in any species concentration is the sum of all elementary reactions, and the rate of reaction is not necessarily determined by the slowest rate of the elementary reactions. Hence, the stoichiometric coefficients may not be used for assessing the overall reaction rate.

$$\left(\frac{dX_i}{dt}\right)_j = (\nu_i'' - \nu_i')[\kappa_{f,i}\Pi(X_i)^{\nu_{i,j}'} - \kappa_{b,i}\Pi)(X_i)^{\nu_{i,j}''}] \tag{2.44b}$$

The continuity equation for nonequilibrium chemically reacting flow becomes

$$\frac{\partial \rho_i}{\partial t} + \nabla \cdot [\rho_i(u + u_i)] = M_i \left(\frac{dX_i}{dt}\right)_i \tag{2.45}$$

In the most recent approaches for analyzing unsteady nonequilibrium ionization, separate energy conservation equations for vibrational and electronic excitation are included with the global energy conservation law to appear as (Shang et al., 2014)

$$\frac{\partial \rho_i e_{iV}}{\partial t} + \nabla \cdot [\rho_i(\vec{u} + \vec{u}_i)e_{iV} + q_{iV})] = e_{iV}\frac{dw_i}{dt} + Q_{V,\Sigma}. \tag{2.46}$$

$$\begin{aligned}\frac{\partial \rho_i e_e}{\partial t} + \nabla \cdot [\rho_i(\vec{u} + \vec{u}_i)e_e + \vec{u} \cdot p_e\vec{I} + q_e)] &= e_e\frac{dw_i}{dt} + \vec{E} \cdot \vec{J} \\ &+ [\rho_e\vec{E} + (\vec{J} \times \vec{B})] \cdot (\vec{u} + \vec{u}_{\rm i}) + Q_{e,\Sigma}.\end{aligned} \tag{2.47}$$

In equation (2.47), $\vec{J}$ is the current density of the partially ionized gas. The vector fields of E and B are sum of applied and induced electric and magnetic field intensities of the electromagnetic field. The net sum of energy transfer between translational, vibration, and electronic excitations is designated as $Q_{V,\Sigma}$ and $Q_{e,\Sigma}$, respectively. These net exchange rates of these internal energies are determined by chemical-physics models.

The typical species concentration distributions of ionized air within a shock layer of blunt body in hypersonic flows are shown in Figure 2.4 to illustrate the current

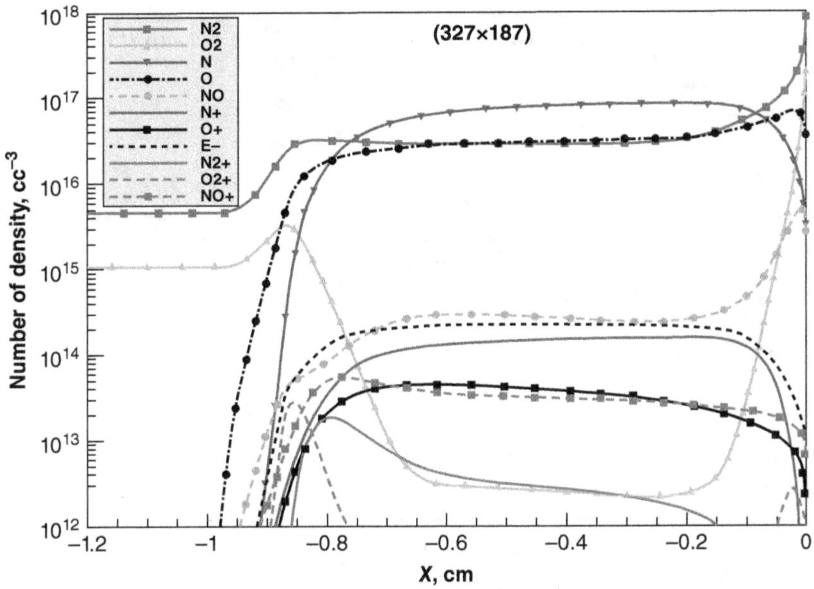

FIGURE 2.4 Species number density along stagnation streamline, $M = 23.9$.

simulation capability in describing the high-temperature air. The numerical results are generated at a characteristic Mach number of 23.9 and the Reynolds number of 1.95×10^4. The maximum translational temperature of the value of 18,750 K is confined within the transient region of the shock jump. The species concentrations in number densities per cubic centimeter along the stagnation streamline consist of 11 molecular, atomic, and ionized species, N_2, O_2, NO, N, O, N^+, O^+, NO^+, $O2^+$, NO^+, and e^-. Only species number densities in the range from 10^{12} to $10^{18}/cm^3$ are included. Two interesting observations can be made in this presentation; first, the species number density spans a very wide range from 10^{18} to 10^{12} and lower but still can be captured by a relatively coarse mesh system of (327 × 187). Second, the sum of the number densities of the positive charged ionized species equals the number density of the free electrons to satisfy the intrinsic globally neutral property of plasma.

The degree of ionization has a strong dependence on the temperature and, to some degree, is also dependent on the rarefication of the gas. This behavior is clearly captured by numerical simulations. Both the flight data and computational simulation indicate a significant change of the thermodynamic states and chemical reactions over the NASA Sample return Probe RAM-CII at the junction of the hemispherical forebody and the conical afterbody. Figure 2.5 describes the chemical species profiles over this junction at three reentry altitudes. The species concentrations are observed to adjust to the rapidly changing thermodynamic condition within the shock layer. The most noticeable adjustments are reflected by the sharpening definition of the bow shock at the lower altitudes and the formation of electrons through the

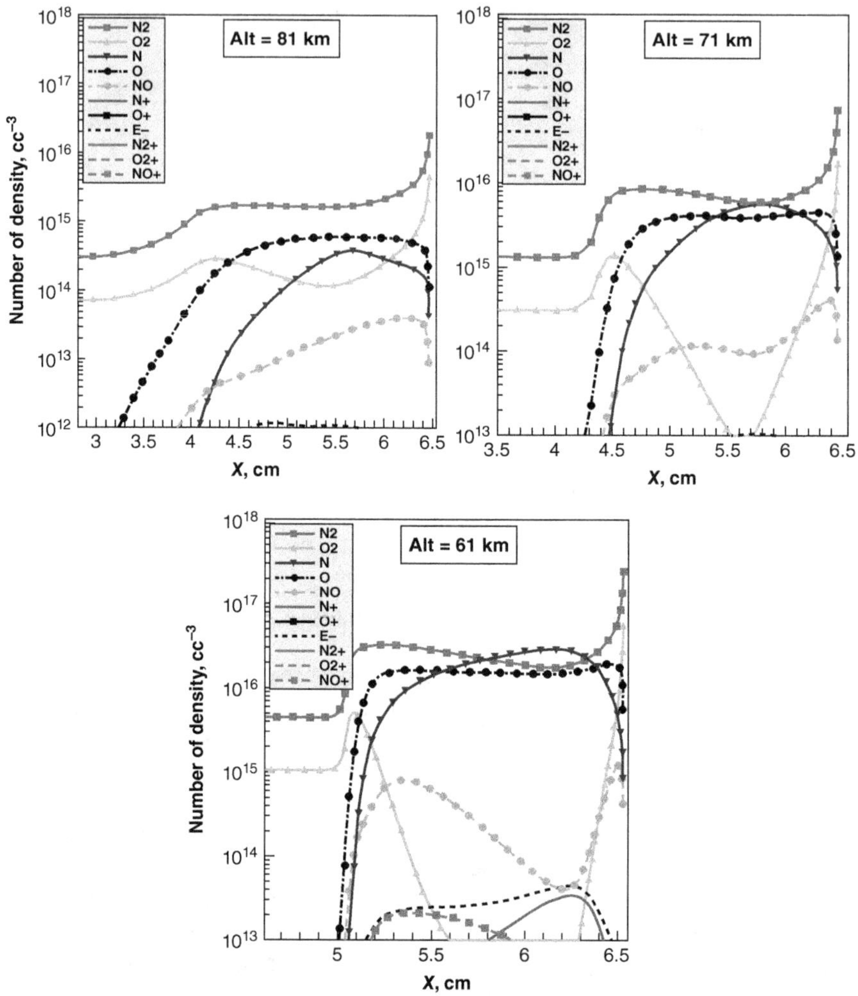

FIGURE 2.5 Species profiles over RAM C-II probe at three altitudes

accumulative collision process. Additional ionized components also are propagated from the stagnation region at the altitude of 61 km, and the maximum concentration of electrons is around 4.0×10^{13} particles per cubic centimeters. The depletion of molecular oxygen and redistribution of nitric oxide also appeared during reentry. At lower altitudes, the species composition profiles develop into an established pattern like that along the stagnation streamline. The shock layer thickness at this juncture increases from 4.3 to 5.0 cm from the altitudes of 71 to 61 km to accommodate the expansion from the stagnation region and the changing internal degrees of excitation of the high-enthalpy gas.

2.8 ELECTRON IMPACT IONIZATION

Electron impact is the most important atomic and molecular ionization mechanism in gas discharge. The energy transfer depends on inelastic and elastic collisions in an externally applied electric field. The temperature of a gas discharge is always substantially lower than the ionization potential. Atoms are ionized by high-energy electron collisions in the outer limits of the Maxwellian distribution. An electron avalanche generated in an electric field evolves not only in time but also in space along the drift direction of free electrons. Therefore, the rate of ionization is not measured by the ionization frequency but the ionization coefficient $\alpha = \nu/u_d$ cm^{-1}.

A unique phenomenon in gas discharge is by secondary emission that produces breakdown for discharge within the electrode gap and sustains a small current. The main mechanism is discovered by Penning in 1928: The accumulated field of positively charged ions next to the cathode creates a narrow potential barrier and is very strong, with an electric intensity on the order of surrounding nucleus. An electron from the cathode, either a conductor or dielectrics, immediately tunnels into the ion layer to neutralize it. The tunneling action leads to electric breakdown, and the process transforms a dielectric medium in a conductor under a sufficient strong electric field. The process usually takes place at time scales from 10^{-4} to 10^{-9} s.

Ionization by electron collision may appear as a spark and the discharge current is limited by the resistance in the external circuit. The primary elements of breakdown are the electron avalanche and the threshold of a steep rate of atomic ionization by electron collisions. The breakdown is controlled by the ignition potential

$$I_{bd} = B(pd)/[c + \ln(pd)] \tag{2.48a}$$

which is essentially the Paschen curve. The product pd (pressure and gap distance between electrodes) is a manifestation of the similar law for gas discharge by Townsend. The minimum value of E/P, which is a measure of energy grain between collisions, corresponds to the Stoletov point. For air, $c = 1.18$, $(pd)_{\min} =$ 0.83 torr-cm, and $(E/P) = 365$ V/(cm-torr), and the minimum voltage for breakdown is around 300 V at low pressure. In general, the degree of ionization in number density of normal discharge is around $10^8 < n_e < 10^{12}/\text{cm}^3$.

The direct current discharge (DCD) has one obvious feature: it is macroscopic time independent. The normal glow discharge has an electric current between 10^{-5} and 10^{-3} atm. At a pressure near atmospheric, the breakdown is likely to result in an arc. The electric potential across the electrodes usually drops to tens of volts. The glow-to-arc transition is not predictable. In the glow discharge, the essential supply of electrons is mainly from secondary emission by bombardment by positive ions onto the cathode. In an arc, another known source is the thermoionic emission from the cathode. The transient discharge or spark can occur over a wide range of electric current. A more detailed description of gas discharge basically can be categorized as follows:

1. Townsend or dark (invisible) discharge occupies the range of a current-voltage curve from A to C in Figure 2.6. Generally, the discharge has a total current up

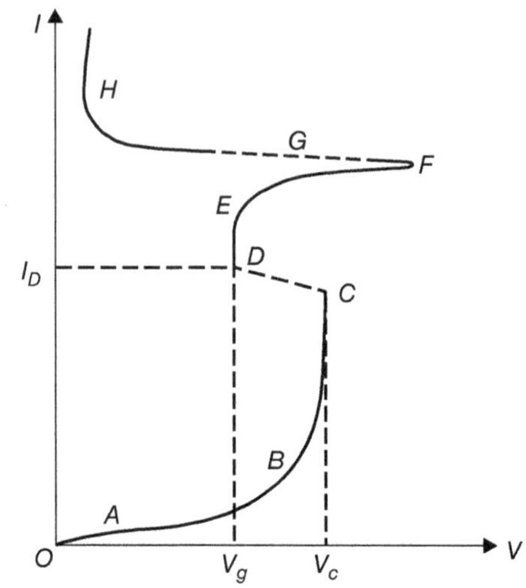

FIGURE 2.6 Voltage–current relationship of gas discharge.

to 10^{-6} A. The notation V_c is the so-called breakdown or sparking potential. Once breakdown occurs, the discharge becomes self-sustaining, taking the form of a glow or an arc discharge and becoming luminous with $pd > 10$ torr cm. The parameter *pd* of Paschen's law for the breakdown voltage is closely related to the collision frequency of the charged particles.

2. Normal and corona discharges take place along the curve from C to D. In this current-voltage range, the discharge has a strong dependence on the shape of electrode and the gas pressure (or correctly the gas density). At high pressure and from a sharp-point electrode, a corona discharge will be emitted, whereas in a low pressure gas, a normal glow discharge sets in. If the power source cannot supply the minimum current I_d, the spark will repetitively ignite in the discharge gap. The current densities of these types of discharge have a range from 10^{-3} to 10 A/cm^2. The glow discharge neither is in thermal equilibrium nor maintains charge neutrality. The atoms have a Maxwellian distribution of energy at a very low temperature, the ions at a slightly higher temperature, and the electron at very high temperature of a few electron volts or 10^4–10^5 K.
3. In the region E-F, the voltage rises sharply across the cathode fall and the higher current density heats the cathode substantially. This electron emission is known as the abnormal discharge.
4. Beyond the critical point F, the glow-arc transition occurs and the more prolific emission mechanisms appear on the heated cathode by the thermoionization and photoionization of electrons. The electric resistance drops and the current therefore surges to 1000 A or more to become arc discharge.

The electric field breakdown at high pressure is similar to that occurring in the low-pressure environment. However, it is not feasible to completely investigate Pschen's law at high pressure, because the minimum breakdown voltage occurs at distance too small to be easily measured. For example, the breakdown field strength for atmospheric air is 33 kV/cm. Breakdown at high pressure in a nonuniform field is markedly dependent on the geometry of the electrodes and on their polarity. If high voltage is applied to the cathode, the discharge is a negative or anode-directed corona; otherwise, it is a positive or cathode-directed corona. Although the corona is basically a Townsend charge, mechanisms of sustaining the continuous ionization depend on the polarity of its electrodes. Generally, the radius of electrodes is much smaller than the gap distance. For parallel wires, it is $d/r > 5.83$. If the ratio is greater than 5.83, the wire produces a spark and transits from corona discharge. The maximum electrical field intensity can be estimated by

$$E_{\max} = (V/2r)\ln(d_g/r) \tag{2.48b}$$

In principle, the ionization of negative corona does not differ from the Townsend breakdown. However, the negative corona generated by a needle-to-plate electrode arrangement is associated with a periodic discharge pattern known as Trichel pulses, which has a repetition rate greater than that in positive coronas. These pulses are connected with streamers starting from the pointed electrode toward the flat counterpart with repetition frequency of about 1 kHz. As the voltage is increased, the pulses vanish and a steady-state corona is sustained until spark breakdown within the discharge gap.

For multiple positive corona discharges, electron reproduction is by the so-called secondary photo process and displays a luminous filament emitting from the electrode as streamers or microdischarges. The streamer is a moderately weak ionized channel formed by the primary avalanche in a sufficiently strong electric field between electrodes. Again in the case of positive corona, the electrons are mostly produced by photoionization and lead to avalanche. The ionization zone has a dimension of a few tenths of a millimeter and display luminous filaments from the anode. The streamer is conveyed by ion drift to have charge number density of 10^9–10^{13}/cm^3. When the current or voltage exceeds a threshold, there will be a corona-to-arc transition.

Electron impact ionization in an AC cycle is not substantially different from a DC field. However, if one electrode is coated by a dielectric film, then the charges accumulated on the electrode surface will reduce the electrical intensity between electrodes and prevent the transition to spark. This discharge is known as the dielectric barrier discharge (DBD) or the silent discharge. The discharge cycle in DBD is dominated by the random streamer, filament, or microdischarge in time and space (Elisson et al., 1991; El-Bahy, 2005). The streamer or the microdischarge is a weakly ionized channel by the primary avalanche in a strong electric field. The streamer can issue from the anode or cathode and lead to spark discharge. The process is random in time of short duration of microseconds and within the electrode gap. The space charges also generate an additional electric field by themselves, which makes it different from corona discharge. When the microdischarge is initiated from anode to cathode, it is known as the cathode-directed or positive streamer; otherwise, it is said to be anode-directed or negative streamer. Negative streamers propagate against the direction of the electric field thus in the same direction of the electron drift velocity, whereas positive streamers propagate in the opposite direction. However, if the electron avalanche transformed into streamer before it reaches electrode, like in a microwave, the streamer grows toward both electrodes. For the positive streamer, the free electrons have to travel a greater distance to the head region; thus, the negative streamer is more diffusive than the positive one. Current computational capability is incapable of a direct simulation of multiple microdischarges within the DBD.

Electron impact discharge is dominated by quantum chemical-physics processes but must still follow the law of mass action. Thus, the finite-rate chemical reaction formulation has been widely applied. Pioneering and luminous work was provided by Elisson and Kogelschatz (1991). They actually displayed the detailed generation rates of the most dominant ionized species in an AC discharge. Chemical kinetics models for low-temperature plasma generation span a very wide range of complexity. The models cover from the detailed description to many simplified approximations. They simplified the chemical kinetic model based on comparable reaction rates of species to the characteristic time scale on the order of 30 ns for atmospheric air. The approach is rational and the fidelity of numerical simulation to nonequilibrium chemical reaction solely depends on the empirical determined rate constants. Unfortunately, chemical kinetic rates are the weakest link of ionization models for both the high-temperature thermal excitation and electron collision ionizations. The complete differential equations system of species conservation is very stiff to become a formidable computational challenge (Shang et al., 2014).

An alternative approach in determining the net rates of generation and depletion of individual ionized species can be achieved by quantum chemical-physics ionization modeling. This approach is also simplified by treating the charge species into the global categories of electrons and positively and negatively charged ions. The justification for this approach is to recognize that the charged species appear only in trace amount in molar or mass fractions and dominated by a few radical components. In fact, ionization by electron collisions occurs at the outer limit of the Maxwell distribution where the gas energy is lower than the ionization potential. The dominant species of DBD are the metastable molecules $N_2(A^3\Sigma_u^+)$ and $O_2(b^1\Sigma_d^+)$. The main ionized processes of a nonequilibrium volumetric discharge consist of four types of reactions: electron/molecular, atomic/molecular, decomposition, and synthesis. At any instance during discharge, the ionized species concentration is the net balance between ionization, detachment, attachment, and recombination processes. The ionization model is, therefore, focused on these mechanisms.

The ionized species can be substantially simplified via multifluid model by grouping species into electrons, neutrals, and positively and negatively charged ions. The physics-base modeling isolates the highly uncertain composition of a weakly ionized gas to collision kinetics. Physical fidelity of ionized species is enforced by identifying the most active ionized species and rates of generation and depletion of ionized species from theory and data of quantum chemical-physics. This ionization model is a specialization from the chemical kinetics for ionization by electron impact and is built on the similarity law by Townsend for discharge (Raizer, 1991). The generation and depletion of ionization processes are derived from collision kinetics in describing ionization, recombination, attachment, and detachment. In fact, this model integrates the drift-diffusion theory and quantum chemical-physics kinetics with the multifluid formulation to achieve a self-consistent approach for simulating electromagnetic–aerodynamics interactions. A detailed derivation and formulation will be provided in the following chapters. The overall formulation for plasma generation and depletion processes can be modeled as the following:

$$\begin{aligned} \left(\frac{dw}{dt}\right)_e &= \alpha(\left|\vec{E}\right|)|\vec{\Gamma}_e| - \beta n_+ n_e - \nu_a n_e + \kappa_d n_n n_- \\ \left(\frac{dw}{dt}\right)_+ &= \alpha(\left|\vec{E}\right|)|\vec{\Gamma}_e| - \beta n_+ n_e - \beta_i n_+ n_- \\ \left(\frac{dw}{dt}\right)_- &- \nabla \cdot (n_- \mu_- \vec{E} + d_- \nabla n_-) = \nu_a n_e - \kappa_d n_n n_- - \beta_i n_+ n_- \end{aligned} \tag{2.49}$$

In these formulations, the ionization coefficient α is described by the Townsend similarity law. The coefficients of the dissociative recombination, and the ion–ion recombination, are designated by β and β_i. The electron attachment and detachment are characterized by the respective coefficients ν_a and κ_d (Shang et al., 2014).

Numerical results from the quantum chemical-physics model in conjunction of the drift-diffusion theory for transport properties of plasma have been demonstrated to be a viable approach in computational simulation of the DCD and DBD discharges. Figure 2.7 illustrates validation for the basic electron impact discharge of a DCD in a parallel electrode configuration. Numerical results capture correctly the structure of the discharge column and the predicted current density reaches a very good agreement with classical solution (Surzhikov et al., 2004). The computed contours of electron and positively charged number density of DCD between two parallel electrodes are

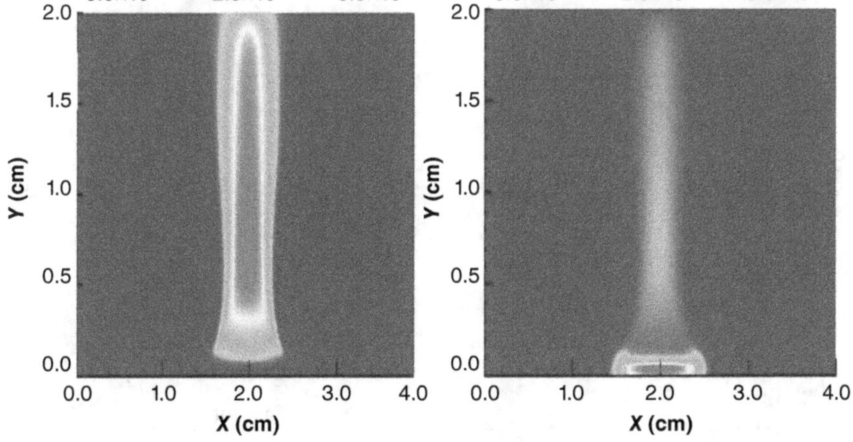

FIGURE 2.7 Electron and positively charged ion number density contours of parallel electrodes, $P = 5$ torr, electromotive force (EMF) = 2.0 kV, gap distance 2.0 cm.

displayed. The diffusive discharge is generated between a gap of 2.0 cm by an EMF of 2.4 kV at an ambient pressure of 5 torr. On the cathode, all the electrons are repelled, but the secondary emission generates a rapid avalanche above it in the positive column. The electric field intensity reaches its maximum value in the cathode layer to have a value that agrees very well with the classical theory. The positively charged ions are clustered in the cathode layer to reach the maximum number density of $3.0 \times 10^{10}/cm^3$. Although the electron temperature is not included in the simulation, the highest value from the classic work is known to exist in the cathode layer and has a value of around 3 eV.

The most important verification of the quantum chemical-physics ionizing model is presented by a direct comparison with a measured conductive current during a DBD. The ionization model is built on the Townsend discharge mechanism together with the bulk and ion–ion recombination, as well as the electron attachment and detachment as described by equation (2.49). For DBD, the verification for modeling and simulation is rest on the prediction of the electric breakdowns during an AC cycle and the conductive electric current density due to charge separation. Therefore, an ionization model by the chemical-physics kinetics must be sufficiently accurate to describe a reasonable discharge composition. From the charge number density of electrons and ions, the current then can be determined. The magnitude of this current component is relatively minuscule and indicates a density of less than 0.008 mA/cm and thus poses a very stringent validation criterion (Stanfield et al., 2009). It is shown that the computation results agree well with data that are collected over a wide group of experiments to indicate that the ionization model by chemical-physics kinetics is credible. The comparing experimental data are collected at an AC cycle of 5 kHz and an electrical potential of 4.0 kV; in Figure 2.8, the calculated conductive current

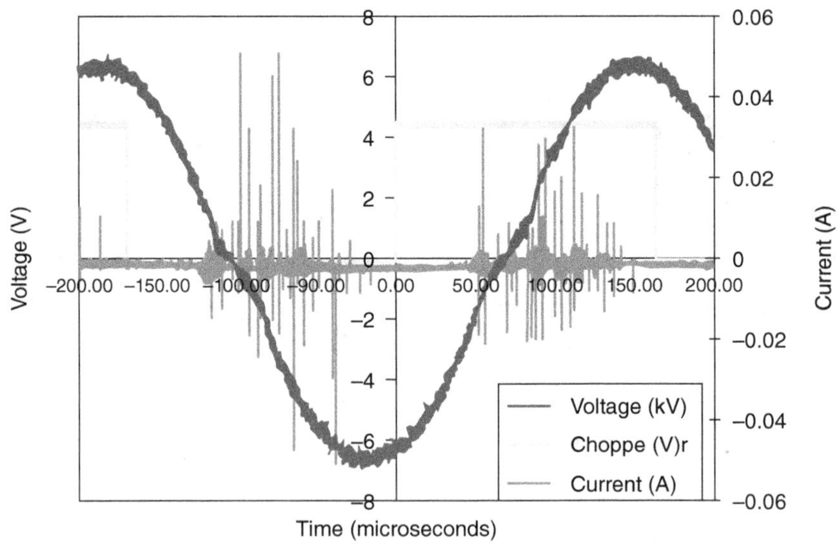

FIGURE 2.8 Comparison of DBD conductive electric current with experimental observation, EMF = 4.0 kV, $f = 5$ kHz.

is designated by the black line that lies within the traces of the experimental data to reflect the correct magnitude and duration of the discharge. Equally important, the effects of different electrical permittivity and secondary emission coefficients on metallic electrode and dielectric surface are also accurately captured.

All ionizing phenomena occur across molecular/atomic scales and beyond the scope of gas kinetics theory. The ionization processes by either the thermal or electron impact have to be modeled. In practices, the thermal excitation is developed by chemical kinetics models, including the energy transfer between translation, vibration, and electronic degrees of freedom. Despite some glaring deficiencies for plasma modeling and simulation, the fundamental formulation based on law of mass action still can be used to describe ionization of partially ionized gas. On the contrary, the ionizations by electron impact have been treated as chemical reactions, but a viable and simple model is developed by a multiple fluid model and with collision kinetic for ionization. A better understanding of electromagnetic–aerodynamics interaction in practical engineering is discussed on the basis of the modeling and computational simulation techniques.

REFERENCES

Anderson, J.D., *Hypersonic and High Temperature Gas Dynamics*, McGraw-Hill, New York, 1989.

Clark, J.F., and McChesney, M., *The Dynamics of Real Gases*, Butterworths, Washington, 1964.

El-Bahy, M.M., A numerical modelling of microdischarge threshold in uniform electric fields, *J. Phys. D. Apply Phys.*, vol. 38, 2005, pp. 103–112.

Elisson, B., and Kogelschatz, U., Nonequilibrium volume plasma chemical processing, *IEEE Trans. Plasma Sci.*, vol. 19, 1991, pp. 1063–1077.

Eyring, H., Walter, J., and Kimball, G.E., *Quantum Chemistry*, John Wiley & Sons, New York, 1944.

Hertzberg, G., *Atomic Spectra and Atomic Structure*, 2nd ed., Dover Publisher, New York, 1944.

Hertzberg, G., *Molecular Spectra and Molecular Structures*, vol. I, Van Nostrand, New York, 1950.

Kuo, K.K., *Principles of Combustion*, 2nd ed., John Wiley and Sons, Hoboken, NJ, 2005.

Leighton, R.B., *Principles of Modern Physics*, McGraw-Hill Inc., New York, 1959.

Moore, W.J., *Physical Chemistry*, Prentice-Hall Inc., Englewood Cliff, NJ, 1963.

Pai, S.-I., *Magnetogasdynamics and Plasma Dynamics*, Springer-Verlag, Wien, 1962.

Park, C., *Nonequilibrium Hypersonic Aerodynamics*, John Wiley & Sons, New York, NY, 1989.

Saha, M.N., Ionization in the solar chromosphere, *Phil. Mag.*, vol. 40, no. 238, 1920, p. 472.

Shang, J.S., and Huang, P.G., Surface plasma actuators modeling for flow control, progress. *In Aerospace Sci.*, vol. 67, 2014, pp. 29–50.

Shang, J.S., Andrienko, D.A., Huang, P.G., and Surzhikov, S.T., A computational approach for hypersonic nonequlibrium radiation utilizing space partition algorithm and Gauss quadrature, *J. Comp. Phys.*, vol. 266, 2014, pp. 1–21.

Shang, J.S., Numerical simulation of hypersonic flows, Chapter 3; *Computational Methods in Hypersonic Aerodynamics*, Ed. Murthy, T., Kluwer Academic Publishers, Boston, MA, 1991, pp. 81–114.

Stanfield, S.A., Menart, J., Dejoseph, C., Kimmel, R.L., Hayes, J.R., Rotational and vibrational temperature distributions for a dielectric barrier discharge in air, *J AIAA*, vol. 47, 2009, pp. 1107–1115.

Surzhikov, S.T., and Shang, J.S., Two-component plasma model for two-dimensional glow discharge in magnetic field, *J. Comput. Phys.*, vol. 199, no. 2, September 2004, pp. 437–464.

Vincenti, W.G., and Kruger, C.H., Introduction to physical gas dynamics, John Wiley & Sons, New York, 1965.

Zel'dovich, Ya. B., and Raizer, Yu. P., *Physics of Shock Waves and High-Temperature Hydrodynamic Phenomena*, Dover Publication, Mineola, 2002.

3

MAGNETOHYDRODYNAMICS FORMULATION

INTRODUCTION

Hannes Alfven initiated the field of magnetohydrodynamics (MHD) that was originally applied mostly to astrophysics and geophysics (Alfven, 1950). For his theoretical works on Van Allen belts, the Earth's magnetic field, and plasma dynamics in the Milky Way, he received the Nobel Prize in physics in 1970.

The distinction between MHD and plasma dynamics is not significant; both analyze the interacting dynamics of an electrically conducting gas or liquid in electromagnetic fields. However, the focuses in studying these phenomena are substantially different; MHD emphasizes mostly on low frequency or quasi-stationary processes and places focus on the global behavior of the conductive medium in the presence of electromagnetic fields. Plasma dynamics is concentrated on the detailed description of the inertial effects of the charged particles and usually describes phenomena in the framework of wave dynamics. For this reason, there are ideas and independent concepts of each in clearly separated domains.

An important feature in MHD is that the motion of collective charged particles is described by an electrically conducting fluid with the usual fluid dynamic variables of velocity, density, and pressure or temperature. In the process, the generalized Ohm's law is adopted to directly relate the electric field strength with current density, even including the Hall Effect and ion slip for a partially ionized gas. This approximation greatly simplifies the formulation at the expense of overlooking the detailed composition of the plasma or the individual behavior of charged particles (Cabannes, 1970;

Computational Electromagnetic-Aerodynamics, First Edition. Joseph J.S. Shang.

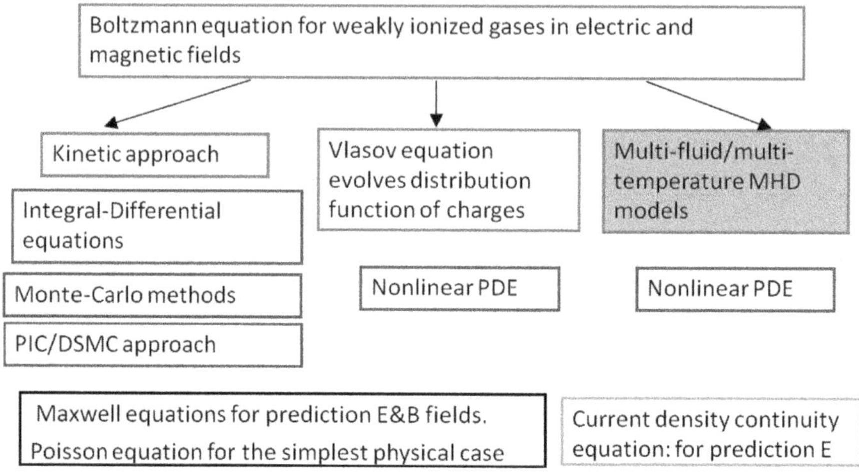

FIGURE 3.1 Hierarchy of MHD formulations.

Jackson, 1975). The MHD formulation in a wide range of physical conditions gives an accurate description of physical phenomena. At the extremely high temperature and low density environment when the link between microscopic and macroscopic dynamics fails, then the accuracy of MHD becomes unreliable.

MHD is a scientific branch for studying the dynamics of fluid motion in the presence of an electromagnetic field, especially for a medium that is electrically conductive because the inductive electric current by the moving medium modifies the field and this coupling must be analyzed simultaneously. The microscopic description of MHD can be carried out by the kinetic approach through the integro-differential Maxwell–Boltzmann equations. By this approach, the particle-in-cell and direct simulation Monte-Carlo methods have been implemented more recently. Another approach based on kinetic theory has also been developed. The Vlasov equation is derived to describe the evolution of the distribution function of plasma consisting of charged particles with long-range force. However, in the framework of continuum mechanics, by neglecting rarified gases, the classical MHD formulation was established by integrating Maxwell's equation for electromagnetics and the Navier–Stokes equations for fluid motion (Figure 3.1).

From a straightforward nondimensionalization of the basic governing equations, two well-known similitudes are revealed. One of the similitudes is the interaction parameter or the Steward number

$$S_{\mathrm{m}} = \sigma B^2 L / \rho_\infty u_\infty \tag{3.1a}$$

and the other is the magnetic Reynolds number

$$R_{\mathrm{m}} = \sigma \mu u_\infty L \tag{3.1b}$$

The former is a measure of the relative importance of the magnetic flux density to the inertia of the flow field or the ratio of Lorentz force $J \times B$ to the inertia of the fluid flow. It may be interesting to point out that the Lorentz force is a first-order relativistic effect. This dimensionless similarity parameter R_{m} becomes a very important indication to highlight the relatively dominant force in an interacting plasma flow field.

The magnetic Reynolds number can be interpreted as the ratio of an induced to total magnetic flux density in plasma field. The plasma exhibits distinctive behaviors in a range of applied magnetic field strength and electric conductivities. At high magnetic Reynolds number or plasma with an infinity electric conductivity, the magnetic field locks on the charge particles that can move freely along the magnetic lines of force. In many flow fields of interest, inertial effects greatly outweigh viscous dissipation and heat transfer in the medium, plasma can be treated as an isotropic medium. Especially at the asymptotic inviscid limit, the strong magnetic field lines will dictate completely the topology of the plasma field structure, including twisting and turning. At the opposite limit of very low magnetic Reynolds number, the convection of the magnetic field line by the fluid is negligible; magnetic induction is therefore insignificant.

In an MHD formulation, the electrostatic effect is essentially ignored in a global neutral condition. The interactions of fluid dynamics and electromagnetics are exclusively derived from the magnetic flux density. As a point of interest, Faraday's induction law can be used to identify different mechanisms for the rate of change of magnetic flux density via the simplified generalized Ohm's law. Through this approach, the mechanisms for temporal change of magnetic flux density are attributable to the production and diffusion processes (Vincenti et al., 1965),

$$\frac{\partial B}{\partial t} = -\nabla \times E \approx \nabla \times \left[u \times B - \nabla \times \frac{B}{\mu\sigma} \right]$$

(production) (diffusion)

Most recently, interdisciplinary scientific formulation has been increasingly and widely applied to flow control through plasma actuators. This emphasis has further evolved from the classic MHD that was originally devised for astrophysics to magneto-aerodynamics by a shift in focus from wave-dominated phenomena to force and energy transfer in electromagnetics and aerodynamics interactions. For this purpose, the electrostatic force due to charge separation becomes the principal mechanism. A systematic derivation of the governing equations of MHD is detailed and delineated in this chapter.

3.1 BASIC ASSUMPTIONS OF MHD

The first and foremost assumption for MHD formulation is that the physical phenomena are describable by macroscopic scales or in the framework of continuum mechanics. All physical properties are resolved as a continuous function by the

Boltzmann distributions. In the basic formulation for classic MHD, another leading approximation is that the displacement current $\partial D/\partial t$ of the Ampere's circuit law is always omitted (Sutton et al., 1965),

$$J = \nabla \times H - \frac{\partial D}{\partial t} \Rightarrow J = \nabla \times H \tag{3.2a}$$

Thus, the electric current density is associated only with the curl of the magnetic field strength. In other words, the electromagnetic interrelationship is described by the original time-independent Ampere's circuit law. The underlying assumption is based on an order of magnitude analysis that indicates that the displacement current can be estimated as $O[(\partial D/\partial t)/J] = O(\varepsilon/t\sigma)$. The electric conductivity in plasma is $\sigma \doteq n_e e^2/m_e v_{eh}$, and v_{eh} characterizes the averaged collision frequency between electrons and heavy particles. Recall that the plasma frequency is defined as $\omega_\mathrm{p} = (n_e e^2/m_e \varepsilon)^{1/2}$. By imposing this restriction, the characteristic time scale for MHD simulations shall always be much greater than the inverse of plasma frequency

$$t \gg \frac{\varepsilon}{\sigma} \gg \frac{1}{\omega_\mathrm{p}} \tag{3.2b}$$

Therefore, the time scale for accurate MHD analysis is limited to microwave frequencies, around tens of gigahertz. Specially, the spectrum of microwaves is bonded at upper limit by visible and infrared waves. The lower limit is the radio frequency; for 100-channel AM radios, the spectrum range is from 0.535 to 1.505 MHz. The range of FM radio is 88 to 108 MHz. The microwave spectrum is actually defined by seven bands from L, S, C, X, K_u, K, to K_a to span a range from 1 to 40 GHz. Beyond this frequency range, the error by MHD equation may become significant.

By neglecting the displacement current is equivalent to set the vanishing gradient of electric current density

$$\nabla \cdot J = 0\,(\nabla \cdot J = \nabla \cdot \nabla \times H \equiv 0) \tag{3.2c}$$

It follows from the charged conservation equation

$$\partial \rho/\partial t + \nabla \cdot J = 0 \tag{3.2d}$$

that the charged number density is therefore independent of time. According to Faraday's induction law, the time derivative of divergence of magnetic flux density also vanishes,

$$\partial(\nabla \cdot B)/\partial t = 0\,(\nabla \cdot \partial B/\partial t = -\nabla \cdot \nabla \times E \equiv 0) \tag{3.2e}$$

The requirement of $\nabla \cdot B = 0$ is imposed only as an initial value, but in practice it is applied as a consistent constraint.

A consequence of the globally neutral electric field of plasma, it means that there is no free space charge. The Lorentz acceleration is, therefore, far greater than the electrostatic force $|J \times B| \gg |\rho \cdot E|$. This condition generally holds except when charge separation occurs in plasma sheath and especially near electrodes. In the anode or cathode layer of a discharge domain, the electrostatic force is no longer negligible.

A usual approximation in MHD is the generalized Ohm's law for electrical conductivity. The kinetic foundation of this law for charged particles must be started from the conservation law of mass and momentum of charged particles, namely,

$$\begin{aligned}&\frac{\partial n_i m_i}{\partial t} + \nabla \cdot (n_i m_i u_i) = s_i \\ &\frac{\partial n_i m_i u_i}{\partial t} + \nabla \cdot (n_i m_i u_i u_j) = p_i^* + n_i e(E + u_i \times B) - M_{ij}\end{aligned} \tag{3.3}$$

where the average momentum exchange between colliding particles i and j,

$$M_{ij} = n_i \nu_{ij} \left(\frac{m_i m_j}{m_i + m_j} \right) (u_i - u_j) \tag{3.4}$$

In expression (3.4), the average momentum transfer collision frequency is

$$\nu_{ij} = n_i \sigma_{ij} \iiint |u_p - u| \, f_i(u_p) f_j(u)(d^3 u_p)(d^3 u) \tag{3.5a}$$

The notation u_p is the peculiar velocity of the Boltzmann distribution, and the average collision cross section can be given as

$$\sigma_{ij} = 2/3 \int r^2 e^{-x} Q_{ij}^{(1)} \left(\sqrt{2kT/m_{ij} r} \right) dr \tag{3.5b}$$

The global momentum conservation equation, as it has been shown in Section 1.6, can be obtained by summing all species in plasma to get

$$\begin{aligned}&\frac{m_e m_i}{(\rho_e + \rho_i)e^2} \left[\frac{\partial}{\partial t}(\rho_e U + J) + \nabla \cdot (\rho_e UU + UJ + JU) \right] \\ &= E - \left[\frac{m_e \nu_{ei}}{n_i e^2} + \frac{m_e m_{ei}}{(\rho_e + \rho_i)e^2} \left(\nu_{en} + \frac{m_e m_{in}}{m_i m_i} \nu_{in} \right) \right] J - \frac{m_i - m_e}{(\rho_e + \rho_i)e} (\rho_e U + J \times B) \\ &+ \frac{\nabla \cdot (m_i P_e - m_e P_i)}{(\rho_e + \rho_i)e} \\ &- \frac{\rho_e m_e \nu_{ei}}{n_i e^2} \frac{\rho_n U_n}{(\rho_e + \rho_i)} - \frac{\rho_n}{(\rho_e + \rho_i)} (U_n \times B) - \frac{\rho_n m_{ei}}{(\rho_e + \rho_i)e} \left[\left(\nu_{en} - \frac{m_{in}}{m_i} \nu_{in} \right) \right. \\ &\left. - \frac{\rho_i m_i}{\rho_e} \left(\nu_{en} + \frac{m_e m_{in}}{m_i m_i} \right) \right] U_n\end{aligned} \tag{3.6a}$$

Since masses of neutrals and ions are much greater than those of electrons, the reduced mass becomes $m_{ei} = m_e m_i / (m_e + m_i) \approx m_e$. The reduced mass $m_{ei} \approx m_e$ will also be neglected. After some algebraic manipulations and by using the definition of the charge particle number density, current density $J = \sigma E$, and electric conductivity $\sigma = n_e e^2 / m_e \nu_{ei}$, it can be shown for a fully ionized gas that we shall have

$$\left(\frac{m_e}{n_e e^2}\right)\left[\frac{\partial J}{\partial t} + \nabla \cdot (UJ + JU)\right] = E - \frac{J}{\sigma} + \left(\frac{\rho_n}{\rho_i}\right) U_n \times B$$
$$-\left(\frac{1}{en_e}\right)[(J \times B) - \nabla \cdot P^*] + \left(\frac{\rho_n}{\rho_i}\right)\left(\frac{m_e}{e}\right)\left[\nu_{en} - \left(\frac{m_n}{m_i}\right)\nu_{in}\right] U_n \tag{3.6b}$$

From the global momentum equation, it has been argued that $J \times B$ is on the same order of magnitude as the gradient of the partial pressure in the partially and fully ionized limits, and these terms can be neglected (Mitchner et al., 1973). Using the argument of the great disparity in mass of electrons and the mass of ions and neutrals between electron, ion, and neutral

$$0 = E - \frac{J}{\sigma} + U_n \times B - \left(\frac{1}{en_e}\right)[(J \times B) - \nabla \cdot P^*] \tag{3.6c}$$

In addition, the much lower characteristic frequency of MHD than the plasma frequency allows the simplified generalized Ohm's law to be written as

$$J = \sigma(E + U \times B) \tag{3.7}$$

For partially ionized plasma, the simplified Ohm's law can also be used to show a unique ion slip effect (Mitchner et al., 1973). Most important, it is clearly shown that the simplified generalized Ohm's law has reduced the electric conductivity from a tensor to a simple scalar property.

Again, the globally neutral behavior of plasma leads to the consequence that the Lorentz acceleration overwhelms the electrostatic force $|J \times B| \gg \rho_e |E|$ except near electrodes or within the plasma sheath where charge separation occurs (Shang et al., 2014). For flow control using surface plasma actuators, this approximation is invalid, because the electrostatic force arises from charge separation near electrodes. The magnitude of this force is proportional to the product of applied electric field intensity and the difference in number density of positively and negatively charged particles. This force becomes the principal mechanism for accelerating neutral particles by collisions with ions.

3.2 IDEAL MHD EQUATIONS

The classic ideal MHD formulation is an integrated equation system that combined the simplified Maxwell equations and the Euler equations. The approximated Maxwell equation is the result by the basic MHD approximations in that the displacement current is considered to be negligible, and current density depends exclusively on the magnetic field strength $J = \nabla \times H$. The second major assumption that is the adoption of the generalized Ohm's law on a moving frame of reference $J = \sigma(E + u \times B)$, which establishes a relationship between the electric field strength and the current density. The Gauss law for magnetic field $\nabla \cdot B = 0$, which eliminates the existence of nonphysical magnetic dipole in electromagnetic field is invoked repeatedly in derivation but not included explicitly in the equation system. A reinforcement of this zero divergence of magnetic field flux density constraint has often been imposed on computational simulations.

The rationale of replacing the Navier–Stokes equations by the Euler equations is the fact that the transport property in terms of the molecular viscosity, thermal conductivity, and diffusion is relatively insignificant in comparison with the inertia of the fluid motion. This assumption is valid for original study of astrophysics and geophysics in the absence of a solid boundary. From the viewpoint of fluid dynamics, the approximation is by setting the characteristic Reynolds number of the flow field to approach infinity, and this asymptotic limitation can be easily removed. The interdisciplinary governing equations have been given as (Brio and Wu, 1988; Shang, 2001)

$$\frac{\partial \rho}{\partial t} + \nabla \cdot (\rho u) = 0 \tag{3.8a}$$

$$\frac{\partial B}{\partial t} + \nabla \times H = 0 \tag{3.8b}$$

$$\frac{\partial \rho u}{\partial t} + \nabla \cdot (\rho uu + p\bar{\bar{I}}) = J \times B \tag{3.8c}$$

$$\frac{\partial \rho e}{\partial t} + \nabla \cdot (\rho eu + u \cdot p\bar{\bar{I}}) = (E + u \times B) \cdot J \tag{3.8d}$$

where the symbol $\bar{\bar{I}}$ denotes the identity matrix. The system of equations is not closed because there is an unbalanced number of unknowns and equations. Closure is achieved by the generalized Ohm's law, and by the constitute relationship

$$B = \mu H \tag{3.8e}$$

The magnetic permeability μ is treated as a slow variation parameter so it can be moved outside of the spatial differential operators. It has been mentioned previously that Ohm's law can also explicitly include the Hall current

$$J = \sigma[E + u \times B - \beta(J \times B)] \tag{3.8f}$$

and the $(u \times B)$ term is required for describing the electric field strength on a moving frame (Mitchner et al., 1973). It is an induced electric current contributed by the moving charged particles in a magnetic field.

The associated initial value and boundary condition can be briefly summarized. For the gas dynamics variable, the nonpermeable condition is required for the velocity components on the fluid–solid interface $\vec{n} \cdot u = 0$. The normal component of the magnetic flux density is continuous across the boundary $\vec{n} \cdot \nabla B = 0$, and the discontinuity of the magnetic field intensity must be balanced by the surface current; $\vec{n} \times \nabla H = J$.

After substituting the reduced Ampere's electric law $\nabla \times H = J$ and the generalized Ohm's law,

$$E = \nabla \times B/\sigma\mu - u \times B \tag{3.9a}$$

Faraday induction law becomes

$$\frac{\partial B}{\partial t} + \frac{1}{\sigma\mu}(\nabla \times \nabla \times B) - (\nabla \times u \times B) = 0 \tag{3.9b}$$

By a vector identity, the second term becomes

$$\nabla \times \nabla \times B \equiv \nabla \cdot (\nabla \cdot B) - \nabla \cdot (\nabla B) \tag{3.9c}$$

After applying Gauss's law, it yields

$$\nabla \times \nabla \times B = -\nabla \cdot (\nabla B) \tag{3.9d}$$

Expanding the curl operators of the third term, we shall have

$$\nabla \times u \times B \equiv (B \cdot \nabla)u - (\nabla \cdot u)B - (u \cdot \nabla)B + (\nabla \cdot B)u \tag{3.9e}$$

By regrouping the curl and gradient operators, it gives

$$\nabla \times u \times B = \nabla \cdot (Bu - uB) \tag{3.9f}$$

Faraday's induction law now acquires the following expression:

$$\frac{\partial B}{\partial t} + \nabla \cdot (uB - Bu) = \frac{\nabla \cdot (\nabla B)}{\sigma\mu} \tag{3.10}$$

The term on the right-hand side of equation (3.10) is referred to as the Pierre–Simon Laplace operator and is customarily neglected by virtue of a very high value of the electric conductivity in most MHD environment.

The Lorentz acceleration in the momentum conservation equation can be expressed as

$$J \times B = \nabla \times \left(\frac{B}{\mu}\right) \times B = \left(\frac{1}{\mu}\right)[(\nabla \times B) \times B] \tag{3.11a}$$

From two following vector identities, we will have

$$(\nabla \times B) \times B \equiv (B \cdot \nabla)B - (1/2)\nabla(B \cdot B) \tag{3.11b}$$

and

$$(B \cdot \nabla)B \equiv \nabla \cdot (BB) - B(\nabla \cdot B) \tag{3.11c}$$

The second term of the last vector identity vanishes by invoking Gauss's law for magnetic field. The Lorentz acceleration in magnetic flux density is then

$$J \times B = \left(\frac{1}{\mu}\right)\left[\nabla \cdot BB - \frac{1}{2}\nabla(B \cdot B)\right] \tag{3.12}$$

The term BB is the well-known Maxwell tensor or the magnetic tensor of rank 2, and the term $(B \cdot B)$ is its normal stress component. It is also called the magnetic pressure that is the counterpart of hydrodynamic pressure of the molecular shear stress tensor. Substituting into the momentum equation (3.8c) yields

$$\frac{\partial \rho u}{\partial t} + \nabla \cdot [\rho uu - BB/\mu + (p + B \cdot B/2\mu)\bar{\bar{I}}] = 0 \tag{3.13}$$

The Joule heating in the energy conservation equation is rearranged by substituting the electric current by the approximate Ampere's circuit law, $J = \nabla \times H$ and the electric field intensity by the generalized Ohm's law, $E = \nabla \times H/\sigma - u \times B$ to appear as

$$\begin{aligned} E \cdot J &= \left[\frac{\nabla \times H}{\sigma} - u \times B\right] \cdot (\nabla \times H) \\ &= \frac{[\nabla \times (B/\mu)]^2}{\sigma} - (u \times B) \cdot (\nabla \times H) \end{aligned} \tag{3.14a}$$

From the vector identity,

$$(u \times B) \cdot (\nabla \times H) \equiv \nabla \cdot (H \times u \times B) + H \cdot (\nabla \times u \times B) \tag{3.14b}$$

By further applying the following two vector identities to the first and second terms

$$\nabla \cdot (H \times u \times B) \equiv \nabla \cdot [u(H \cdot B) - B(H \cdot u)] \tag{3.14c}$$

and

$$\nabla \times u \times B \equiv (B \cdot \nabla)u - B(\nabla \cdot u) - (u \cdot \nabla)B + u(\nabla \cdot B) \tag{3.14d}$$

$$\begin{aligned} (u \times B) \cdot (\nabla \times B) = \nabla \cdot &\left\{\left[u\left(\frac{B}{\mu}\right) \cdot B\right] - \left[B\left(\frac{B}{\mu}\right) \cdot u\right]\right\} \\ &+ \left(\frac{B}{\mu}\right) \cdot [\nabla \cdot (Bu) - \nabla \cdot (uB)] \end{aligned} \tag{3.15}$$

Replace the last term of this equation from the derived Faraday's induction law, the Joule heating is expressed as

$$E \cdot J = \frac{\nabla \times (B/\mu)^2}{\sigma} - \nabla \cdot \left[u\left(\frac{B \cdot B}{\mu}\right) - B\frac{B \cdot u}{\mu}\right] - \left(\frac{1}{2\mu}\right)\frac{\partial (B \cdot B)}{\partial t} + \left(\frac{B}{\mu}\right) \cdot \left[\frac{(\nabla \cdot \nabla B)}{\mu\sigma}\right] \tag{3.16}$$

Substituting all the derived expressions in the MHD governing equations and rearranging yields

$$\frac{\partial \rho}{\partial t} + \nabla \cdot (\rho u) = 0 \tag{3.17a}$$

$$\frac{\partial B}{\partial t} + \nabla \cdot (uB - Bu) = 0 \tag{3.17b}$$

$$\frac{\partial \rho u}{\partial t} + \nabla \cdot [\rho uu - BB/\mu + (p + B \cdot B/2\mu)\bar{\bar{I}}] = 0 \tag{3.17c}$$

$$\frac{\partial(\rho e + B^2/2\mu)}{\partial t} + \nabla \cdot [(\rho e + p + B \cdot B/2\mu)u - (B \cdot u)B/\mu] = 0 \tag{3.17d}$$

A few remarks are needed here; first, Gauss's divergence law for magnetic flux density has been consistently invoked in the derivation. Second, the electric conductivity is sufficient large so the terms involving reciprocal conductivity that appear in the derived Faraday induction law and Joule heating are dropped (Zachary et al., 1992; Powell et al., 1999). The ideal MHD equations are not a strictly hyperbolic type of nonlinear partial differential system. Thus, it means that the analytic structure of the weak solution around some point is generally unknown (Brio and Wu, 1988). The coefficient matrix of equation has duplicated eigenvalues; thus, not all the eigenvectors are linearly independent and there are also fields that can be characterized neither as linearly degenerate nor genuinely nonlinear, so the equation system is nonconvex.

More importantly, the electromagnetic field introduces the magnetic pressure $B \cdot B/2\mu$ and the magnetic stress matrix BB to the conservation momentum equation. The magnetic stress matrix is the degenerate Maxwell stress tensor without the contribution from the electric field intensity and singles out the normal stress components $B \cdot B/2\mu$ as the magnetic pressure. The added pressure is resulted from the basic MHD approximation by including the Lorentz force in the conservation momentum equation. The magnetic stress matrix is a second rank tensor similar to the dyadic of the velocity uu. In the Cartesian coordinates, the tensors in matrix forms are

$$BB = \begin{bmatrix} B_xB_x & B_xB_y & B_xB_z \\ B_yB_x & B_yB_y & B_yB_z \\ B_zB_x & B_zB_y & B_zB_z \end{bmatrix} \quad \text{and} \quad uu = \begin{bmatrix} u_xu_x & u_xu_y & u_xu_z \\ u_yu_x & u_yu_y & u_yu_z \\ u_zu_x & u_zu_y & u_zu_z \end{bmatrix} \tag{3.18}$$

The additional components of the magnetic pressure also contribute to the internal energy. The specific enthalpy now consists of four components: the internal energy, the kinetic energy, static hydrodynamic pressure of the plasma, and the magnetic pressure in the presence of a magnetic field,

$$h = e + u \cdot u/2 + P/\rho + B \cdot B/2\mu \tag{3.19}$$

The intrinsic property of the magnetic field is that it acts perpendicularly to the electric field, creates some unique behavior, and challenges analytic formulation. This behavior can be easily appreciated by examining the divergence operator in the derived Faraday induction equation, namely the derivatives of magnetic flux density along the respective coordinate are consciously absent. As a consequence, the corresponding elements in the coefficient matrix will appear as a null value to complicate eigenvalue and eigenvector analyses

$$\begin{aligned} \nabla \cdot (uB - Bu) = {} & [0 + \partial(B_x u_y - B_y u_x)/\partial y + \partial(B_x u_z - B_z u_x)/\partial z]\bar{i} \\ & + [\partial(B_y u_z - B_z u_y)/\partial z + 0 + \partial(B_y u_x - B_x u_y)/\partial x]\bar{j} \\ & + [\partial(B_z u_x - B_x u_z)/\partial x + \partial(B_z u_y - B_y u_z)/\partial y + 0]\bar{k} \end{aligned} \tag{3.20}$$

The effect of the magnetic pressure has been indisputably demonstrated by the Ziemer experiment in Figure 3.2 (Ziemer, 1959). In his experiment, a hemispherical-nose cylinder is immersed in an electromagnetic shock tube and the fluid medium

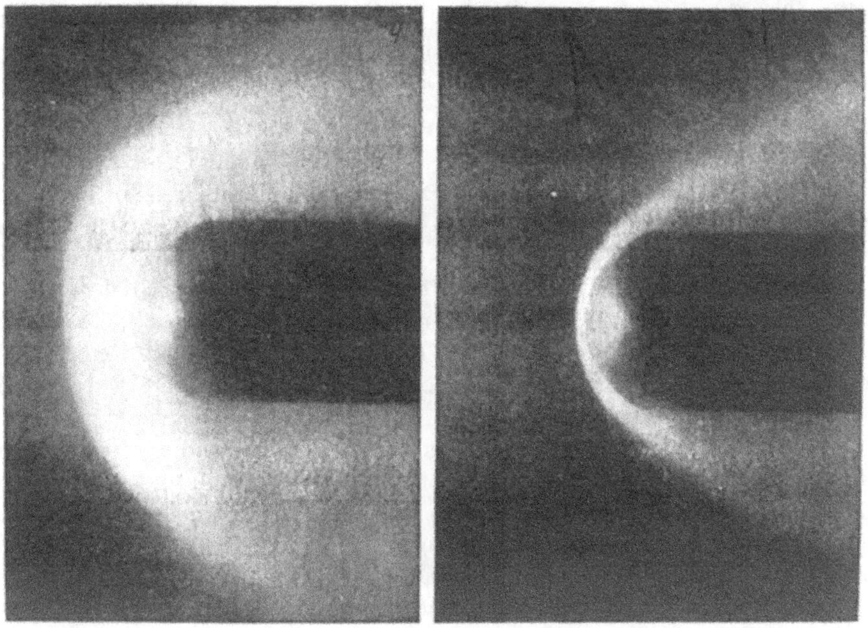

FIGURE 3.2 Magnetic field effect to bow shock standoff distance. From Ziemer, R. *J. Am. Rocket Soc.* vol. 29, 1959, pp. 642–647. Used with permission from AIAA.

is a partially ionized gas moving at a supersonic speed. The electric conductivity of the steady ionized air is estimated in the range of 28–55 mho/cm. A magnetic field is generated by a magnetic coil located coaxially in the hemispherical nose and aligns with the oncoming partially ionized gas. The magnetic field is therefore parallel to the oncoming plasma in the stagnation region. Through this arrangement, the generated maximum magnetic field strength is 40 kilogauss (4.0 Tesla). The activated magnetic field pushes the bow shock wave upstream, but the quantification of the standoff distance is not ascertained because of optical distortion of the emitting glow. When the current supply to the magnetic coil is cut off, the externally applied magnetic field ceases. The bow shock standoff distance retracts and is predicted by the aerodynamics to be proportional to the density jump across the shock and the radius of the hemispherical nose.

When the magnetic field is actuated, a magnetic pressure is generated and combines with the static hydrostatic pressure. An outwardly displaced standoff distance of the bow shock is recorded by the photograph. In the stagnation region, the combined pressure is essentially an invariant. Since the magnetic pressure attains the maximum value at the stagnation point over the magnetic pole, the hydrostatic pressure will decrease proportionally as the flow approaches the solid surface. The increased standoff distance in the plasma is then the result of a nonmonotonic and a lower postshock density in the shock layer. As a consequence, a large stream tube cross section is required to ingest the same amount of mass flow in the shock layer to satisfy the continuity condition. All numerical simulations also duplicate this phenomenon, but consistently underpredict the shock wave standoff distance. It has been pointed out that the observed increased bow shock standoff distance in the experiment is exaggerated by the illumination of the discharge glow.

3.3 EIGENVALUES OF IDEAL MHD EQUATION

The governing equations of an ideal MHD constitute a nonstrictly hyperbolic system; the most effective solving procedure is still the approximate Riemann solver (Powell et al., 1999). This basic approach is a characteristic method that builds on the eigenvector and eigenvalue structure of the equation system. A substantial amount of effort has been devoted in developing a complete set of right and left eigenvectors for the one-dimensional projection of the ideal MHD equations. The objective is to keep the eigenvectors well-defined even in domains where the equations are degenerate or no longer strictly hyperbolic (Brio and Wu, 1988).

The higher-order extension of the Godunov method for the solutions of gas dynamics, including shock jumps, emerges to be an attractive solving procedure for the ideal MHD problem (Zachary and Colella, 1992). Accounting to the number of scalar ideal MHD equations, a complete set of eight eigenvalues and eigenvectors is presented. To analyze the structure, the equation system is first expressed in flux vector form and solves in the one-dimensional time-space domain,

$$\frac{\partial U}{\partial t} + \nabla \cdot F = 0 \tag{3.21}$$

In this equation, the dependent variable and flux vectors are

$$U = \begin{pmatrix} \rho \\ B \\ \rho u \\ \rho(e + B \cdot B/2\mu) \end{pmatrix} \quad F = \begin{bmatrix} \rho u \\ uB - Bu \\ \rho uu + (p + B \cdot B\big/2\mu)\bar{\bar{I}} - BB/\mu \\ \rho eu + (p + B \cdot B\big/2\mu)\bar{\bar{I}} \cdot u - (B \cdot u)B/\mu \end{bmatrix} \tag{3.22}$$

The governing equation in flux vector is best cast in Cartesian coordinates, because scale factors of all coordinates for differential operators have values of unity that simplifies the formulation. On the fundamental Cartesian coordinates, the flux vectors can be described by three one-dimensional time-space domain. The peculiar magnetic field structure highlighted by equation (3.30) is easily displayed by the one-dimensional projection of the flux vector of ideal MHD equations. It becomes obvious that the one-dimensional flux vectors along the three coordinates reveal that the ideal MHD equations must be solved as a multidimensional problem. The temporal advancement of the magnetic flux density is not described in the corresponding spatial coordinate. In other words, there is no physically realistic magnetohydrodynamic problem that can be properly implemented in a single temporal–spatial dimension.

$$F_x = \begin{bmatrix} \rho u_x \\ 0 \\ u_x B_y - u_y B_x \\ u_x B_z - u_z B_x \\ \rho u_x^2 + p + B_x^2\big/2\mu \\ \rho u_x u_y - B_x B_y\big/\mu \\ \rho u_x u_z - B_x B_z\big/\mu \\ (\rho e + p + B \cdot B\big/2\mu)u_x - (B_x u_x + B_y u_y + B_z u_z)B_x\big/\mu \end{bmatrix} \tag{3.23}$$

$$F_y = \begin{bmatrix} \rho u_y \\ u_y B_x - u_x B_y \\ 0 \\ u_y B_z - u_z B_y \\ \rho u_x u_y - B_x B_y\big/\mu \\ \rho u_y^2 + p + B_y^2\big/2\mu \\ \rho u_y u_z - B_y B_z\big/\mu \\ (\rho e + p + B \cdot B\big/2\mu)u_v - (B_x u_x + B_y u_y + B_z u_z)B_y\big/\mu \end{bmatrix} \tag{3.24}$$

$$F_z = \begin{bmatrix} \rho u_z \\ u_z B_x - u_x B_z \\ u_z B_y - u_y B_z \\ 0 \\ \rho u_x u_z - B_x B_z / \mu \\ \rho u_y u_z - B_y B_z / \mu \\ \rho u_z^2 + p + B_z^2 / 2\mu \\ (\rho e + p + B \cdot B / 2\mu) u_x - (B_x u_x + B_y u_y + B_z u_z) B_z / \mu \end{bmatrix} \tag{3.25}$$

To solve the above equations system, equation (3.21) through (3.25) for its eigenvalues, the flux vector form in each Cartesian coordinate is required to be expressed in coefficient matrices,

$$\frac{\partial U}{\partial t} + \left(\frac{\partial F_x}{\partial U}\right)\frac{\partial U}{\partial x} + \left(\frac{\partial F_y}{\partial U}\right)\frac{\partial U}{\partial y} + \left(\frac{\partial F_z}{\partial U}\right)\frac{\partial U}{\partial z} = [A_x]\frac{\partial U}{\partial x} + [A_y]\frac{\partial U}{\partial y} + [A_z]\frac{\partial U}{\partial z} = 0 \tag{3.26a}$$

The typical Jacobian or the coefficient matrix $(\partial F_x / \partial U)$ by a quasi-linearized approximation can be rewritten as (Brio and Wu, 1988; Powell et al., 1999)

$$\frac{\partial F_x}{\partial x} = \begin{bmatrix} u & \rho & 0 & 0 & 0 & 0 & 0 & 0 \\ 0 & u & 0 & 0 & -B_x/\rho & +B_y/\rho & +B_z/\rho & 1/\rho \\ 0 & 0 & u & 0 & -B_y/\rho & -B_x/\rho & 0 & 0 \\ 0 & 0 & 0 & u & -B_z/\rho & 0 & -B_x/\rho & 0 \\ 0 & 0 & 0 & 0 & 0 & 0 & 0 & 0 \\ 0 & B_y & -B_x & 0 & -v & u & 0 & 0 \\ 0 & B_z & 0 & -B_x & -w & 0 & u & 0 \\ 0 & \gamma p & 0 & 0 & -(\gamma - 1) u \cdot B & 0 & 0 & u \end{bmatrix} \tag{3.26b}$$

By counting the number of elements or the eigensystem of the Coefficient matrix, there shall be complete eight sets of right and left eigenvalues. All the eigenvalues are real for the hyperbolic partial differential equations system and if the coefficient matrix [A] = $[\partial F_i / \partial x_i]$ or Jacobian matrix can be diagonalized; $[A]=[S][\lambda][S]^{-1}$. Then, the eigenvalues of the equation system are obtained by solving the coefficient matrix by the determinant,

$$\det\{[A_i] - \lambda \bar{\bar{I}}\} = 0 \tag{3.27}$$

The type of waves and the wave speeds are uniquely determined from the eigenvalues of the governing equations. However, the fifth row of the matrix is 0 and the matrix is singular. The approach is then to modify the governing equations so as to make the matrix nonsingular. The criterion is that the one-dimensional Riemann solver remains unchanged.

For a one-dimensional formulation, the Jacobian becomes a 7×7 matrix and up to five out of the seven eigenvalues may coincide. A set of right eigenvector for the ideal MHD equations has been given by Jeffrey and Taniuti (1964), as well as by Brio and Wu (1988). The eigenvalues are

$$\lambda_1 = u$$

$$\lambda_2 = u + \sqrt{\frac{B_n^2}{\rho\mu}} \quad \lambda_3 = u - \sqrt{\frac{B_n^2}{\rho\mu}}$$

$$\lambda_4 = u + \sqrt{\left\{\left(\frac{\partial p}{\partial \rho}\right)_s + \frac{B_n^2}{\rho\mu} + \sqrt{\left[\left(\frac{\partial p}{\partial \rho}\right)_s + \frac{B_n^2}{\rho\mu}\right]^2 - 4\left(\frac{\partial p}{\partial \rho}\right)_s\left(\frac{B_n^2}{\rho\mu}\right)}\right\}\Big/2}$$

$$\lambda_5 = u - \sqrt{\left\{\left(\frac{\partial p}{\partial \rho}\right)_s + \frac{B_n^2}{\rho\mu} + \sqrt{\left[\left(\frac{\partial p}{\partial \rho}\right)_s + \frac{B_n^2}{\rho\mu}\right]^2 - 4\left(\frac{\partial p}{\partial \rho}\right)_s\left(\frac{B_n^2}{\rho\mu}\right)}\right\}\Big/2}$$

$$\lambda_6 = u + \sqrt{\left\{\left(\frac{\partial p}{\partial \rho}\right)_s + \frac{B_n^2}{\rho\mu} - \sqrt{\left[\left(\frac{\partial p}{\partial \rho}\right)_s + \frac{B_n^2}{\rho\mu}\right]^2 - 4\left(\frac{\partial p}{\partial \rho}\right)_s\left(\frac{B_n^2}{\rho\mu}\right)}\right\}\Big/2}$$

$$\lambda_7 = u + \sqrt{\left\{\left(\frac{\partial p}{\partial \rho}\right)_s + \frac{B_n^2}{\rho\mu} - \sqrt{\left[\left(\frac{\partial p}{\partial \rho}\right)_s + \frac{B_n^2}{\rho\mu}\right]^2 - 4\left(\frac{\partial p}{\partial \rho}\right)_s\left(\frac{B_n^2}{\rho\mu}\right)}\right\}\Big/2} \tag{3.28a}$$

$$\lambda_8 = u \tag{3.28b}$$

The eighth eigenvalue λ_8 is known as a duplicity of λ_1. They are interpreted as the entropy and the contact surface waves, respectively. The eigenvalues λ_2 and λ_3 are identified with the Alfven waves. The fast electromagnetic waves are associated with eigenvalues λ_4 and λ_5. The slow electromagnetic waves are tied to the eigenvalues λ_6 and λ_7. One needs to note in the these equations that the symbol B_n is the magnetic flux density that is perpendicular to the direction of wave propagation.

In Figure 3.3, the four electromagnetic wave speeds versus the phase angle of the magnetic flux density are displayed from 1 to 180°, measured between the direction of wave propagation and the magnetic field (Shang, 2002). The presented results are obtained by assuming a homogeneous plasma field and all wave speeds are normalized by the speed of sound $c = \sqrt{(\partial p/\partial \rho)_s}$. The result are presented for the parameter $(B_n^2/\rho\mu)/c^2 = 0.5$. Based on this condition, the fast electromagnetic wave can propagate at 1.225 times the speed of sound in a transverse magnetic field, and

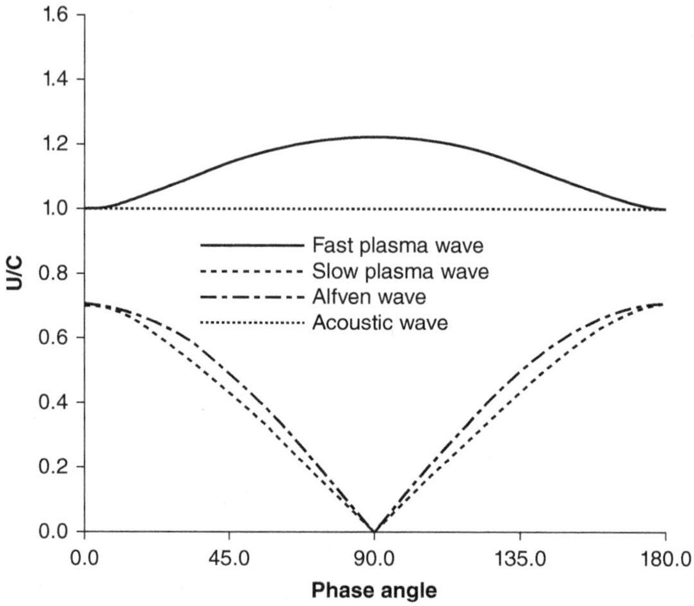

FIGURE 3.3 Different MHD wave speeds $(B_n^2/\rho\mu)/c^2 = 0.5$.

this wave speed is greater than or equal to the speed of sound over the entire range of the phase angles. The slow electromagnetic wave, on the contrary, propagates at a rate that is slower than the speed of sound over the complete phase angle range. The highest ratio of the slow electromagnetic wave to speed of sound is 0.7071 when the wave motion is parallel to the magnetic field. The slow wave ceases as the transverse magnetic field vanishes. The fast and slow electromagnetic wave speeds represent the upper and lower bounds for the propagating speeds of the Alfven and acoustic waves (Figure 3.4).

The electromagnetic waves are compound waves for which the compression and rarefaction components can coexist and the direct consequence of the nonconvexity of the ideal MHD equations. The drastic different behavior has been illustrated and identified by a numerical simulation of a shock tube by Gaitonde (1999). The degenerate electromagnetic wave is produced by reversing the polarity of an externally applied transverse magnetic field at the midpoint of a shock tube. The computational results of the density distribution are presented when the diaphragm is ruptured between the two sections of different pressures. The complex electromagnetic wave system is presented in direct contrast to the pure longitudinal acoustic wave. The fast rarefaction waves propagate at speeds greater than the acoustic speed toward both ends of the shock tube. The precursor of the slow right-running shock is identified as the fast electromagnetic wave, moving into the compressing section of the tube. The similar left-running fast rarefaction wave, moving into the expanding section of the

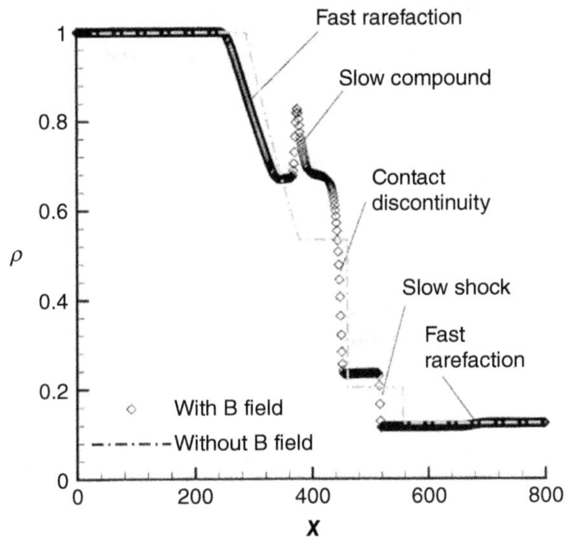

FIGURE 3.4 Density distribution in an ideal MHD shock tube.

shock tube, reveals a more complex wave system that propagates at different speed than the gas dynamic solution.

At the location where the transverse magnetic field reverses its polarity, an intermediate or compound shock emerges at the instance of null value. The compound wave actually shows that a slow rarefaction wave can be attached to a slow shock temporally in the plasma. This unique phenomenon occurs when the transverse magnetic field vanished during a polarity switches and the distinct electromagnetic wave degenerated. However, the existence and physical significance of the compound wave are still uncertain because there is no known experimental observation that supports the theoretical result, but all independent computational simulations of ideal MHD equations have consistently exhibited the distinct behavior when a transverse magnetic field presents in the plasma field (Brio and Wu, 1988; Gaitonde, 1999).

A distinct and unique feature of the plasma shock structure is demonstrated by the behavior of normal component of velocity u by the shock tube solution to ideal MHD equations. A varying magnetic field component B_y is applying perpendicular to the shock and undergoes polarization reversal between the contact discontinuity and the fast rarefaction wave. At the instant of vanishing transverse magnetic field, the multiple plasma waves degenerate into one hydrodynamic wave and create a condition for an intermediate or compound shock wave. Meanwhile, the streamwise component of the magnetic flux density B_x steadfastly remains at a constant value to ensure Gauss's law for the magnetic field $\nabla \cdot B = 0$. The solution depicted in Figure 3.5 represents the region bounded by the right-running, rarefaction precursor and left-running fast expansion waves. In the bounded region, the electromagnetic force induces a large and negative normal velocity component. The altered velocity

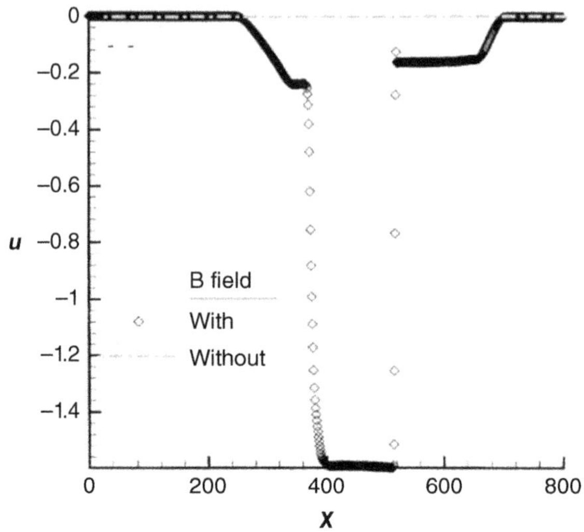

FIGURE 3.5 Normal velocity component across normal shock with varying transverse magnetic field.

distribution markedly differs from the corresponding gas dynamic shock, which leaves the zero normal velocity component unperturbed. From gas dynamic theory, it is well known that a straight normal will not produce any vorticity. However, the computational simulation of an ideal MHD shock shows that the two slow plasma shock waves generate strong normal stress component of $(\lambda + \mu)(\partial u/\partial x)$ in the direction normal to the planar shock.

The added dilution generation mechanism of a straight plasma shock and its more complex wave structure not only provide clear possibility that an applied electromagnetic field can alter the entropy production mechanism, but it also points out a new approach for flow field manipulation. Physics validation for this phenomenon is not easily achievable because the simulation is conducted for plasma with a perfect electric conductivity, but all numerical simulations have produced the identical results (Shang et al., 2014).

3.4 FULL MHD EQUATIONS

The full MHD equations are a set of nonlinear partial differential system of equations consisting of the reduced Maxwell equations with the classic MHD assumptions and the compressible Navier–Stokes equations. The difference from ideal MHD equations is that the flow medium now included viscosity and heat transfer to expanding the capability for engineering application. From a pure fluid dynamic viewpoint, the assumption of infinity Reynolds number is removed. A wide spectrum description for the electromagnetic–fluid–dynamics interaction has been put forth, and a rich variety of the formulations are derived as the consequence (Mitchner et al., 1973; Gaitonde, 1999; Shang, 2001; 2002). The reduced Maxwell equations are consistently adopted like that by the ideal MHD formulation. The approximation includes Faraday's induction law, by neglecting the displacement current in the Ampere's circuit law so the electric field intensity and current relationship are provided by the generalized Ohm's law (Sutton et al., 1965). The constitutive relation between magnetic flux density and magnetic field intensity is also introduced, and finally by applying Gauss's law in equation derivations but not explicitly imposed. In short, the difference between the ideal MHD and full MHD equation lies only in the description of transport properties for the flow medium; when the transport properties in molecular viscosity, conductivity, and diffusion are fully incorporated. The governing equations acquire the following form:

$$\frac{\partial \rho}{\partial t} + \nabla \cdot (\rho u) = 0 \tag{3.29a}$$

$$\frac{\partial B}{\partial t} + \nabla \cdot (uB - Bu) = \frac{\nabla \cdot (\nabla B)}{\mu \sigma} \tag{3.29b}$$

$$\frac{\partial \rho u}{\partial t} + \nabla \cdot [\rho uu - \bar{\bar{\tau}} - BB/\mu + (p + B \cdot B/2\mu)\bar{\bar{I}}] = 0 \tag{3.29c}$$

$$\frac{\partial (\rho e + B \cdot B/2\mu)}{\partial t} + \nabla \cdot [(\rho e + p + B \cdot B/2\mu)u - \kappa \nabla T - u \cdot \bar{\bar{\tau}} - (B \cdot u)B/\mu]$$
$$= \frac{(\nabla \times B/\mu)^2}{\sigma} + \left(\frac{B}{\mu}\right) \cdot \left[\frac{(\nabla \cdot \nabla B)}{\mu \sigma}\right] \tag{3.29d}$$

When dependent variables are normalized with respect to the free stream conditions in the these equations, the magnetic Reynolds number $\mathrm{Re_m} = \mu \sigma u L$ will explicitly appear in Faraday's induction and global energy conservation equations (3.19c) and (3.29d). Another similitude in the full MHD equations is the magnetic interaction parameter or the Steward number, $S = L\sigma B^2/\rho u$ (Table 3.1). Some of the typical values for these MHD similitudes were provided by Mitchner and Kruger (Mitchner et al., 1973).

TABLE 3.1 Typical values of electric conductivity, magnetic Reynolds, and Steward Numbers

	σ (mhos/m)	$\mathrm{Re_m}$	S
Glow discharge	10^2	10^{-3}	10^{-1}
Experimental MHD generator	10	10^{-3}	10^{-1}
Arc heater	5×10^3	6×10^{-2}	5×10^{-2}
MHD thruster	5×10^3	3	10
Controlled thermonuclear reaction	10^5	10^2	10^6
Ionosphere	10	10^3	10^3
Solar atmosphere	10^3	10^8	10^8

The specific components of flux vectors written in flux vector form in each Cartesian coordinate are

$$\frac{\partial U}{\partial t} + \frac{\partial(F_x + F_{x,v})}{\partial x} + \frac{\partial(F_y + F_{y,v})}{\partial y} + \frac{\partial(F_z + F_{z,v})}{\partial z} = S \tag{3.30}$$

$$U = U(\rho, B_x, B_y, B_z, \rho u_x, \rho u_y, \rho u_z, \rho e + p - B \cdot B/\mu) \tag{3.31}$$

$$F_x = \begin{bmatrix} \rho u_x \\ 0 \\ u_x B_y - u_y B_x \\ u_x B_z - u_z B_x \\ \rho u_x^2 + p + B_x^2/2\mu \\ \rho u_x u_y - B_x B_y/\mu \\ \rho u_x u_z - B_x B_z/\mu \\ (\rho e + p + B \cdot B/2\mu)u_x - (B_x u_x + B_y u_y + B_z u_z)B_x/\mu \end{bmatrix} \tag{3.32}$$

$$F_{x,v} = \begin{bmatrix} 0 \\ 0 \\ [(\partial B_y/\partial x) - (\partial B_x/\partial y)]/\mu\sigma \\ [(\partial B_z/\partial x) - (\partial B_x/\partial z)]/\mu\sigma \\ \tau_{xx} \\ \tau_{xy} \\ \tau_{xz} \\ u_x \tau_{xx} + u_y \tau_{xy} + u_z \tau_{xz} + \kappa \partial T/\partial x \\ \{B_y[(\partial B_y/\partial x) - (\partial B_x/\partial y)] + B_z[(\partial B_z/\partial x) - (\partial B_x/\partial z)]\}/\mu^2\sigma \end{bmatrix} \tag{3.33}$$

$$F_y = \begin{bmatrix} \rho u_y \\ u_y B_x - u_x B_y \\ 0 \\ u_y B_z - u_z B_y \\ \rho u_x u_y - B_x B_y/\mu \\ \rho u_y^2 + p + B^2/2\mu \\ \rho u_x u_z - B_x B_z/\mu \\ (\rho e + p + B \cdot B/2\mu)u_v - (B_x u_x + B_y u_y + B_z u_z)B_y/\mu \end{bmatrix} \tag{3.34}$$

$$F_{y,v} = \begin{bmatrix} 0 \\ [(\partial B_x/\partial y) - (\partial B_y/\partial x)]/\mu\sigma \\ 0 \\ [(\partial B_z/\partial y) - (\partial B_y/\partial z)]/\mu\sigma \\ \tau_{xy} \\ \tau_{yy} \\ \tau_{yz} \\ u_x \tau_{xy} + u_y \tau_{yy} + u_z \tau_{yz} + \kappa \partial T/\partial y \\ \{B_x[(\partial B_x/\partial y) - (\partial B_y/\partial x)] + B_z[(\partial B_z/\partial y) - (\partial B_y/\partial z)]\}/\mu^2\sigma \end{bmatrix} \tag{3.35}$$

$$F_z = \begin{bmatrix} \rho u_z \\ u_z B_x - u_x B_z \\ u_z B_y - u_y B_z \\ 0 \\ \rho u_x u_z - B_x B_z / \mu \\ \rho u_y u_z - B_y B_z / \mu \\ \rho u_z^2 + p + B_z^2 / 2\mu \\ (\rho e + p + B \cdot B / 2\mu) u_x - (B_x u_x + B_y u_y + B_z u_z) B_z / \mu \end{bmatrix} \tag{3.36}$$

$$F_{z,v} = \begin{bmatrix} 0 \\ [(\partial B_x / \partial z) - (\partial B_z / \partial x)] / \mu\sigma \\ [(\partial B_y / \partial z) - (\partial B_z / \partial y)] / \mu\sigma \\ 0 \\ \tau_{xz} \\ \tau_{yz} \\ \tau_{zz} \\ u_x \tau_{xz} + u_y \tau_{yz} + u_z \tau_{zz} + \kappa \partial T / \partial z \\ \{B_x[(\partial B_x / \partial z) - (\partial B_z / \partial x)] + B_y[(\partial B_y / \partial z) - (\partial B_z / \partial y)]\} / \mu^2 \sigma \end{bmatrix} \tag{3.37}$$

The source terms on the right-hand side of the governing equation (3.30) are results of the Pierre–Simon Laplacian of the magnetic flux density from Ampere's electric circuit law, and the Joule heating of the energy conservation equations,

$$\mathrm{S} = \begin{bmatrix} 0 \\ \partial^2 B_x / \partial x^2 + \partial^2 B_y / \partial y \partial x + \partial^2 B_z / \partial z \partial x \\ \partial^2 B_x / \partial x \partial y + \partial^2 B_y / \partial y^2 + \partial^2 B_z / \partial z \partial y \\ \partial^2 B_x / \partial x \partial z + \partial^2 B_y / \partial y \partial z + \partial^2 B_z / \partial z^2 \\ 0 \\ 0 \\ 0 \\ (\partial B_z / \partial y - \partial B_y / \partial z)^2 + (\partial B_x / \partial z - \partial B_z / \partial x)^2 + (\partial B_y / \partial x - \partial B_x / \partial y)^2 \\ +B_x[\partial^2 B_x / \partial x^2 + \partial^2 B_y / \partial y \partial x + \partial^2 B_z / \partial z \partial x] \\ +B_y[\partial^2 B_x / \partial x \partial y + \partial^2 B_y / \partial y^2 + \partial^2 B_z / \partial z \partial y] \\ +B_z[\partial^2 B_x / \partial x \partial z + \partial^2 B_y / \partial y \partial z + \partial^2 B_z / \partial z^2] \end{bmatrix} \tag{3.38}$$

The associated initial values and boundary conditions for the fluid dynamics variables are identical to the Navier–Stokes equations but with additional conditions for the electromagnetic variables. The nonslip condition is imposed for all velocity components on the fluid and solid interface. The density on the solid surface is derived from the intrinsic zero-pressure gradient condition of the boundary inner layer $\bar{n} \cdot \nabla p = 0$. The surface temperature can be specified as either a given value or by the adiabatic condition. The boundary conditions for the electric variables are identical to that of the ideal MHD equations.

In most MHD applications, the assumption of negligible electric conductivity or low magnetic Reynolds number is imposed by dropping off all the source terms. The governing equations acquire the following form (Shang, 2002):

$$\begin{aligned}
&\frac{\partial \rho}{\partial t} + \nabla \cdot (\rho u) = 0 \\
&\frac{\partial B}{\partial t} + \nabla \cdot (uB - Bu) = 0 \\
&\frac{\partial \rho u}{\partial t} + \nabla \cdot [\rho uu - \bar{\bar{\tau}} - BB/\mu + (p + B \cdot B/2\mu)\bar{\bar{I}}] = 0 \\
&\frac{\partial(\rho e + B \cdot B/2\mu)}{\partial t} + \nabla \cdot [(\rho e + p + B \cdot B/2\mu)u - \kappa \nabla T - u \cdot \bar{\bar{\tau}} - (B \cdot u)B/\mu] = 0
\end{aligned} \tag{3.39}$$

The assumption is based on the fact that the inducted magnetic field is negligible in comparison with the externally applied magnetic field. In numerical simulation, this simplification to computational stability and efficiency is noticeable by eliminating a stiff condition for evaluating a fractional term with numerator and denominator that approach vanishing values simultaneously (Gaitonde, 1999; Poggie and Gaitonde, 2002). Meanwhile, a good approximation of numerical simulations of high physical fidelity is retained. Finally, some of the formulations are based on the low magnetic Reynolds number approximation and limit the externally applied magnetic field to a steady state $\partial B/\partial t = 0$. The permutation between the components of the magnetic flux density and the velocity, uB and Bu, becomes irrelevant, and thus the magnetic and flow fields can be decoupled (Sheriff, 1965). The MHD governing equations are further simplified to become the interdisciplinary Navier–Stokes equations. The simplifications are achieved by appended source terms of the Lorentz acceleration and Joule heating to the global conservation laws

$$\begin{aligned}
&\frac{\partial \rho}{\partial t} + \nabla \cdot (\rho u) = 0 \\
&\frac{\partial \rho u}{\partial t} + \nabla \cdot (\rho uu + p\bar{\bar{I}}) = J \times B \\
&\frac{\partial \rho e}{\partial t} + \nabla \cdot (\rho eu + u \cdot p\bar{\bar{I}}) = E \cdot J
\end{aligned} \tag{3.40}$$

or

$$\begin{aligned}
&\frac{\partial \rho}{\partial t} + \nabla \cdot (\rho u) = 0 \\
&\frac{\partial \rho u}{\partial t} + \nabla \cdot (\rho uu + p\bar{\bar{I}}) = \nabla \left(\frac{B}{\mu}\right) \times B \\
&\frac{\partial \rho e}{\partial t} + \nabla \cdot (\rho eu + u \cdot p\bar{\bar{I}}) = -(u \times B) \cdot \left(\frac{\nabla \times B}{\mu}\right)
\end{aligned} \tag{3.41}$$

The relationship between the electric field intensity and electric current is explicitly described by the reduced generalized Ohm's law and Ampere's electric circuit law. Gauss's divergence law for the magnetic flux density is implicitly included but must be imposed explicitly in most numerical simulations. These formulations tend to focus on a specific application under additional restrictions for applicability. A common feature also stands out that the detailed and complex plasma property between the electric field intensity and the current density is approximated by generalized Ohms Law. This approximation provides a compact formulation while frequently leads to plausible physical fidelity concerns by an oversimplification.

3.5 SHOCK JUMP CONDITION IN PLASMA

The shock discontinuity in perfect gas is well understood and summarized by the classic Rankine–Hugoniot relationship in gas dynamics. The added mechanisms such as Joule heating and Lorentz acceleration of an electrically conducting medium in an actuated electromagnetic field certainly will alter the discontinuous behavior. Specially, plasma properties change rapidly across a shock that may involve localized charge separation to produce a double layer with a large electric potential difference. The electric field is constrained within the layer, and the double structures are often found in current-carrying plasma.

However, MHD shock velocities are close to the speed of light and thus must include a relativistic treatment. The simplest possible approach is to assume that the electric conductivity is approaching infinity, which is equivalent to assume that the magnetic Reynolds number is infinite (De Hoffmann et al., 1950). By this assumption, the fluid particles are free to move parallel to the magnetic field, but the motion perpendicular to the magnetic field is firmly attached to lines magnetic force. The discontinuity of fluid behavior is describable by a two-dimensional shock wave. In order to derive the equivalent shock jump condition, the energy equation needs to be rewritten in terms of the specific enthalpy h rather than the specific internal energy e,

$$\frac{\partial \rho h}{\partial t} - \frac{\partial p}{\partial t} + \nabla \cdot (\rho h u + q - u \cdot \bar{\bar{\tau}}) = (E + u \times B) \cdot J \tag{3.42}$$

Consider the simplest shock wave structure in steady state across the x coordinate and recall the electromagnetic field will not permit a realistic one-dimensional phenomenon. Therefore, the shock jump condition must be formulated by the tangential and normal components across a shock wave. The shock relation is essentially an oblique wave in the three-dimensional electromagnetic field. The macroscopic conservative equation can then be integrated across the shock. From the Maxwell equations, the normal components of the magnetic flux density B and the tangential component of the electric field intensity E must be continuously across any interface. The general shock jump conditions for plasma become

$$d(\rho u_x)/dx = 0$$

$$dB_x/dx = 0$$

$$
\begin{aligned}
&dE_y/dx = 0\\
&dE_z/dx = 0\\
&d\left(\rho u_x^2 + p - \tau_{xx}\right)/dx - (J \times B)_x = 0\\
&d(\rho u_x u_y - \tau_{xy})/dx - (J \times B)_y = 0\\
&d(\rho u_x u_z - \tau_{xz})/dx - (J \times B)_z = 0\\
&d[\rho u_x h + q_x - (u \cdot \bar{\bar{\tau}})_x]/dx - [E + (u \times B)] \cdot J = 0
\end{aligned}
\tag{3.43a}
$$

In the MHD formulation, the globally neutral condition implies that there are no free space charges. A fundamental law of Gauss divergent for electric field becomes; $\nabla \cdot E = 0$. By the generalized Ohm's law, $E_y = J_y/\sigma - u_w B_x + u_x B_z$ and $E_z = J_z/\sigma - u_x B_y + u_y B_x$. The cross product $J \times B$ can be manipulated to be a perfect divergence operator of magnetic flux density. Finally by a perfect electric conductivity assumption, the Joule heating also achieves a perfect divergent form in steady state

$$
[E + (u \times B)] \cdot J \doteq -\nabla \cdot [u(B \cdot B/\mu) - B(B \cdot u/\mu)]
\tag{3.43b}
$$

The conservation laws now appear as

$$
\begin{aligned}
&\nabla \cdot \rho u = 0\\
&\nabla \cdot (uB - Bu) = 0\\
&\nabla \cdot [\rho uu - BB - \bar{\bar{\tau}} + (p + B \cdot B/2\mu)\bar{\bar{I}}] = 0\\
&\nabla \cdot [\rho hu + \kappa \nabla T - u \cdot \bar{\bar{\tau}} - (B \cdot u)B/\mu] = 0
\end{aligned}
\tag{3.44}
$$

Meanwhile, the shock jump relationship under these assumptions can be integrated across the shock. Viscous dissipation and heat transfer have no influence on the jump condition. This observation can be understood on the base of an equilibrium thermodynamic state. Since entropy is a function of the local thermodynamic state, the path of fluid particle motion is irrelevant. At the integral limits, up- and downstream of the shockwave, the normal shear stress, and heat transfer are negligible to yield

$$
\begin{aligned}
&[\rho u_x] = 0\\
&[B_x] = 0\\
&[u_x B_y - u_y B_x] = 0\\
&[u_x B_z - u_z B_y] = 0\\
&\left[\rho u_x^2 + p + \left(B_y^2 + B_z^2\right)/2\mu\right] = 0\\
&[\rho u_x u_y - B_x B_y/\mu] = 0\\
&[\rho u_x u_z - B_x Bz/\mu] = 0\\
&\left[\rho u_x h + u_x \left(B_y^2 + B_z^2\right)/2\mu - B_x(u_y B_y + u_z B_z)/\mu\right] = 0
\end{aligned}
\tag{3.45}
$$

The square brackets indicate the jumps of the end states across the shock wave. For a plasma with a perfect electric conductivity and a negligible Hall current, the above jump conditions are identical to these derived by Sutton and Sherman (Sutton et al., 1965).

The above equations summarize the changes to the Rankine–Hugoniot relationship of gas dynamics across a shock, and it then will revert to the classic relationship of gas dynamics when the applied magnetic field vanishes. They have been demonstrated by numerical simulations that the degenerate shocks occur either at the instances the polarity of the magnetic field is switched or the applied magnetic field vanishes (Brio and Wu, 1988; Gaitonde, 1999).

The entropy increases across the shock can be evaluated by the second law of thermodynamics

$$\frac{dS}{dt} = \frac{1}{T}\left(\frac{dh}{dt} - \frac{1}{\rho}\frac{dp}{dt}\right) \tag{3.46}$$

Here, $d/dt = \partial/\partial t + \rho u_x \partial/\partial x$ is the substantial derivative in the Eulerian frame and along the x coordinate. In steady state, $d/dt = \rho u_x \partial/\partial x$. Thus, the change in entropy per unit mass is

$$\rho u_x \frac{dS}{dx} = \frac{1}{T}\left(\rho u_x \frac{dh}{dx} - u_x \frac{dp}{dx}\right) \tag{3.47}$$

Substitute the derivatives from the momentum and energy equations and invoke the fact that ρu_x is an invariant across the shock to get

$$\rho u_s \Delta S = \int \frac{\kappa}{T^2}\left(\frac{dT}{dx}\right)^2 dx + \int \frac{(\mu_m + \lambda)}{T}\left(\frac{du_x}{dx}\right)^2 + \int \frac{J^2}{(\sigma T)}dx - \int \frac{u_x (J \times B)_x}{T} dx \tag{3.48}$$

The modification of entropy increment rests on two additional entropy change mechanisms. Joule heating and the work performed by the moving charged particles contribute to the change across the shock. The conductive heat transfer, viscous dissipation, and Joule heating constitute the positive and definitive terms that contribute to entropy increase across the shock. On the contrary, the work done by electromagnetic force depends on the polarity of the electromagnetic field and the orientation of the moving charged particles is not. This mechanism, if realizable, may lead to some innovative application for wave drag reduction. Some unique insights may also shed light on the shock structure of a plasma. Additional and sustained efforts by both experimental and computational are critical in this area of research for validating the modified shock wave structures in an electromagnetic field.

Canupp (2000) used the Runge–Kutta–Fehlberg method to solve the conservation laws for the fine structure of a shock by assuming $B_x = B_z = 0$, the sole component

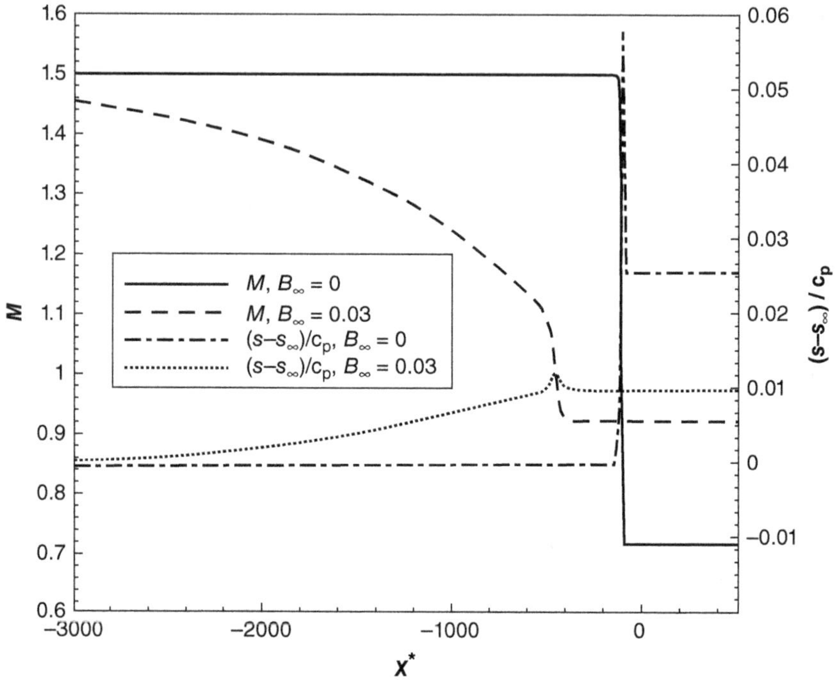

FIGURE 3.6 Shock structure in a transverse magnetic field.

being $B_y = 0.03T$(300 gauss). The electric conductivity has a large but finite value of 10^5 mho/m.

$$\begin{aligned} &d[\rho u_x]dx = 0 \\ &d[(1/\mu\sigma)(dB_x/dx) - u_x B_y]/dx = 0 \\ &d[\rho u_x^2 + p + B_y^2/2\mu - \tau_{xx}]/dx = 0 \\ &d[\rho u h + u_\infty B_\infty B_y/\mu - u_x \tau_{xx} - \kappa dT/dx]/dx = 0 \end{aligned} \tag{3.49}$$

The thin shock structure is stretched by the shock wave Reynolds number $x^* = \rho_\infty u_\infty x/\mu_\infty$. The shock structures in the electromagnetic and acoustic fields are presented together in Figure 3.6 with the range of $x^* = -3000$ to highlight the internal shock structure. In order to maintain computational stability, the calculation proceeded from the downstream saddle point in the temperature–velocity phase space. For each case, the calculated Mach number and the entropy variations correctly reached the identical upstream asymptotes near $x^* = -7500$ beyond the presented computational domain (Canupp, 2001).

The computational results show that the entropy of the gas dynamic shock rapidly rises in the thin shock structure to a maximum value at the sonic point and then quickly relaxes to the final value downstream. The entropy of the shock in the plasma exhibits a similar overall behavior, but the changes are much more gradual and less intense. The increased entropy initiates farther upstream in the magnetic field than that of the gas dynamic counterpart. This phenomenon is understandable on the basis of great disparity in signal propagation speed between the two gas mediums. The entropy increment in the shock structure of plasma is a factor of 2.6 times lower than the acoustic field. The entropy variation may derive from the strong dependence of the polarity of the transverse magnetic field, but this simplified one-dimensional shock structure simulation does not illustrate all possible magneto-aerodynamic interaction. However, it still reveals an interesting glance of the basic magnetic field effects to gas dynamics discontinuity.

3.6 SOLUTIONS OF MHD EQUATIONS

The magnetic field interaction with fluid dynamics intrinsically altered the flow behavior is presented in this chapter. The classic Hartman flow of an incompressible medium and Ziemer's experiment on shock standoff distance are examples for this purpose. Additional computational simulations will also be discussed to show the lack of detailed plasma composition that is required for determining electric conductivity of plasma in the MHD formulation. This simplification can limit the fidelity to physics.

The Harman solution is a generalization of the incompressible steady plane Poiseuille flow within a rectangular channel with an imposed uniform and constant transverse magnetic flux density B_z (Hartman, 1937; Hartman et al., 1937). From the eigenvalue structure of the MHD equations, a physically meaningful solution must involve multiple spatial dimensions by any numerical simulation. The electrode surfaces are thus prescribed on the constant y coordinate (height) and the insulated walls of the channel are imposed on the constant value of the z coordinate (width). The height and the width of the channel are resolved by a 100-point mesh system over the channel. By this arrangement, all solutions are function of z, and the induced electric field intensity by the generalized Ohm's law is $J_y = \sigma(E_y - uB_z)$. From the prescribed vanishing electric field intensity at the isolator surface E_y becomes a constant. By the low magnetic Reynolds number and the symmetrical condition along the centerline of the channel, the induced magnetic flux density B_x is negligible. Hartman showed that the pressure must be balanced by the Lorentz force $J \times B$.

The ratio of the Lorentz force to the viscous force defines the Hartman number that is related to the two similarity parameters of Magnetic Reynolds and the Steward numbers given by equations (3.1a) and (3.1b). In fact, the relationship between three similarity parameters is given as (Mitchner et al., 1973)

$$H = (R_{\mathrm{m}}S)^{1/2} = Bl(\sigma/\nu)^{1/2} \tag{3.50a}$$

For the Hartman flow with the half channel width z to be characteristic length, the Hartman is

$$H = (w/2)B(\sigma/\nu)^{1/2} \tag{3.50b}$$

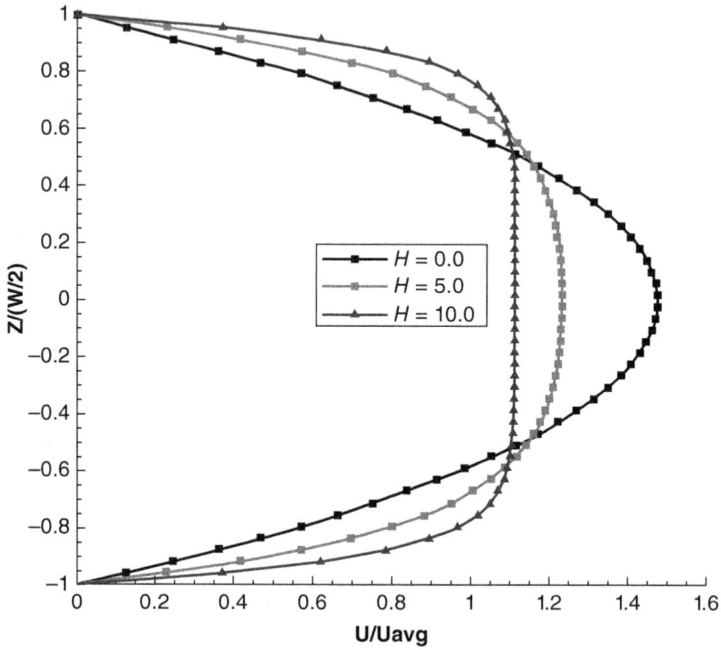

FIGURE 3.7 Velocity profiles of Hartman flow, $0.0 < H < 1.0$.

The direct comparison of the classic Poiseuille velocity profile and MHD solution with Hartman numbers of 5 and 10 is depicted in Figure 3.7. The effect of a uniform and constant transverse magnetic field to the incompressible channel flow is retarding the flow near the centerline of the channel by the decelerating force $J \times B = J_y B_z < 0$. Close to the isolator surface, the force becomes positive to accelerate the fluid to steepen the slope of the velocity profile. The net result of an applied transverse magnetic flux density to the channel flow is decelerating the flow and flattening the velocity profile in the rectangular channel. The effect is intensified and much more pronounced as the Hartman number is increased. The increased surface shear stress is substantial. At a Hartman number of 5, the increment value is more than double that of the Poiseuille flow. Experimental data support the general trend of the classic Hartman result but reveal a greater sensitivity to the surface temperature and thermodynamic condition of plasma. It was found that a decreased surface temperature also reduced electric conductivity thereby diminishing the electromagnetics and fluid interaction by the transverse magnetic field.

The modification to and accommodation of the electromagnetic field by the fluid motion is reflected by a close relationship with the kinematics of the flow. The analytic result is given by the simple equation related to the electric current density normalized by the product of electric conductivity, averaged velocity, and the applied magnetic flux density.

$$J_y(z)\big/(\sigma\bar{u}B_z) = 1 - u(z)/\bar{u} \tag{3.51}$$

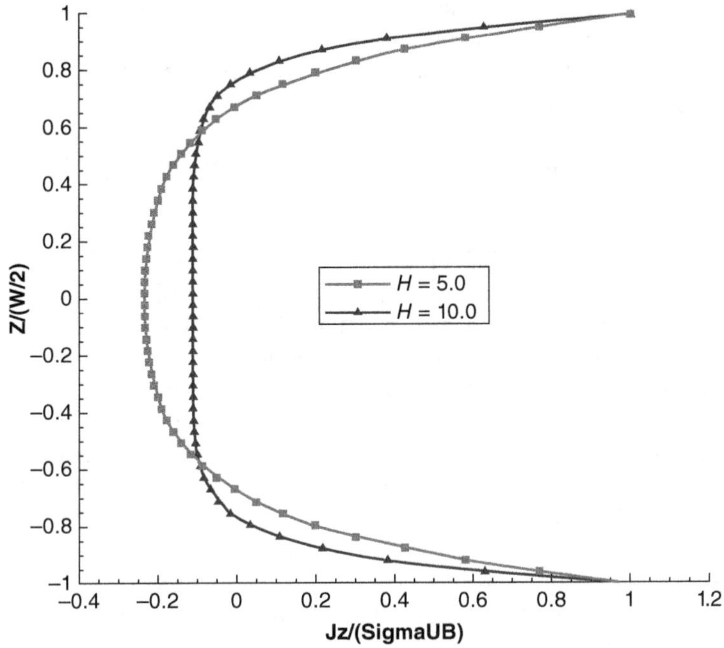

FIGURE 3.8 Electric current profiles of Hartman flow, $5.0 < H < 10.0$.

This interdependence of velocity profile and electric current distribution across the channel is easily displayed in Figure 3.8. The maximum electric current is located at the isolated channel surface for all Hartman numbers and reverses its direction to become negative near the central portion of the channel. The gradient of electric current in the near wall region is seen to steepen with the increased Hartman number. The Hartman current is actually an eddy current in the *yz* cross-sectional plane and loops between the electrode surfaces to form a closed circuit.

An interesting observation is noted but without any in-depth study. In principle, the Hartman flow is laminar and most computational simulation is restricted to a Reynolds number less than 1200. However, it has been conjectured that the externally applied magnetic field may reduce the turbulent shear stress by damping the fluctuating eddying motion (Sonju and Kruger, 1969; Mitchner et al., 1973).

The MHD theory and formulation show that the magnetic–fluid dynamic interaction in an applied magnetic flux density around a blunt body at supersonic speed leads to an increase in bow shock wave standoff distance, a deceleration of flow over a solid surface, and a mitigation of stagnation point heat transfer (Bush, 1961; Shang, 2001, 2002). These attractive and unique features of MHD offer a new physical dimension for high-speed flights such as space travel launch and reentry. However, a more realistic electrical conductivity distribution in plasma has been shown to diminish the effectiveness of the electromagnetic–fluid–dynamics interaction (Locke et al., 1965). The lack of physical fidelity for plasma description in MHD by using generalized Ohms law can be alleviated only by removing this fundamental and simplifying assumption that a scalar relationship between electric current density and an applied electric field is an oversimplification.

Theoretical studies have indicated that the force produced by the interaction of the electrically conducting fluid with an applied magnetic field can produce appreciable changes in the supersonic flow pattern about aerodynamic shapes. Ziemer (1959) conducted an experimental investigation in an electromagnetic shock tube. The thermodynamic state of the ionized gas is equivalent to a hypervelocity velocity up to 12 km/s and a stagnation point temperature of up to 25,000 K. Therefore, the estimated free stream Mach number has an equivalent range from 15 to 40, but the free stream Mach number in the shock tube only spans a range from 2.9 to 3.9 in the shock tube. A magnetic field was generated by a magnet coil at 40 kilogauss (4.0 Tesla) at the stagnation point of a hemispherical blunt body, and the standoff distance of the bow shock showed an increment of 7.5 times to that of a perfect gas at $Q = \sigma B^2 r / \rho_\infty u_\infty^2 = 69.0$. The data also show that the shock standoff distance increases with the magnetic flux density and correlates with the parameter Q.

The altered bow shock standoff distance has been simulated by solving the ideal MHD and the MHD equations. The difference between numerical results is anticipated to be small, because the standoff distance is the direct consequence of the continuity condition across the shock, and the weaker dependence to the detailed thermodynamic state and chemical composition of plasma is consistently excluded in these formulations. The magnetic field distribution was also modeled either by a dipole at the stagnation point similar to the field distribution of a solenoid or by a simple surface outward normal distribution of constant magnetic flux density. A typical result is presented in Figure 3.9, and the numerical results are generated for a

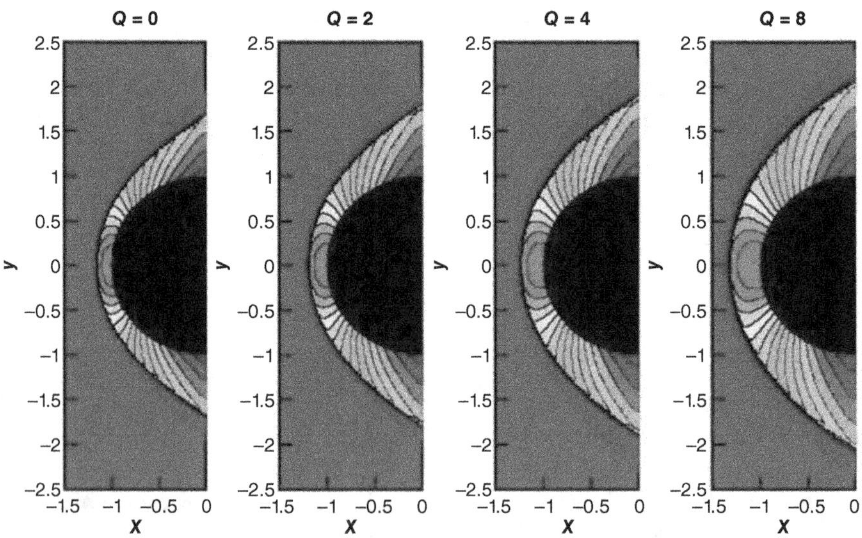

FIGURE 3.9 Computed standoff distances at different magnetic field flux density. From Poggie, J., and Gaitonde, D., *Phys. Fluids*, vol. 14, no. 5, 2002, pp. 1720–1731. Used with permission.

Mach 5 flow over a hemisphere (Poggie and Gaitonde, 2002). The simulated solution does not have a specific comparative validation with experimental observation, but the solution is consistent to all computational results. The standoff distance over the hemisphere is seen to increase with the elevating magnitude of the electromagnetic interaction parameter Q. The predicted behavior is in general agreement to that of the pioneering efforts by Ziemer.

The numerical result of a Mach number of 5.85 convincingly elucidates how the magnetic pressure or the normal stress component of the Maxwell tensor affects the bow shock standoff distance in supersonic plasma (Canupy, 2001). In a magnetic field, the pressure exerting on a plasma must include the magnetic pressure for momentum balance but excluded from the equation of state for determining thermodynamic variables. The difference in density distribution along the stagnation streamline with and without the presence of a magnetic field is depicted in Figure 3.10. The numerical result in the absence of an externally applied magnetic field shows the density jump across the shock wave according to the Rankine–Hugoniot relationship and a continuous compression toward the stagnation point in the shock layer. In contrast to that, the density decreases along the stagnation streamline as the flow decelerated to stop by an increased contribution from the magnetic pressure. This particular phenomenon is supported by experimental observation in that the density behind the shock wave drops quickly through the shock layer. At a plasma interaction parameter

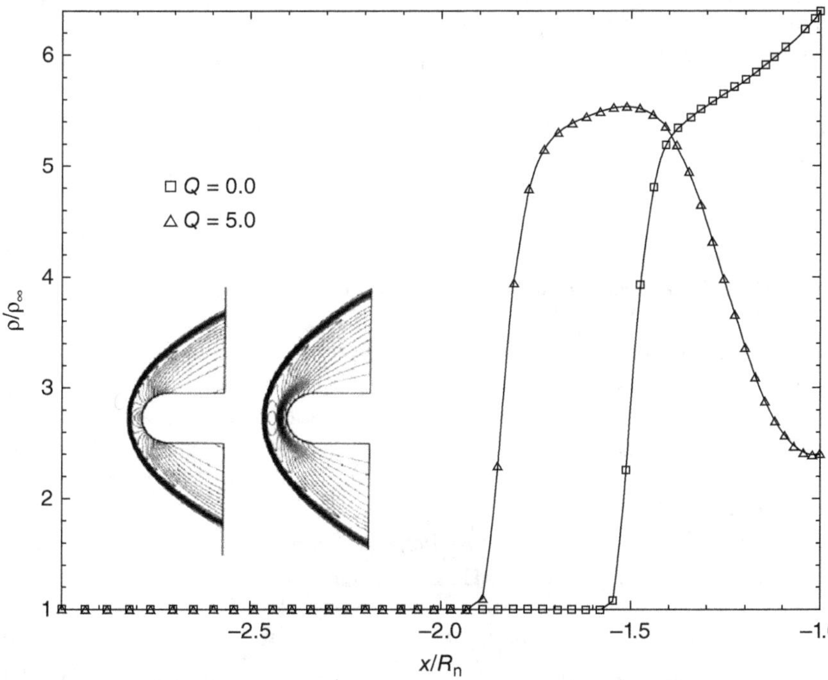

FIGURE 3.10 Effect of Maxwell's normal stress or magnetic pressure to shock standoff distance.

of $Q = 5.0$, the density distribution along the stagnation streamline follows a non-monotonic path, and more importantly, it yields a lower value in the shock layer. The standoff distance of a bow shock is inversely proportional to the density jump across the shock by satisfying the continuity condition. The magnetic pressure decreases the density jump across the bow shock; thus, an expanding shock layer thickness or standoff distance of the bow shock is required to accommodate the same amount of mass flow. This phenomenon is a canonical behavior for the supersonic MHD flow over a blunt body and fully verified by experimental observations and computational simulations.

From the increase shock layer thickness and the decelerating Lorentz force of an externally applied magnetic field normal to solid surface, mitigation of heat transfer and modification to surface shear stress can be anticipated for plasma applications. The potential benefit to aerospace science and engineering technology is extremely attractive. However, a detailed and accurate analysis can be achieved only by a specific description of the ionized gas medium, which is beyond the reach of the MHD formulation based on the elegant and simplified transport property. The MHD discipline is and will remain at the frontier of scientific research.

REFERENCES

Alfven, H., *Cosmical Electrodynamics*, Clarendon Press, Oxford, 1950.

Bush, W.B., The stagnation-point boundary layer in the presence of an applied magnetic field, *J. Aerospace Sci.* vol. 28, 1961, pp. 610.

Brio, M., and Wu, C.C., An upwind differencing scheme for the equations of ideal magnetohydrodynamics, *J. Comp. Phys.*, vol. 75, 1988, pp. 400–422.

Cabannes, H., *Theoretical Magnetofluid dynamics*, Academic Press, New York, 1970.

Canupp, P.W., *The Influence of Magnetic Fields for Shock Waves and Hypersonic Flows*, AIAA 2000-2260, Denver, CO, June 2000.

Canupp, P.W. *Influence of Magnetic Field Divergence Errors in Magneto-Aerodynamic Computations*, AIAA 2001-0197, Reno, NV, 2001.

Gaitonde, D.V., *Development of a Solver for 3-D Non-ideal Magnetogasdynamics*, AIAA 1999-3610, Norfolk, VA, 1999.

Hartman, J., Hg-dynamics I: theory of the laminar flow of an electric conductive liquid in a homogeneous magnetic field. Danske Videnskab, *Mat-fys. Medd.* vol. 15, no. 6, 1937.

Jackson, J.D., *Classical Electrodynamics*, 2nd ed., John Wiley & Sons, New York, 1975.

Hartman, J., and Lararrus, F., Hg dynamics II: experimental investigations on the flow of mercury in a homogeneous magnetic field. Danske Videnskab, *Mat-fys. Medd.* vol. 15, no. 7, 1937.

Jeffrey, A., and Taniuti, T., *Non-linear Wave Propagation*, Academic Press, New York, 1964.

Locke, E., Petschek, H.E., and Rose, P.H., Experiments with magnetodrodynamically supported shock layers, *Phys. Fluids*, vol. 8, no. 1, 1965, pp. 872–885.

Mitchner, M., and Kruger, C.H., *Partially Ionized Gases*, John Wiley & Sons, New York, 1973.

Poggie, J., and Gaitonde, D.V., Magnetic control of flow past a blunt body: numerical validation and exploration, *Phys. Fluids*, vol. 14, no. 5, 2002, pp. 1720–1731.

Powell, K.G., Roe, P. L., Linde, T.J., Gombosi, T.I., and Zeeuw, D.L., A solution-adaptive upwind scheme for ideal magnetohydrodynamics, *J. Comp. Phys.*, vol. 154, 1999, pp. 284–309.

Shang, J.S., Recent research in magneto-aerodynamics, *Prog. Aerosp. Sci.*, vol. 37, no. 1, 2001, pp. 1–20.

Shang, J.S., Shared knowledge in computational fluid dynamics, electromagnetics, and magneto-aerodynamics, *Prog. Aerosp. Sci.*, vol. 38, no. 6-7, 2002, pp. 449–467.

Shang, J.S., and Huang, P.G., Surface plasma actuators modeling for flow control, progress. *In Aerospace Sci.*, vol. 67, 2014, pp. 29–50.

Sheriff, J.A., *A Textbook of Magnetohydrodynamics*, Pergamon Press, Oxford, 1965.

Sonju, O.K., and Kruger, C.H., Experiments on Hartman channel flow in plasma, *Phys. Fluids*, vol. 12, no. 12, 1969, pp. 2548.

Sutton, G.W., and Sherman, A., *Engineering Magnetohydrodynamics*, McGraw-Hill, New York, 1965.

Vincenti, W.G., and Kruger, C.H., *Introduction to Physical Gas Dynamics*, John Wiley & Sons, New York, 1965.

Zachary, A., and Colella, A., A high-order Godunov method for the equations of ideal magnetohydrodynamics, *J. Comp. Phys.*, vol. 99, 1992, pp. 341–347.

Ziemer, R.W., Experimental investigation in magneto-aerodynamics, *J. Am. Rocket Soc.*, vol. 29, 1959, pp. 642–647.

4

COMPUTATIONAL ELECTROMAGNETICS

INTRODUCTION

All phenomena of electromagnetics are governed by the time-dependent Maxwell equations; computational simulations of electromagnetic–aerodynamic interactions must be built on the capability of computational electromagnetics (CEM) that has a tier structure for solving the Maxwell equations. Physics-based simulations fall naturally into groups according to the frequency spectrum of the dominant physical phenomena, depending on whether they occur in the Raleigh, Resonance, or the Optical region (Shang, 2002). Ray tracing is suitable for high-frequency applications where the wavelength is so much shorter than the characteristic dimension of the electromagnetic field, such as in wideband antenna design, real-time range profiling, and synthetic aperture radar imaging. The underlying principle is based on physical optics, theory of diffraction, or a combination of both. The scattered far field is derived from an integration of induced surface currents through physical optics; the physical optics is therefore an asymptotic method and bypasses direct simulation by solving fundamental governing equations of electromagnetics.

The collective modeling and simulation tools for solving the Maxwell equations span a range from the method of moments in the frequency domain, finite-difference/volume approximation in time domain, and a more recent development of a hybrid technique. The numerical simulation in the frequency domain is solving the second-order curl–curl formulation of Maxwell equations. The temporal variation

Computational Electromagnetic-Aerodynamics, First Edition. Joseph J.S. Shang.

is represented by a simple harmonic function using dependent variable separation technique. The resultant partial differential equations are transformed to the frequency domain as elliptic type and usually solved by fast multipole methods. These solving schemes rely on a hierarchal subdivision of space to process the integral kernel by a selected basis function (Harrington, 1968; Taflove, 1995). The computational efficiency is dominated by matrix–vector multiplications and is strongly depended on the spatial distribution point density and methods of implementation. Impressive improvements to both computing efficiency and memory address reduction have been achieved through the adaptive integral method (Luebber et al., 1991; Chew et al., 2001).

A relatively new approach in CEM for full-scale dynamic simulation is the hybrid method. This method is designed to simulate physics involving interaction of wave discontinuities on large structures. Three levels of hybridization are made possible by consistently combining computational tools of the time, frequency, and asymptotic procedures for scattering and radiation. The hybrid formulation in CEM melds the best of the high-frequency asymptotic with rigorous low-frequency approach that is based on first principles, but this approach is not necessarily the best choice for interdisciplinary simulation.

CEM in the time domain solves the time-dependent Maxwell equations. This approach represents the most general and versatile approach for analyzing electromagnetic phenomena and is compatible with MHD or computational electromagnetic-aerodynamics. Since the pioneering CEM effort in the time domain by Yee (1966), an ever-increasing range of engineering applications in broad bandwidth and dynamic electromagnetic phenomena have been realized. CEM in the time domain has become an effective tool for electromagnetics research (Taflove, 1995). However, maturation of this scientific discipline for design and evaluation is still hampered by issues in numerical resolution and computational efficiency. A fundamental limitation emerges from describing wave motion via the solutions of hyperbolic partial differential equations in discrete space. This limitation is imposed by the basic and absolute numerical accuracy constraint for simulating wave motion by the Nyquist frequency limit. In short, a minimum of two discretized points per wavelength is necessary to achieve a physical meaningful simulation. This theoretical limit defines the upper bound of practical application in physical dimension of a transient phenomenon that can be simulated. The computational resource requirement for three-dimensional simulations over an extensive distance measured in wavelength can be truly daunting.

Another formidable challenge for CEM in the time domain arises from the eigenvalue structure of governing equations. The Maxwell equations are a system of partial differential equations and its associated initial values and boundary conditions constitute the hyperbolic system. Solutions to the Maxwell equations must satisfy compatibility conditions to be well posed. Most of all, simulations of an initial-value problem in a truncated computational domain will avoidably introduce spurious wave reflections at the artificial boundary. A range of numerical procedures have been developed to overcome this fundamental dilemma with different degrees of success, and which still remains as a challenge to simulate time-dependent electromagnetic phenomena. A continuous wave motion of simple electric dipole simulation

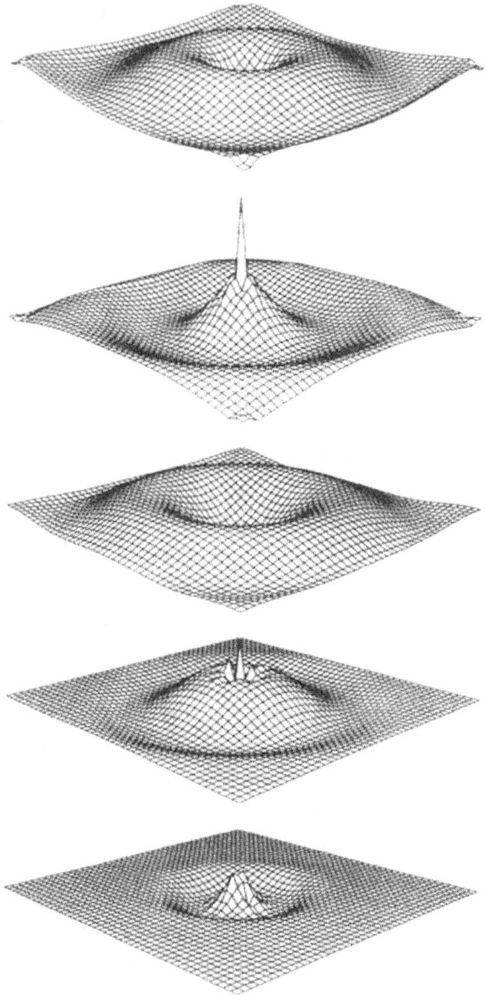

FIGURE 4.1 Electromagnetic wave propagation by a dipole.

is depicted by Figure 4.1 to illustrate the resolution requirement for CEM simulation and the need to eliminate reflected waves from computational domain.

Nevertheless, the dynamic nature of electromagnetic wave propagation and the interaction with fluid motion require accurate time integration of the governing equations. The detailed formulation includes an appropriate initial/boundary condition and the no-reflection far field treatments, and concurrent computational techniques will be included in the following two chapters. Interesting applications for the electric and magnetic energy distributions in scattering far field fields are demonstrated first as a validation of the CEM in the time domain versus frequency domain methodologies. The study of electromagnetic wave propagation in dipole radiation, waveguide with and without an embedded plasma region, and a pyramidal-horn antenna for plasma diagnostics is also presented.

4.1 TIME-DEPENDENT MAXWELL EQUATIONS

Computational electromagnetics in the present context is focused on solving Maxwell equations in the time domain. First of all, the time-dependent formulation is the basic form of the Maxwell equations that describes dynamic electromagnetic fields not confined as a single frequency, time-harmonic phenomena (Krause, 1953; Harrington, 1961). CEM in the time domain has the widest range of engineering applications. In addition, several innovative numerical algorithms for solving the first-order hyperbolic partial differential equations, as well as coordinate transformation techniques, have been introduced to the CEM community. These finite-difference and finite-volume numerical algorithms were devised specifically to mimic physics involving directional wave propagation. Meanwhile, very complex shapes associated with an electromagnetic field can be easily accommodated by incorporating the coordinate transformation technique.

In order to use CEM effectively, it will be beneficial to understand the fundamentals of numerical simulation and its limitations. An inaccurate result by a numerical simulation is attributable to a mathematical model for the physics, numerical algorithm, and computational accuracy. Numerical accuracy is also controlled by the algorithm and computing system adopted. The error induced by the discretization consists of the round-off and the truncation error. The round-off error is the result of a computing system and is problem-size-dependent. Since this error is random, it is the most difficult to evaluate. One anticipates that this type of error will be a concern for solving procedures involving a large-scale matrix manipulation such as the method of moments and the implicit numerical algorithm for finite-difference or finite-volume methods. The truncation error for time-dependent calculations appears as dissipation and dispersion, and the multidimensional aliasing can be assessed and alleviated by better selection of numerical algorithms and mesh-system refinements (Figure 4.2).

The Maxwell equations in the time domain consist of a first-order divergence curl system that is classified as hyperbolic partial differential equations. The solutions need not be analytic functions (Courant and Hilbert, 1965; Shang, 1995b). More importantly, the initial values, together with any possible discontinuities, are continued along a time–space trajectory, which is commonly referred to as the characteristic (Sommerfeld, 1949). A series of numerical schemes has been devised to duplicate the directional information-propagation feature. These numerical procedures are collectively designated as the characteristic-based method, which in its most elementary form is identical to the Riemann problem or the Riemann invariant (Shang, 2002). The characteristic-based method, when applied to solve the Maxwell equations in the time domain, exhibits many attractive attributes. A synergism of the new numerical procedures and scalable parallel-computing capability has opened up a new frontier in electromagnetics research. For this reason, a major portion of this chapter is

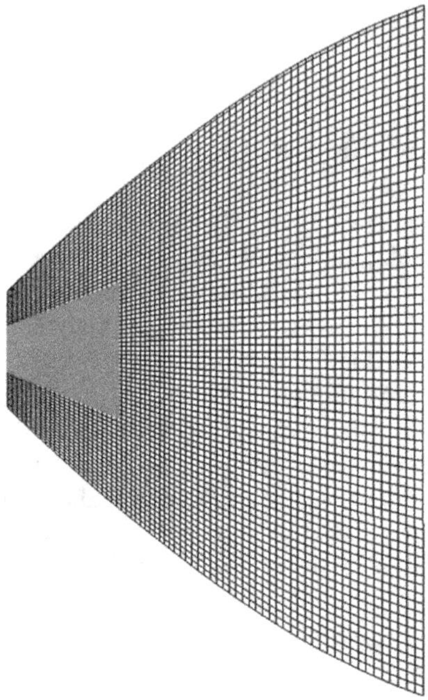

FIGURE 4.2 Mesh system for pyramidal-horn antenna (148 × 5 × 142).

focused on introducing the characteristic-based finite-volume and finite-difference methods.

Faraday's induction law and the generalized Ampere's electric circuit law of the time-dependent Maxwell equations in integral form can be given as (Kong, 1986)

$$\begin{aligned} \oint E \cdot dl &= -\oiint \frac{\partial B}{\partial t} \cdot ds \\ \oint H \cdot dl &= \oiint \left(J + \frac{\partial D}{\partial t}\right) \cdot ds \end{aligned} \tag{4.1}$$

Apply the Stokes divergence theorem, the line and surface integrals, equation (4.1) can be expressed as volume integrals, and we have then the complete Maxwell equations, including Gauss's law for electric displacement and the magnetic flux density. These integral relationships hold only if the first derivatives of the electric displacement D and the magnetic flux density B are continuous throughout the control volume:

$$\begin{aligned} &\oiint (\nabla \times E) \cdot ds + \iiint \frac{\partial B}{\partial t} dv = 0 \\ &\oiint (\nabla \times H) \cdot ds + \iiint \frac{\partial D}{\partial t} dv = \iiint J dv \end{aligned}$$

$$
\begin{aligned}
&\oiint B \cdot ds = 0 \\
&\oiint D \cdot ds = \oiiint \rho dv
\end{aligned}
\tag{4.2}
$$

The description of the differential system is complete only after appropriated initial values and boundary conditions are prescribed. For the associated boundary condition on a stationary boundary, the partial time derivative operator is exchangeable with the spatial integral. Since the integration over the control volume and the results are functions of time only, the partial becomes the total derivative in time. We have

$$
\begin{aligned}
&\oiint (\nabla \times E) \cdot ds = -\frac{d}{dt} \oiiint Bdv \\
&\oiint (\nabla \times H) \cdot ds + \frac{d}{dt} \oiiint Ddv = \oiiint Jdv \\
&\oiint B \cdot ds = 0 \\
&\oiint D \cdot ds = \oiiint \rho dv
\end{aligned}
\tag{4.3}
$$

By shrinking the control volume to an infinitesimal space and assuming the field vectors B, D, E, and H to be finite but that may be discontinuous across boundaries, the system of equations transforms to differential form. The Maxwell equations and the associated boundary conditions are the governing equations of CEM in the time domain.

$$
\begin{aligned}
&\frac{\partial B}{\partial t} + \nabla \times E = 0 \\
&\frac{\partial D}{\partial t} - \nabla \times H = -J \\
&\nabla \cdot B = 0 \\
&\nabla \cdot D = \rho
\end{aligned}
\tag{4.4}
$$

The associated boundary conditions are

$$
\begin{aligned}
n \times (E_1 - E_2) &= 0 \\
n \times (H_1 - H_2) &= J_s \\
n \cdot (B_1 - B_2) &= 0 \\
n \cdot (D_1 - D_2) &= \rho_s
\end{aligned}
\tag{4.5a}
$$

Although the current and charge number density may be infinite on the surface of a perfect conductor, the surface current density is still definable as

$$J_s = \Delta J \tag{4.5b}$$

in the limit as $\Delta \to 0$ and $J \to \infty$.

In equation (4.5b), the surface current density has a dimension of amp/m and the surface charge density has a dimension of coulomb/m^2.

The second-order curl–curl form of the Maxwell equations is derived by applying the differentiation with respect to time to get (Harrington, 1968),

$$\begin{aligned}\nabla \times \nabla \times B + \frac{1}{c^2}\frac{\partial^2 B}{\partial t^2} &= \nabla \times (\mu J)\\ \nabla \times \nabla \times E + \frac{1}{c^2}\frac{\partial^2 E}{\partial t^2} &= -\frac{\partial(\mu J)}{\partial t}\end{aligned} \tag{4.6a}$$

where c is the phase velocity of the electromagnetic wave. In an unbounded medium, it is a function of the magnetic permeability and electric permittivity

$$c = 1/\sqrt{\varepsilon\mu} \tag{4.6b}$$

In free space, the velocity is a constant and is the speed of light. The outstanding feature of the curl–curl formulation of the Maxwell equations is that the electric and magnetic fields are decoupled. The second-order equations can be further simplified for harmonic fields. The time-dependent behavior can be represented by a harmonic function $e^{i\omega t}$. Separation-of-variables transforms the Maxwell equations into the frequency domain. The resultant partial differential equations in spatial variables become elliptic type

$$\begin{aligned}\nabla \times \nabla \times B - k^2 B &= -i\omega(\mu J)\\ \nabla \times \nabla \times E - k^2 E &= \nabla \times (\mu J)\end{aligned} \tag{4.7}$$

where $k = \omega/c$ is the so-called propagation constant or the wave number and is the angular frequency of a component of the Fourier series. The above equations are also frequently the basis for finite-element approaches (Rahman et al., 1991).

4.2 CHARACTERISTIC-BASED FORMULATION

The fundamental idea of the characteristic-based method for solving the hyperbolic system of equations is derived from the eigenvalue–eigenvector analyses of the governing equations (Sommerfeld, 1949; Courant et al., 1965). For Maxwell equations in the time domain, every eigenvalue is real but not all of them are distinct. In a time–space plane, the eigenvalue defines the slope of the characteristic or equivalently the phase velocity of the wave motion. All dependent variables within the time–space domain bounded by two intersecting characteristics are completely determined by the values along these characteristics and by the compatibility relationship (Richtmyer and Morton, 1967). The direction of information propagation is clearly described by these two characteristics. In numerical simulation, the well-posedness requirement on initial or boundary conditions and the stability of a numerical approximation are also ultimately linked to the eigenvalues of the governing equation. Therefore, characteristic-based methods have demonstrated superior numerical stability and accuracy than all other schemes. However, characteristic-based algorithms also have an inherent limitation in that the governing equation can be diagonalized only in one spatial dimension at a time (Roe, 1981). The multidimensional equations are required to be split into multiple one-dimensional space-time formulations. This limitation is not unusual for numerical algorithms that are always adopted in CFD and CEM such as the approximate factored and the fractional-step schemes (Shang, 1995b; Shang et al., 1995 and 1996). A consequence of this restriction is that solutions of the characteristic-based procedure may exhibit some degree of sensitivity to the orientation of the coordinate selected. This numerical behavior is consistent with the concept of optimal coordinates.

In the characteristic formulation, data on the wave motion are first split according to the direction of phase velocity and then information can transmit in both orientations. In each one-dimensional time–space domain, the direction of the phase velocity degenerates into either a positive or a negative direction. They are commonly referred to as the right-running and left-running wave components. The sign of the eigenvalue is thus an indicator of the direction of signal transmission. The corresponding eigenvectors are the essential elements for diagonalizing the coefficient matrices and for formulating the approximate Riemann problem (Shang et al., 1995a, 1996). In essence, knowledge of eigenvalues and eigenvectors of the Maxwell equations in the time domain becomes the first prerequisite of the present formulation.

In CEM, only the Faraday induction law and the generalized Ampere circuit law are included. The Gauss laws for magnetic and electric fields are imposed as constraints. The system of equations adopts the magnetic flux density B and the electric displacement D as the dependent variables, and the electric field strength E and magnetic field strength H are eliminated by constitutive relationships $B = \mu H$ and $D = \varepsilon E$. The governing equations with a reduced number of dependent variables in Cartesian frame can be given as

$$
\begin{aligned}
&\frac{\partial B_x}{\partial t} + \frac{1}{\varepsilon}\frac{\partial D_z}{\partial y} - \frac{1}{\varepsilon}\frac{\partial D_y}{\partial z} = 0 \\
&\frac{\partial B_y}{\partial t} - \frac{1}{\varepsilon}\frac{\partial D_z}{\partial x} + \frac{1}{\varepsilon}\frac{\partial D_x}{\partial z} = 0 \\
&\frac{\partial B_z}{\partial t} + \frac{1}{\varepsilon}\frac{\partial D_z}{\partial x} - \frac{1}{\varepsilon}\frac{\partial D_x}{\partial y} = 0 \\
&\frac{\partial D_x}{\partial t} - \frac{1}{\mu}\frac{\partial B_z}{\partial y} + \frac{1}{\mu}\frac{\partial B_y}{\partial z} = -J_x \\
&\frac{\partial D_y}{\partial t} + \frac{1}{\mu}\frac{\partial B_z}{\partial x} - \frac{1}{\mu}\frac{\partial B_x}{\partial z} = -J_y \\
&\frac{\partial D_z}{\partial t} - \frac{1}{\mu}\frac{\partial B_y}{\partial x} + \frac{1}{\mu}\frac{\partial B_x}{\partial y} = -J_z
\end{aligned}
\tag{4.8}
$$

When three-dimensional, time-dependent Maxwell equations are cast in the flux-vector form, them appear as

$$\frac{\partial U}{\partial t} + \frac{\partial F_x}{\partial x} + \frac{\partial F_y}{\partial y} + \frac{\partial F_z}{\partial z} = R \tag{4.9a}$$

where U is the vector of dependent variables. The flux vectors are formed by the inner product of the coefficient matrix and the dependent variable

$$U = \{B_x, B_y, B_z, D_x, D_y, D_z\}^T \tag{4.9b}$$

with

$$F_x = C_x U, \quad F_y = C_y U \quad \text{and} \quad F_z = C_z U \tag{4.9c}$$

The coefficient matrices C_x, C_y, and C_z can be given in a contracted form as

$$\begin{aligned} F_x &= \{0, -D_z/\varepsilon, D_y/\varepsilon, 0, B_z/\mu, -B_y/\mu\}^T \\ F_y &= \{D_z/\varepsilon, 0, -D_x/\varepsilon, -B_z/\mu, 0, B_x/\mu\}^T, \\ F_z &= \{-D_y/\varepsilon, D_x/\varepsilon, 0, B_y/\mu, -B_x/\mu, 0\}^T \end{aligned} \tag{4.9d}$$

The typical coefficient matrices or the Jacobians of the flux vectors C_x, C_y, and C_z are

$$C_x = \begin{bmatrix} 0 & 0 & 0 & 0 & 0 & 0 \\ 0 & 0 & 0 & 0 & 0 & -1/\varepsilon \\ 0 & 0 & 0 & 0 & 1/\varepsilon & 0 \\ 0 & 0 & 0 & 0 & 0 & 0 \\ 0 & 0 & 1/\mu & 0 & 0 & 0 \\ 0 & -1/\mu & 0 & 0 & 0 & 0 \end{bmatrix} \tag{4.10}$$

where ε and μ are the electric permittivity and magnetic permeability, which relate the electric displacement to the electric field intensity ($D = \varepsilon E$) and the magnetic flux density to the magnetic field intensity ($B = \mu H$) by constitutive relationships.

The eigenvalues of coefficient matrices C_x, C_y, and C_z in the Cartesian frame are identical and contain multiplicities. Care must be exercised to ensure that all associated eigenvectors are linearly independent. The linearly independent eigenvectors associated with each eigenvalue are found by reducing the determinant, $\text{Det}(C - \bar{\bar{I}}\lambda) = 0$, to the Jordan normal form, yield

$$\lambda = \left\{ +\frac{1}{\sqrt{\varepsilon\mu}}, -\frac{1}{\sqrt{\varepsilon\mu}}, 0, +\frac{1}{\sqrt{\varepsilon\mu}}, -\frac{1}{\sqrt{\varepsilon\mu}}, 0 \right\} \tag{4.11}$$

As has been mentioned earlier, the value of the eigenvalues or the phase velocity of wave propagation is the speed of light, $c = 1/\sqrt{\varepsilon\mu}$ in vacuum (Krause, 1953). All the eigenvalues are real numbers for all hyperbolic partial differential equations. The coefficient matrices C_x, C_y, and C_z can be diagonalized if there exist nonsingular similar matrices S_x, S_y, and S_z such that

$$\begin{aligned}\Lambda_x &= S_x^{-1} C_x S_x \\ \Lambda_y &= S_y^{-1} C_y S_y \\ \Lambda_z &= S_z^{-1} C_z S_z\end{aligned} \tag{4.12}$$

where the Λ's are the diagonalized coefficient matrices. The columns of similar matrices S_x, S_y, and S_z are simply the linearly independent eigenvectors of the coefficient matrices C_x, C_y, and C_z, respectively.

The fundamental relationship between the characteristic-based formulation and the Riemann problem can be best demonstrated by the Cartesian frame. For the Maxwell equations on this frame of reference and in an isotropic medium, all similar matrices are invariant with respect to temporal and spatial variables. In each time–space plane (*x–t*, *y–t*, and *z–t*), the one-dimensional governing equation can be given just as in the *x–t* plane, for example,

$$\frac{\partial U}{\partial t} + C_x \frac{\partial U}{\partial x} = 0 \tag{4.13}$$

Substitute the diagonalized coefficient matrix to get

$$\frac{\partial U}{\partial t} + S_x \Lambda_x S_x^{-1} \frac{\partial U}{\partial x} = 0 \tag{4.14}$$

Since the similar matrix under the present consideration is invariant with respect to time and space, it can be brought into the differential operator. Multiplying equation (4.14) by the left-hand inverse of S_x, $S_x{}^{-1}$, we have

$$\frac{\partial (S_x^{-1} U)}{\partial t} + \Lambda_x \frac{\partial (S_x^{-1} U)}{\partial x} = 0 \tag{4.15a}$$

or

$$\frac{\partial W_x}{\partial t} + \Lambda_x \frac{\partial W_x}{\partial x} = 0 \tag{4.15b}$$

One immediately recognizes the group of variables $S_x{}^{-1} U$ as the characteristics, and the system of equations is decoupled. In a scalar-variable form and with appropriate initial values, this is the classic Riemann problem (Courant and Hilbert, 1965). This differential system is specialized for studying single discontinuity, and the piecewise continuous solutions separated by the singular point are also invariant along characteristics. Equally important, stable numerical operators can now be easily devised to solve the split equations according to the sign of the eigenvalues. In practice, it has been found that if the multidimensional problem can be split into a sequence of one-dimensional equations, this numerical technique is applicable to all of these one-dimensional equations in all coordinates.

The gist of the characteristic-based formulation is also clearly revealed by the decomposition of the flux vector into positive and negative components corresponding to the sign of the eigenvalue:

$$\begin{aligned} \Lambda &= \Lambda^+ + \Lambda^-, \qquad F = F^+ + F^- \\ F^+ &= S\Lambda^+ S^{-1}, \qquad F^- = S\Lambda^- S^{-1} \end{aligned} \tag{4.16}$$

The superscripts + and − denote the split vectors associated with positive and negative eigenvalues, respectively. The governing equation in flux-vector form can be given as

$$\frac{\partial U}{\partial t} + \frac{\partial}{\partial x}(F_x^+ + F_x^-) + \frac{\partial}{\partial y}(F_y^+ + F_y^-) + \frac{\partial}{\partial z}(F_z^+ + F_z^-) = R \tag{4.17}$$

Characteristic-based algorithms have a deep-rooted theoretical basis for describing wave dynamics. An unconditionally stable windward numerical algorithm can be devised to solve the time-dependent Maxwell equations and to effectively treat the piecewise continuous solution like the Riemann problem. They also have, however, an inherent limitation in that the diagonalized formulation is achievable only in one spatial dimension and time. All multidimensional equations are required to be split into multiple one-dimensional formulations. The approach yields accurate results so long as discontinuous waves remain aligned with one of the computational coordinates. This limitation is also the state-of-the-art constraint in solving partial differential equations.

4.3 GOVERNING EQUATIONS ON CURVILINEAR COORDINATES

In order to develop a versatile modeling and simulation tool for CEM for a wide range of applications, the Maxwell equations need to be cast in a general curvilinear frame of reference. For efficient simulation of complex electromagnetic field configurations, the adoption of a general curvilinear mesh system becomes necessary. The system of equations in general curvilinear coordinates is derived by a coordinate transformation (Thompson, 1982). The mesh system of the transformed space can be obtained by numerous grid generation procedures, and computational advantages in the transformed space are realizable. For a body-oriented coordinate system, the interface between two different media is easily defined by one of the coordinate surfaces. Along this coordinate parametric plane, all discretized nodes on a curve interface are

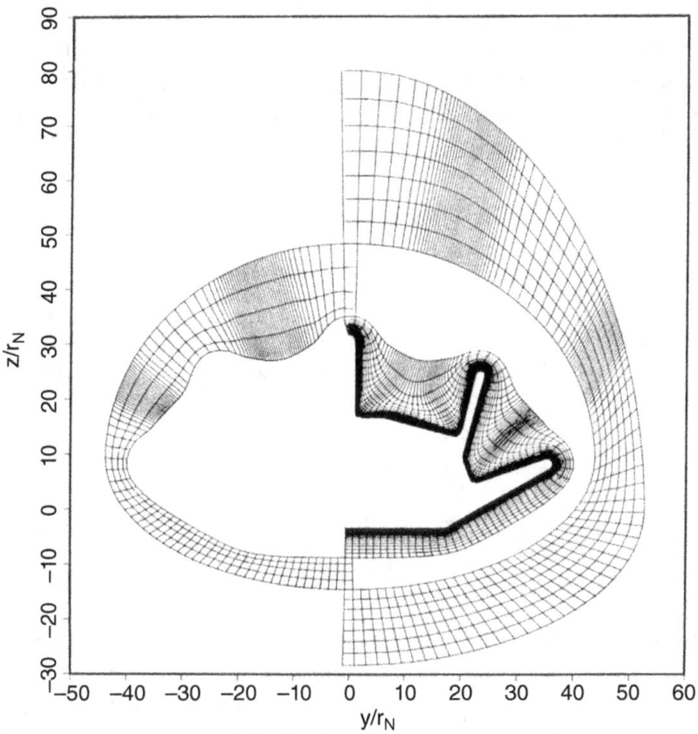

FIGURE 4.3 X24C-10D aerospace vehicle (181 × 59 × 62).

precisely prescribed without the need of an interpolating procedure from a rectilinear frame. The outward normal of the interface, which is essential for the boundary-value implementation, can be computed easily by $n = \Delta S/\|\Delta S\|$ (Sommerfeld, 1949). In the transformed space, computations are performed on a uniform mesh, but the corresponding physical spacing can be highly clustered at boundary region to enhance the numerical resolution.

As an illustration of the numerical advantage of solving the Maxwell equations on nonorthogonal curvilinear, body-oriented coordinates, the simulations of wave propagation within a simple pyramidal-horn antenna, and the scattered electromagnetic field from a complicated reentry vehicle are presented. For the antenna, the curved divergent channel surface is transformed into a straight edge so the staircase formation is completely avoided. For the X24C-10D aerospace vehicle, its complex geometrical shape, including the strakes and fins, can be transformed by body-conformal coordinates (Figure 4.3). A single-block mesh system enveloping the configuration has been accomplished. The numerical grid system is generated by using a hyperbolic grid generator (Steger and Warming, 1987) for the near-field mesh adjacent to the solid surface, and a transfinite technique for the far field.

The most general coordinate transformation of the Maxwell equations in three-dimensional space is a one-to-one relationship between two sets of spatial independent variables. The coordinate transformation can easily extend to include temporal variation. However, for most practical applications, the spatial coordinate transformation alone is sufficient:

$$\begin{aligned} \xi &= \xi(x, y, z) \\ \eta &= \eta(x, y, z) \\ \zeta &= \zeta(x, y, z) \end{aligned} \tag{4.18a}$$

In Euclidian space, the derivatives of independent variables are uniquely related to each other,

$$\begin{bmatrix} d\xi \\ d\eta \\ d\zeta \end{bmatrix} = \begin{bmatrix} \xi_x & \xi_y & \xi_z \\ \eta_x & \eta_y & \eta_z \\ \zeta_x & \zeta_y & \zeta_z \end{bmatrix} \begin{bmatrix} dx \\ dy \\ dz \end{bmatrix} \tag{4.18b}$$

The partial derivatives of the independent variables ξ_x, η_x, and ζ_x, ($\xi_x = \partial\xi/\partial x$, $\xi_y = \partial\xi/\partial y$, and $\xi_z = \partial\xi/\partial z$) etc. are the metrics of the coordinate transformation and can be computed easily from the definition of the general coordinate transformation by the chain rule. The governing equation in the strong conservation form is obtained by dividing the chain-rule differentiated equations by the Jacobian of coordinate transformation and by invoking metric identities (Vinokur, 1974),

$$\begin{aligned} &[(\xi_x/V)_\xi + (\eta_x/V)_\eta + (\zeta_x/V)_\zeta] \equiv 0 \\ &[(\xi_y/V)_\xi + (\eta_y/V)_\eta + (\zeta_y/V)_\zeta] \equiv 0 \\ &[(\xi_z/V)_\xi + (\eta_z/V)_\eta + (\zeta_z/V)_\zeta] \equiv 0 \end{aligned} \tag{4.18c}$$

Here V is the Jacobian of the coordinate transformation and is also the inverse local cell volume. If the Jacobian has nonzero values in the computational domain, the correspondence between the physical and the transformed space is uniquely defined. Since systematic procedures from the fundamental Cartesian frame to a body-oriented coordinate system have been developed to ensure this property of coordinate transformations. The procedure is applied the chain-rule of differentiations between the group of independent variables equation (4.18b); thus, detailed information on this point is not repeated here (Anderson et al., 1984). We have

$$V = \begin{bmatrix} \xi_x & \eta_x & \zeta_x \\ \xi_y & \eta_y & \zeta_y \\ \xi_z & \eta_z & \zeta_z \end{bmatrix} \tag{4.19}$$

The time-dependent Maxwell equations on a general curvilinear frame of reference and in the strong conservative form become (Elliott, 1966; Shang et al., 1996);

$$\frac{\partial U}{\partial t} + \frac{\partial F_\xi}{\partial \xi} + \frac{\partial F_\eta}{\partial \eta} + \frac{\partial F_\zeta}{\partial \zeta} = R \tag{4.20}$$

where the dependent variables are now defined as

$$U = U(B_x V,\ B_y V,\ B_z V,\ D_x V,\ D_y V,\ D_z V) \tag{4.21}$$

The flux-vector components in the transformed space achieve the following form. The flux vector in the transformed coordinate ξ is

$$F_\xi = \frac{1}{V}\begin{bmatrix} 0 & 0 & 0 & 0 & -\xi_z/\varepsilon & \xi_y/\varepsilon \\ 0 & 0 & 0 & \xi_z/\varepsilon & 0 & -\xi_x/\varepsilon \\ 0 & 0 & 0 & -\xi_y/\varepsilon & \xi_x/\varepsilon & 0 \\ 0 & \xi_z/\mu & -\xi_y/\mu & 0 & 0 & 0 \\ -\xi_z/\mu & 0 & \xi_x/\mu & 0 & 0 & 0 \\ \xi_y/\mu & -\xi_x/\mu & 0 & 0 & 0 & 0 \end{bmatrix} \begin{Bmatrix} B_x \\ B_y \\ B_z \\ D_x \\ D_y \\ D_z \end{Bmatrix} \tag{4.22a}$$

In the transformed coordinate η

$$F_\eta = \frac{1}{V}\begin{bmatrix} 0 & 0 & 0 & 0 & -\eta_z/\varepsilon & \eta_y/\varepsilon \\ 0 & 0 & 0 & \eta_z/\varepsilon & 0 & -\eta_x/\varepsilon \\ 0 & 0 & 0 & -\eta_y/\varepsilon & \eta_x/\varepsilon & 0 \\ 0 & \eta_z/\mu & -\eta_y/\mu & 0 & 0 & 0 \\ -\eta_z/\mu & 0 & \eta_x/\mu & 0 & 0 & 0 \\ \eta_y/\mu & -\eta_x/\mu & 0 & 0 & 0 & 0 \end{bmatrix} \begin{Bmatrix} B_x \\ B_y \\ B_z \\ D_x \\ D_y \\ D_z \end{Bmatrix} \tag{4.22b}$$

And in the transformed coordinate ζ

$$F_\zeta = \frac{1}{V}\begin{bmatrix} 0 & 0 & 0 & 0 & -\zeta_z/\varepsilon & \zeta_y/\varepsilon \\ 0 & 0 & 0 & \zeta_z/\varepsilon & 0 & -\zeta_x/\varepsilon \\ 0 & 0 & 0 & -\zeta_y/\varepsilon & \zeta_x/\varepsilon & 0 \\ 0 & \zeta_z/\mu & -\zeta_y/\mu & 0 & 0 & 0 \\ -\zeta_z/\mu & 0 & \zeta_x/\mu & 0 & 0 & 0 \\ \zeta_y/\mu & -\zeta_x/\mu & 0 & 0 & 0 & 0 \end{bmatrix} \begin{Bmatrix} B_x \\ B_y \\ B_z \\ D_x \\ D_y \\ D_z \end{Bmatrix} \tag{4.22c}$$

After the coordinate transformation, all coefficient matrices now contain metrics that are position dependent, and the system of equations in the most general frame of reference possesses variable coefficients (Shang, 1995; Shang and Fithen, 1996). This added complexity of the characteristic formulation of the Maxwell equations no longer permits the system of time-dependent equations to be decoupled into six scalar equations and reduced to the true Riemann problem like that on the Cartesian frame.

As previously mentioned, knowledge of eigenvalues and the eigenvectors is prerequisite for developing characteristic-based algorithms. The analytic process to obtain the eigenvalues and the corresponding eigenvectors of the Maxwell equations in general curvilinear coordinates is identical to that in the Cartesian frame. In each of the temporal–spatial planes t–ξ, t–η, and t–ζ, eigenvalues are easily found by solving the sixth-degree characteristic equation associated with the coefficient matrices to get

$$\begin{aligned}
\lambda_\xi &= \left\{ -\frac{\alpha}{V\sqrt{\varepsilon\mu}}, -\frac{\alpha}{V\sqrt{\varepsilon\mu}}, \frac{\alpha}{V\sqrt{\varepsilon\mu}}, \frac{\alpha}{V\sqrt{\varepsilon\mu}}, 0, 0 \right\} \\
\lambda_\eta &= \left\{ -\frac{\beta}{V\sqrt{\varepsilon\mu}}, -\frac{\beta}{V\sqrt{\varepsilon\mu}}, \frac{\beta}{V\sqrt{\varepsilon\mu}}, \frac{\beta}{V\sqrt{\varepsilon\mu}}, 0, 0 \right\} \\
\lambda_\zeta &= \left\{ -\frac{\gamma}{V\sqrt{\varepsilon\mu}}, -\frac{\gamma}{V\sqrt{\varepsilon\mu}}, \frac{\gamma}{V\sqrt{\varepsilon\mu}}, \frac{\gamma}{V\sqrt{\varepsilon\mu}}, 0, 0 \right\}
\end{aligned} \tag{4.23}$$

where $\alpha = \sqrt{\xi_x^2 + \xi_y^2 + \xi_z^2}$, $\beta = \sqrt{\eta_x^2 + \eta_y^2 + \eta_z^2}$, and $\gamma = \sqrt{\zeta_x^2 + \zeta_y^2 + \zeta_z^2}$.

One recognizes that the eigenvalues in each one-dimensional time–space plane contain multiplicities, and hence the eigenvectors do not necessarily have unique elements. Nevertheless, linearly independent eigenvectors associated with each eigenvalue still can be found by reducing the coefficient matrix to the Jordan normal form. For reasons of wide applicability and internal consistency, eigenvectors are selected in such a fashion that similar matrices of diagonalization will reduce to the same form as in the Cartesian frame. Furthermore, in order to accommodate a wide range of electromagnetic field configurations such as antennas, wave guides, and scatterers, the eigenvalues are no longer identical in the three time–space planes. This complexity of formulation is essential to facilitate boundary-condition implementation on the interfaces of media with different characteristic impedances.

From eigenvector analysis, similarity transformation matrices for diagonalization in each time–space (t–ξ, t–η, and t–ζ) plane are formed by using eigenvectors as the column arrays. These eigenvectors are associated with the eigenvalue $\lambda_\xi = -\alpha/V\sqrt{\varepsilon\mu} = -\alpha c/V$ of the coefficient matrix. For example, the first column of the similar matrix of diagonalization is $[\sqrt{\mu/\varepsilon}\xi_y/\alpha,\ \sqrt{\mu/\varepsilon}(\xi_x^2 + \xi_y^2)/\xi_x),\ \sqrt{\mu/\varepsilon}\xi_y\xi_z/\xi_x,\ \xi_y/\xi_x,\ 0,\ 1]^T$. It is recognized, $\sigma = \sqrt{\mu/\varepsilon}$ has the dimension of electric resistance. In free space, it has the value of 120 π or approximately 367.7 ohms and is often referred to as the intrinsic impedance of a lossless dielectric media (Krause, 1953). The eigenvector in the similar matrix of the t–ξ plane can, therefore, be expressed simply as $[-\sigma\xi_y/\alpha, \sigma(\xi_x^2 + \xi_z^2)/\alpha\xi_x, \sigma\xi_y\xi_z/\alpha\xi_x, \xi_y/\varsigma_x, 0, 1]^T$.

Similarly, the eigenvectors corresponding to the eigenvalues in the $t-\eta$ and the $t-\zeta$ plane $\lambda_\eta = -\beta c/V$ and $\lambda_\zeta = -\gamma c/V$ are the column arrays of the similar matrices for diagonalization. Since the similar matrices are nonsingular, the left-hand inverse matrices are easily found. We have the similar (S_ξ, S_η, S_ζ) matrices and its left-hand inverse matrices $(S_\xi^{-1}, S_\eta^{-1}, S_\zeta^{-1})$ in each (ξ, η, ζ) coordinate as

$$S_\xi = \begin{bmatrix} -\sigma\xi_y/\alpha & \sigma\xi_z/\alpha & \sigma\xi_y/\alpha & -\sigma\xi_z/\alpha & 1 & 0 \\ \sigma\left(\xi_x^2+\xi_z^2\right)/\alpha\xi_x & \sigma\xi_y\xi_z/\alpha\xi_x & -\sigma\left(\xi_x^2+\xi_z^2\right)/\alpha\xi_x & \sigma\xi_y\xi_z/\alpha\xi_x & \xi_y/\xi_x & 0 \\ -\sigma\xi_y\xi_z/\alpha\xi_x & -\sigma\left(\xi_x^2+\xi_y^2\right)/\alpha\xi_x & \sigma\xi_y\xi_z/\alpha\xi_x & \sigma\left(\xi_x^2+\xi_y^2\right)/\alpha\xi_x & \xi_z/\xi_x & 0 \\ -\xi_z/\xi_x & -\xi_y/\xi_x & -\xi_z/\xi_x & -\xi_y/\xi_x & 0 & 1 \\ 0 & 1 & 0 & 1 & 0 & \xi_y/\xi_x \\ 1 & 0 & 1 & 0 & 0 & \xi_z/\xi_x \end{bmatrix} \tag{4.24a}$$

$$S_\xi^{-1} = \begin{bmatrix} -\sigma\xi_y/2\alpha & \sigma\xi_x/2\alpha & 0 & -\xi_x\xi_z/2\alpha^2 & \xi_y\xi_z/2\alpha^2 & \left(\xi_x^2+\xi_y^2\right)/2\alpha^2 \\ \sigma\xi_z/2\alpha & 0 & -\sigma\xi_x/2\alpha & -\xi_x\xi_y/2\alpha^2 & \left(\xi_x^2+\xi_z^2\right)/2\alpha^2 & -\xi_y\xi_z/2\alpha^2 \\ \sigma\xi_y/2\alpha & -\sigma\xi_x/2\alpha & 0 & -\xi_x\xi_z/2\alpha^2 & -\xi_y\xi_z/2\alpha^2 & \left(\xi_x^2+\xi_y^2\right)/2\alpha^2 \\ -\sigma\xi_z/2\alpha & 0 & \sigma\xi_x/2\alpha & -\xi_x\xi_y/2\alpha^2 & \left(\xi_x^2+\xi_y^2\right)/2\alpha^2 & -\xi_y\xi_z/2\alpha^2 \\ \xi_x^2/\alpha^2 & \xi_x\xi_y/\alpha^2 & \xi_x\xi_z/\alpha^2 & 0 & 0 & 0 \\ 0 & 0 & 0 & \xi_x^2/\alpha^2 & \xi_x\xi_y/\alpha^2 & \xi_x\xi_z/\alpha^2 \end{bmatrix} \tag{4.24b}$$

$$S_\eta = \begin{bmatrix} -\sigma\left(\eta_y^2+\xi_z^2\right)/\beta\eta_y & -\sigma\eta_x\eta_z/\beta\eta_y & \left(\eta_y^2+\xi_z^2\right)/\beta\eta_y & \sigma\eta_x\eta_z/\beta\eta_y & \eta_x/\eta_y & 0 \\ \sigma\eta_x/\beta & -\sigma\eta_z/\beta\eta_y & -\sigma\eta_x/\beta & \sigma\eta_z/\beta\eta_y & 1 & 0 \\ -\sigma\eta_x\eta_z/\alpha\eta_y & \sigma\left(\eta_x^2+\eta_y^2\right)/\alpha\eta_y & \sigma\eta_x\eta_z/\alpha\eta_y & -\sigma\left(\eta_x^2+\eta_y^2\right)/\alpha\eta_y & \eta_z/\eta_y & 0 \\ 0 & 1 & 0 & 1 & 0 & \eta_x/\eta_y \\ -\eta_z/\eta_y & -\eta_x/\eta_y & -\eta_x/\eta_y & -\eta_x/\eta_y & 0 & 1 \\ 1 & 0 & 1 & 0 & 0 & \eta_z/\eta_y \end{bmatrix} \tag{4.25a}$$

$$S_\eta^{-1} = \begin{bmatrix} -\sigma\eta_y/2\beta & \sigma\eta_x/2\beta & 0 & -\eta_x\eta_y/2\beta^2 & -\eta_y\eta_z/2\beta^2 & \left(\eta_x^2+\eta_y^2\right)/2\beta^2 \\ 0 & -\sigma\eta_z/2\beta & \sigma\eta_y/2\beta & \left(\eta_y^2+\eta_z^2\right)/2\beta^2 & -\eta_x\eta_y/2\beta^2 & -\eta_x\eta_y/2\beta^2 \\ \sigma\eta_y/2\beta & -\sigma\eta_x/2\beta & 0 & -\eta_x\eta_y/2\beta^2 & -\eta_y\eta_z/2\beta^2 & \left(\eta_x^2+\eta_y^2\right)/2\beta^2 \\ 0 & \sigma\eta_z/2\beta & -\sigma\eta_y/2\beta & \left(\eta_y^2+\eta_z^2\right)/2\beta^2 & -\eta_x\eta_y/2\beta^2 & -\eta_x\eta_z/2\beta^2 \\ \eta_x\eta_y/\beta^2 & \eta_y^2/\beta^2 & \eta_y\eta_z/\beta^2 & 0 & 0 & 0 \\ 0 & 0 & 0 & \eta_x\eta_y/\beta^2 & \eta_y^2/\beta^2 & \eta_y\eta_z/\beta^2 \end{bmatrix} \tag{4.25b}$$

$$S_{\zeta} = \begin{bmatrix} \sigma\left(\zeta_y^2+\zeta_z^2\right)\Big/\gamma\zeta_z & \sigma\zeta_x\zeta_y/\gamma\zeta_z & -\sigma\left(\zeta_y^2+\zeta_z^2\right)\Big/\gamma\zeta_z & -\sigma\zeta_x\zeta_y/\gamma\zeta_z & \zeta_x/\zeta_z & 0 \\ -\sigma\zeta_x\zeta_y/\gamma\zeta_z & -\sigma\left(\zeta_x^2+\zeta_z^2\right)/\gamma\zeta_z & \sigma\zeta_x\zeta_y/\gamma\zeta_z & \sigma\left(\zeta_x^2+\zeta_z^2\right)/\gamma\zeta_z & \zeta_y/\zeta_z & 0 \\ -\sigma\zeta_x/\gamma & \sigma\zeta_y/\gamma & \sigma\zeta_x/\gamma & -\sigma\zeta_y/\gamma & 1 & 0 \\ 0 & 1 & 0 & 1 & 0 & \zeta_x/\zeta_z \\ 1 & 0 & 1 & 0 & 0 & \begin{matrix} 1 \\ \zeta_y/\zeta_z \end{matrix} \\ -\zeta_y/\zeta_z & -\zeta_x/\zeta_z & -\zeta_y/\zeta_z & -\zeta_x/\zeta_z & 0 & 1 \end{bmatrix} \tag{4.26a}$$

$$S_{\zeta} = \begin{bmatrix} \sigma\zeta_z/2\gamma & 0 & -\sigma\zeta_x/2\gamma & -\zeta_x\zeta_y/2\gamma^2 & \left(\zeta_x^2+\zeta_z^2\right)/2\gamma^2 & \left(\zeta_y\zeta_z\right)/2\gamma^2 \\ 0 & -\sigma\zeta_z/2\gamma & \sigma\zeta_y/2\gamma & \left(\zeta_y^2+\zeta_z^2\right)\Big/2\gamma^2 & -\zeta_x\zeta_y/2\gamma^2 & -\zeta_x\zeta_z/2\gamma^2 \\ -\sigma\zeta_z/2\gamma & 0 & \sigma\zeta_x/2\gamma & -\zeta_x\zeta_y/2\gamma^2 & -\left(\zeta_x^2+\zeta_z^2\right)/2\gamma^2 & -\left(\zeta_y\zeta_z\right)/2\gamma^2 \\ 0 & \sigma\zeta_z/2\gamma & -\sigma\zeta_y/2\gamma & \left(\zeta_y^2+\zeta_z^2\right)\Big/2\gamma^2 & -\zeta_x\zeta_y/2\gamma^2 & -\zeta_x\zeta_z/2\gamma^2 \\ \zeta_x\zeta_z/\gamma^2 & \zeta_y\zeta_z/\gamma^2 & \zeta_z^2/\gamma^2 & 0 & 0 & 0 \\ 0 & 0 & 0 & \zeta_x\zeta_z/\gamma^2 & \left(\zeta_y\zeta_z\right)/\gamma^2 & \zeta_z^2/\gamma^2 \end{bmatrix} \tag{4.26b}$$

An efficient and computationally stable flux-vector splitting algorithm for solving the Euler equations has been developed by Steger and Warming (1987). The basic concept is equally applicable to any hyperbolic differential system for which the solution need not be analytic (Roe, 1981, 1986). In most CEM applications, discontinuous behavior of a solution is associated only with the wave crossing an interface between different medium to become a piecewise continuous solution. Even if a jump condition exists, the magnitude of the finite jump across the interface is much less drastic than the shock waves encountered in supersonic flows. Nevertheless, the salient feature of the piecewise continuous solution stands out for the hyperbolic partial differential equation. On general curvilinear coordinates, the coefficient matrices of the time-dependent, three-dimensional Maxwell equations contain metrics of coordinate transformation. Therefore, the equation system no longer has constant coefficients even in an isotropic and homogeneous medium. Under this circumstance, eigenvalues may change sign at any given field location due to the metric variations of the coordinate transformation.

The basic idea of the Steger–Warming flux-vector splitting (Steger and Warming, 1987) is to process data according to the direction of information propagation. Since diagonalization is achievable only in each time–space plane, the direction of wave propagation degenerates into either the positive or the negative orientation. This designation is consistent with the notion of the right- and left-running wave components. The flux vectors are computed from the point value, including the metrics at the node of interest. This formulation for solving hyperbolic partial differential equations ensures not only the well-posedness of the differential system but also the stability of the numerical procedure. Specifically, the flux-vectors F_{ξ}, F_{η}, and F_{ζ} are split according to the sign of their corresponding eigenvalues. The split fluxes are differenced by an upwind algorithm to allow for the zone of dependence of an initial-value problem.

From the previous discussion, it is clear that the eigenvalues contain multiplicities, and hence the split flux of the three-dimensional Maxwell equations is not unique. All flux vectors in each time–space plane are split according to the signs of the local eigenvalues:

$$\begin{aligned} F_\xi &= F_\xi^+ + F_\xi^- \\ F_\eta &= F_\eta^+ + F_\eta^- \\ F_\zeta &= F_\zeta^+ + F_\zeta^- \end{aligned} \tag{4.27}$$

The flux-vector components associated with the positive and negative eigenvalues are obtainable by a straightforward matrix multiplication:

$$\begin{aligned} F_\xi^+ &= S_\xi \lambda_\xi^+ S_\xi^{-1} U, \quad & F_\xi^- &= S_\xi \lambda_\xi^- S_\xi^{-1} U \\ F_\eta^+ &= S_\eta \lambda_\eta^+ S_\eta^{-1} U, \quad & F_\eta^- &= S_\eta \lambda_\eta^+ S_\eta^{-1} U \\ F_\zeta^+ &= S_\zeta \lambda_\zeta^+ S_\zeta^{-1} U, \quad & F_\zeta^- &= S_\zeta \lambda_\zeta^- S_\zeta^{-1} U \end{aligned} \tag{4.28}$$

It is important to recognize that even if the split flux vectors in each time–space plane are not unique, the sum of the split components must be unambiguously identical to the original flux vector of the governing equation. This fact is easily verifiable by performing the addition of the split matrices to reach the identities in the original flux vector. In addition, if one sets the diagonal elements of metrics, ξ_x, η_y, and ζ_z equal to unity and the all off-diagonal elements equal to zero, the coefficient matrices will recover the form in Cartesian coordinate:

$$F_\varepsilon^+ = \frac{c}{2V}\begin{bmatrix} (\xi_y^2+\xi_z^2)/\alpha & -\xi_x\xi_y/\alpha & -\xi_x\xi_z/\alpha & 0 & -\xi_z\sigma & \xi_y\sigma \\ -\xi_x\xi_y c/\alpha & (\xi_x^2+\xi_z^2)c/\alpha & -\xi_y\xi_z c/\alpha & \xi_z\sigma & 0 & -\xi_x\sigma \\ -\xi_x\xi_z c/\alpha & -\xi_y\xi_z c/\alpha & (\xi_x^2+\xi_y^2)c/\alpha & -\xi_y\sigma & \xi_x\sigma & 0 \\ 0 & \xi_z/\sigma & -\xi_y/\sigma & (\xi_y^2+\xi_z^2)/\alpha & -\xi_x\xi_y/\alpha & -\xi_x\xi_z/\alpha \\ -\xi_z/\sigma & 0 & \xi_x/\sigma & -\xi_x\xi_y/\alpha & (\xi_x^2+\xi_z^2)/\alpha & -\xi_y\xi_z/\alpha \\ \xi_y/\sigma & -\xi_x/\sigma & 0 & -\xi_x\xi_z/\alpha & -\xi_y\xi_z/\alpha & (\xi_x^2+\xi_y^2)/\alpha \end{bmatrix}\begin{Bmatrix} B_x \\ B_y \\ B_z \\ D_x \\ D_y \\ D_z \end{Bmatrix} \tag{4.29a}$$

$$F_\varepsilon^- = \frac{c}{2V}\begin{bmatrix} -(\xi_y^2+\xi_z^2)/\alpha & \xi_x\xi_y/\alpha & \xi_x\xi_z/\alpha & 0 & -\xi_z\sigma & \xi_y\sigma \\ \xi_x\xi_y/\alpha & -(\xi_x^2+\xi_z^2)/\alpha & \xi_y\xi_z/\alpha & \xi_z\sigma & 0 & -\xi_x\sigma \\ \xi_x\xi_z/\alpha & \xi_y\xi_z/\alpha & -(\xi_x^2+\xi_y^2)/\alpha & -\xi_y\sigma & \xi_x\sigma & 0 \\ 0 & \xi_z/\sigma & -\xi_y/\sigma & -(\xi_y^2+\xi_z^2)/\alpha & \xi_x\xi_y/\alpha & \xi_x\xi_z/\alpha \\ -\xi_z/\sigma & 0 & \xi_x/\sigma & \xi_x\xi_y/\alpha & -(\xi_x^2+\xi_z^2)/\alpha & \xi_y\xi_z/\alpha \\ \xi_y/\sigma & -\xi_x/\sigma & 0 & \xi_x\xi_z/\alpha & \xi_y\xi_z/\alpha & -(\xi_x^2+\xi_y^2)/\alpha \end{bmatrix}\begin{Bmatrix} B_x \\ B_y \\ B_z \\ D_x \\ D_y \\ D_z \end{Bmatrix} \tag{4.29b}$$

$$F_\eta^+ = \frac{c}{2V}\begin{bmatrix} \left(\eta_y^2+\eta_z^2\right)/\alpha & -\eta_x\eta_y/\alpha & -\eta_x\eta_z/\alpha & 0 & -\eta_z\sigma & \eta_y\sigma \\ -\eta_x\eta_y/\alpha & \left(\eta_x^2+\eta_z^2\right)/\alpha & -\eta_y\eta_z/\alpha & \eta_z\sigma & 0 & -\eta_x\sigma \\ -\eta_x\eta_z/\alpha & -\eta_y\eta_z/\alpha & \left(\eta_x^2+\eta_y^2\right)/\alpha & -\eta_y\sigma & \eta_x\sigma & 0 \\ 0 & \eta_z/\sigma & -\eta_y/\sigma & \left(\eta_y^2+\eta_z^2\right)/\alpha & -\eta_x\eta_y/\alpha & -\eta_x\eta_z/\alpha \\ -\eta_z/\sigma & 0 & \eta_x/\sigma & -\eta_x\eta_y/\alpha & \left(\eta_x^2+\eta_z^2\right)/\alpha & -\eta_y\eta_z/\alpha \\ \eta_y/\sigma & -\eta_x/\sigma & 0 & -\eta_x\eta_z/\alpha & -\eta_y\eta_z/\alpha & \left(\eta_x^2+\eta_y^2\right)/\alpha \end{bmatrix}\begin{Bmatrix} B_x \\ B_y \\ B_z \\ D_x \\ D_y \\ D_z \end{Bmatrix} \tag{4.30a}$$

$$F_\eta^- = \frac{c}{2V}\begin{bmatrix} -\left(\eta_y^2+\eta_z^2\right)/\alpha & \eta_x\eta_y/\alpha & \eta_x\eta_z/\alpha & 0 & -\eta_z\sigma & \eta_y\sigma \\ \eta_x\eta_y/\alpha & -\left(\eta_x^2+\eta_z^2\right)/\alpha & \eta_y\eta_z/\alpha & \eta_z\sigma & 0 & -\eta_x\sigma \\ \eta_x\eta_z/\alpha & \eta_y\eta_z/\alpha & -\left(\eta_x^2+\eta_y^2\right)/\alpha & -\eta_y\sigma & \eta_x\sigma & 0 \\ 0 & \eta_z/\sigma & -\eta_y/\sigma & -\left(\eta_y^2+\eta_z^2\right)/\alpha & \eta_x\eta_y/\alpha & \eta_x\eta_z/\alpha \\ -\eta_z/\sigma & 0 & \eta_x/\sigma & \eta_x\eta_y/\alpha & -\left(\eta_x^2+\eta_z^2\right)/\alpha & \eta_y\eta_z/\alpha \\ \eta_y/\sigma & -\eta_x/\sigma & 0 & \eta_x\eta_z/\alpha & \eta_y\eta_z/\alpha & -\left(\eta_x^2+\eta_y^2\right)/\alpha \end{bmatrix}\begin{Bmatrix} B_x \\ B_y \\ B_z \\ D_x \\ D_y \\ D_z \end{Bmatrix} \tag{4.30b}$$

$$F_\zeta^+ = \frac{c}{2V}\begin{bmatrix} \left(\zeta_y^2+\zeta_z^2\right)/\alpha & -\zeta_x\zeta_y/\alpha & -\zeta_y\zeta_z/\alpha & 0 & -\zeta_z\sigma & \zeta_y\sigma \\ -\zeta_x\zeta_y/\alpha & \left(\zeta_x^2+\zeta_z^2\right)/\alpha & -\zeta_y\zeta_z/\alpha & \zeta_z\sigma & 0 & -\zeta_x\sigma \\ -\zeta_x\zeta_z/\alpha & -\zeta_y\zeta_z/\alpha & \left(\zeta_x^2+\zeta_y^2\right)/\alpha & -\zeta_y\sigma & \zeta_x\sigma & 0 \\ 0 & \zeta_z/\sigma & -\zeta_y/\sigma & \left(\zeta_y^2+\zeta_z^2\right)/\alpha & -\zeta_x\zeta_y/\alpha & -\zeta_x\zeta_z/\alpha \\ -\zeta_z/\sigma & 0 & \zeta_x/\sigma & -\zeta_x\zeta_y/\alpha & \left(\zeta_x^2+\zeta_z^2\right)/\alpha & -\zeta_x\zeta_z/\alpha \\ \zeta_y/\sigma & -\zeta_x/\sigma & 0 & -\zeta_x\zeta_z/\alpha & -\zeta_x\zeta_z/\alpha & \left(\zeta_x^2+\zeta_y^2\right)/\alpha \end{bmatrix}\begin{Bmatrix} B_x \\ B_y \\ B_z \\ D_x \\ D_y \\ D_z \end{Bmatrix} \tag{4.31a}$$

$$F_\zeta^- = \frac{c}{2V}\begin{bmatrix} -\left(\zeta_y^2+\zeta_z^2\right)/\alpha & \zeta_x\zeta_y/\alpha & \zeta_y\zeta_z/\alpha & 0 & -\zeta_z\sigma & \zeta_y\sigma \\ \zeta_x\zeta_y/\alpha & -\left(\zeta_x^2+\zeta_z^2\right)/\alpha & \zeta_y\zeta_z/\alpha & \zeta_z\sigma & 0 & -\zeta_x\sigma \\ \zeta_x\zeta_z/\alpha & \zeta_y\zeta_z/\alpha & -\left(\zeta_x^2+\zeta_y^2\right)/\alpha & -\zeta_y\sigma & \zeta_x\sigma & 0 \\ 0 & \zeta_z/\sigma & -\zeta_y/\sigma & -\left(\zeta_y^2+\zeta_z^2\right)/\alpha & \zeta_x\zeta_y/\alpha & \zeta_x\zeta_z/\alpha \\ -\zeta_z/\sigma & 0 & \zeta_x/\sigma & \zeta_x\zeta_y/\alpha & -\left(\zeta_x^2+\zeta_z^2\right)/\alpha & \zeta_x\zeta_z/\alpha \\ \zeta_y/\sigma & -\zeta_x/\sigma & 0 & \zeta_x\zeta_z/\alpha & \zeta_x\zeta_z/\alpha & -\left(\zeta_x^2+\zeta_y^2\right)/\alpha \end{bmatrix}\begin{Bmatrix} B_x \\ B_y \\ B_z \\ D_x \\ D_y \\ D_z \end{Bmatrix} \tag{4.31b}$$

The governing equation in flux-vector form becomes

$$\frac{\partial U}{\partial t} + \frac{\partial(F_\xi^+ + F_\xi^-)}{\partial \xi} + \frac{\partial(F_\eta^+ + F_\eta^-)}{\partial \eta} + \frac{\partial(F_\zeta^+ + F_\zeta^-)}{\partial \zeta} = R \tag{4.32}$$

The characteristic-based methods have several clear advantages over the classic approach. First, the characteristics formulation naturally enforces the well-posedness condition for specifying the initial values to the differential system (Hadmard, 1932). Second, the spatial discretization is windward (directional biased). It follows the direction of the phase velocity of the wave propagation (Anderson et al., 1984). This discretization mimics the physics in directional-signal motion and therefore provides a favorable stability property than a central differencing scheme for solving the Maxwell equations in the time domain. The appropriately posed initial-value problems, combined with a consistent and computational stable approximation, are the necessary and sufficient conditions for numerical convergence.

The most important attribute of the characteristics formulation is the ability to remove a fundamental dilemma in CEM (Shang et al., 1995a, 1996). The finite size of computer memory compels all computations to be conducted in a truncated computational domain, thus introducing an artificial boundary to the numerical simulation. Spurious waves reflecting from the computational boundary will invariably be induced. The undesirable disturbances will propagate to and from the interior computational domain. In addition to causing poor numerical accuracy, the reflecting waves also incur unrealistic amplitude modulations. Characteristics-based procedures can eliminate the undesirable wave reflections, if one of the transformed coordinates is aligned with the direction of the electromagnetic wave motion $E \times H$. Under this circumstance, the characteristics formulation satisfies the compatibility condition identical at the far field boundary. Furthermore, this formulation is also capable of resolving the discontinuous interface boundary across different media by accommodating a finite jump of the dependent variables.

4.4 FAR FIELD BOUNDARY CONDITIONS

Well-posed and stable boundary conditions are paramount to accurate numerical simulation. A problem of mathematical physics is considered well posed if a unique solution exists, and depends continuously on the initial values and boundary conditions. In short, the well-posed condition is a physical requirement of a mathematical problem (Hadmard, 1932). The well-posed boundary conditions are needed both at the media interface and at the truncated computational domain. The boundary conditions across the medium interface are rigorously derived from the integral form of Maxwell equations to supplement wherever the spatial derivatives do not exist.

The fundamental dilemma in solving the time-dependent Maxwell equations is that the solution in a discretized domain whose sizes must be restrained. The theoretically boundless space is required to impose, an artificial condition on the computational outer boundary to absorb or to pass the outgoing wave without reflection. Numerous boundary conditions have been developed to annihilate the nonphysical reflected waves from the numerical boundary. Two theoretical frameworks have been established by either the absorbing boundary conditions or the characteristic formulation. The absorbing boundary condition cannot be directly obtained from the numerical algorithm because the conventional differential approximation requires information from the fields form points beyond the computational domain. Absorbing boundary condition is usually implemented by the partial derivatives on the computational boundary by expanding the differential approximation about an auxiliary discretized point (Enquist and Majda, 1977; Mur, 1981). This type of approach has achieved some degree of success.

The perfectly matched layer by Berenger (1994) as an absorbing far field boundary condition has proven to be an attractive answer to suppress the reflected wave from the numerical boundary. The basic idea consists of surrounding the computational domain with an absorbing medium whose impedance matches that free space. This approach also has its root in characteristics theory and the method is best illustrated by applying it to a two-dimensional transverse electromagnetic wave. A cornerstone of the perfectly matched layer for the transverse electric wave is further breaking the magnetic field into two subcomponents, $H_z = H_{zx} + H_{zy}$, so Faraday's law is separated in two components, the x and y coordinates.

$$
\begin{aligned}
\varepsilon \frac{\partial E_x}{\partial t} + \sigma_x E_x &= \frac{\partial (H_{zx} + H_{zy})}{\partial y} \\
\varepsilon \frac{\partial E_y}{\partial t} + \sigma_y E_y &= \frac{\partial (H_{zx} + H_{zy})}{\partial x} \\
\mu \frac{\partial H_{zx}}{\partial t} + \sigma_x^* H_{zx} &= -\frac{\partial H_y}{\partial x} \\
\mu \frac{\partial H_{zy}}{\partial t} + \sigma_y^* H_{zy} &= \frac{\partial H_x}{\partial y}
\end{aligned}
\tag{4.33}
$$

In the formulation (4.33), σ^* designates the so-called magnetic conductivity. The electric and magnetic conductivities become vectors with the x and y components. When $\sigma_x = \sigma_y$ and $\sigma_x^* = \sigma_y^*$, the formulation represents the absorbing medium.

For the perfectly matched medium, if the condition of $\sigma/\varepsilon = \sigma^*/\mu$ is satisfied, then the impedance of the medium equals that of vacuum and no reflection occurs when a plane wave propagates normally across a vacuum–medium interface. Under this condition, the wave components and the impedance σ become, respectively,

$$
\begin{aligned}
&\psi = a \exp\{i\omega[t - (x \cos\varphi + y \sin\varphi)/c] - [\sigma_x \cos\varphi/\varepsilon c]x - [\sigma_y \sin\varphi/\varepsilon c]y\} \\
&\sigma = \sqrt{\mu/\varepsilon}
\end{aligned}
\tag{4.34}
$$

In a two-dimensional transverse electric wave, the magnitude of the wave will decrease exponentially along the coordinates x and y. The layer of attenuation can be reduced to within one mesh spacing. Interestingly, the corner of plane intersections does not add any numerical reflection; the numerical results are consistent with the computed reflection factor of a wave striking a plane boundary. In summary, the computational results generated by a finite difference procedure have shown a small amount of numerical reflection, and whose magnitude can be further reduced by tuning the so-called magnetic conductivity of the layer. The basic formulation of the perfectly matched layer can be derived either from the mathematical construct or by the Lorentz material model. In this aspect, the perfectly matched layer can be more effective than the characteristic formulation for wave motion without knowledge of the principal axis of propagation (Zhao et al., 1966).

The characteristic formulation of Maxwell equations in the time domain has a rigorous theoretical foundation to treat the no-reflection far field boundary condition. The reflected waves can be completely eliminated by simply specifying a null value to the incoming wave, $F^{-}(r) = 0$ (Shang, 1995b; Shang et al., 1995). The outer boundary of a computational domain on the transformed general curvilinear coordinates $\eta \to \infty$, so that the perfect no-reflection condition is simply

$$F_{\eta}^{-} = \lim_{\eta \to \infty} \begin{bmatrix} -(\eta_y^2+\eta_z^2)/\alpha & \eta_x\eta_y/\alpha & \eta_x\eta_z/\alpha & 0 & -\eta_z\sigma & \eta_y\sigma \\ \eta_x\eta_y/\alpha & -(\eta_x^2+\eta_z^2)/\alpha & \eta_y\eta_z/\alpha & \eta_z\sigma & 0 & -\eta_x\sigma \\ \eta_x\eta_z/\alpha & \eta_y\eta_z/\alpha & -(\eta_x^2+\eta_y^2)/\alpha & -\eta_y\sigma & \eta_x\sigma & 0 \\ 0 & \eta_z/\sigma & -\eta_y/\sigma & -(\eta_y^2+\eta_z^2)/\alpha & \eta_x\eta_y/\alpha & \eta_x\eta_z/\alpha \\ -\eta_z/\sigma & 0 & \eta_x/\sigma & \eta_x\eta_y/\alpha & -(\eta_x^2+\eta_z^2)/\alpha & \eta_y\eta_z/\alpha \\ \eta_y/\sigma & -\eta_x/\sigma & 0 & \eta_x\eta_z/\alpha & \eta_y\eta_z/\alpha & -(\eta_x^2+\eta_y^2)/\alpha \end{bmatrix} \begin{Bmatrix} B_x \\ B_y \\ B_z \\ D_x \\ D_y \\ D_z \end{Bmatrix} = 0 \tag{4.35}$$

However, this accurate and simple far field boundary implementation must be placed on the boundary where the orientation of wave motion is unambiguously identifiable to be perpendicular to the boundary surface. This is the consequence of the fact that there are no genuine characteristics for two- and three-dimensional problems. For this reason, the placement of the truncated numerical domain still remains as a fundamental issue in the far field boundary condition implementation.

4.5 FINITE-DIFFERENCE APPROXIMATION

The classic Yee algorithm in CEM is a second-order explicit leapfrog scheme that is non-dissipative but exhibits predominantly dispersive error, and this method is neutrally stable (Yee, 1966). Yee's ingenuity is derived from solving the electric and magnetic fields on a staggered grid in that every electric component is surrounded by magnetic components and vice versa. The central idea is interlinking array of Faraday and Ampere laws. His approach proves to be very successful and opened an avenue for others to follow. Another insight gleaned by him for analyzing electromagnetic wave phenomena in the time domain is that more computational efficient implicit schemes may not be necessary because the temporal and spatial evolutions are equally important. The explicit scheme brings additional concern on computational stability that can be investigated through the amplification matrix or the amplification factor by a Fourier analysis and leads to the well-known Courant–Friedrichs–Levy (CFL) condition (Anderson et al., 1984).

Numerical procedures using the characteristic formulation are well established. Once the detailed split fluxes are known, the governing equation in characteristic variable $W_\xi = S_\xi^{-1}U$, $W_\eta = S_\eta^{-1}U$, and $W_\zeta = S_\zeta^{-1}U$ in each coordinate for the finite-difference approximation is straightforward. The difference operators in three coordinate directions for each dependent variable $(D_x, D_y, D_z, B_x, B_y, B_z)$ can be given as

$$\begin{aligned}
L_\xi &: \frac{\partial(W_\xi)}{\partial t} + \Lambda_\xi \frac{\partial(W_\xi)}{\partial \xi} = 0 \\
L_\eta &: \frac{\partial(W_\eta)}{\partial t} + \Lambda_\eta \frac{\partial(W_\eta)}{\partial \eta} = 0 \\
L_\zeta &: \frac{\partial(W_\zeta)}{\partial t} + \Lambda_\zeta \frac{\partial(W_\zeta)}{\partial \zeta} = 0
\end{aligned} \tag{4.36}$$

From the sign of an eigenvalue, the stencil of a spatially second- or higher-order accurate windward differencing can be easily constructed to form multiple one-dimensional difference operators. In this regard, the forward and backward difference approximations are used for the negative and positive eigenvalues, respectively. The split flux vectors are evaluated at each discretized point of the field according to the signs of the eigenvalues.

The merits of characteristic-based methods for solving the time-dependent, three-dimensional Maxwell equations can best be illustrated by the following two examples in Figure 4.4. In the first example, the exact electrical field of a traveling wave is compared with numerical results. The numerical results were generated at the

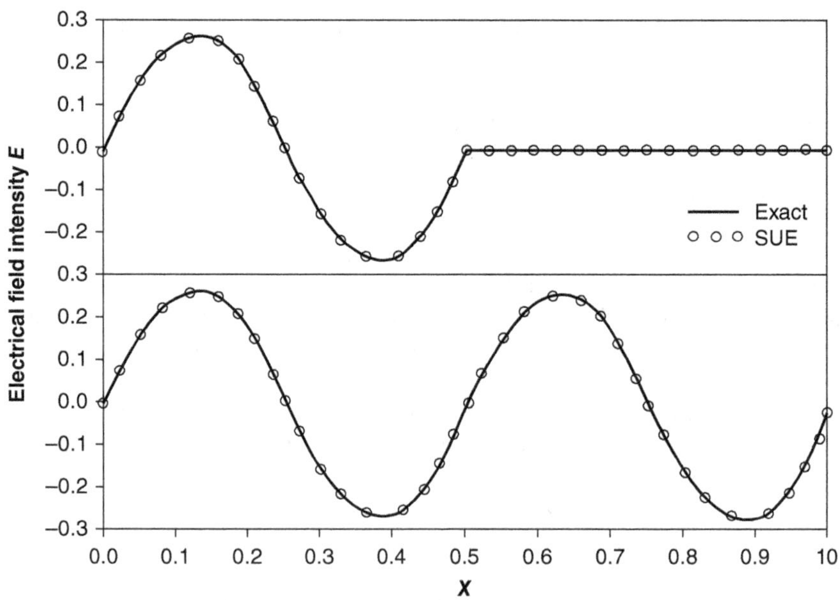

FIGURE 4.4 Perfect-shift solution of scalar wave equation.

maximum allowable time-step size defined by the CFL number of 2 ($\lambda\Delta x/\Delta t = 2$) by a single-step upwind explicit scheme. The numerical solutions presented are at instants when a right-running wave reaches the midpoint of the computational domain and exits the numerical boundary, respectively. For this one-dimensional simulation, the characteristic-based scheme using the single-step upwind explicit algorithm exhibits the perfect shift property, which indicates a perfect translation of the initial value in space (Anderson et al., 1984). As the impulse wave moves through the initially quiescent environment, the numerical result duplicates the exact solution at each and every discretized point, including the discontinuous incoming wave front. Although this highly desirable property of a numerical solution is achievable only under very restrictive conditions and may not be preserved for multidimensional problems, the ability to simulate the nonanalytic solution behavior in the limit is clearly illustrated.

For this purpose, a second-order accurate procedure in spatial derivatives is given as follows:

$$\begin{aligned} &\textit{If} \quad \lambda < 0 \quad \Delta W_{i,j,k} = [-3W_{i,j,k} + 4W_{i+1,j,k} - W_{i+2,j,k}]/2 \\ &\textit{If} \quad \lambda > 0 \quad \Delta W_{i,j,k} = [3W_{i,j,k} - 4W_{i+1,j,k} + W_{i+2j,k}]/2 \end{aligned} \tag{4.37}$$

The necessary metrics of the coordinate transformation are calculated by central differencing, except at the edges of computational domain, where one-sided differences are used. Although the fractional step or the time-splitting algorithm has demonstrated greater efficiency in data storage and a higher data-processing rate than predictor–corrector time integration procedures, it is limited to second-order accuracy in time and space. In the following fractional-step method, the temporal second order result is obtained by a sequence of symmetrically cyclic operators (Shang, 1995b):

$$W_{i,j,k}^{n+1} = L_\xi L_\eta L_\zeta L_\zeta L_\mu L_\xi W_{i,j,k}^n \tag{4.38}$$

where L_ξ, L_η, and L_ζ are the difference operators for one-dimensional equations in the ξ, η, and ζ coordinates, respectively. The symmetric cyclic spatial difference operators are applied to eliminate the numerical sweep bias. The overall numerical stability of the factorized scheme is the product of all individual amplification factors,

$$G = G_\xi G_\eta G_\zeta G_\zeta G_\eta G_\xi \tag{4.39}$$

where $G_\xi = W_\xi^{n+1}/W_\xi^n$, $G_\eta = W_\eta^{n+1}/W_\eta^n$ and $G_\zeta = W_\zeta^{n+1}/W_\zeta^n$.

In general, second-order and higher temporal resolution are achievable through multiple-time step schemes. However, one-step schemes are more attractive because they have less memory storage requirements and do not need special startup procedures. For the higher-order temporal accurate solution procedure, the single-step, multistage Runge–Kutta procedures are frequently adopted. This choice is also consistent with the accompanying characteristic-based finite-volume method.

The two-stage Runge–Kutta scheme is a formal second-order temporal accurate scheme:

$$
\begin{aligned}
W^0 &= W^n_{i,j,k} \\
W^1 &= W^0 - W'(W^0) \\
W^2 &= 0.5[W'(W^1 + W'(W^0)] \\
W^{n+1}_{i,j,k} &= W^2
\end{aligned} \tag{4.40}
$$

The four-stage, formally fourth-order accurate scheme can be presented as

$$
\begin{aligned}
W^{n+1}_{i,j,k} &= W^n_{i,j,k} + (\Delta t/6)\left(W'_1 + 2W'_2 + 2W'_3 + W'_4\right) \\
W'_1 &= W'_1\left(t, W^n_{i,j,k}\right) \\
W'_2 &= W'_2\left(t + \Delta t/2, W^n_{i,j,k} + W'_1\Delta t/2\right) \\
W'_3 &= W'_3\left(t + \Delta t/2, W^n_{i,j,k} + W'_2\Delta t/2\right) \\
W'_4 &= W'_2\left(t + \Delta t, W^n_{i,j,k} + W'_3\Delta t\right)
\end{aligned} \tag{4.41}
$$

where W comprises the incremental values of dependent variables during each temporal sweep. The resultant characteristics-based finite-difference scheme for solving the three-dimensional Maxwell equations in the time domain is second-order accurate in both time and space. For the high-resolution procedure, a consistent fourth-order accurate solution in time and space is achievable by combining the four-stage Runge–Kutta and a compacting differencing procedures.

In practical applications, a second-order accurate computational procedure is mostly popular, because the boundary-conditions implementation is straightforward. The second-order implicit procedure in delta formulation uses the ADI (alternative direction implicit) scheme (Peaceman and Rachford, 1955; Richtmyer and Morton, 1967) and has been adopted. The formulation breaks down a three-dimensional problem into three consecutive one-dimensional formulations as

$$
\begin{aligned}
[I + \alpha\Delta t D_\xi \partial/\partial\xi]\Delta W^*_\xi &= -\Delta t\{D_\xi \partial W^n_\xi/\partial\xi + S^{-1}_\xi S_\eta D_\eta \partial W^n_\eta/\partial\eta \\
&\quad + S^{-1}_\xi S_\zeta D_\eta \partial W^n_\zeta/\partial\zeta] + J\} \\
[I + \alpha\Delta t D_\eta \partial/\partial\eta]\Delta W^{**}_\eta &= \Delta W^*_\eta \\
[I + \alpha\Delta t D_\zeta \partial/\partial\zeta]\Delta W^{n+1}_\zeta &= \Delta W^{**}_\zeta
\end{aligned} \tag{4.42}
$$

where α is the numerical parameter that splits the implicit and explicit portions of the calculation. The Crank–Nicolson scheme (Crank and Nicolson, 1947) is recovered by assigning a value of 1/2. The diagonalized structure of the coefficient matrices are $D_\xi = S^{-1}_\xi AS_\xi$, $D_\eta = S^{-1}_\eta BS_\eta$, and $D_\zeta = S^{-1}_\zeta AS_\zeta$. The six characteristics in the differential equation system are completely decoupled from each other. Therefore, intensive matrix manipulations that are required by solving these equations are substantially reduced.

From the analysis of a simple scalar wave equation, the explicit numerical algorithm always has a conditional stability constraint for time step taken but does not apply to implicit algorithms. A greater numerical stability of an implicit over explicit scheme is one reason that the implicit method is favored for time-dependent simulations. The numerical stability of the ADI scheme in delta formulation applied to three-dimensional hyperbolic system is known. Using a model equation with constant coefficients, it can be given as follows (Richtmyer and Morton, 1967):

$$\frac{\partial u}{\partial t} + c_1 \frac{\partial u}{\partial x} + c_2 \frac{\partial u}{\partial y} + c3 \frac{\partial u}{\partial z} = 0 \tag{4.43}$$

The computational stability is studied by amplification factor of Fourier components in solutions of a consecutive temporal evolution, $G = (u^{n+1}/u^n)$. The modulus of amplification factors that determine numerical stability for equation (4.43) in the discretized space is

$$\begin{aligned}(GG^*)^{1/2} = &\left\{1 + (1-\alpha)\sum G_i + \alpha^2 \sum (1-\delta_{ij})G_iG_j - \alpha^3 G_1G_2G_3\right\} / \\ &\left\{1 - \alpha\sum G_i + \alpha^2 \sum (1-\delta_{ij})G_iG_j - \alpha^3 G_1G_2G_3\right\}\end{aligned} \tag{4.44}$$

where the amplification factors in each coordinate can be given as

$$G_i = \left(\frac{c_i\Delta t}{\Delta c_i}\right)\left[\frac{3}{2} - 2\exp(-i\kappa_i\Delta x_i) + \frac{1}{2}\frac{c_i\Delta t}{\Delta c_i}\right]\left[\frac{3}{2} - 2\exp(-i\kappa_i\Delta x_i)\right], \; i = 1, 2, 3$$

The ADI scheme, when applied to two-dimensional hyperbolic equations, yields an unconditionally stable approximation. Numerical stability is retained regardless of variations in the adopted spatial differencing schemes, whether they are one-sided and spatially central algorithms. However, for a three-dimensional hyperbolic partial differential equation, the spatially central differencing scheme will lead to an unconditionally unstable condition. The three-dimensional factored numerical procedure has been mended to become a conditional stable procedure by using a windward or one-sided discretization with numerical artificial viscosities, but the constraint is more restrictive for the three-point than the two-point windward differencing. In short, the ADI scheme for solving the three-dimensional hyperbolic equation is subjected to a conditional stability criterion (Beam and Warming, 1976). However, the instability condition can be removed by the Gauss–Seidel relaxation algorithm (Anderson et al., 1984). In fact, the relaxation schemes are mainstays for solving the flux splitting Navier–Stokes equations. Simulating three-dimensional wave motion in CEM using ADI algorithm is not advocated, because of the unconditionally stable property for solving the time-dependent Maxwell equations.

In general, the implicit factored scheme is unconditionally stable when applied to a simple scalar wave equation, a feature that is valuable for understanding and improving numerical efficiency to computational technique in electromagnetics. The particular features of numerical dissipation and phase errors are critically important

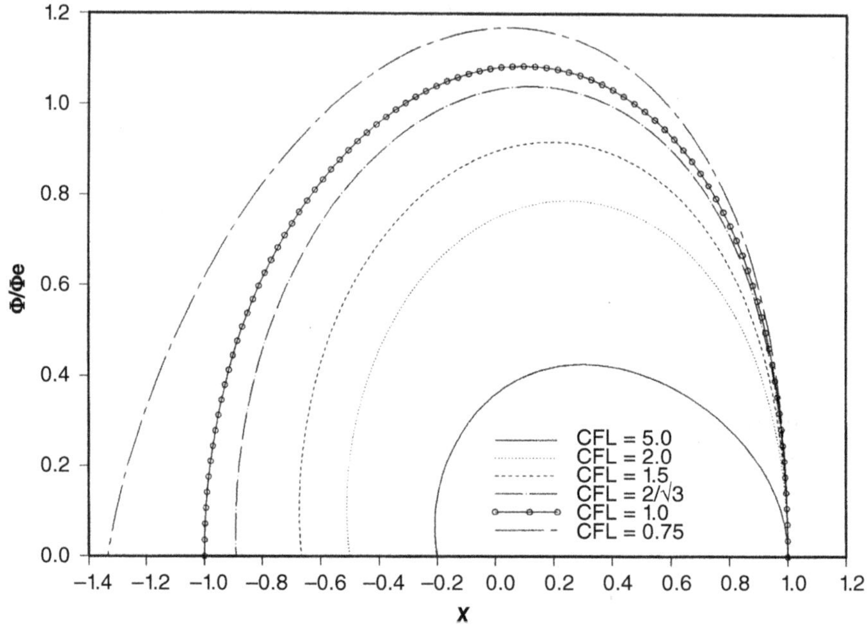

FIGURE 4.5 Relative phase error of implicit factor scheme.

for simulating wave motions. Therefore, the elementary stability analysis of this algorithm is performed and displayed by studying the modulus of amplification factor and the relative phase error in the CFL numbers ranging from 0.75 to 5.0. The numerical results are presented in Figure 4.5. This numerical procedure is dissipative and a numerical error is diminished monotonically as the CFL number is increased toward unity. On the contrary, the relative phase angle exhibits a predominant leading error for CFL values of less than unity and then switches to a persistent lagging error responding to the increasing CFL number. The cross-over of relative phase error from the leading to lagging error takes place at the unity CFL number and constrains the discrepancy to a relatively low magnitude in a wide spectrum. Most importantly, the relative phase error is significantly less than most time- and space-centered implicit scheme in the high wave number domain. The lesson learned from solutions of a simple wave equation is that numerical method must honor the physics to reproduce wave motions.

The most significant feature of the flux-vector splitting scheme lies in its ability to easily suppress reflected waves from the truncated computational domain. For simulating wave motion, the compatibility condition at any point in space is described by the split flux vector. In other words, the wave propagation in a time-space domain has a rigidly defined zone of dependence. The information is allowed only in the bounded regime by characteristic lines and signal can travel unimpeded along characteristics. In split flux-vector formulation, an approximated no-reflection condition can be achieved by setting either the incoming or the outgoing flux-vector

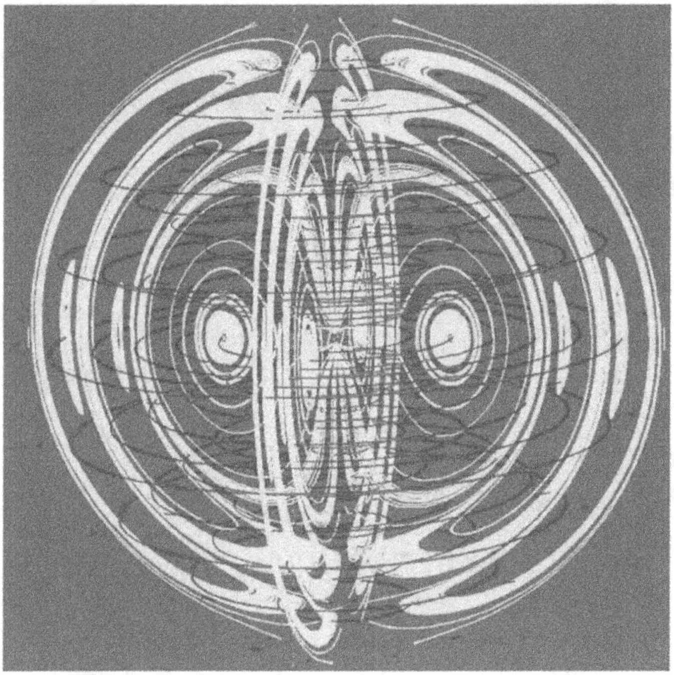

FIGURE 4.6 Electromagnetic field structure of a dipole.

component equal to zero as shown by equation (4.35): either $\lim_{r\to\infty} F^+(\xi, \eta, \zeta) = 0$ or $\lim_{r\to\infty} F^-(\xi, \eta, \zeta) = 0$. The one-dimensional compatibility condition is exact when the wave motion is aligned with one of the coordinates. This unique attribute of the characteristics-based numerical procedure in removing a fundamental dilemma in CEM will be demonstrated repetitively.

The electric dipole is a benchmark problem in CEM because it contains all these challenging attributes for numerical simulations and exemplifies the behavior of wave propagation in the near field. The near-field solution of an electromagnetic wave generated by an oscillating electric dipole is depicted in Figure 4.6. The closed form solution to the time-dependent, three-dimensional Maxwell equation is known and will be used as the base for validating the numerical solutions. This selection is based on the fact the solution contains a singular point at the dipole and the leading term singularity of the field variables appears as the inverse cubic power of radial distance from the pole. The numerical solution is generated by a relatively spare discretized mesh of $(25 \times 24 \times 48)$ and at CFL = 1.0, but all the salient features of the contour of the electromagnetic field have been captured to reveal the orthogonal structure of the electric and magnetic field intensity. The detailed and specific comparison is presented in the following text.

The characteristics-based finite-difference solution to the oscillating electric dipole is presented in Figure 4.7. For the radiating electric dipole, the depicted temporal

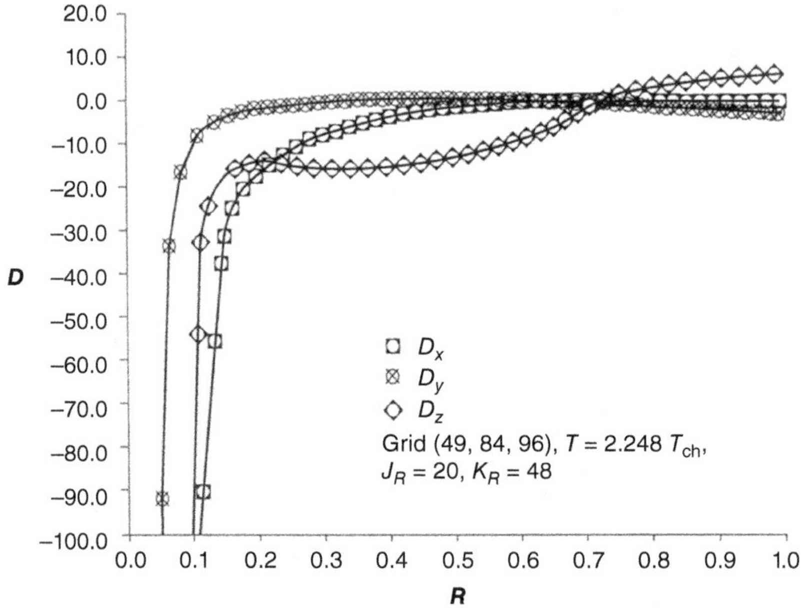

FIGURE 4.7 Instantaneous electric field of oscillating dipole.

calculations are sampled at the instant when the initial pulse has traveled a distance of 2.24 wavelengths from the dipole. The numerical results are generated on a (48 × 48 × 96) mesh system by a second-order scheme. The three electric displacement components D_x, D_y, and D_z are resolved in each wavelength by 15 mesh points. Supporting by sufficient grid-point density, the numerical solutions yield a uniformly excellent comparison with the theoretical result. Significant error appeared only in the immediate region of the coordinate origin, $R \leq 1.0$. At the coordinate origin, the solution has a singularity that behaves as the inverse cubic power of the radial distance. The most interesting numerical behavior, however, is revealed in the truncated far field. The no-reflection condition at the numerical boundary is observed to be satisfied within the order of the truncation error. For a spherically symmetric radiating field, the orientation of the wave is aligned with the radial coordinate, and the suppression of the reflected wave within the numerical domain is the best achievable by the characteristic formulation.

The corresponding magnetic field intensity computed by the second-order accurate finite-difference is also given in Figure 4.8. For the oscillating electric dipole, only the x and y components of the magnetic field exist. Numerical results attain excellent agreement with theoretical results, in that the singular behavior near the dipole origin is described reasonably well and the numerical oscillation is not discernible. The most important feature from the characteristics-based, finite-difference approximation is again reflected by the unimpeded exit condition at the far field computational boundary.

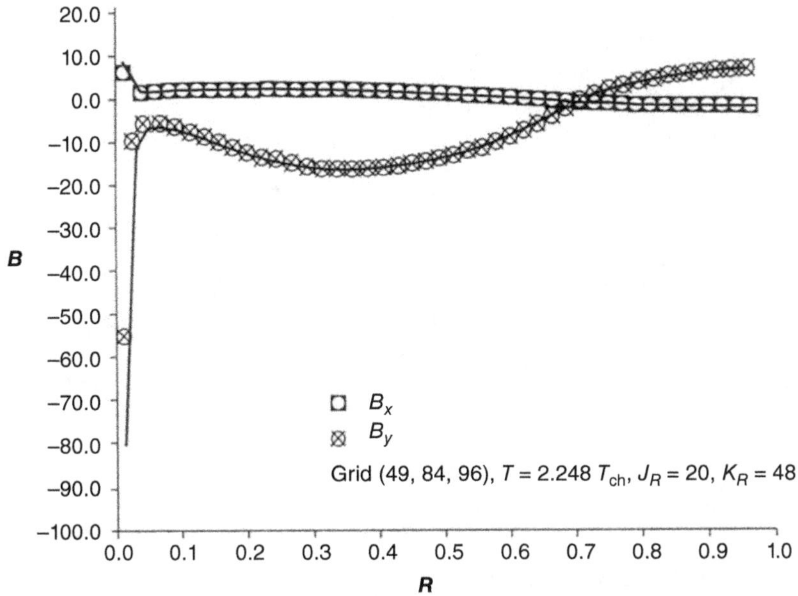

FIGURE 4.8 Instantaneous magnetic field of oscillating dipole.

4.6 FINITE-VOLUME APPROXIMATION

The finite-volume approximation solves the governing equations via integral form over a control volume by discretizing the physical space into contiguous cells and balancing the flux vectors on the cell control surfaces (Anderson et al., 1984; Gaitonde and Shang, 1997). The schematic of the finite volume structure is depicted in Figure 4.9. For the cell-center scheme, the control surfaces are located between grid points along each coordinate and perpendicular to the coordinate axis. The flux vectors are further split into normal and tangential component with respect to the control surfaces. All dependent variables contained within the elementary volume are assigned the value at the cell center (i, j, k). Therefore, the entire computational domains are consisting of contagious elementary volumes from boundaries to boundaries. In a discretized form, the integration procedure reduces to evaluation of the sum of all fluxes aligned with surface-area vectors.

$$\frac{\Delta U}{\Delta t} + \frac{\Delta F}{\Delta \xi} + \frac{\Delta G}{\Delta \eta} + \frac{\Delta H}{\Delta \zeta} - R = 0 \tag{4.45}$$

In this approach, the continuous differential operators have been replaced by discrete operators. The finite-volume approximation satisfies the governing equation over an elementary volume that bounded by six control surface perpendicular to three coordinates. In essence, the numerical procedure needs only to evaluate the sum of all flux vectors aligned with surface-area vectors. Only one of the vectors is required

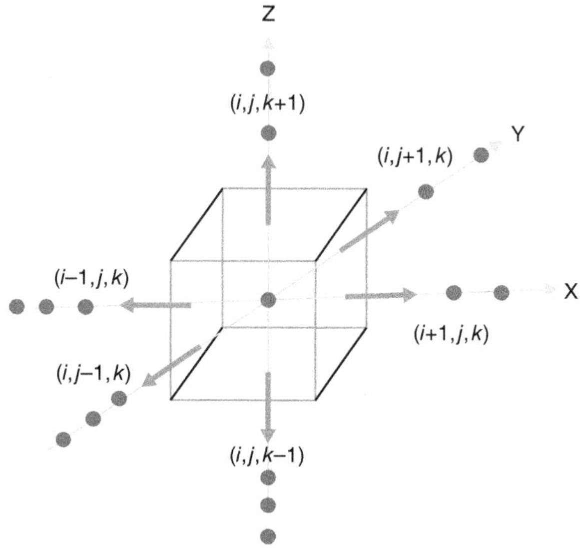

FIGURE 4.9 Control volume and control surface of finite volume approximation.

to coincide with the outward normal to the cell surface, and the rest of the orthogonal triad can be made to lie on the same surface. The metrics, or more appropriately the direction cosines, on the cell surface are uniquely determined by the nodes and edges of the elementary volume. This feature is distinct from the finite-difference approximation. The shape of the cell under consideration and the stretching ratio of neighbor cells can lead to a significant deterioration of the accuracy of finite-volume schemes.

The most outstanding aspect of finite-volume schemes is the elegance of its flux-splitting process. The flux-difference splitting for equation (4.45) is greatly facilitated by a locally orthogonal system in the transformed space (ξ, η, ς). In this new frame of reference, eigenvalues and eigenvectors, as well as metrics of the coordinate transformation between two orthogonal systems, are known. The inverse transformation is simply the transpose of the forward mapping. In particular, the flux vectors in the transformed space have the same functional form as that in the Cartesian frame. The difference between the flux vectors in the transformed and the Cartesian coordinates is a known quantity and is given by the product of the surface outward normal $(\nabla S/||\nabla S||)$ and the cell volume V. Therefore, the flux vectors can be split in the transformed space according to the signs of the eigenvalues but without detailed knowledge of the associated eigenvectors in the transformed space. This feature of the finite-volume approach provides a tremendous advantage over the finite-difference approximation in solving complex problems in physics.

The widely used formulation to calculate the flux vector on control surface often adopts Van Leer's kappa scheme (Van Leer, 1982) in which the solution vectors are reconstructed on the cell surface from the piecewise data of neighboring cells

$U^+_{i+1/2}$, $U^-_{i+1/2}$. The reconstructed data on the control surface are used to compute the flux vectors and determined the accuracy of the finite-volume schemes. The spatial accuracy of this scheme spans a range from first- to third-order upwind biased approximations,

$$\begin{aligned} U^+_{i+\frac{1}{2}} &= U_i + \frac{\phi}{4}[(1-\kappa)\nabla + (1+\kappa)\Delta]U_i \\ U^-_{i+\frac{1}{2}} &= U_i - \frac{\phi}{4}[(1+\kappa)\nabla + (1-\kappa)\Delta]U_i \end{aligned} \tag{4.46}$$

where $\Delta U_i = U_i - U_{i-1}$ and $\nabla U_i = U_{i+1} - U_i$ are the forward and backward differencing discretizations. The parameters φ and κ control the accuracy of the numerical results. For $\varphi = 1$, $\kappa = -1$, a two-point windward scheme is obtained. This method has an odd-order leading truncation-error term; the dispersive error is expected to dominate. If $\kappa = 1/3$, a third-order upwind-biased scheme will emerge. In fact, both upwind procedures have discernible leading phase error. This behavior is a consequence of using the two-stage time integration algorithm. The dispersive error can be alleviated by increasing the temporal resolution. For $\varphi = 1$, $\kappa = 0$, the formulation recovers the Fromm scheme. If $\kappa = 1$, the formulation yields the spatially central scheme. Since the fourth-order dissipative term is suppressed, however, the central scheme is susceptible to parasitic odd–even point decoupling.

Time integration is carried out by the same two-stage Runge–Kutta method as in the finite difference procedure. The finite-volume procedure is therefore second-order accurate in time and up to third-order accurate in space. For the present purpose, only the second-order windward and the third-order upwind biased options are exercised. The second-order windward schemes in the form of the flux-vector splitting finite-difference and the flux-difference splitting finite-volume scheme are formally equivalent.

In order to compare the relative merit of finite-difference and finite-volume approximations for solving the Maxwell equations, numerical results of the dipole by these two approximations are examined at selected locations in space, the time trace of field variable, and a point spectral analysis. For the radiating problem, the z component of the magnetic field intensity is identical to zero; thus, not all electric and magnetic field components are presented. From comparative studies, the similar basic concept of the two algorithms and the vastly different detailed formulation are illustrated and validated against theoretical results (Shang et al., 1996). On the spherical coordinates, the second-order windward numerical simulations for the radiating phenomenon are compared in terms of accuracy and efficiency. The ability of suppressing the reflected wave from the truncated boundary is also demonstrated. Three mesh systems, $(25 \times 24 \times 48)$, $(37 \times 36 \times 72)$, and $(49 \times 48 \times 96)$, are used to perform the investigations. By scaling the independent and dependent variables within the range of unity, the coarsest mesh system does not satisfy the marginal point per wavelength density of 10 points for wave motion; a significant error up to a few percent is anticipated. The finite-difference calculations are performed on the nodes that are located at the cell

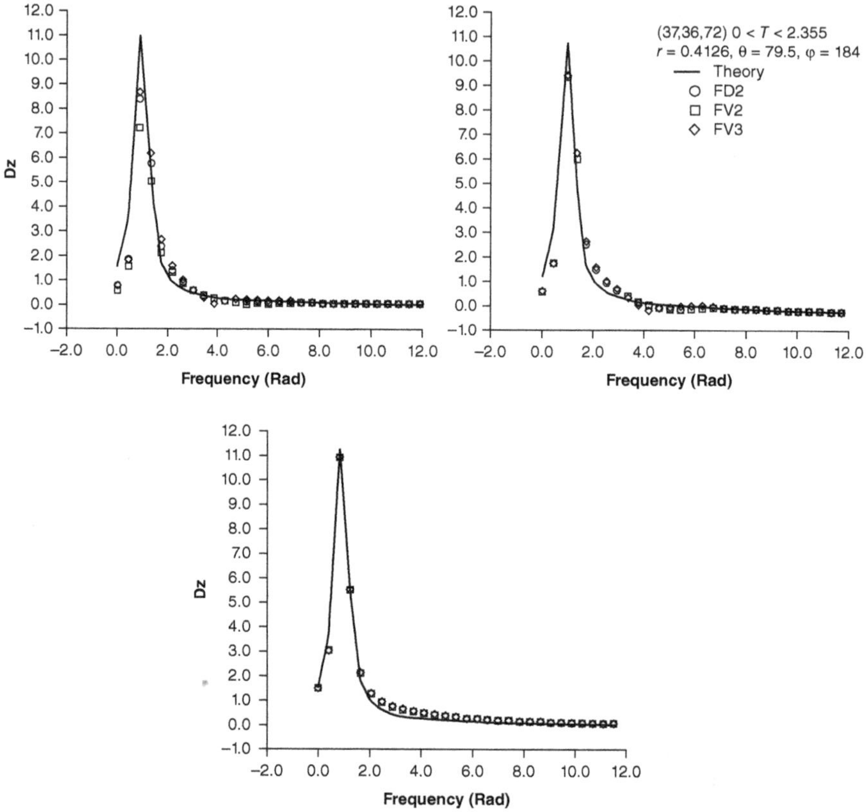

FIGURE 4.10 Spectral analysis of electric field intensity D_z on different grid-point densities.

center of the finite-volume formulation. Therefore, the mesh systems between two algorithms are slightly different but within the uncertainty limit of discretization.

The spectral analysis of the z component of the electric displacement resolved by the three mesh system is displayed first in Figure 4.10. The electric component of the dipole D_z is singled out, because the singular behavior of the induced electric field at the pole has steeper inverse radius dependence than the magnetic field. Spectral analysis is also performed for a few randomly selected points from the studied mesh systems. Since the mesh systems are shared between the definite-difference and the finite-volume scheme, the comparing results are not able to scale onto identical location. Instead the point spectral analyses are conducted along the radial ray defined by $\theta = 79.5$, $\phi = 184$, and $0.4107 < r < 0.4144$. A two-hundred-terms Fourier series is used to perform the spectral analysis. The data sampling includes the transient state to reflect the worst possible error and provides the longest period for numerical resolution.

For the time-step sizes used ($0.9 < \text{CFL} < 1.0$), the numerical behavior indicates a leading phase error. On the coarsest mesh system ($25 \times 24 \times 48$), the solution of

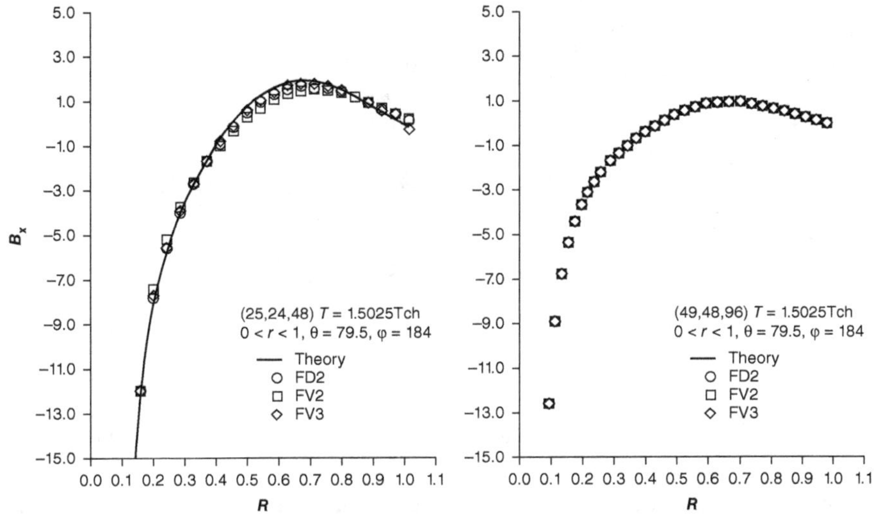

FIGURE 4.11 Numerical resolution of magnetic field intensity *Bx* with different grid-point densities.

the z-component electric field, D_z by the second-order finite-volume scheme, FV2 contains the largest error in prediction of the local extreme. Away from the dipole origin, the maximum error is located at $r = 0.6268$ and as a relative magnitude of 9.09%. Refined to a (37 × 36 × 72) grid system, the specific comparison of numerical results yields little additional insight. Once sufficient numerical resolution is secure on the (49 × 48 × 96) mesh, the differences between numerical and theoretical results are confined within the plotting accuracy. The relative phase errors by the three numerical procedures are reduced significantly from the maximum of 1.1044 (FV2) to less than 1.0044. Among the three numerical results, the calculated amplitudes at the principal frequency have maximum difference of 0.37%.

In Figure 4.11, the comparisons of computed x-component of the magnetic intensity, B_x with theory on the coarse (25 × 24 × 48), and the fine grid (49 × 48 × 96) are depicted. Under the condition of merely 10 cells per wavelength, the maximum relative error is still confined to within 1% by the spatial third-order scheme, FV3. The solution produced by the second-order windward finite-volume scheme shows the worst discrepancy from the theoretical result. The relative error is 5.0% at $r = 0.7133$. The most important feature of eliminating reflected wave from the truncated boundary by the characteristics-based numerical procedure is repeatedly demonstrated. On the (49 × 48 × 96) mesh system, the second-order procedures, either by the finite-volume FV2 or by the finite-difference approximation, have the precise formulation in enforcing the compatibility condition. At the truncated boundary, the vanishing incoming flux components derived from characteristics theory are imposed. Within the limit of truncation error, no reflected wave from the far field is detected. On the contrary, the third-order windward biased scheme reconstructs the

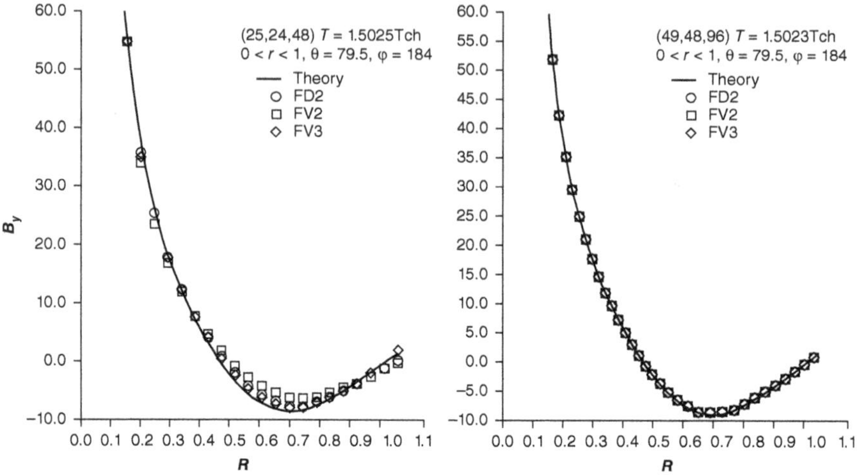

FIGURE 4.12 Numerical resolution of magnetic field intensity B_y with different grid-point densities.

split flux on the interface by an interpolation process that includes data beyond the physical domain of dependence: the perfect no-reflection condition is not achieved. The resulting wave reflection is observable from both computational results on the coarse and fine mesh systems. Numerical results reflect the fact that an accurate simulation for wave propagation can be achieved only by both sufficient numerical resolution and appropriately specified boundary condition.

Direct comparison of three numerical results of the y-component of the magnetic field intensity with the theoretical value is given by Figure 4.12 on two mesh systems. The maximum error is revealed by the solution of the second-order windward-volume scheme FV2 on the coarse grid ($25 \times 24 \times 48$). The relative error has a magnitude of 8.39% at $r = 0.6268$. The counterpart of the second-order finite-difference approximation FD2 performs much better. The accuracy of the solution by the FD2 scheme approaches that of the formally third-order finite-volume scheme, FV3. On an irregular mesh like the present spherical polar system, the finite-volume procedure shows a greater numerical error than the finite-difference procedure on highly stretched grid systems. When a sufficient numerical resolution is provided by the fine mesh system ($49 \times 48 \times 96$), the difference in computational results of the three comparing numerical scheme becomes negligible. Under the present computational conditions, both numerical procedures yield uniformly excellent comparison with the theoretical result. Significant error appears only in the immediate region of the coordinate origin. At that location, the solution has a singularity that behaves as the inverse cubic power of the radial distance. The most interesting numerical behavior, however, is revealed in the truncated far field. The no-reflection condition at the numerical boundary is observed to be satisfied within the order of the truncation error. For a spherically symmetric radiating field, the orientation of the wave is aligned with the radial coordinate, and the suppression of the reflected wave within the numerical domain is the best achievable by the characteristic formulation.

The computational efficiency of a numerical procedure is another consideration for selecting simulation tools. The database is established by a direct comparison between finite-volume and finite-difference approximations that are carried on vector processors. All computations were processed on either a Cray Y-Mp8/8128 or C916/19256 system. For the numerical efficiency comparison, the date procession rates and the central processing time to advance the solution for a characteristic time period were recorded. For the relative computational efficiency, the consistent second-order finite-difference scheme, FD2 has a less central processing time requirement at ratios of 0.7534, 0.6808, and 0.6460 than that of the second-order finite volume scheme, FV2 on the coarse, medium, and fine mesh systems.

4.7 HIGH-RESOLUTION ALGORITHMS

Time-dependent Maxwell equations were established in 1873 but with only a limited number of analytic solutions. In computational simulation, the accuracy requirement for high-frequency or transient computational simulations is severe both in dynamic range and in absolute value. Improved numerical efficiency has always been a pacing item for expanding the application envelope of CEM in the time domain. For multidimensional computations, the truncation error of a numerical result manifests in dissipative, dispersive, and anisotropic errors. The accumulative numerical error during a sustained period of calculation or in an extensive computational domain may result in unacceptable wave modulation and phase error.

Numerical resolution is naturally quantifiable by Fourier analysis in terms of normalized wavenumber. In application, the quantification is measured by the grid-point density per wavelength. Thus, a direct correlation between the grid-point density and the wavenumber exists. All numerical schemes have a limited wavenumber range for accurate computations. For most popular second-order accurate methods using an unstaggered mesh system, the rule of thumb for acceptable accuracy is on the order of 20 nodes per wavelength (Shang, 1999, 2002). However, a simple grid density criterion becomes insufficient when the computational domain contains multiple media with a wide range of characteristic impedances. For a high frequency, CEM simulation or a large electrical configuration is often beyond the reach of conventional computing systems. The need to develop high-resolution schemes for extending the present CEM simulating capability to high-frequency spectra is urgent.

Using the formal order of accuracy of truncation error exclusively to select an algorithm for time-dependent simulation is an oversimplification (Lele, 1992; Tam et al., 1993). A simple illustration is offered by the dispersion behavior of semidiscrete schemes by Figure 4.13. The third-order MUSCL scheme has nearly identical dispersive error to that of the explicit fourth-order scheme, and the different fourth-order schemes can have different dispersive attributes over the entire wavenumber range. Thus, the desired feature of a numerical algorithm may be derived from

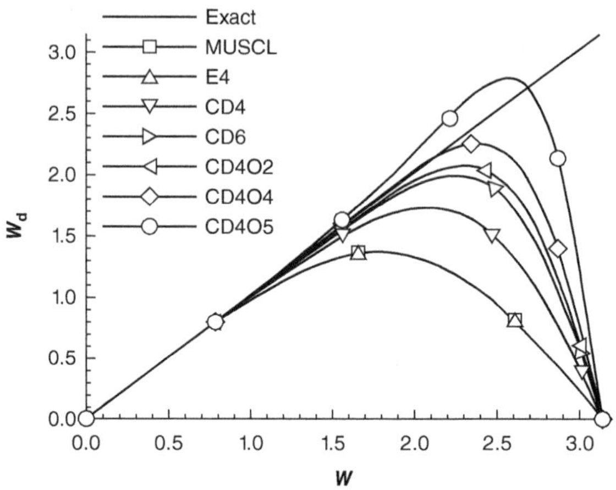

FIGURE 4.13 Dispersion behavior of high-order algorithms.

optimization in the Fourier space rather by focusing on the lowest possible truncation error (Gaitonde et al., 1997).

In light of the above line of reasoning, compact differencing and optimized numerical procedures are viable approaches to achieve high resolution. Both approaches seek algorithms that have a small stencil dimension and maintain a lower level of dispersive and dissipative error than conventional numerical schemes. Collatz (1966) has pointed out that the compact difference is based on Hermite's generalization of the Taylor series. The basic formulation is an implicit procedure for evaluating derivatives of the dependent variables. The stencil structure for most compact-difference formulas is spatially central and requires additional derivative values at the boundaries. The spatially central scheme is inapplicable to the immediately adjacent grid point at a boundary for a stencil greater than 3. Additional transitional boundary operator to interior domain is required. This transitory boundary implementation not only requires transmitting data from the boundary but must also preserve the stability and accuracy of interior domain for the global order of resolution. This unique feature of compact-difference schemes is also the source of spurious high-frequency numerical oscillations.

The most general compact difference or the Hermitian formulation has been written for the first and second derivatives by Lele (1992) as

$$\begin{aligned}\beta U'_{i-2} + \alpha U'_{i-1} + U'_i + \alpha U'_{i+1} + \beta U'_{i+2} &= c(U_{i+3} - U_{i-3})/6\Delta\xi \\ b(U_{i+2} - U_{i-2})/4\Delta\xi &+ a(U_{i+1} - U_{i-1})/2\Delta\xi\end{aligned} \tag{4.47}$$

$$\begin{aligned}\beta U''_{i-2} + \alpha U''_{i-1} + U''_i + \alpha U''_{i+1} + \beta U''_{i+2} &= c(U_{i+3} - 2U_i + U_{i-3}/9\Delta\xi^2 \\ +b(U_{i+2} - 2U_i + U_{i-2})/4\Delta\xi^2 &+ a(U_{i+1} - 2U_i + U_{i-1})/\Delta\xi^2\end{aligned} \tag{4.48}$$

If one limits the formulation to tridiagonal form, the fourth-order Pade' and six-order first derivatives can be obtained by setting $\alpha = 1/4, \beta = 0$, $a = 3/2b = 0$,

$c = 0$ and $\alpha = 1/3, \beta = 0, \ a = 14/9, b = 1/9, \ c = 0$. Similarly, the fourth-order and six-order second derivatives are achieved by letting $\alpha = 1/10, \beta = 0, \ a = 6/5, b = 0, \ c = 0t$ and $\alpha = 2/11, \beta = 0, \ a = 12/11, b = 3/11, c = 0$.

The consistent fourth-order, first and second derivatives by one-sided difference operators to be used at the boundary are

$$\begin{aligned} U_i' &= (-25U_i + 48U_{i+1} - 36U_{i+2} + 16U_{i+3} - 3U_{i+4})/12\Delta\xi \\ U_i'' &= (45U_i - 154U_{i+1} + 214U_{i+2} - 156U_{i+3} + 61U_{i+4} - 10U_{i+5})/12\Delta\xi^2 \end{aligned} \quad (4.49)$$

An undesirable feature of compact-difference schemes is the temporal instability during a long-time integration process. Although high-resolution schemes are stable in the classic sense, numerical results have frequently exhibited nonphysical growth in time. Research efforts have shown that even satisfying the generalized summation-by-parts energy norm formulation is not sufficient for preventing this instability, which is incurred by a positive real eigenvalue component of the spatial operator matrix (Carpenter et al., 1994). This eigenvalue will dominate the numerical results after a long-time evolution. A very effective remedy originated by Gaitonde and Shang (1997) is by a low-pass filter (Vichnevetsky and Bowles, 1982) that eliminates the destabilizing Fourier components from the numerical results.

A low-pass filter is ideal for eliminating the undesirable high-frequency Fourier components and does not affect the remaining components of a numerical solution. The adopted spatial filter modifies only the amplitude but not the phase relation among all Fourier components of the solution. The spectral function is a symmetric numerical filter that contains no imaginary part and has the low-pass amplitude response. For example, the tridiagonal filter is given as

$$\beta U_{i-1} + U_i + \beta U_{i+1} = \Sigma a_n(u_{i+n} + u_{i-n})/2 \quad n = 0, 1, 2 \ldots N \quad (4.50a)$$

where the variables U and u represent the filtered and raw data of the numerical solution. The only free parameter is bounded by $-0.5 < \beta < 0.5$. When the value approaches 0.5, the coefficients of the symmetric filter can be obtained by expanding and matching the spectral function,

$$SF(\omega) = \Sigma a_n \cos(n\omega)/[1 + 2\beta\cos(\omega)] \quad (4.50b)$$

by a Maclaurin series.

The numerical behavior of the spatial filters is depicted over the entire normalized wavenumber range of $(2\pi\kappa\Delta x/L)$ form aero to π in Figure 4.14. The unique characteristic of a low pass is displayed by filtering out only the high-frequency components and leaves the low-frequency components unaffected. It is immediately recognized that the frequency responses to the sixth- and eight-order filters suppress only the amplitude of the highest-frequency components. The effectiveness of the parameter β in regulating the filter performance is also clearly exhibited. For the eight-order

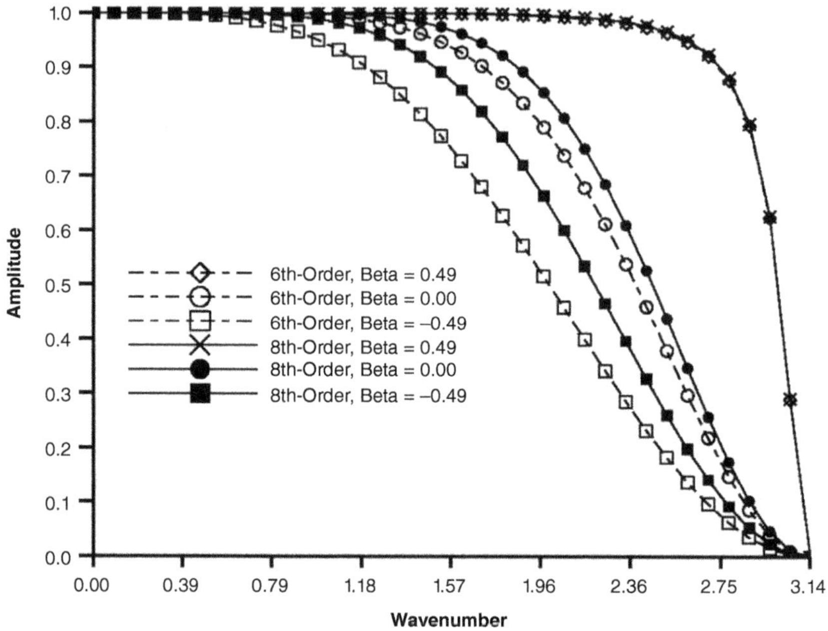

FIGURE 4.14 High-frequency suppression characteristic of low-pass filter.

filter, the performance is similar to the six-order result except that the amplitude of the Fourier components with a wavenumber greater than $\pi/4$ (wavenumber $= 0.785$) is gradually suppressed. The Fourier component at the extremely high frequency is completely suppressed. The depicted numerical results do not include the phase information of the filtered data. However, the complete analysis reveals that the filter does not affect the phase relationship among components of the filtered final solution.

For finite volume, the flux vectors collapse telescopically by construction; thus, the flux conservation is easily enforced even on arbitrary meshes (Harten et al., 1997). The telescopic effect is meant that the internal cancellation is complete between elementary control surfaces along any coordinates (Figure 4.9). Analysis of cell-centered formulation shows superior performance in the high wavenumber range and exhibits lower truncation error. The basic formulation combines the primitive function approach with 5-point spatially fourth- and sixth-order methods. The optimization is achieved by minimizing dispersion and isotropic error in the semidiscrete space (Casper et al., 1994). The temporal accuracy is determined by the classic fourth-order Rung–Kutta technique. In the flux-reconstruction process, a primitive function is defined according to Harten et al., as $\bar{v} = \int_0^x u dx$. By the second fundamental theorem of calculus, smooth solutions of the desired pointwise values are obtainable as

$$\bar{u}_{i+1/2} = \frac{d\bar{v}}{dx} + O(\Delta x^n) \tag{4.50}$$

A 5-point compact stencil is employed to evaluate the term $v' = dv/dx = u$;

$$\alpha v'_{i-1/2} + v'_{i+1/2} + \alpha v'_{i+3/2} = b\frac{v_{i+5/2} - v_{i-3/2}}{4\Delta x} + a\frac{v_{i+3/2} - v_{i-1/2}}{4\Delta x} \tag{4.51}$$

where α. a, and b are constants that determine the spatial properties of the algorithm. Equation (4.51) is a tridiagonal system if $\alpha \neq 0$. A similar stencil is adopted by the finite-difference approximation.

The optimized scheme for the improved wideband wave propagation performance in the high wavenumber range can be achieved by manipulating a spectral function that modifies the parameter for compact-difference formulation as

$$\hat{A}(\omega) = \frac{i[a\sin(\omega) + b\sin(2\omega)/2]}{1 + 2\alpha\cos(\omega)} \tag{4.52}$$

The computed dispersion wave number ω_{d} is related to that of the imposed trial solution ω as

$$\omega_{\mathrm{d}} = \mathrm{Im}[\hat{A}(\omega)] \tag{4.53}$$

A scaled isotropy wave number ω_i may be similarly defined for a wave traveling at an angle θ to the coordinate,

$$\omega_i(\omega, \theta) = \cos\theta\omega_{\mathrm{d}}(\omega\cos\theta) + \sin\theta\omega_{\mathrm{d}}(\omega\sin\theta) \tag{4.54}$$

The merit of high-resolution scheme in solving a benchmark problem of dipole oscillation for computational electromagnetics is revealed by comparing solutions via a conventional second-order and high-resolution compact-differencing scheme on the identical grid system. The improved numerical accuracy is clearly illustrated by the improved numerical accuracy (Shang, 1999). The quantification of reduced error for the oscillating dipole is presented by the L2 norm of the z component of the magnetic flux density B_z in Figure 4.15. For this radiating problem, the theoretical value of this field component is identically zero. Therefore, a nonzero value is an indication of the magnitude of the numerical error. In this figure, the numerical result generated by a fourth-order spatial filter is recorded after a duration of eight periods. The accuracy improvement is clearly demonstrated over that of the conventional second-order scheme. The numerical errors of the L2 norm are reduced from the band around 10^{-5} to 10^{-11} for the second-order to high-resolution algorithms, respectively. A total of five calculations are performed using different parameter β of the low-pass filter. As a reference of comparison, the time-dependent calculation diverges after a time elapse of 3.4 periods without the use of the spatial filter. All computations are sustained over 20 periods when the low-pass filter is activated.

The further specific comparison of microwave propagations in a waveguide via the fourth-order, compact-difference-based high-resolution scheme and the MUSCL scheme is generated on a $(25 \times 24 \times 96)$ grid. All the x components of the magnetic

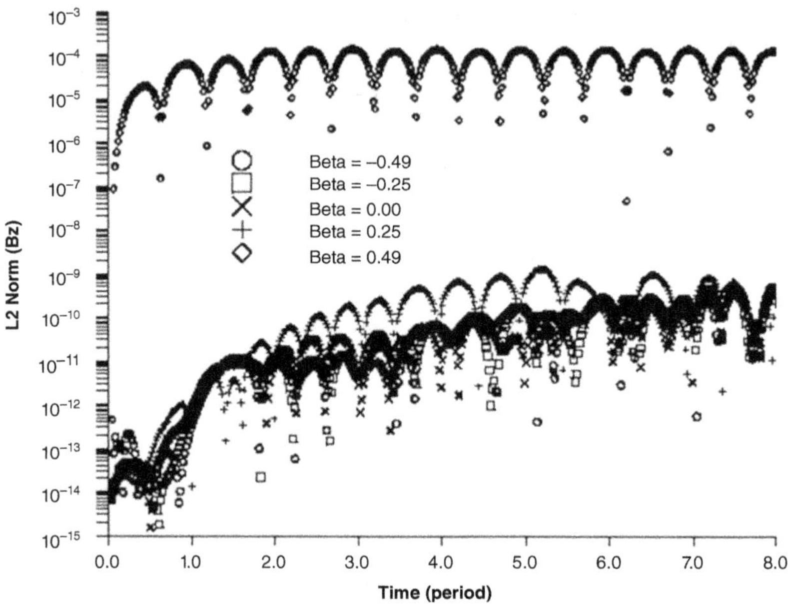

FIGURE 4.15 Comparison of errors in L2 norm of compact-differencing and central schemes.

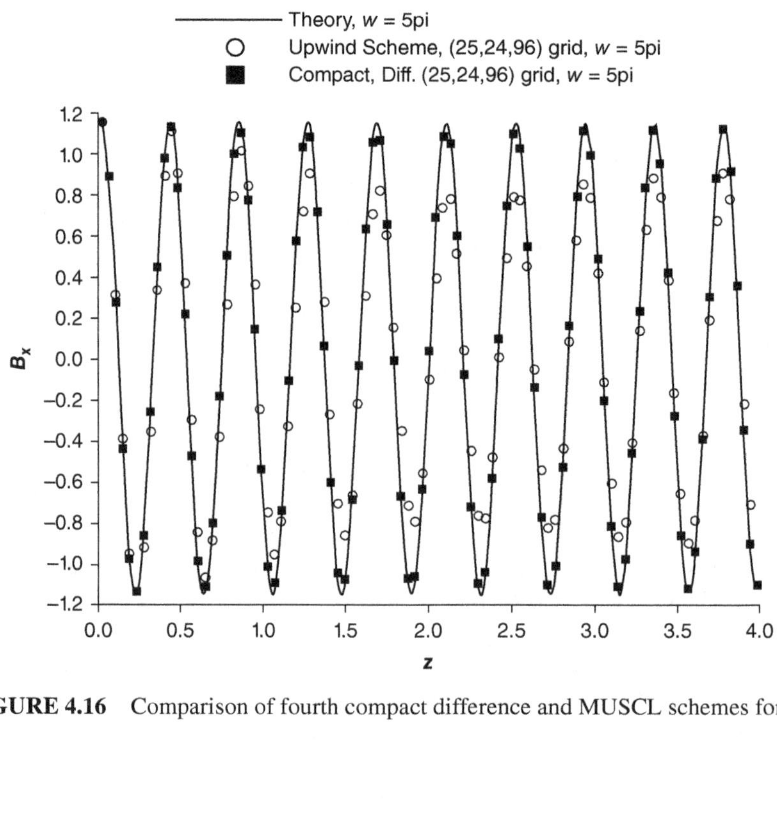

FIGURE 4.16 Comparison of fourth compact difference and MUSCL schemes for B_x.

flux density B_x of a $TE_{1,1}$ wave in a rectangular guide are calculated at a wavenumber of $\omega = 5\pi$, which corresponds to a grid-point density per wavelength of 9.15. The superior numerical resolution of the high-resolution scheme is highlighted by the negligible dissipative error and relative phase relationship over the entire waveguide in comparison with the theoretic result as shown by Figure 4.16. The high-resolution results by compact-difference scheme are designated by solid black points and the hollow circles depicted solutions by the second-order method. The high-resolution results maintained an excellent agreement with the theoretical solution of waveguide in both wave amplitude and phase relationship over 10 wavelengths, whereas the solution by the second-order scheme loses amplitude by numerical dissipation and shows shifted phase angles by numerical dispersion. From a series of studies, the L2 norm error decreases approximately linearly with time and a substantially improvement in spatial resolution. In fact, for each 2 grid-point density per wavelength increased, the resolution lowers the error by nearly two thirds for the wave number range of $3\pi < \omega < 7\pi$.

REFERENCES

Anderson, D.A., Tannehill, J.C., and Pletcher, R.H. 1984. *Computational Fluid Mechanics and Heat Transfer*. Hemisphere, New York, 1984.

Beam, R.M., and Warming, R.F., An implicit finite-difference algorithm for hyperbolic systems in conservation law form, *J. Comp. Phys.*, vol. 22, 1976, pp. 87–110.

Berenger, J.A., A perfectly match layer for the absorption of electromagnetic waves, *J. Comp. Physics*, vol. 114, 1994, pp. 185–200.

Casper, J., Shu, C.W., and Atkins, H., Comparison of two formulations for high-order accurate essentially non-oscillatory schemes, *AIAA J.*, vol. 32, 1994, pp. 1970–1977.

Carpenter, M.K., Gottlieb, D., and Abarbanel, S., Time-stable boundary conditions for finite-difference scheme solving hyperbolic systems: methodology and application to high-order compact schemes, *J. Comp. Phys.*, vol. 111, 1994, pp. 220–236.

Chew, W.C., Jin, J.M., Michielssen, E., and Song, J.M., *Fast and Efficient Algorithms in Computational Electromagnetics*, Artech House, Boston, 2001.

Collatz, L., *The Numerical Treatment of Differential Equations*, Springer-Verlag, New York, 1966, p. 538.

Courant, R., and Hilbert, D., *Methods of Mathematical Physics*, vol. II, Interscience, New York, 1965.

Crank, J., and Nicolson, P., A practical method for numerical calculation of solutions of partial differential equations of the heat-conduction type. *Proc. Camb. Philos. Soc.*, vol. 43, 1947, pp. 50–67.

Elliott, R.A., *Electromagnetics*, Ch. 5. McGraw-Hill, New York, 1966.

Enquist, B., and Majda, A., Absorbing boundary conditions for the numerical simulations of waves, *Math. Comp.* vol. 31, 1977, pp. 629–651.

Gaitonde, D., and Shang, J.S., Optimized compact-difference-based finite-volume schemes for linear wave phenomena, *J Comp. Phys.* vol. 138, 1997, pp. 617–643.

Hadmard, J., *Lecture on Cauchy's Problem in Linear Partial Differential Equation*, Yale University Press, New haven, CT, 1932.

Harrington, R.R., *Time-Harmonic Electromagnetic Fields*, McGraw-Hill, New York, 1961.

Harrington, R.F., *Field Computation by Moment Methods*, 4th ed., Robert E. Krieger, Malabar, FL, 1968.

Harten, A., Enquist, B., Osher, S., and Chakravarthy, S.R., Uniformly high order accurate essentially non-oscillatory Scheme III, *J. Comp. Phys.*, vol. 131, 1997, pp. 3–47.

Krause, J.D., *Electromagnetics*, 1st ed., McGraw-Hill, New York, 1953.

Lele, S.K., Compact finite difference schemes with spectral-like resolution. *J. Comp. Phys.*, vol. 103, 1992, pp. 16–42.

Luebbers, R.J., Kunz, K.S., Schneider, M., and Funsberger, F., A finite-difference time-domain near zone transformation, *IEEE Trans. Antennas Propag.*, vol. 39, no. 4, 1991, pp. 429–433.

Mur, G., Absorbing boundary conditions for the finite difference approximation of the time domain electromagnetic field equations, *IEEE Trans. Electromagn. Compat.*, vol. 23, 1981, pp. 377–382.

Peaceman, D.C., and Rachford, H.H., The numerical solution of parabolic and elliptic differential equations, *J. Soc. Ind. Appl. Math.*, vol. 3, 1955, pp. 38–41.

Rahman, B.M.A., Fernandez, F.A., and Davies, J.B., Review of finite element methods for microwave and optical waveguide. *Proc. IEEE*, vol. 79, 1991, pp. 1442–1448.

Richmyer, R.D., and Morton, K.W., *Difference Methods for Initial-Value Problem*. Interscience, New York, 1967.

Roe, P.L., Approximate Riemann solvers, parameter vectors and difference scheme, *J. Comp. Phys.*, vol. 43, 1981, pp. 357–372.

Roe, P.L., Characteristic-based schemes for the Euler equations. *Ann. Rev. Fluid Mech.*, vol. 18, 1986, pp. 337–365.

Shang, J. S., A fractional-step method for solving 3-D, time-domain Maxwell equations. *J. Comp. Phys.* vol. 118, no. 1, April 1995a, pp. 109–119.

Shang, J.S., Characteristic-based algorithms for solving the Maxwell equations in the time domain, *IEEE Antenna Propagat. Mag.*, vol. 37, no. 4, 1995b, pp. 15–25.

Shang, J.S., and Fithen, R.M., A comparative study of characteristic-based algorithms for the Maxwell equations, *J. Comput. Phys.*, vol. 125, 1996, pp. 378–394.

Shang, J.S., and Gaitonde, D.V., Characteristic-based, time-dependent Maxwell equation solvers on a general curvilinear frame, *AIAA J.*, vol. 33, no. 3, 1995a, pp. 491–498.

Shang, J.S., and Gaitonde, D., Scattered electromagnetic field of a reentry vehicle, *J. Spacecraft Rockets*, vol. 32, no. 2, 1995b, pp. 294–301.

Shang, J.S., High-order compact-difference schemes for time-dependent Maxwell equations, *J. Comp. Phys.*, vol. 153, 1999, pp. 312–333.

Shang, J.S., Shared knowledge in computational fluid dynamics, electromagnetics, and magneto-aerodynamics, *Prog. Aerosp. Sci.*, vol. 38, no. 6-7, 2002, pp. 449–467.

Sommerfeld, A., *Partial Differential Equations in Physics*, Ch. 2. Academic Press, New York, 1949.

Steger, J.L., and Warming, R.F., Flux vector splitting of the inviscid gas dynamics equations with application to finite difference methods. *J. Comput. Phys.*, vol. 20, no. 2, 1987, pp. 263–293.

Thompson, J.F., *Numerical Grid Generation*, Elsevier Science, New York, 1982.

Tam, C.K.W., and Weber, J.C., Dispersion-relation-preserving finite difference schemes for computational acoustics, *J. Comp. Phys.*, vol. 107, 1993, pp. 262–281.

Taflove, A., *Computational Electromagnetics, the Finite-Difference Time Domain Method*, Artech House, Boston, 1995.

Van Leer, B., Flux-vector splitting for the Euler equations. *TR* 82-30, ICASE, Sept., *Lect. Notes Phys.*, vol. 170, 1982, pp. 507–512.

Vichnevetsky, R., and Bowles, J.B., *Fourier Analysis of Numerical Approximations of Hyperbolic Equations*, SIAM Studies in Applied Mathematics, SIAM, Philadelphia, 1982.

Vinokur, M., Conservation equations of gas-dynamics in curvilinear coordinate system, *J. Comput. Phys.*, vol. 14, 1974, pp. 105–125.

Yee, K.S., Numerical solution of initial boundary value problems involving Maxwell's equation in isotropic media, *IEEE Trans. Antennas Propag.*, vol. 14, 1966, pp. 302–307.

Zhao, L., and Cagellaries, A.C., A general approach for the development of unsplit-filed time-domain implementations of perfect match layer for FD-FV grid truncation, *IEEE Microwave Guided Wave Lett.*, vol. 6, 1966, pp. 209–211.

5

ELECTROMAGNETIC WAVE PROPAGATION AND SCATTERING

INTRODUCTION

An electromagnetic wave is a transverse wave and is generated by electric charge excitation. The wave is associated with synchronized electric and magnetic fields that oscillate perpendicularly to each other. The direction of wave propagation is perpendicular to the direction of energy transfer. In free space, its phase velocity has a constant value of speed of light. According to the Maxwell equations, a spatially varying electric field is directly linked to a temporally varying magnetic field. Conversely, the magnetic field change in space is associated with an altered electric field in time. It is important to know that the electric and magnetic fields are vibrating orthogonally from each other in phase but both reaching maxima and minima at the same points in place. The coexisting fields occur simultaneously and are linked by the theory of special relativity. In fact, the magnetic field is considered as a relativistic distortion of the electric field.

The wave speed is the product of wavelength and frequency, $u = f\lambda$. For practical purposes, the electromagnetic spectrum is divided into radio waves, microwaves, infrared, visible light, ultraviolet, x-rays, and gamma rays. The very long wavelength of radio wave is in the tens of meters and the very short wavelength gamma ray is smaller than an atomic nucleus. The solutions to the Maxwell electromagnetic wave equations have often been classified as the plane waves and spherical waves (Jordan, 1960). The former is not but a limiting case of the latter at a very large distance from the source of the wave generator. Like all waves, an electromagnetic wave can be

Computational Electromagnetic-Aerodynamics, First Edition. Joseph J.S. Shang.

constructed by any number of monochromatic wave components with a single frequency, amplitude, and phase. Therefore, it is amiable to be treated by Fourier analysis.

Since the electric and magnetic fields have the properties of superposition, an electromagnetic wave contributes accumulatively to the total field together with other waves in the same space. Furthermore, the waves are vector fields, so they may interact through constructive or destructive interference. Electromagnetic waves also possess properties of refraction, diffraction, and refection, and they cross from one medium to another or are scattered from an interface of media (Harrington, 1961). Furthermore, the driven mechanisms of electromagnetic waves are dependent on the electric and magnetic properties of the medium that have the controlling influence of how the waves dispersed and dissipated particularly, when the medium is electrically conducting or contains charged particles that can alter the electromagnetic field. Therefore, in a semiconductor, the wave can be completely reflected from a medium interface or the wave amplitude of the incident wave can be attenuated in a very short distance within the medium. This phenomenon is known as communication blackout in microwave communication.

Another unique feature of the electromagnetic wave in free space is the conservation of transmitted energy through any spherical boundary surface surrounding its source (Balanis, 2005; Crispin et al., 1968). The total area of the spherical surface is defined by the square of the distance from the origin. Therefore, the power or the energy of electromagnetic radiation always obeys the inverse–square law, which is in the direct contrast to the behavior of the near field of a dipole. The electromagnetic field of a dipole decays rapidly with distance to exhibit clear distinctive structures of the far field form the near field. The wave reflection feature is exemplified by the radar cross section (RCS) of a reflected wave from electric scatterers. The quantity of primary interest of RCS is placed on the distribution of electromagnetic energy from the far field. All these behaviors of propagating electromagnetic waves are studied in this chapter.

5.1 PLANE ELECTROMAGNETIC WAVES

Simple plane waves in dielectric and electric conducting medium will be presented to show the interdependence of electric and magnetic fields and the drastic different propagating patterns between them. These basic formulations, some key pertaining parameters, and unique feature of electromagnetic waves can be easily pointed out.

The linearly polarized transverse electromagnetic or a simple harmonic wave in a lossless dielectric medium is the principal or the zero-order electromagnetic wave traveling through a nonconducting medium in which the conductive electrical current $J = 0$. Hence, the generalized Ampere law gives

$$i\left(\frac{\partial H_z}{\partial y} - \frac{\partial H_y}{\partial z}\right) + j\left(\frac{\partial H_x}{\partial z} - \frac{\partial H_z}{\partial x}\right) + k\left(\frac{\partial H_y}{\partial x} - \frac{\partial H_x}{\partial y}\right) = \frac{\partial}{\partial t}(iD_x + jD_y + kD_z) \tag{5.1}$$

Following the right-hand rule in the positive x direction, the wave equation of a plane wave is

$$-j\frac{\partial H_z}{\partial x} = j\frac{\partial D_y}{\partial t} = j\varepsilon\frac{\partial E_y}{\partial t} \tag{5.2}$$

Faraday law gives

$$i\left(\frac{\partial E_z}{\partial y} - \frac{\partial E_y}{\partial z}\right) + j\left(\frac{\partial E_x}{\partial z} - \frac{\partial E_z}{\partial x}\right) + k\left(\frac{\partial E_y}{\partial x} - \frac{\partial E_x}{\partial y}\right) = \frac{\partial}{\partial t}(iB_x + jB_y + kB_z) \tag{5.3}$$

For a plane wave in the positive x direction, we have

$$-k\frac{\partial E_y}{\partial x} = k\frac{\partial B_z}{\partial t} = k\mu\frac{\partial H_z}{\partial t} \tag{5.4}$$

In a dielectric medium, equations (5.2) and (5.4) can be manipulated by taking the temporal and spatial derivatives, respectively, to get the plane electromagnetic wave equations in separated electric and magnetic components. These equations are also referred to as the D'Alembert equations:

$$\begin{aligned} \frac{\partial^2 E_y}{\partial t^2} &= \frac{1}{\mu\varepsilon}\frac{\partial^2 E_y}{\partial x^2} \\ \frac{\partial^2 E_z}{\partial t^2} &= \frac{1}{\mu\varepsilon}\frac{\partial^2 H_z}{\partial x^2} \end{aligned} \tag{5.5}$$

The general solutions to the equation (5.5) are simply

$$\begin{aligned} E_y &= \sin\beta(x - vt) + \sin\beta(x - vt) \\ H_z &= \sin\beta(x - vt) + \sin\beta(x - vt) \end{aligned} \quad \text{or} \quad \begin{aligned} E_y &= E_0 e^{i(\omega t \pm \beta x)} \\ H_z &= H_0 e^{i(\omega t \pm \beta x)} \end{aligned} \tag{5.6}$$

Since $v = \lambda f$ then $\beta v = (2\pi/\lambda)f\lambda = 2\pi f = \omega$. From the elementary solution, some key parameters for the propagating wave are found. By the constant value of $\omega t - \beta x$, the phase velocity of the wave is

$$v = \omega/\beta = \lambda f = 1/\sqrt{\mu\varepsilon} \tag{5.6a}$$

and the group velocity can be given as

$$u = d\omega/d\beta = \beta(dv/d\beta) = v - \lambda dv/d\lambda \tag{5.6b}$$

From the general solutions of the plane electromagnetic wave, $E_y/H_z = \sqrt{\mu/\varepsilon}$ is the so-called intrinsic impedance of the lossless medium.

The energy transfer by a traveling electromagnetic plane wave can be evaluated by the electric and magnetic energy density $D_e = \varepsilon E^2/2$ and $D_m = \mu H^2/2$, both has the physical unit of Joules per cubic meter. The energy increase within a control volume must be balanced by an inflow of energy. Therefore, the energy per unit area passing per unit time at any location is required to satisfy a divergent condition:

$$\nabla \cdot S = -\partial/\partial t(D_e + D_m) \tag{5.6c}$$

It can be shown from the plane electromagnetic wave equations that

$$S = E \times H \tag{5.6d}$$

is the Poynting vector that gives the rate of energy flow per unit area in a wave and perpendicular to the electric and magnetic fields.

The electromagnetic plane wave described by the generalized Ampere law in electric conducting medium yields

$$\nabla \times H = j + \frac{\partial D}{\partial t} = \sigma E + \frac{\partial D}{\partial t} \tag{5.7}$$

Assume that E_y is a harmonic function in time $E_y = E_o e^{i\omega t}$. We have then

$$-\frac{\partial H_z}{\partial x} = \sigma E_y(\text{conductive}) + i\omega\varepsilon E_y(\text{displacement}) \tag{5.8}$$

By convention, the relative magnitude of the current components determines whether the medium is a dielectric, quasi-conductor, or a conductor. As a reference for the magnitude of electric conductivity, copper has value of 5.8×10^7 and fresh water of 10^{-3} mho/m.

The electric complement of the plane wave equation in electrically conducting medium is

$$\frac{\partial^2 E_y}{\partial x^2} - (i\mu\sigma\omega - \mu\varepsilon\omega^2)E_y = 0 \tag{5.9}$$

Letting $\gamma^2 = i\mu\sigma\omega - \mu\varepsilon\omega^2$, the equation of wave motion in a conducting medium becomes

$$\frac{\partial^2 E_y}{\partial x^2} - \gamma^2 E_y = 0 \tag{5.10}$$

Again, assume that the temporal varying of the electric field can be represented by a simple harmonic motion $E_y = E_o e^{i\gamma x}$. For a conductor, then

$$\gamma = (1 + i)\sqrt{\omega\mu\varepsilon/2} = \alpha + i\beta \tag{5.10a}$$

The real part α is associated with wave attenuation and the imaginary component β describes the phase relation. The attenuation factor is given as $e^{-\sqrt{\omega\mu\sigma/2}}x$, and the depth of wave penetration in a conducting medium is defined by the magnitude of E_y decreasing to the original value of *1/e* = 0.368 times (Krause, 1953). The penetration depth is

$$\delta = \frac{1}{\sqrt{\pi\mu\sigma f}} \tag{5.11}$$

It is clear that an electromagnetic wave will be attenuated when propagating in an electrically conducting medium and a high-frequency wave will be damped out quicker than a low-frequency counterpart.

5.2 MOTION IN WAVEGUIDE

Electromagnetic energy transmission is a very important subject in microwaves for telecommunication and plasma diagnostic applications. The waveguide can transmit energy in high-order modes, and practical waveguides take the form of rectangular or circular cylinders. The electromagnetic field structure within the waveguide is obtainable by solving the Maxwell equation by imposing null values to either the normal E or H component at the surface of the conductor. Assuming that temporal variations of a microwave are given by $e^{i\omega t}$, and the variations in the z direction may be expressed as $e^{-\gamma z}$, where $\gamma = \alpha + i\beta$, the Maxwell equations for E_z and H_z are

$$\frac{\partial^2 E_z}{\partial x^2} + \frac{\partial^2 E_z}{\partial y^2} + \gamma^2 E_z = -\omega^2 \mu\varepsilon E_z$$

$$\frac{\partial^2 H_z}{\partial x^2} + \frac{\partial^2 H_z}{\partial y^2} + \gamma^2 Z_z = -\omega^2 \mu\varepsilon H_z \tag{5.12}$$

Because the electric and magnetic fields are mutually perpendicular in an electromagnetic field, not all electric and magnetic components of a transverse electromagnetic wave can exist simultaneously inside a conducting wave guide. It is usual to designate the possible field structures into the transverse magnetic (TM) waves for which H_z is identical to zero, and the transverse electric (TE) waves for which E_z vanishes.

By the separation of variable technique, two constants in the governing equations appear and tie directly to the width and height of the waveguide (*a,b*); $A = m\pi/a$ and $B = n\pi/b$. The propagation constant is

$$\gamma = \sqrt{\left(\frac{m\pi}{a}\right)^2 + \left(\frac{n\pi}{b}\right)^2 - \omega^2\mu\varepsilon} \tag{5.13}$$

From the ordinary transmission line theory, γ is a complex number $\gamma = \alpha + i\beta$. If γ is real, β must be zero, and there cannot be a phase shift along the guide that means there is no wave motion at low frequency. Therefore, at a frequency below this value, wave propagation will not occur in the wave guide. This gives rise to the definition of the cutoff frequency

$$f_c = \frac{1}{2\pi\sqrt{\mu\varepsilon}}\sqrt{\left(\frac{m\pi}{a}\right)^2 + \left(\frac{n\pi}{b}\right)^2} \tag{5.14}$$

Another peculiarity of a waveguide is that the velocity of wave propagation in a guide can be greater than the phase velocity in free space

$$v = \omega/\beta = w/\sqrt{\omega^2\mu\varepsilon - (m\pi/a)^2 - (n\pi/b)^2} \tag{5.15a}$$

The transverse waves are designated as $TE_{m,n}$ or $TM_{m,n}$ and the lowest cutoff frequency will occur for m and n equal to unity. In fact, the lowest TE wave in a rectangular guide is the $TE_{1,0}$ wave, which has the lowest cutoff frequency and is called the dominant wave. The higher-order waves (larger values of m and n) are obtained at higher frequencies in order to be propagated within a waveguide of a given physical dimension.

The analytic solution for a lossless dielectric medium in a rectangular wave guide is well known and the electromagnetic wave can propagate unimpeded. However, an electromagnetic wave rapidly attenuates in a conducting medium. In a conductor, the microwave attenuation is so rapid that the wave may only penetrate the conductor by a shallow depth. To validate numerical simulations of the waveguide, a theoretical reference is invaluable. The generalized TE wave traveling through the lossy medium along the z coordinate has been derived as

$$\begin{aligned}
H_x &= (H_o/\kappa^2)(n\pi/x_o)e^{-\alpha z}[\alpha\cos(\omega t - \beta z) - \beta\sin(\omega t - \beta z)] \\
&\quad \sin(n\pi x)\cos(m\pi y) \\
H_y &= (H_o/\kappa^2)(m\pi/y_o)e^{-\alpha z}[\alpha\cos(\omega t - \beta z) - \beta\sin(\omega t - \beta z)] \\
&\quad \cos(n\pi x)\sin(m\pi y) \\
H_z &= H_o e^{-\alpha z}\cos(n\pi x)\cos(m\pi y) \\
E_x &= (H_o\omega v/\kappa^2)(n\pi/y_o)e^{-\alpha z}\sin(\omega t - \beta z) \\
&\quad \cos(n\pi x)\sin(m\pi y) \\
E_y &= (H_o\omega v/\kappa^2)(n\pi/x_o)e^{-\alpha z}\sin(\omega t - \beta z) \\
&\quad sin(n\pi x)\cos(m\pi y) \\
E_z &= 0
\end{aligned} \tag{5.15b}$$

A composite pictorial depiction of the field structure of an electromagnetic wave is presented. Both waves in the $TE_{1,0}$ and $TE_{1,1}$ modes were computed on a (25 × 25 × 197) mesh system at a microwave frequency of 4 GHz. The basic topology of the TE waves are similar; thus, only the result of the TE1,0 is depicted in Figure 5.1. The surface magnetic field is projected on upper surface, and electric surface fields of the electric transverse wave are projected on the vertical and lower surfaces of the waveguide, respectively. For clarity only the far-side wall of the

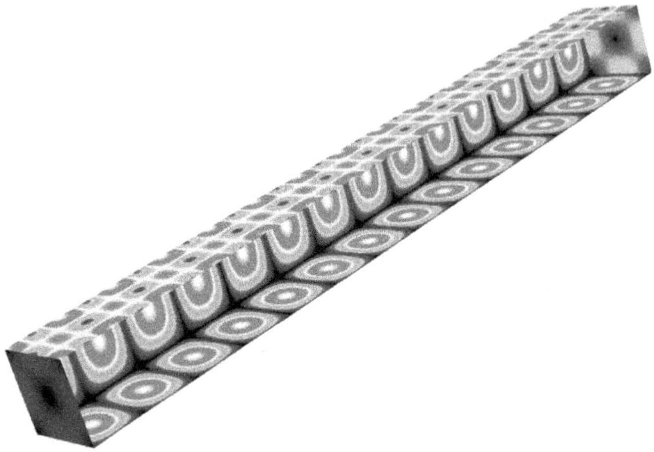

FIGURE 5.1 Electromagnetic field structure on waveguide surface.

waveguide is shown in this graph. From these numerical results, it is clearly revealed that the electromagnetic wave in the dominant wave is essentially a two-dimensional motion that is displayed in the next few figures. It is interesting to point out that the wave cutoff frequency of the $TE_{1,1}$ mode in the rectangular guide is 2.84 GHz (wave number of 1.41 π) in contrast to the lowest cutoff frequency for the TE1,0 wave in the same rectangular guide is 2.42 GHz (wave number of 1.21 π).

The validation of the computed electrical transverse wave of the $TE_{1,0}$ mode is accomplished by comparing with theoretical results; see equation (5.14) in Figure 5.2. The z components of the magnetic field intensity propagating along the waveguide are given versus the value of $\sigma/\omega\varepsilon$ from 0,0 to 0,1. This parameter actually represents the ratio of the conductive to displacement current that the electromagnetic wave encounters in propagation. For the waveguide simulations, all numerical results are generated on a uniform mesh ($25 \times 25 \times 197$) in a square cross-sectional configuration. The aspect ratio between the length and the width is 10, and the electromagnetic wave is propagating at a frequency of 4 GHz. There is no detectable difference between computational and theoretical results in wave profile as it propagates in the dielectric medium. The computational simulations are accurate and show that numerical dispersion and dissipation are negligible. The microwave attenuation in a quasi conductor is significant and is captured by numerical simulations. The rate of attenuation is exponential and this behavior is best demonstrated by a higher value of the electric conductivity. The energy conversion is known to be controlled by the Joule heating mechanism but is not modeled by the governing equations (Swason, 1989; Snoot et al., 1966).

Additional verification of the numerical simulation of the waveguide filled with dielectric and quasi conductor is carried out by comparison of numerical solutions with theoretical results on the x component of the electric displacement (Figure 5.3). The agreements with the theoretical results are excellent to reflect the fact that the numerical procedure, including the boundary condition implementation, is correct.

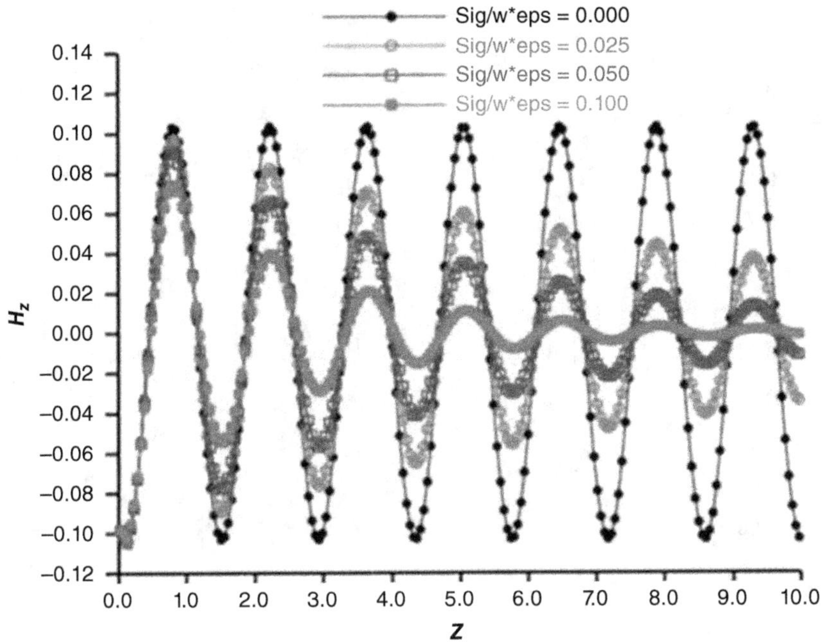

FIGURE 5.2 Attenuation of magnetic flux density Bz in lossy medium, $0.0 < \sigma/\omega\varepsilon < 0.1$.

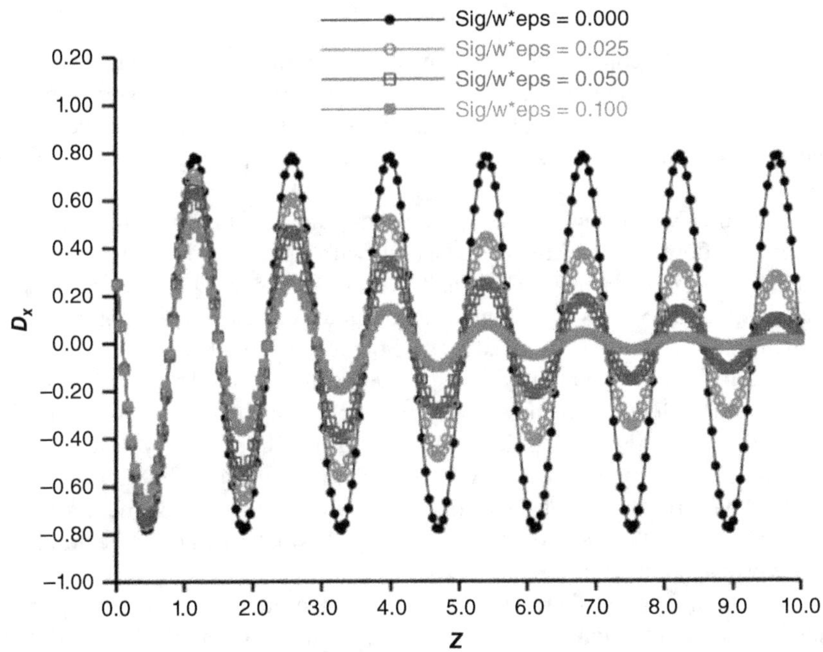

FIGURE 5.3 Attenuation of electric displacement Dx in lossy medium, $0.0 < \sigma/\omega\varepsilon < 0.1$.

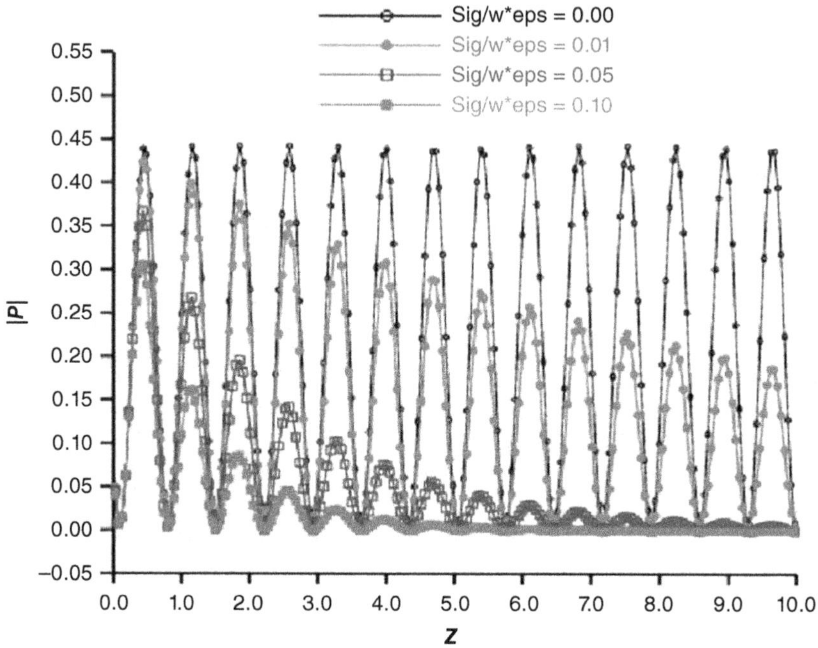

FIGURE 5.4 Poynting vector distributions in lossy medium, $0.0 < \sigma/\omega\varepsilon < 0.1$.

The effect of electric conductivity of the medium within the wave guide on the attenuated electric displacement is analogous to that of the magnetic intensity. Microwave attenuation through the electrically conducting medium is clearly indicated in both the electric and magnetic fields as electromagnetic waves propagating in rectangular waveguide.

The Poynting vector $S = E \times H$ in an electromagnetic field is a measure of the rate of energy flow per unit area at that point, with the direction of the energy flow perpendicular to both E and H. The magnitude of the Poynting vector is depicted in Figure 5.4. Only the values at the centerline of the wave guide are computed over the range of the parameter $\sigma/\omega\varepsilon$ from 0.0 to 0.1. For the dielectric medium, the magnitude of the Poynting vector maintains a constant value as it shall. The magnitude of the rate of the energy flow diminishes rapidly as the electric conductivity of the medium is increased. At the highest electric conductivity studied, the energy flow essentially ceased after the electromagnetic wave has propagated a distance of 8.2 wavelengths. These displayed results of the Poynting vector of an electric conducting medium describe pictorially the electromagnetic energy dissipation rate of the microwave along the propagation path.

It can be summarized that electromagnetic wave attenuation in an electrically conducting medium or a plasma is governed by parameters of the electric conductivity and permittivity of the medium and the incident wave frequency $\sigma/\omega\varepsilon$. In a weakly ionized gas or partially ionized plasma, the numerical simulations are validated by the well-known theory. However, the basic phenomenon of amplitude attenuation in microscopic scale is governed by the electron number density, the momentum transfer collision frequency, and the electron mobility of plasma.

5.3 WAVE PASSES THROUGH PLASMA SHEET

The spectra of electron collision, plasma, and incident microwave frequencies for microwave diagnostics are generally $\nu \ll \omega_p \ll \omega$. The wave diagnostic techniques are developed by changes of phase angle and amplitude of a wave when propagating in different media. The measured phase shift depends only upon the electron density, but the attenuation of wave amplitude depends on both the electron density and the collision frequency for momentum transfer. For a transverse plane electromagnetic wave in the $TE_{1,0}$ mode that travels in the positive z direction, the motion can be described by

$$E = E_o e^{[-\alpha z + i(\beta z - \omega t)]} \tag{5.16a}$$

where α and β are the attenuation and phase constants. The phase constant is given by the relationship $\beta = 2\pi/\lambda$, and the phase velocity is defined as $v = \omega/\beta$. The constants of the transverse wave can be calculated by the following formulas:

$$\begin{aligned} \alpha &= (n\pi/a)^2 \\ \beta &= \sqrt{\beta_o^2 + \sqrt{\beta_o^4 - (\sigma/c\varepsilon)^2} \Big/ 2} \end{aligned} \tag{5.16b}$$

where $\beta_o = (\omega/c)^2 - (n\pi/a)^2$. When the incident microwave has a shorter wavelength than the plasma, the plasma behaves as a relatively low-loss dielectric medium. Then the wave amplitude attenuates and the phase shifts linear proportional to the distance propagated. However, in practical applications, the plasma domain always has a finite, irregular, and fluctuating dimension. There are serious limitations of applying a one-dimensional, plane-wave model for plasma diagnostics. Added to these complications, the wave diffraction and reflection at the media interface from a divergent wave are problem dependent. In general, these effects are inaccessible to make the microwave diagnostic very uncertain.

The computational simulation has been developed only for a traveling microwave in homogeneous, isotopic plasma. The beam path for microwave diagnostics usually passes across multiple media. The interferences at media interface on the transmitting microwave are dominated, especially when the difference of intrinsic impedance $\sqrt{\mu/\varepsilon}$ is large and the plasma is only a few wavelengths thick. This type of plasma formation is frequently encountered in most plasma channel in which the interference effect can easily obscure the plasma absorption. In addition, if the microwave wavelength is comparable with the dimension of a plasma region, diffraction and surface waves around the discharge are also presented. These effects can be more pronounced than the absorption.

Therefore, the accuracy of computational simulation needs to assess the simulation capability for analyzing a finite plasma domain of a small dimension like a sheet. For plasma sheet of a finite thickness and with a sharp boundary, the reflection of the microwave at the boundary always takes place because of the mismatched intrinsic impedance of two media. Refraction occurs at both the incident and exiting interfaces and affects measurements. This phenomenon is clearly demonstrated by numerical simulation of a guided $TE_{1,0}$ microwave with a frequency of 4 GHz. All the following numerical results are obtained on a (25 × 25 × 197) grid system. The entire computational domain consists of 10 wavelengths and the square waveguide has sidewall dimensions of one wavelength each. The plasma sheet strides the guided wave at the middle point of the waveguide and has a thickness of two wavelengths. The plasma sheet is characterized by a uniform and greater electrical conductivity, $\sigma = 0.75\ \varpi\varepsilon$.

The microwave attenuation, when it propagates through a thin plasma sheet, is a frequently encountered environment for plasma diagnostics. Thus, the overall behavior of the guided microwave, as it passes through a plasma sheet with a thickness of two wavelengths, is given by Figure 5.5. For the purpose of a simple illustration, different components of the guided wave are projected onto the upper (electric displacement D) and lower (magnetic flux density B) surface. The electric displacement is also included on the far-side wall to show the planer structure of the $TE_{1,0}$ wave. The overall wave field structure substantiates the observation that the reflections from the media interfaces create a very complex wave pattern within the plasma sheet, but the attenuated microwave still retains most of its incident wave characteristics. Since the microwave measurement for plasma diagnostics is an integrated result from the beam path, it reflects only an accumulated effect. The measurement is therefore sampling condition dependent; thus, there is no general conclusion that can be drawn

FIGURE 5.5 Electromagnetic field projection containing a plasma sheet on waveguide surfaces.

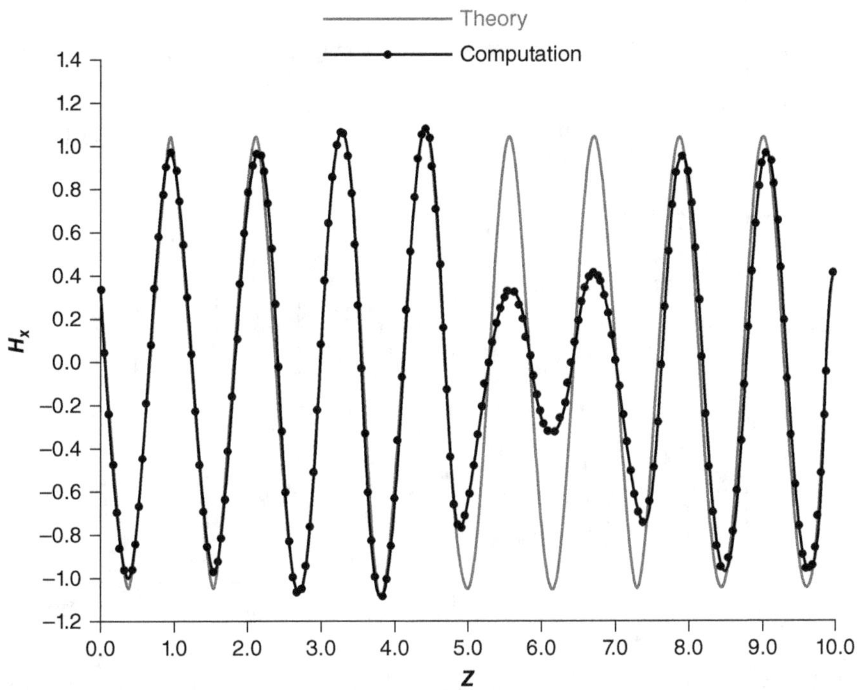

FIGURE 5.6 Magnetic field intensities H_x of $TE_{1,0}$ wave in vacuum and passed a plasma sheet.

from the microwave pattern within the waveguide and to compensate the effects of reflection.

The $TE_{1,0}$ wave has only two magnetic field components H_x, H_z, and a sole electric displacement component E_y. The x component of the magnetic field strength H_x, along the microwave path, is presented in Figure 5.6. For purpose of comparison, the same variable in free space is pended in the figure as a reference. Both waves travel from the coordinate origin and sample at the identical time interval. From the numerical results, the magnetic field is substantially distorted by the reflections from the media interfaces. This behavior is drastically different from the numerical results, describing the microwave propagation that is completely confined in the plasma. Under that circumstance, the reflected wave does not exist and the wave attenuation is continuous and persistent through the entire computational domain.

The reflected wave from the incident interface produces a wave cancellation toward its origin. In the plasma domain, the microwave magnetic component exhibits a significant distortion and remerges at the exit interface to show attenuation in wave amplitude. This interference has been known to obscure the plasma absorption measurements and is demonstrated by the present numerical simulation. The microwave propagated in a single medium and used as a reference beam for comparison actually indicates a phase shift and amplitude damping of the microwave that travels through the plasma with only a distance of two wavelengths.

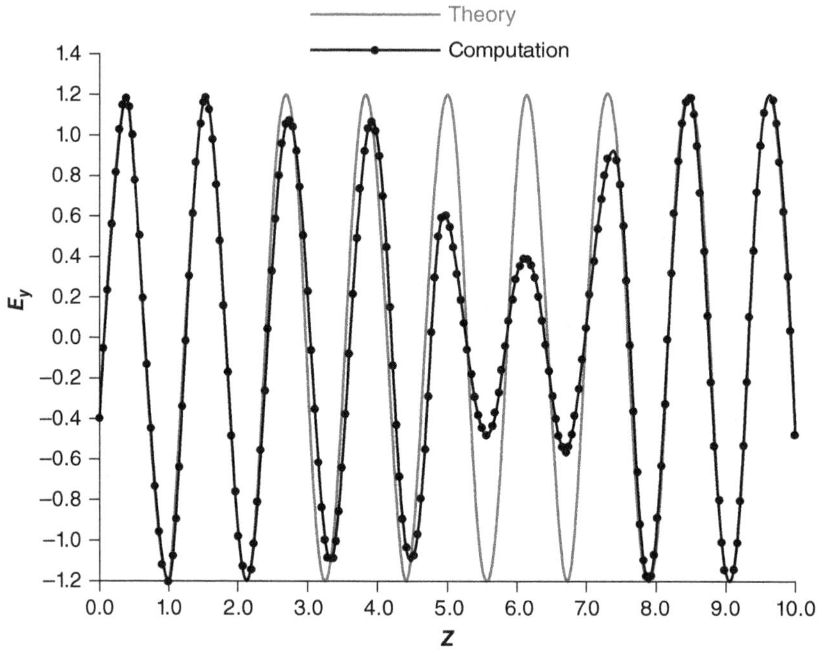

FIGURE 5.7 Electric field intensities E_y of $TE_{1,0}$ wave in vacuum and passed a plasma sheet.

The y component of the electrical field intensity E_y of the $TE_{1,0}$ wave is depicted in Figure 5.7. Again the reflected wave from the media interfaces, $z = 0.5$, substantially distorts the electrical field of the microwave in direct contrast to the microwave propagating in the plasma without interface boundaries. For the electrical field component, the reflected wave from the incident media interface exhibits a much more pronounced phase shift and amplitude reduction in comparing with a homogeneous medium. These distortions also propagate with and partially reflect from the media interface with respect to the incident microwave.

The Poynting vector is the basic measurement of the microwave attenuation. For the $TE_{1,0}$ wave mode, the Poynting vector is given as

$$S = -(E_z H_y)i - (E_x H_z)j + (E_x H_y - E_y H_x)k \tag{5.17}$$

The magnitudes of the Poynting vector of $TE_{1,0}$ electromagnetic waves at an instant are presented in Figure 5.8. This quantity in a constant cross-sectional square waveguide can be viewed as the dynamic range of the electromagnetic energy flux. As before, this figure presents only the results of two values of the parameters $\sigma/\omega\varepsilon$ of 0.0 and 0.25. The former represents theoretical result for the dielectric medium ($\sigma = 0$), and the magnitude of the Poynting vector maintains a constant value, as

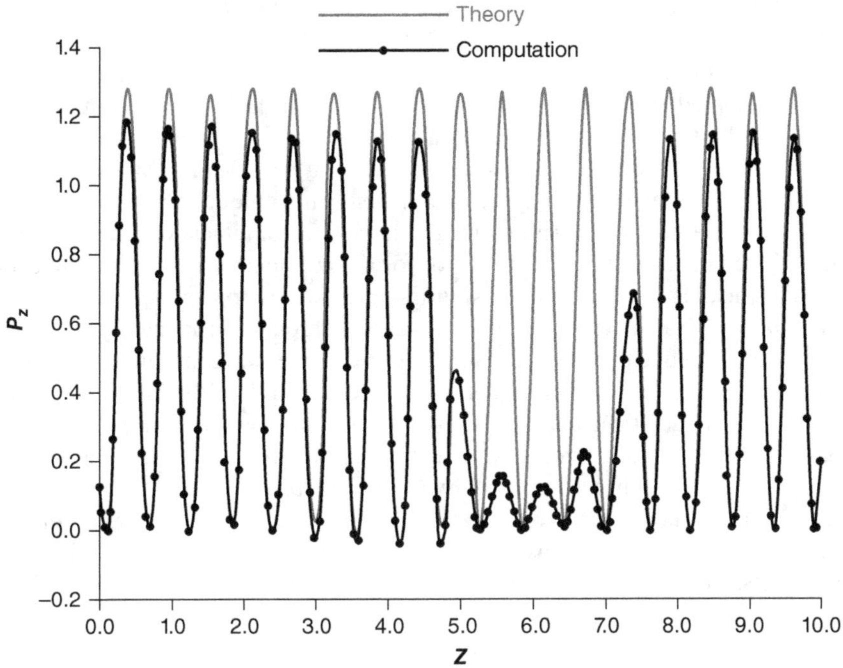

FIGURE 5.8 Poynting vector distribution in waveguide with/without thin plasma sheet.

it should. The magnitude of the energy flow diminishes by dissipation when the electric conductivity of the medium has a finite value. In the thin plasma sheet, the Poynting vector exhibits drastic adjustments to the refractions from the media interfaces and yields uniform magnitude attenuation toward the exiting computational domain. The total energy conveyed by the microwave is the integrated result over the entire waveguide and over at least one period of wave motion, only a sampling along the guided wave is depicted to highlight the discussion.

In summary, the dynamic range of the spatially averaged Poynting vector reduces sharply along the electromagnetic wave path when the probing wave passes through thin plasma sheets. The reduced magnitude is proportional to the thickness of plasma sheets. At a probing microwave frequency of 12.5 GHz and a plasma sheet characterized by a parameter of $\sigma/\varpi\varepsilon = 0.25$, the range of the Poynting vector reduces by a value of 65% with a 55% thinner plasma sheet. The reflected waves from the front and the back interfaces of a plasma sheet induced by the mismatched intrinsic impedances generate a drastic distortion of the electromagnetic wave pattern within the waveguide. All numerical simulations are conducted on a simple parameter $\sigma/\varpi\varepsilon$, either with a value 0.25 or 0.75. The attenuation of a passing microwave at a lower microwave frequency is still strongly dependent on the electrical conductivity and less on a nearly constant value of electric permittivity of plasma.

5.4 PYRAMIDAL HORN ANTENNA

A nonintrusive plasma diagnostic tool is microwave probing. The microwave system is equally adopted for both plasma diagnostics and in deep-space communication (Fedrick et al., 1995). For plasma diagnostics, the number density of the charge particles and its collision frequency with the neutral particles are measured on the basis of the microwave attenuation phenomenon. This unique microwave behavior in weakly ionized air is also known for the famous communication blackout in the reentry phase either for an earth orbital or for an interplanet flight. Communication blackout is the consequence of a transmitting microwave propagating at a frequency lower than the plasma frequency when it passes through an ionized gas domain. When the two frequencies are equal, the propagating wave becomes evanescent and the transmission of electromagnetic energy also ceases. When the transmission bandwidth is greater than that of the plasma, the microwave will attenuate as it propagates through the ionized gas. The dissipated energy along the wave path is proportional to the electrical conductivity of the medium.

An electromagnetic wave propagating in an electrically conducting medium depends strongly on its electrical conductivity and the transmission frequency (Golant, 1983; Lochte-Holtgreven, 1968). For a linearly polarized plane wave traveling in a conducting medium, the current density consists of the conductive and displacement components. For a dielectric medium, the electric conductivity σ and the conductive current vanish, and the wave may travel without any impedance (Meuth et al., 1989). Otherwise, the relative magnitude of σ and the product of wave frequency and electrical permittivity $\omega\epsilon$ will regulate the behavior of the propagating wave. In the absence of an external magnetic field, the partially ionized gas can be studied as an isotropic medium. The plasma behaves as a simple quasi conductor and will support high-frequency wave motion through the response of electrons.

This salient feature of microwave attenuation in plasma has been used to measure the plasma transport property of a rocket exhaust plume and the weakly ionized air stream in a hypersonic wind tunnel (Shang et al., 2000). For plasma diagnostics, added complications arise when a microwave impinges on a boundary of two media. At the interface, a part of the incident wave is reflected and another part is transmitted into the second medium. The portion of the transmitted wave is governed by the continuity of the tangential components of the electric and magnetic fields. The intrinsic impedance of the media ultimately controls the behavior of this electromagnetic field at the interface. For a plane wave traveling in air impacts normally to a perfect conductor, the incident magnetic field doubles its intensity at the media interface. On the exit interface, the electric field can also double its strength at the boundary. In both cases, a standing wave originates at the boundary. In all circumstances, there will be a reflected wave from the interface, except when the two media have identical impedances, and then the incident wave will transmit uninterrupted.

In most microwaves diagnostics, the wave produced by the generator is directed by a waveguide and transmitted from an antenna; the basic arrangement is depicted in Figure 5.9. The widely used antennae usually are a pair of pyramidal horns whose main function is to collimate the wave and reduce diffraction. The behavior of these

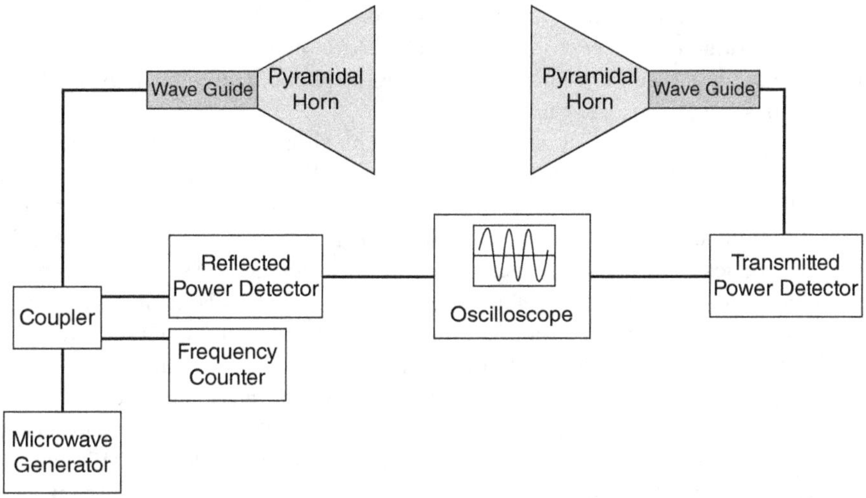

FIGURE 5.9 Typical plasma diagnostic setup using microwave antenna.

components in plasma diagnostics is also the least understood. The sources of data disparity are many; they include the explicit assumptions of the outer boundary of the plasma and diffraction/refraction of the microwave. For this reason, some fundamental and data accuracy uncertainty for plasma diagnostics remains. The evidence is clearly demonstrated by the disparity of independent measurements using Langmuir probe and microwave absorption techniques. The measurement discrepancy on charged number density of a direct current glow discharge in a hypersonic stream is greater than a factor of 2 and in practical applications the difference can be as high as one order of magnitude. Even for a simple glow discharge simulation, these simplifying assumptions of a weakly ionized gas are seriously challenged. It is therefore natural to explore a numerical simulation technique to obtain a better understanding for the basics of applying microwave diagnostics.

For the application range of microwave diagnostics, the plasma encountered is mostly in the so-called ideal classical domain. According to the classification by charged particle number density and electron temperature, the thermal kinetic energy in this domain is large in comparison with the Coulomb interaction energy, $k_b T \gg e^2/n^{-1/3}$. This condition is equivalent to $n\lambda_D^3 \gg 1$, where λ_D is the Debye length; under this condition, the weak electromagnetic–aerodynamics interaction model prevails.

The governing equations for microwave diagnostics are built around the solution to the three-dimensional Maxwell equations in the time domain; see equation (4.4). The closure of the partial differential equations requires additional constitutive relationships to describe the electrical current density J, which is defined by the average electron collision frequency and the charged number density; see equation (1.31). The charged species number density in theory is obtainable from chemical kinetics. The kinetic foundation of a many-component model for the partially ionized gas is the Boltzmann equation. In a dilute gas, the collision terms in the Boltzmann equation are short-range, so the contribution of the intermolecular collisions is practically instantaneous and localized at a point in space. However, there is no unique formulation to bridge the microscopic and macroscopic quantity. Some of the required knowledge in these disciplines is known only at the phenomenological level, and the fidelity to physics is thus uncertain.

To alleviate these presently irresolvable complexities, nonequilibrium chemical kinetics is required but is not included in the Maxwell equations; see equation (5.7) (Shang, 2006a, 2006b; Katz et al., 1991). In fact, the description of the transport property of low-temperature plasma is possible by drift-diffusion theory. The complex chemical kinetics for electron-impact ionizations are modeled by simplified approximation, the classic Thompson similar law. This approximation is justified by the fact that the charged particles, electrons and ions, are presented only in a trace amount of mass fraction – O (10^{-6}). The partial ionized gas is thereby treated as an isotropic and homogeneous lossy medium. An alternative and appropriate approximation that relates the electric current density and electric field intensity must be adopted to solve the Maxwell equations. Following the traditional approach by MHD, the rate of change of the electric current density is derived from the generalized Ohm's law (Mitchner et al., 1973).

$$\left(\frac{1}{\nu_{\text{eh}}}\right)\frac{\partial J}{\partial t} - \sigma E + J = 0 \tag{5.18}$$

where ν_{eh} is the average collision frequency between electrons and heavy particles and $\sigma = n_e e^2/m_e \nu$ is the electrical conductivity. The governing equations for simulating plasma diagnostic using microwave absorption are built around computational solutions to the three-dimensional Maxwell equations in the time domain.

The following discussion is focused on the development of a numerical simulation capability for microwave diagnostics using a pyramidal horn. From this physics-based simulation, the coordinate origin is moved to the axis of symmetry to take advantage of the particular feature of the configuration. The simulated pyramidal horn has a rectangular cross section, and the smallest physical dimension is located at the juncture with waveguide with a height, $b = 9.5$ mm. The minimum width is 2.3 times of its height. At the exit plane of the horn or antenna aperture, the height and width increase linearly by a factor of 7.6 and 9.2, respectively. At the operating frequency of 14.5 GHz, the total length of the antenna is 3.31 wavelengths of the $TE_{1,0}$ wave, and the aperture has the dimensions in wavelength of (3.045×3.640). A mesh system of $(75 \times 25 \times 71)$ is implemented for the numerical simulation. The entire three-dimensional computational domain is generated by a surface-oriented coordinate transformation on which the outward normal over antenna surface can be easily defined. While the normalized physical dimensions measured by wavelength yield $(15.67 \times 6.51 \times 18.80)$. A validation of the computed electromagnetic components and the theoretical results within the pyramidal horn has been performed in an earlier effort. Over the entire range of the parameter $0.0 < \sigma/\omega\varepsilon < 0.25$, the agreement between the computational and theoretical results is reasonable while the numerical results slightly underpredict the theory by 1.22%.

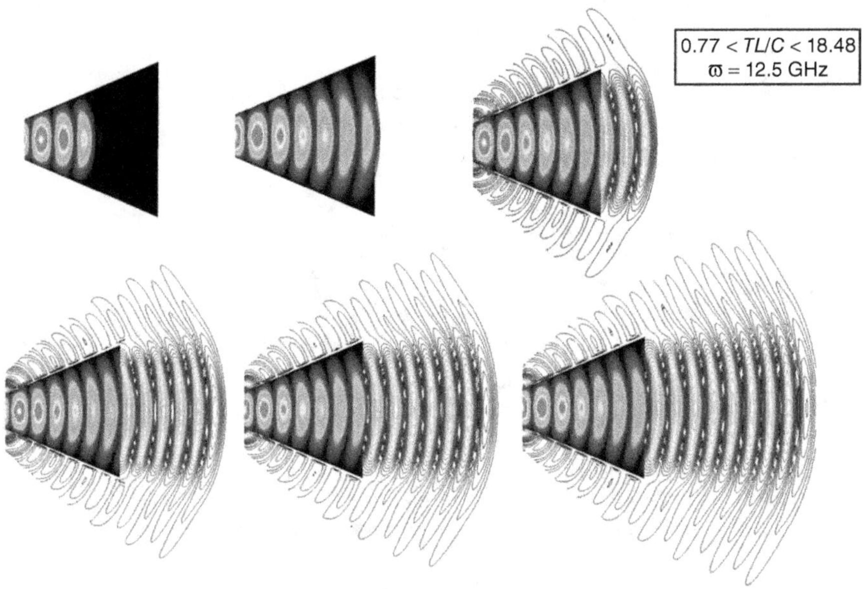

FIGURE 5.10 Sequential electromagnetic wave propagation patterns from the pyramidal horn exiting to free space.

The microwave propagation is strictly a time-dependent phenomenon that may have a periodic asymptote. The temporal sequence of the microwave, as it propagates from the waveguide to the far field boundary, is presented by its electrical displacement component, D_y by a sequential contours insert in Figure 5.10. The sequence shows the microwave propagation exits from the wave guide and begins to emerge from the pyramidal horn into a quiescent free space. The $TE_{1,0}$ wave transmitting from the waveguide exhibits the development of a spherical wave front within the pyramidal horn and the wavelength deceases as the wave approaches the antenna exit. As the wave completely exits into the free space, the edge diffraction at the pyramidal horn is clearly indicated. Complex wave refraction also takes place just outside the horn. The numerical results also show that the reflected wave is not in phase with the emitting microwave and has a wavelength different from that within the waveguide. The final graphic depicts the microwave as it reaches the halfway point and the outer edge of the computational domain. Three features stand out: first, the edge diffraction is persistently presented and the wavelength remains to be a constant and shorter value than that within the antenna. Second, at the far field boundary, the microwave front becomes spherical. Finally, the simulated electrical field correctly captures the discontinuous behavior of the electrical displacement D_y, on the outer surface of an electrically perfect conducting pyramidal horn.

The instantaneous electric displacements D_y along a z coordinate are presented in Figure 5.11. The computational result of microwave radiates into the free space ($\sigma = 0$) by propagating through uniform and sharp-edge plasma sheets of different thicknesses

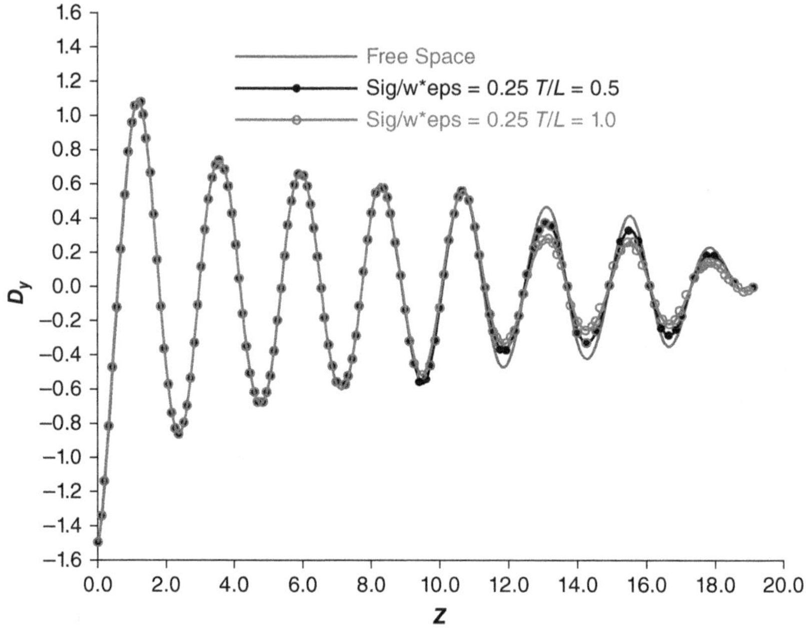

FIGURE 5.11 Electric displacements distributions, D_z with different plasma sheet thicknesses, $\sigma/\varpi\varepsilon = 0.25$.

presented together. The plasma sheets are assigned to have an identical transport property by a value of $\sigma/\varpi\varepsilon = 0.25$, but passing through plasma sheets of different thicknesses. In the radiating microwave field where the waves expand continuously, the amplitude of the electromagnetic field intensities decreases accordingly. With all simulated waves exiting the antenna at the location $z/a = 7.704$, on this plotting scale, the edge diffraction of the antenna is not discernible. The reflected wave from the front edge of the plasma sheet is detected by decreased wave amplitude through wave cancellation. The effect of reflection to the incident wave appears to be minuscule, because there is no evidence of wave phase shift. For both plasma sheets simulated, the front edge of the plasma sheet is located at the distance of $z/a = 10.955$ from the pyramidal horn entrance. The attenuation of the electric displacement of the microwave is greater for the wave penetrating through the thicker plasma sheet.

The instantaneous Poynting vectors along a z coordinate of the three simulations are depicted in Figure 5.12. In the continuously expanding antenna cross-sectional area, the Poynting vector no longer represents the dynamic range of the transmitting energy as in a constant cross-sectional waveguide, because the energy flux density decreases with the increasing cross section of beam path. Nevertheless, the different attenuations of the microwaves along different beam paths are demonstrated. The microwave that passes through the thicker plasma sheet suffers the greater loss of electromagnetic energy than the thinner sheet. In all plasma diagnostics using microwaves, even if the transport property remains unaltered, the fluctuating plasma

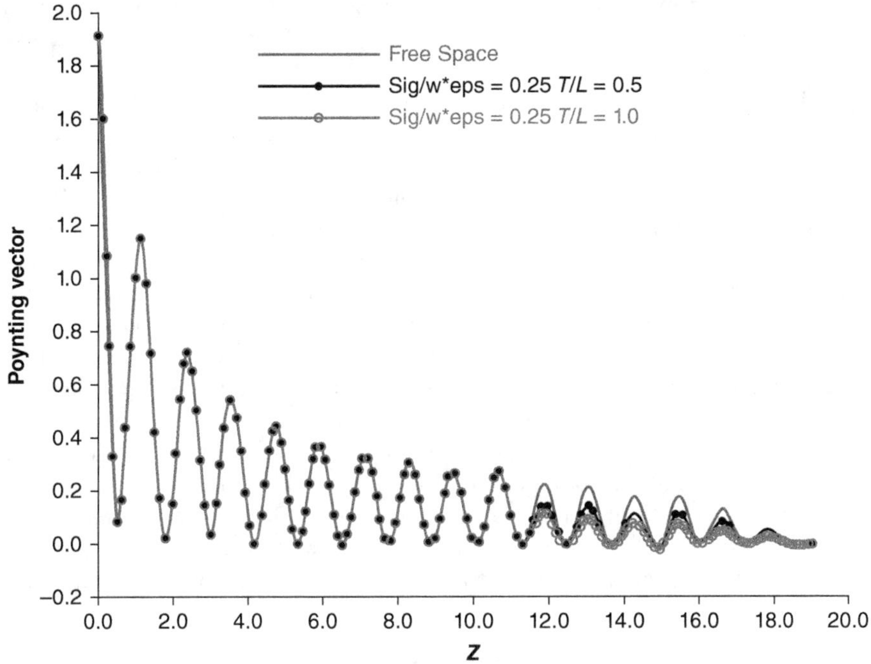

FIGURE 5.12 Poynting vector distributions, D_z with different plasma sheet thicknesses, $\sigma/\varpi\varepsilon = 0.25$.

still can change its dimension and geometric boundary constantly. Under this circumstance, the measurement accuracy can be seriously challenged.

The principle for determining the electron number density of plasma by a microwave diagnostic tool measured the amount of microwave energy absorbed by the plasma relative to the amount of energy carried by the incident microwave. Another technique for electron number density determination is used by analyzing the phase shift or the scattering data of the incident microwave as it propagating through plasma domain. However, the most widely diagnostic technique is based on the attenuation of the probing microwave traveling through plasma. According to the Poynting theorem, the numerical evaluation for the microwave attenuation is straightforward as the following. The power flowing out of a control volume is equal to the time rate of decrease of the energy stored within the volume minus the conduction losses. The instantaneous power transmitted by the microwave through a plasma at any point along the beam path can be calculated by the surface integral

$$\iint (E \times H)dA = \frac{1}{2}\frac{\partial}{\partial t}\iiint (\varepsilon E^2 + \mu H^2)dV + \iiint \sigma E^2 dV \tag{5.19}$$

where dA is the surface vector of a control volume associated with an elementary volume dV. This instantaneous surface integral of the Poynting vector bounded by surfaces of the pyramidal horns and the cross sections along the beam path will be performed to quantify the microwave attenuation when propagating through a lossy medium.

The key element of plasma diagnostics using microwave probing is a pair of pyramidal horns. Their main function is to collimate the microwave beam and to transmit and receive a microwave across a plasma medium. At the frequency of 12.5 GHz ($\lambda = 2.398$ cm), the microwave is required to travel only 10.71 wavelengths to cover the entire distance from the entrance of the transmitting horn to the exit of the receiving pyramidal horn. The simulated radiation field reveals a very complex wave interference pattern, including the diffractions at the edges of the transmitting and receiving horns, and the multiple reflecting waves originating from the receiving horn. The dissipated microwave amplitude when propagates through plasma sheets of different thicknesses is made visible by computational simulations and cannot be obtained in any other means. Equally interesting, the phase shift between microwave propagations in the transmitting antenna and those across the horn surface in free space along the side of the horn are also clearly observable in Figure 5.13.

The interacting field of a $TE_{1,0}$ microwave at 12.5 GHz and a plasma sheet of uniform electrical conductivity of $\sigma/\varpi\varepsilon = 0.25$ between the antenna can be best visualized by the contours of the electrical field intensity. The plasma sheet of four

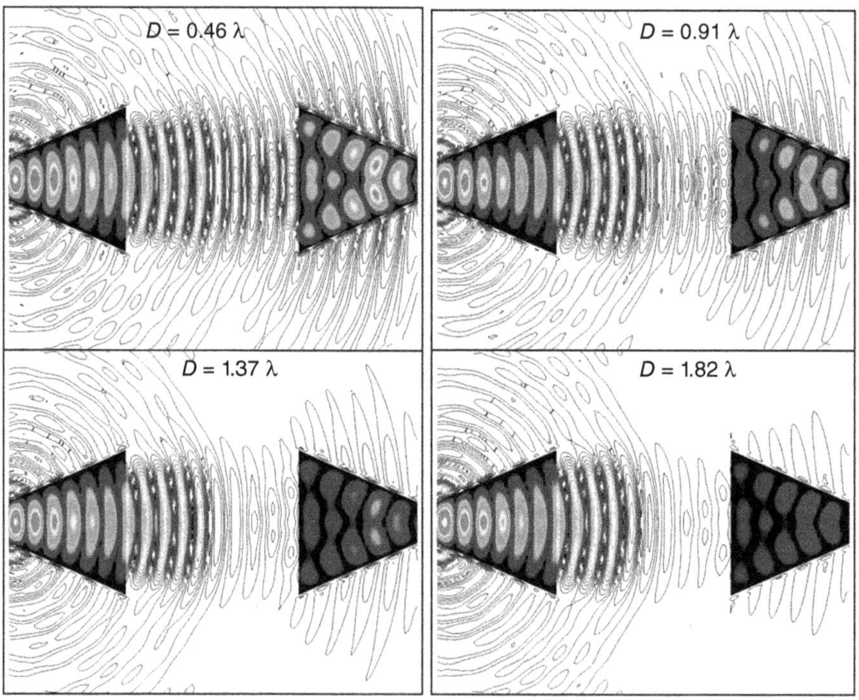

FIGURE 5.13 Electromagnetic waves between pyramidal horns passing through plasma sheet of different thickness, $0.46\lambda < D < 1.82\lambda$.

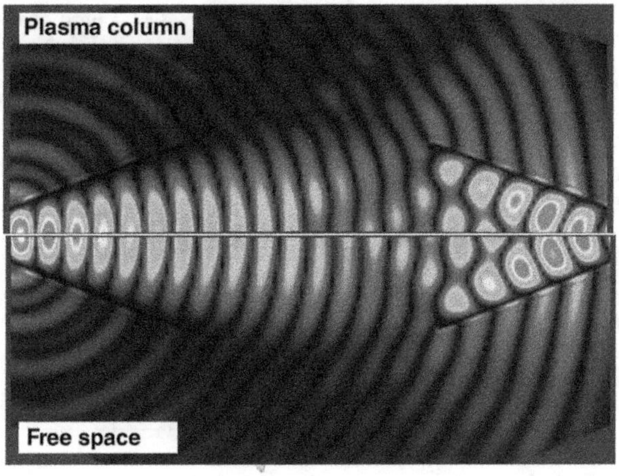

FIGURE 5.14 Contrast of electromagnetic field patterns with/without the presence of plasma medium.

different thicknesses of 0.46, 0.91, 1.37, and 1.82 wavelengths are simulated. The numerical result correctly displayed that the wave amplitude attenuation increases as the wave propagates through a longer distance in a lossy medium. The results of thinner plasma sheets reveal a small difference from the wave interaction pattern in the free space; thus, only the computed result of the plasma sheet with a thickness of 1.823 wavelengths is further discussed in detail. The instantaneous contour plot is recorded at $t = 28.52$ and has 40 equal increments from the maximum magnitude of the electrical field intensity. The leading and trailing edge of the plasma sheet is located at $z/L = 11.09$ and $z/L = 15.68$, respectively. Substantial microwave attenuation is clearly visible in the plasma sheet and the reflected wave from the walls of the receiving pyramidal horn is barely identifiable.

The computational simulation of the thickest plasma sheet is depicted in Figure 5.14 in contours of electric field intensity. The diffraction and reflection of a microwave encountering a plasma column of a finite physical dimension are simulated by inserting a plasma column on the centerline of the beam path. The $TE_{1,0}$ microwave propagates at a frequency of 12.5 GHz ($\lambda = 2.398$ cm). The rectangular plasma column has the dimension of 1 wavelength in width and 1.82 wavelengths in depth, and the uniform plasma is characterized by an electrical conductivity of 69.5 mho/m. The radiating field in free space is depicted in the lower half plane and the interacting field of the microwave and the uniform plasma column is inserted in the upper half plane. The contours of the radiating fields are generated to highlight the structural differences between the microwave propagation in the presence of a plasma column and without. The diffracted and reflected wave patterns are clearly discernible by the contrast of the continuous contours of electrical field intensity with and without the presence of the plasma column. The diffraction and reflection by the plasma column are detectable by the distorted wave fringe pattern and a perceptible phase shift in comparison.

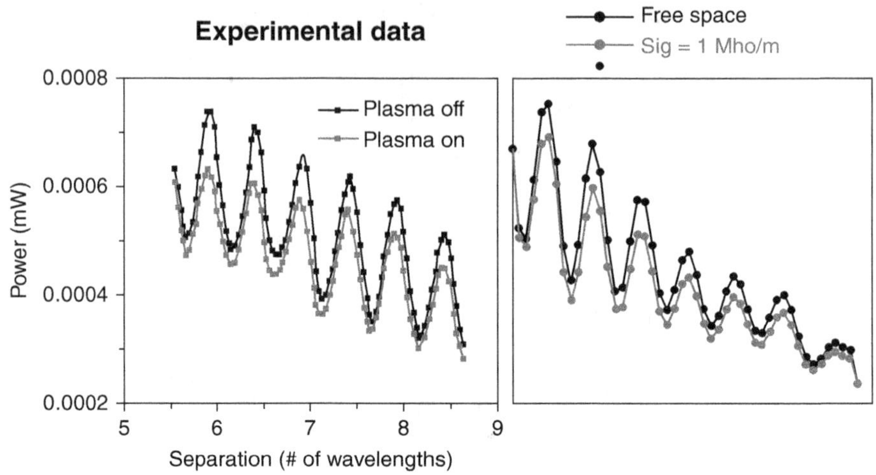

FIGURE 5.15 Comparison of computational results with experiments in measured transmittal microwave power.

The comparison with experimental data is presented in Figure 5.15. The measured transmittal power at a frequency of 12.5 GHz is presented as a function of the separation distance between pyramidal horns with and without the presence of a plasma sheet. First, the dissipated power pattern of the incident microwave is similar, in that when the microwave propagates in free space as black connected lines, the diminished control volume and the reflected wave from the surface of the antenna lead to steadily decreasing power transmission. In the receiving pyramidal horn, the transmitted microwave power is reduced by an order of magnitude by the wave attenuation when it propagates through plasma sheets of finite thicknesses. The difference between the free-space reference and the measured data then determines the number density of electrons for the plasma diagnostics.

A definitive quantification of the numerical simulation by comparing with experimental measurements is short. The diffraction and reflection of the probing beam and plasma are unaccounted for in most experimental investigations. These issues of accurate assessment of experimental measurements have been persistently pursued but unresolved (Heald et al., 1965). A major concern is that all reflected microwaves could be a source of error to determine the transmitted power, particularly in a nonuniform weakly ionized air. In addition, fluctuating plasma can change its geometric configuration constantly. In this environment, the measurement accuracy can be seriously in error. For this reason, numerical simulations are carried out for the purposes of better understanding the pertinent physics and wave motion visualization for insight.

5.5 WAVE REFLECTION AND SCATTERING

Electromagnetic wave reflection at a media interface is complex and depends exclusively on the intrinsic impedance of the medium. The electric and magnetic fields of a plane wave are related to the impedance of the medium, *E/H*. An incident electromagnetic wave at the boundary between two media consists of two components: A part of the incident wave will be reflected while another part is transmitted into the second medium. If incident wave is propagating from air to a conductor, for which $(E/H)_a \gg (E/H)_c$, then the magnetic field intensity is nearly double at the metallic side of interface and the incident magnetic field intensity is nearly equal to the reflected intensity and leads to a standing wave on the air side boundary. At the opposite condition of impedance difference, when an electromagnetic wave exits a conductor into air, the electric field intensity is nearly doubled at the conductor side of boundary but the standing wave will attenuate quickly in a short distance from the boundary.

The electromagnetic wave traveling in air is completely reflected when it encounters a perfect conductor, the magnitude of the reflected electric intensity is exactly equal to that the of the incident wave but in opposite orientation, while the magnetic field intensity is precisely doubled on the media interface. If the impedances of the media are equal, the electromagnetic wave will pass through the interface without any reflection. It has been mentioned before that this is the premise of the perfect-match-layer approach for developing a no-reflection far field boundary condition. Therefore, a wave-terminating interface can also be realized for an incident electromagnetic wave that will not be transmitted or reflected. This concept of a space sheet is treated a surface uniformly by a resistance of 376.7 Ω/m^2, and on this surface no reflected wave exists (Krause, 1953). In all, absorption or termination of an electromagnetic wave can be achieved only by sheets of suitable resistance and spacing extending to an appreciable distance measured in wavelength.

Electromagnetic wave reflection or scattering phenomenon from the far field is the underlining principle for RCS in detecting and ranging of an electromagnetic wave scatterer. The basic technique was established by physical and geometric optics. A perfect conductor is analogous to an optical reflector. The interface boundary condition of a specular reflection requires the incident and reflected wave to be bisected by the surface outward normal. From the optical analogy, this location is referred to as the specular point and the RCS arises mostly from the specular backscatter. The monostatic RCS is obtained from the principal scattering field that when the observation angle coincides with the surface normal. On the contrary, if the radar transmitter and receiver are placed at different locations, the RCS is called bistatic. Some of the complexity in electromagnetic wave propagation and scattering is induced by the general relationship between field and boundary conditions that are not independent.

The quantity of primary interest is the distribution of electromagnetic energy at a large distance from the scatterer (Stratton et al., 1939). For a three-dimensional scatterer in an isotropic homogeneous medium, the RCS is defined as

$$\sigma(\theta, \varphi) = \lim_{r\to\infty} 4\pi r^2 [H_s/H_i]^2 \tag{5.20}$$

or

$$\sigma(\theta, \varphi) = \lim_{r\to\infty} 4\pi r^2 [E_s/E_i]^2 \tag{5.21}$$

where the subscript i and s denote the quantities of incident and the scattered fields, respectively.

The scattered far field can be defined as $E_s = \lim_{r\to\infty}(E - E_i)$ and $H_s = \lim_{r\to\infty}(H - H_i)$. The scattered electric field and magnetic field strengths in general take the form of

$$\begin{aligned} E_s(r) &\sim E_s(\kappa, r)e^{ik\bar{r}}/r \\ H_s(r) &\sim H_s(\kappa, r)e^{ik\bar{r}}/r \end{aligned} \tag{5.22}$$

The cross-sectional σ is a function of the direction to the observer relative to the propagation direction of the incident field that is assumed to be a plane wave. The above equations also imply that the electric and magnetic energies in the scattering field are asymptotically equal at a large distance from the scatterer.

The far field asymptotes for RCS computation H_s and E_s are expressible in terms of parameters associated with the near field by a near-field transformation. These transformed vectors are called the scattering amplitudes and depend upon the propagation vector κ and the direction of the point of observation r in the far field (Crispin et al., 1968).

$$E(\kappa, r) = (i\kappa/4\pi) \oiint \left[\sqrt{\mu/\varepsilon}(n \times H) - (n \times E) \times r - (n \cdot E)r\right] e^{-\kappa r \cdot \bar{r}} ds \tag{5.23}$$

$$H(\kappa, r) = (i\kappa/4\pi) \oiint \left[\sqrt{\mu/\varepsilon}(n \times E) - (n \times H) \times r - (n \cdot H)r\right] e^{-\kappa r \cdot \bar{r}} ds \tag{5.24}$$

In these equations, the symbols n and r denote the surface outward normal of the scatterer and the unit vector of the observation direction. The individual terms $(n \times H)$ and $(n \cdot E)$ are the electric surface current and surface charge. The terms $(n \times E)$ and $(n \cdot H)$ represent the magnetic surface current and surface charge, respectively. For a perfect conductor, both the tangential component of the electric field and the normal component of the magnetic of the magnetic field vanish, and the scattering magnitudes are simplified. Furthermore, the dependent variables of the integrand are now the Fourier transform variables in the frequency domain.

In early engineering applications, two techniques commonly adopted for radar signature evaluation are the method of moments and the geometric optics theory of diffraction. The former is an accurate boundary-value analysis of the electromagnetic-wave interaction for incident waves at a fixed frequency. The latter has provided an asymptotic analysis in the high-frequency regime. Computational electromagnetics (CEM) in the time domain is developed to meet the technique need and to examine the bistatic RCS in the complete range of the scattering phenomena in a wide-range, multifrequency band. In solving the time-dependent Maxwell equations in a discretized space for an initial and boundary value problem, the first concern is the placement of the far field computational boundary. A systematic verification procedure is conducted first by comparing the numerical results with the theoretical Mie series (Bowman et al., 1987) for perfectly electrical conducting spheres of different formulation, dimensions of computational domain, and incident wave frequencies.

The pioneering effort in RCS calculation by CEM usually employed the total-field variable on a staggered mesh system (Yee, 1966). The physical-based numerical procedure proved to be very effective. The total-field formulation was also utilized in conjunction with the characteristic method in earlier RCS calculations. However, accuracy of RCS prediction can be improved significantly by a different formulation. An alternative approach using the scattered-field variable has been demonstrated to be accurate and it is easily achieved just by subtracting the well-defined incident wave component from the total electromagnetic field:

$$U_s(x, y, z) = U_t(x, y, z) - U_i(x, y, z) \tag{5.25}$$

Since the incident field $U_i(x,y,z)$ must satisfy the time-dependent Maxwell equation and is the necessary initial value, the formulation using the scattered variable to solve the Maxwell equations are thus unaltered from the total field variable. The scattered variable is just a simple dependent variable transformation. For this reason, no different notations are used between the formulations by the two dependent variables in the subsequent derivation and computations.

The relative merit of the scattered-field dependent variable is presented in Figure 5.16, by comparing the horizontal polarized $\sigma(\theta, 0)$ at the wavenumber of 9.4, which corresponds to a parameter for the Mie series of $ka = 4.7$ (where a is the radius

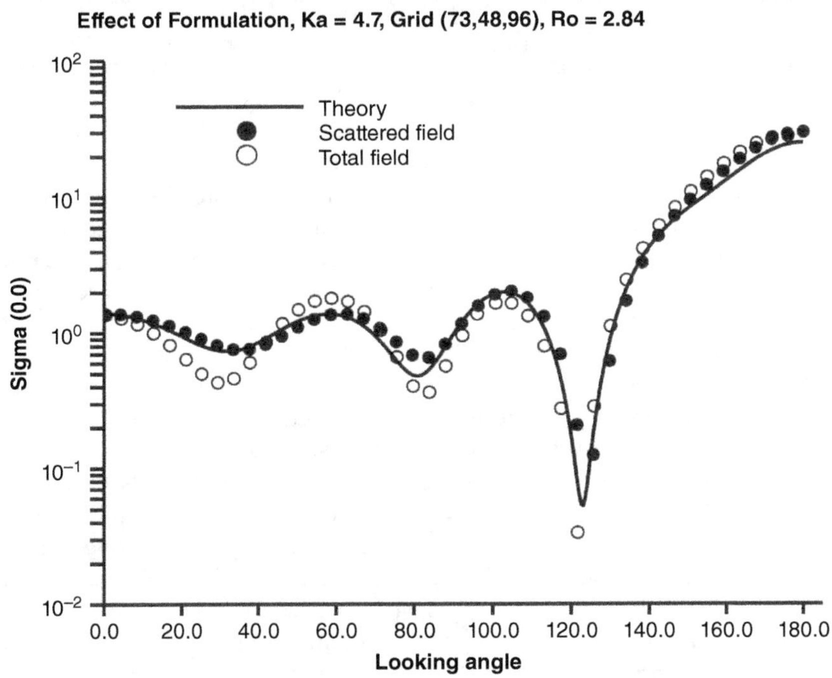

FIGURE 5.16 RCS results in HH polarity by total and scattering field formulations.

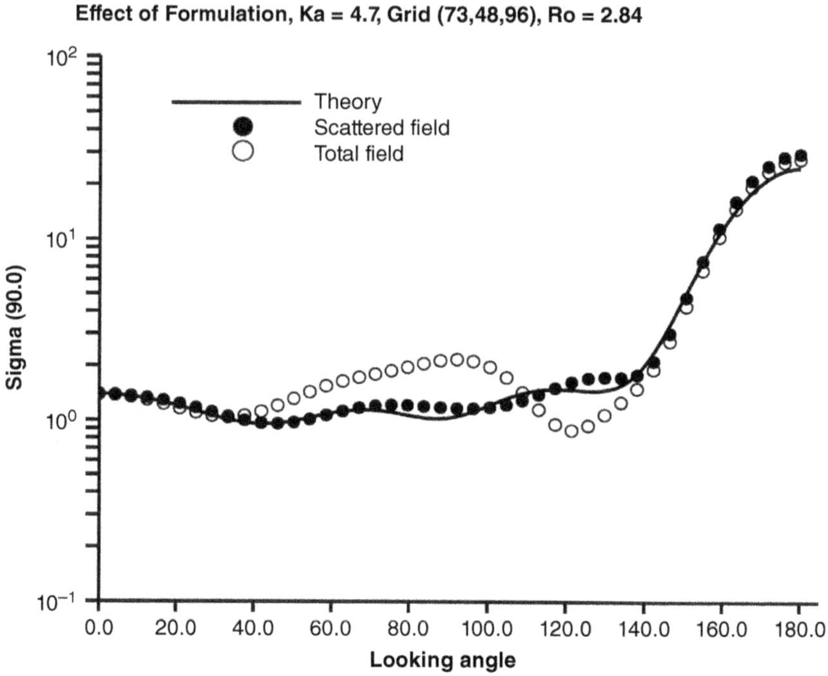

FIGURE 5.17 RCS results in VV polarity by total and scattering field formulations.

of the sphere). Both numerical results in the total and scattered field variables are generated at identical computational conditions. The numerical results of the total-field formulation reveal far greater error than the scattered field counterpart. An additional source of error is incurred when the incident wave must propagate from the far field to the scatterer. In the scatter-field formulation, the incident field data are described precisely on the scatterer surface. Since the far field electromagnetic energy distribution is derived from the near-field input, the advantage of describing the incident data on the scatterer is tremendous without error. The accumulated numerical error in RCS computation by the total field variable is manifested by the predicted value over the entire range of viewing angles of the bistatic RCS.

Figure 5.17 displays the bistatic vertically polarized RCS $\sigma(\theta, \pi/2)$ at the same wavenumber. The computational result fully substantiates the previous observation that the formulation by the scatter field variable is superior in predictive RCS accuracy. The numerical error of the result by the total-field variable is excessive in comparison with that by the scattered-field formulation, and the deviation from the theoretical result becomes unacceptable. In addition, computations using the total-field variable exhibit a stronger sensitivity to the placement of the far field boundary. A small perturbation of the far field boundary placement leads to drastic change in RCS prediction: a behavior resembling the ill-posedness condition that is extremely undesirable for numerical simulation. Therefore, the scattered-field variable of CEM is strongly recommended for RCS simulations.

In the validation process for RCS calculations, it has also been found out that the placement far field boundary is dependent on the dimension of the scatterer and the wave length of the incident wave. In general, the far field boundary of computational domain is required to place at a diameter away from a spherical scatterer in achieving a negligible disparity with the theoretical results. At very high incident wave frequency, the placement of far field can be much smaller than stated. The much smaller computational domain at high frequency even eliminates the creeping surface wave. Thus, the placement of the far field boundary is problem dependent in CEM for RCS prediction. The appropriate location of an imposed far field is actually a conflicting requirement for numerical simulation: a larger computational domain also needs a greater numerical of grid points to maintain a suitable grid point density for numerical resolution. To assess the requirement for a specific numerical algorithm, a grid refinement study is conducted for the second-order in-time and third-order finite-volume scheme.

The effect of grid density per wavelength for the RCS computational resolution is presented in Figures 5.18 and 5.19. For a perfectly conducting sphere illuminated by a linearly polarized incident harmonic wave with a wave number of 10.6, the *ka* parameter of the Mie series is 5.3. The far field boundary is located at a distance of 2.53 wavelengths away from the scatterer. Three grid systems ($49 \times 48 \times 96$), ($61 \times 54 \times 96$), and ($73 \times 60 \times 96$) are investigated, resulting in a grid point number per wavelength of about 10.0, 12.0, and 15.0. The varying grid spacing is

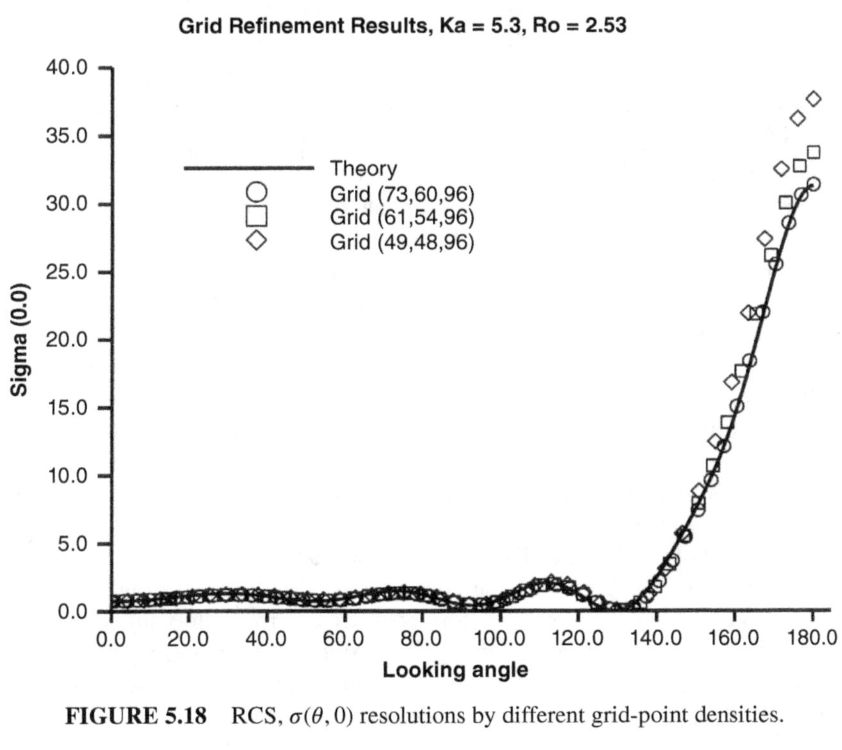

FIGURE 5.18 RCS, $\sigma(\theta, 0)$ resolutions by different grid-point densities.

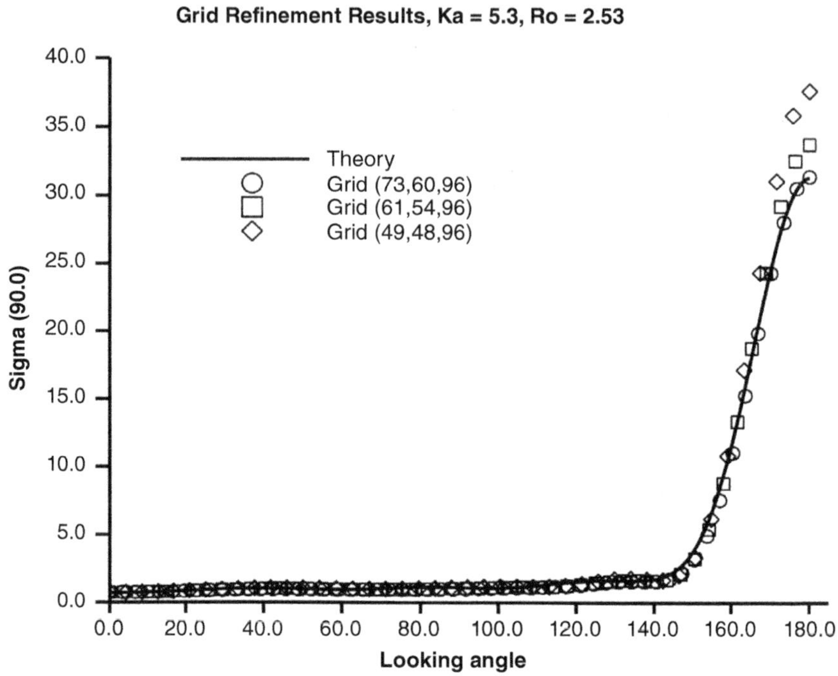

FIGURE 5.19 RCS, $\sigma(\theta, \pi/2)$ accuracy resolutions by different grid-point densities.

accommodated in the radial and azimuthal (θ) coordinates, while the circumferential angular advancement remains at a constant value of $\nabla\varphi = 4°$. Figure 5.18 depicts the normalized horizontal polarized bistatic RC $\sigma(\theta, 0)$ for the three different simulations. It clearly demonstrates that the calculation generated by a grid-point density of 15 points per wavelength attains perfect agreement with the analytical result. On the linear scale, the numerical discrepancies of the coarser mesh computations appear to concentrate in the shadow region of the scatterer, but the deviations are spread nearly uniformly over the entire viewing angles.

The normalized bistatic RCS in vertical polarization $\sigma(\theta, \pi/2)$ is presented in Figure 5.19. The numerical error pattern is identical to that of the horizontal polarization. At the grid point density of 10.0 and 12.0 per wavenumber, the maximum numerical errors are 19.1% and 7.7%, respectively. The accuracy improvement is rather drastic as the grid-point density is enriched. The Maximum numerical error is reduced to less than one-quarter of 1% by the result generated on the mesh of 15 points per wavelength. Although all studied simulations are produced by uniform grid spacing on a spherical coordinate system, the contiguous cell volume still highly stretched according to its radial location. The former order of any finite-volume scheme is known to degrade on an irregular and highly clustered mesh system. Thus, for most spatially second-order accurate schemes still require an even higher grid-point density per wavelength to yield a comparable solution in verification.

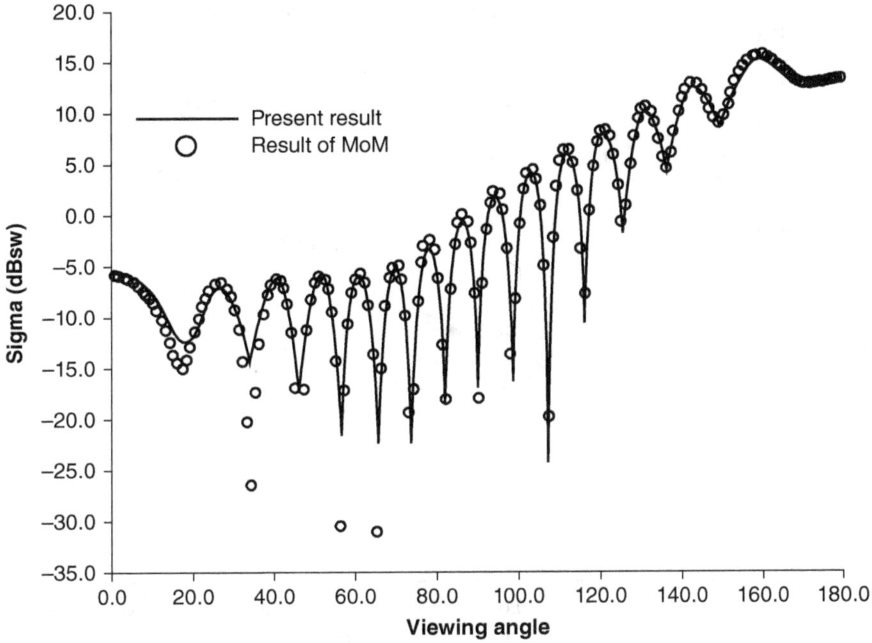

FIGURE 5.20 Bistatic RCS of ogive cylinder in HH polarization, 4 GHz.

Validating computations are also required by implementing an overset multiblock grid procedure for rather complex configurations that are unable to be described by a single-connected grid system. It is especially challenged for RCS simulation because a scatterer contains discontinuous surface elements that appear as multiple diffraction points. At the same time, a direct comparison of solutions by entirely different time- and phase-domains (method of moment) approaches for RCS prediction is examined. In Figures 5.20 and 5.21, the RCSs of an ogive cylinder in HH and VV polarizations are displayed. These computation results are generated by a single incident harmonic wave at 4 GHz and accompanied by results of method of moment. The simulated CEM results in the time domain are supported grid density per wavelength of 16.5. The agreement between results of entirely different numerical methods is excellent in decibel per square wavelength (dBsw) over a dynamic range of 50 (Luebbers et al., 1991). The disparity in all predicted lobes is less than a 10th of one dBsw, and the discrepancy appears only in the nulls among the predicted results and has no physical significance for the extremely low RCS value. The nearly perfect agreement is noticeable for calculations of the VV polarization. Additional comparisons have also been performed at different incident wave frequencies and the excellent agreement has reached for all cases considered. In short, the computational results by solving the Maxwell equations for RCS by a simple frequency incident harmonic wave are nearly identical via either the method of moment or CEM in the time domain.

The comparison of predicted RCS and data from a compact radar range is depicted in Figure 5.22. The monostatic RCS of a four-fin missile configuration illuminated

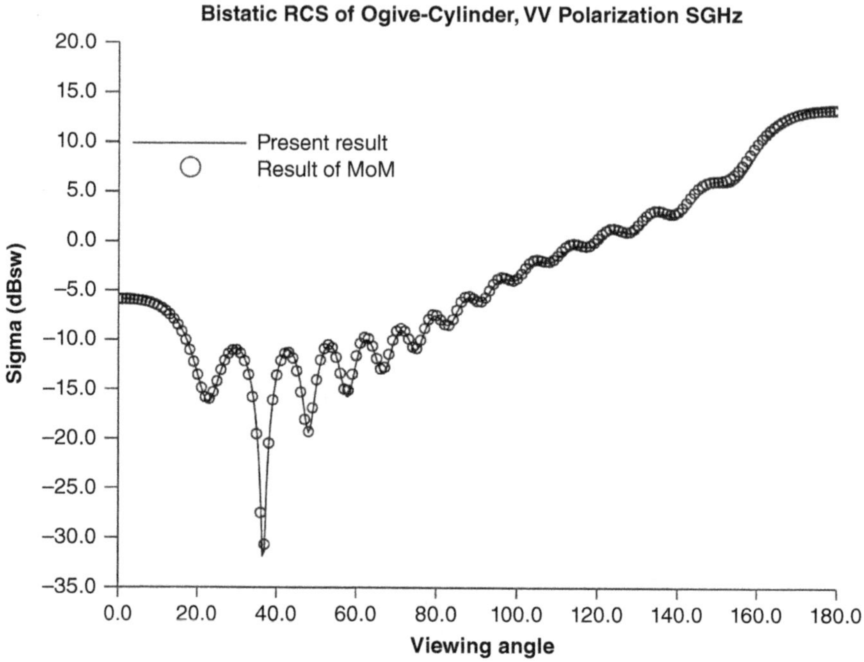

FIGURE 5.21 RCS of ogive cylinder in VV polarization, 4 GHz.

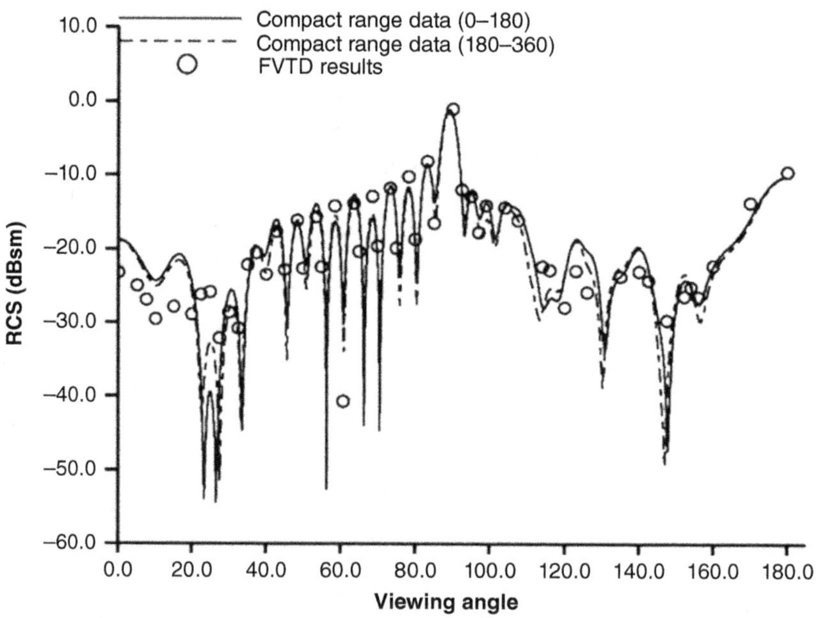

FIGURE 5.22 Comparing computational results with compact range data.

by a head-on incident microwave at 4 GHz is compared with data collected in a compact range. The basic scatterer consisting of ogive forebody and four equal-spaced triangular fins is described by an eight-block composite grid slightly over 2 million cells. The measured data are continuously sampled until the consecutive data over one period of the incident wave has converged to less than one-half of a decibel per square meter (dBsm). For all incident angles considered, the time elapsed is consistently within 20 periods. The agreement with the data is very good; in general, it falls within the data-scattering band. The maximum disparity between data and computational simulations is concentrated mostly in the region of the head-on incidences and the maximum discrepancy is 7 dBsm. In view of the excellent affinities among results of the basic ogive cylinder without the protruding fins, the error is attributed to the insufficient numerical resolution in capturing all diffraction points on the scatterer. Meanwhile the numerical simulation reveals that the resonating surface waves induced by multiple diffraction points can still be significant and even approach the optical region. In this regard, the simple criterion for numerical resolution requirement based on the grid-point density per wave length is insufficient.

The characteristic-based CEM methods have been successfully applied for antenna analysis, microstrip patch antenna, and the scattering by coated objective (White et al., 2000; Jiao et al., 2000a, 2000b, 2001). However, the wider range of applications of CEM in the time domain is still constrained by the numerical resolution limit for computational simulations at high frequency or complex configurations. In order to overcome this formidable challenge, either the innovative numerical algorithm developments or multiple orders of magnitude enhanced computational capability are required to break the barrier for CEM to enter the forbidden optical frequency regime.

5.6 RADAR SIGNATURE REDUCTION

The traditional classification of electrostatic, electromagnetic, and hydrodynamic waves as distinct groups in a plasma is an oversimplification. In most circumstances, a propagating microwave in plasma is always a mixture of several groups. Fundamental works in the 1960s have provided a thorough understanding of small-amplitude wave propagation and fluctuation in plasmas. In the absence of applied electric or magnetic fields, the plasma can be studied as an isotropic medium. Under this condition, the plasma behaves as a simple dielectric and will support high-frequency electromagnetic wave propagation through the response of electrons. However, electromagnetic waves may not propagate when the frequency is lower than the plasma frequency. The frequency at which propagating waves started to become evanescent is the cutoff frequency.

The scattering of electromagnetic wave in stationery plasma with an isotropic and uniform electric conductivity can be analyzed without an externally applied magnetic field. Based on this simplified premise, the electron pressure gradient is negligible. The time-dependent generalized Ohm's law, equation (5.18), is solved coupled with the Maxwell equations. The dissipation and refraction of a microwave are measured by the far field electromagnetic energy distribution and the result is presented in

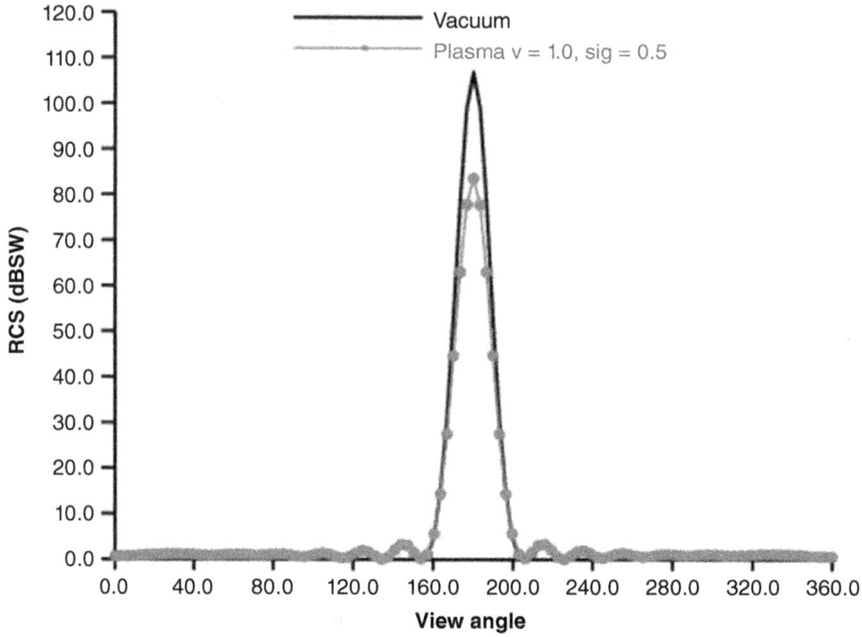

FIGURE 5.23 Bistatic RCS in plasma, $v_c = 1.0, \sigma = 0.5$, HH polarity.

Figure 5.23. The numerical result is obtained by assuming the ratio of the plasma and microwave frequencies to be equal. In this sense, the frequency of the microwave is set to be on the same as the plasma frequency to permit the wave propagation. The physical dimension for the computations is Ohm/m and leads to a normalized value from 0.0 to 0.5. An excellent agreement with Mie's series has been achieved as previously presented in this chapter. From the computational results, the bistatic RCS over a perfect electrically conducting (PEC) sphere is decreased when propagated through plasma. The horizontal polarity bistatic RCSs of a PEC sphere in vacuum and plasma are given on a linear ordinate to reveal the relative small reduction by dissipation in a low electric conductivity and electron number density of a glow discharge (10^{12}/cm^3). The magnitude of the main lobe of the bistatic RCS in plasma decreases 19.1% in comparison with that in vacuum. From theory, the mean collision frequency increased relative to the microwave frequency, the time dependency of the current density diminishes. In the asymptotic limit, the value of the computed RCS approaches that of the lossy medium.

The comparison of RCS over a PEC sphere in vacuum and a partially ionized plasma is depicted in Figure 5.24. Both RCSs are recorded for a horizontal polarity; the results are similar for the vertical polarity. The RCS measured in dBsw is significantly decreased by a factor as great as 2.47. The dBsw value of the calculated RCS is lossy medium and is reduced by a factor of 1.7 times in the main lobe and reduces even more in the shade region or the backside of the scatterer to reach a value of 2.47. The general numerical behavior is similar to the calculation using complex electric

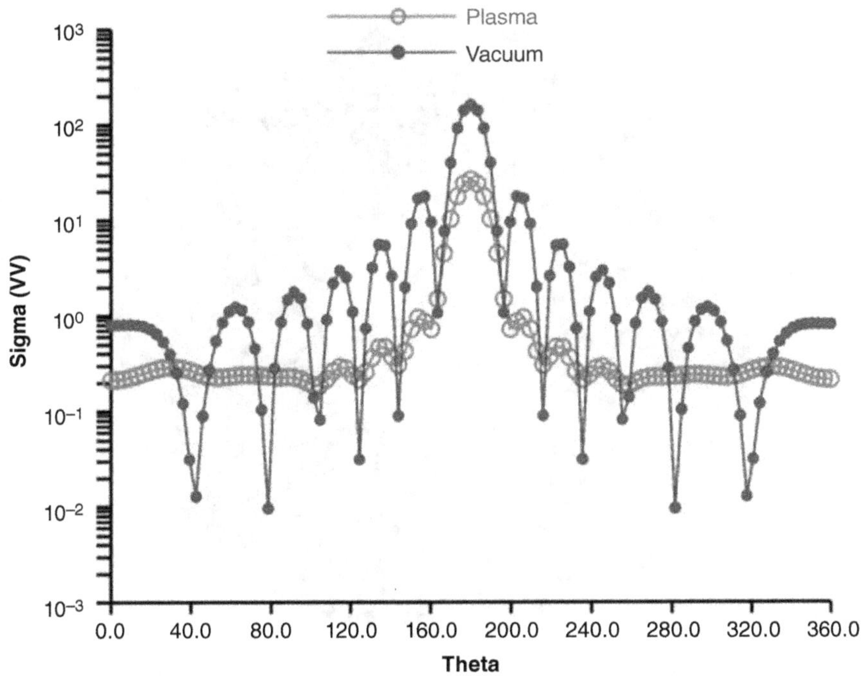

FIGURE 5.24 Bistatic RCS in lossy medium, $\sigma = 0.5$, and vacuum, $\sigma = 0.5$ HH polarity.

conductivity for a coated sphere (Jiao et al., 2000, 2001). It is anticipated that the electromagnetic energy dissipates much more rapidly in a lossy medium than in a nearly isolator.

The scattered electromagnetic fields in a lossy medium and vacuum are given side by side as the composited contours in Figure 5.25. The electric field intensity is obtained from the scattered field formulation; see equation (5.23). The formulation by the scattered field variable, the scatterer, appears as an emitter. On the left-hand side of the figure, the electric field in the vacuum shows the back scattering pattern toward the incident wave in the up portion of the contour plot. The scattering field on the right-hand side of the figure depicts the electric field in a lossy medium. The levels of plotting contours are identical for the two different computational results, so the attenuation of the electric field intensity is easily recognized; the dissipation of the electromagnetic wave amplitude is substantial. As soon as the reflected wave leaves the PEC sphere, the microwave dissipates rapidly and the scattered field vanishes from the same contour levels of the vacuum. The near field pattern in the lossy medium next to the PEC sphere appears as a creeping wave over the surface. This phenomenon is consistent to the reduced RCS, which is presented in Figure 5.23.

By a direct coupling of the generalized Ohm's law to Maxwell's equations in the time domain, approximate numerical simulations have been obtained for the dispersion of microwave in stationary plasma. The far field electric energy distributions in

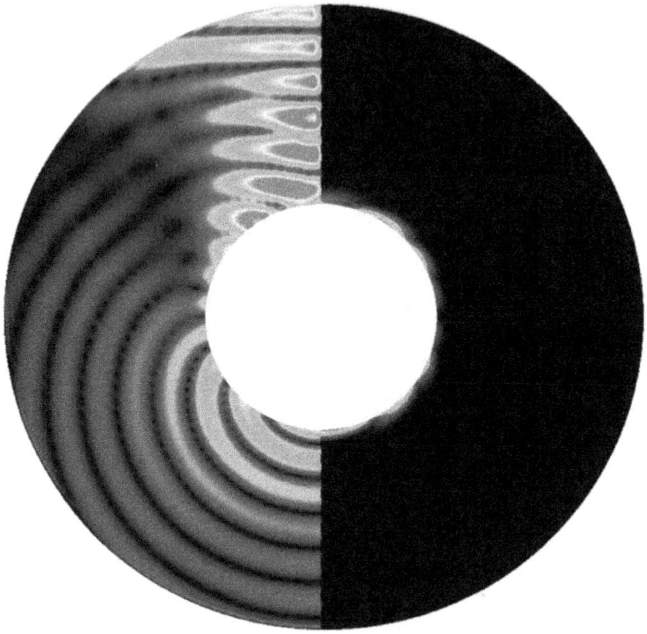

FIGURE 5.25 Comparison of scattered electric field in lossy and vacuum.

the form of a bistatic RCS are achieved in a partially ionized plasma and vacuum. The numerical result shows a drastic dissipation of the microwave by reducing the main lobe of RCS by a factor of 1.7. In a short propagation distance of 2.5 wavelengths, the scattered energy reduced by 10% and is accompanied by a phase shift up to a quarter of wavelength.

5.7 A PROSPECTIVE OF CEM IN THE TIME DOMAIN

There are two inherent limitations for CEM in the time domain by solving the three-dimensional, time-dependent Maxwell equations. One of them is the wave resolution requirement imposed by the Enquist criterion of two grid points per wavelength that requires processing a huge amount of data in high-frequency spectrum. The other constraint is imposed by the no-reflection wave conditions on a finite-size computational domain. In addition, the wave interference is associated with reflected waves from complex geometric configurations that need innovative grid generation techniques to resolve detailed electromagnetic wave patterns. In the following discussions, the research direction and needed progress are presented in numerical algorithm developments, massively parallel computational technology, and over-set grid mapping concepts. These critical scientific areas are essential requirements for additional basic research. At the same time, these innovative approaches naturally recommend the best practices for engineering applications.

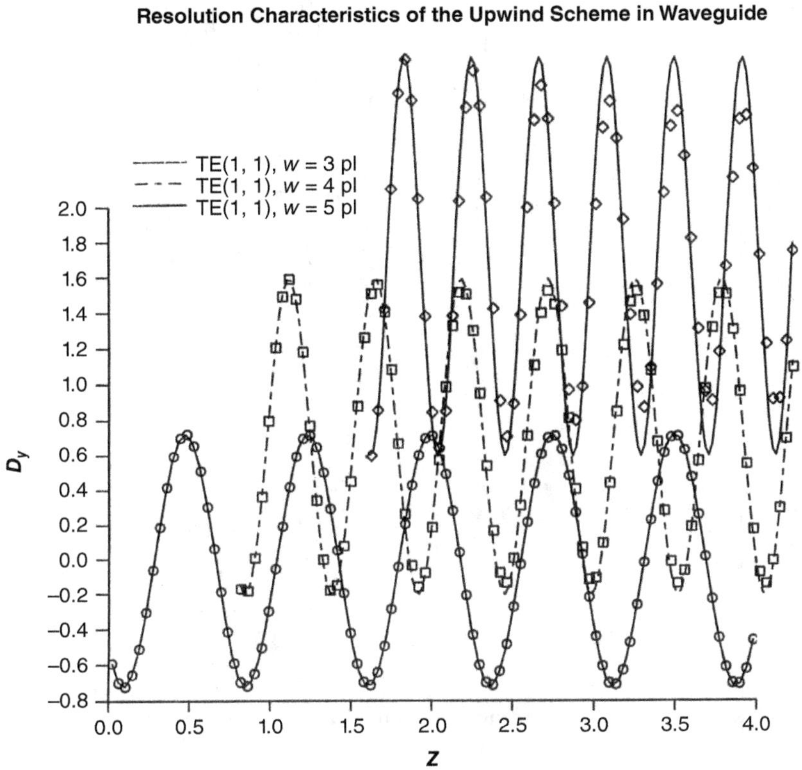

FIGURE 5.26 Numerical resolution of second-order windward scheme.

There are a large variety of second-order accurate finite-difference and finite-volume algorithms in CEM. A group of high-resolution compact-difference schemes has also been introduced to CEM. Indeed, these approximations have shown the ability to extend predictive accuracy to a higher wavenumber range but still not able to approach a truly spectral-like resolution. In fact, the most advanced high-resolution schemes still need a grid/cell density around 6–8 points per wavelength to keep the normalized truncation error within 10^{-4}. The simulated y component of the electric displacement of a $TE_{1,1}$ wave in a rectangular guide at three different frequencies offers the best illustration in Figure 5.26. All numerical solutions are generated by a second-order accurate upwind scheme on a uniformly spaced grid ($25 \times 24 \times 96$). The instantaneous distributions along an entire waveguide are accompanied by the theoretical values for a frequency spectrum from 3π to 5π, which correspond to grid-point density per wavelength from 18.3 to 9.15. At the lowest frequency, the dissipative error is confined to be less than 1%. As the wavenumber or frequency is increased, numerical errors in wave amplitude and phase relation become discernible and unacceptable.

Figure 5.27 displays the resolution feature of solutions by the fourth-order high compact difference for the same waveguide. These numerical accuracies are far

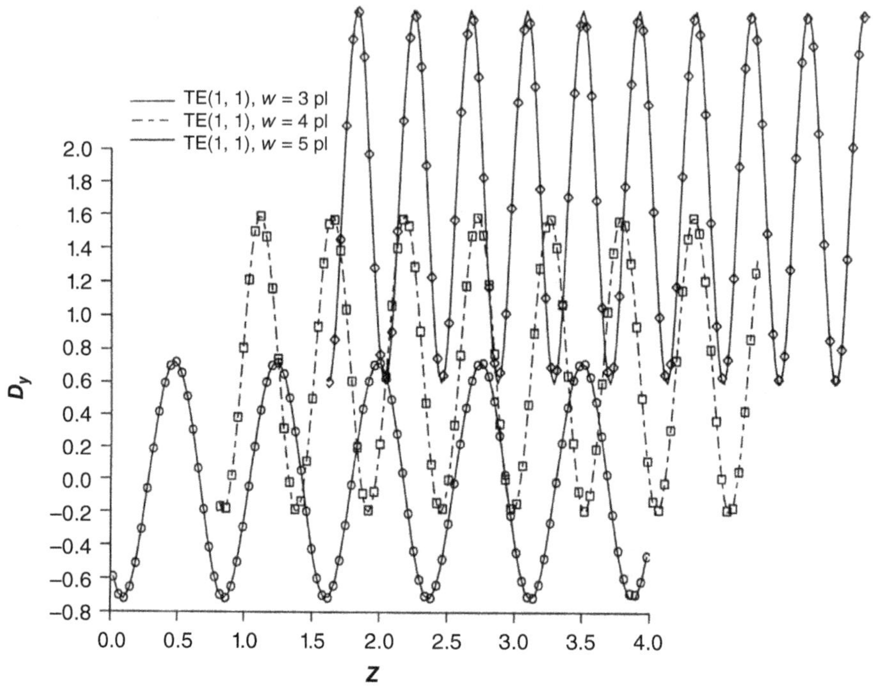

FIGURE 5.27 Resolution features of fourth-order compact difference schemes.

superior to the second-order method. The numerical results are generated on the identical coordinate with grid-point densities of 18.3 and 13.7 per wavelength and are equally accurate. As the grid-point density is further decreased to the value of 9.15 (wavenumber $\omega = 5\pi$), the numerical resolution degrades and yields a maximum error of 0.331% from the theoretical value near the middle of the waveguide. Higher-resolution schemes have also been evaluated, but the accuracy is significantly deteriorated at an angular frequency greater than 6π, which corresponds to a grid-point density per wavelength of a value of 8.34. For diffraction and refraction dominant electromagnetic phenomena, the computational accuracy requirement is more stringent. It is well-known and accepted in practical applications that a predictive dynamic range of 60 dB is required over a broad angular frequency range. In order to meet this criterion, an innovative algorithm requires almost three-order magnitude reduction in grid/cell density per wavelength. The resolution requirement is achievable only by the exact performance level of spectral methods. All these observations point out two important facts: first, like any other computational simulation methodology, the dimensionality of the simulation is critical. When CEM involves more than one-dimensional calculation, numerical simulation must have additional source than that just proportional to a multiple one-dimensional analyses to control anisotropy and stability property. Second, all numerical algorithms have a limited range of accurate resolution on a given time and space scales or a wavenumber.

The numerical resolution of an algorithm limits the computational accuracy and application range of practical simulations by imposing a minimum grid-point density per wave length. Similarly, the stability property of the algorithm controls the computational efficiency by restricting the time-step advancement for temporal evolution. Therefore, the computational memory requirement for a three-dimensional simulation is proportional to the cubic power of number of grid points, and the computational time is proportional to the quadruple power of the grid-point density for an acceptable numerical simulation. For either a high-frequency or a large electrical configuration, the required computational resource is therefore substantial. The numerical algorithm improvement alone is insufficient to overcome the large computational resource needed. The huge amount of computational resource needed for complex simulation can be achieved only by a combination of algorithm research and scalable concurrent computing technique (Shang et al., 2000; Pan et al., 2003).

The memory size, processor clock speed, and data storage of a computing system usually impose the upper limit for the scales of computational simulations. The memory requirement is derived from the addressable locations for each discrete point of the dependent variables of a simulation. The data-processing rate is usually directly proportional to the system clock speed, data transfer rate to and from the memory unit and the central processor, as well as the arithmetic operational count of a numerical procedure. Therefore, the architecture of a computing system impacts greatly on the computational performance.

The advent of multicomputers has made substantial progress and opportunity in computational simulation technology. In theory, distributed-memory computational systems can provide a platform for scalable computations. In turn, it also allows computational simulations that require large addressable memory for rapid turnaround. Research has thoroughly explored the realizable parallel algorithm in CEM and CEM-aerodynamics. These algorithms rely on the underlying approximation to physics according to the range of frequency spectra. For hardware development, the foremost obstacle is the lack of compatible high-speed intercommunication networks to support the advanced single processor for computational simulations in the time domain, while the most urgent software need lies in the porting of existing sequential computer programs to multiprocessor.

Balancing individual processor workloads and minimizing internodal communication are essential for effective use of distributed-memory multicomputers. Using one-dimensional domain decomposition strategy, the most successful approach is by adopting the parallel virtual machine and the message passing interface (MPI) (Gropp et al., 1994). Although the parallel efficiency of the ported software is acceptable but relatively low. Another key and yet often overlooked elements in achieving a high concurrent data process rate (DPR) are managing the cache and the memory hierarchy. A specific example is demonstrated by the actual normalized DPR from both the MPI and power FORTRAN (PFA) programming models (Koelbel, 1994). The actual Mflop (million floating point per second) counts for the now small memory requirements of 429,300 cells; 256 Mbyte computations are 119.93 and 36.08 by the power Fortran and MPI, respectively, on an SGI R1000 processor. Again, this demonstration depicted in Figure 5.28 shows that the efficient cache utilization for high-performance

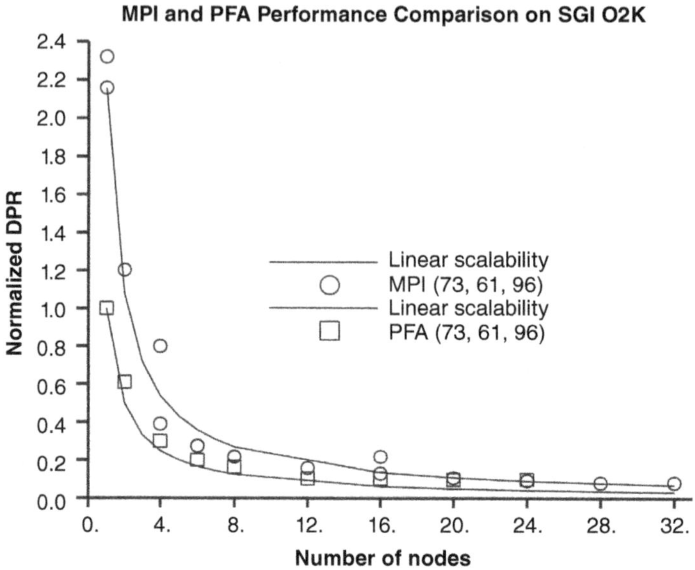

FIGURE 5.28 Scalability performance by MPI and PFA.

computing is critical. Obviously, there are still numerous possibilities for extracting higher computing efficiency from each individual reduced instruction set computing.

The improvement is most noticeable when the multicomputer memory size is significant for memory-size bound computations. As a consequence, based on message-passing paradigm, an approach other than domain decomposition, such as the data-dimension partition, can be useful. MPI is an attractive message-passing procedure because it is one of the high-level languages that emphasize portability over efficiency. In particular, the MPI does not impose semantic restrictions on efficiency; it also makes easy-to-implement spatial domain decomposition.

Three-dimensional decomposition is the most flexible and efficient data partition; it implicitly contains the one- and two-dimensional data partitions as subsets. In CEM, the numerical procedure requires specifications of the global domain dimensions in each coordinate, the number of processors desired, and the overlapping domain of the adjacent subdomains that depend uniquely on the adopted numerical algorithm. Each distributed processor owns the overlapping domains defined by the inclusive boundaries. The treatment of the data within the overlapping region of the processor is critical for data update and communication between processors. At the six interface subdomains, only the local processor reads the data of its immediately adjacent processor. In this sense, the overlapping region is actually the zonal boundary between computational subdomains and must transmit information between processors.

Figure 5.29 indicates the scalability of the solving procedure by the MPI on the IBM SP system. The RCS calculation performed for a perfectly electrical conducting sphere on a (91 × 241 × 384) mesh system, resulting in a memory size of 4.596 Gbytes. A baseline of scalable performance is established by a minimum number

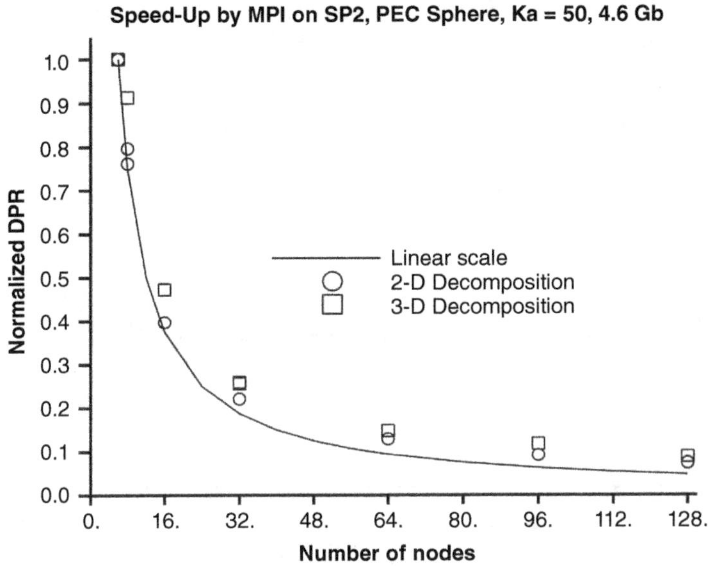

FIGURE 5.29 Scalability performance by MPI on IBM SP system.

of six nodes. The DPR of the six nodes is 1.664×10^{-6} seconds per grid point per time step which the computing rate is estimated at 853 Mflops. Both the two- and three-dimensional data decomposition options are exercised. Over the range from 6 to 128 nodes, the MPI model demonstrated a consistent and reasonable scalable performance. For the 128-node calculation, the computing rate is around 11.65 Gflops (Giga floating point per second). The speedup from the minimum to the maximum number of nodes is a factor of 13.66 over a possible linear scale of 23.33, and the parallel computing efficiency drops from a value from 98.4% to 64.7%. For the 128-node calculation, the computing rate is around 11.65 Gflops, more recently the similar parallel computational efficiency has been sustained for a thousand-node multicomputer. At the same time, the two-dimensional data partition algorithm reveals a higher computational and parallel efficiency than the three-dimensional counterpart.

The massively parallel computing capability has enabled at the least a hundred-fold speedup and an immense increase in memory size for CEM. The enhanced computational efficiency permits a benchmark RCS calculation of a perfectly electrical conducting sphere at a wavenumber beyond 100 (ka = 50.0). The calculation is carried out on a $(91 \times 241 \times 384)$ mesh system, and the bistatic horizontal polarized RCS was sampled for the elapse time of three periods. The RCS computation requires 2887.6 s on an IBM SP 256 node system that is the first-generation commercially available, massive parallel computing system. The calculated horizontal polarized bistatic RCS distribution $\sigma(\theta, 0.0)$ is presented in Figure 5.30 to exhibit an excellent agreement with the theoretical results by the Mie series over a dynamic range of six decades (Bowman et al., 1987).

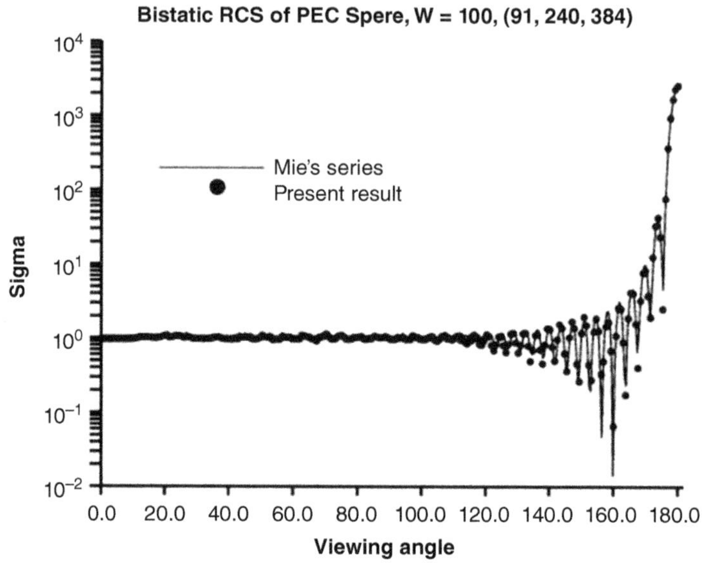

FIGURE 5.30 Comparing computational bistatic RCS of PEC sphere with Mie's series, Ka = 50.0, HH polarity.

The vertically polarized RCS $\sigma(\theta, \pi/2)$, as shown in Figure 5.31, substantiates the previous observation. The numerical result agrees well with theory. The numerical error of the scattered-field calculation is limited mostly as spurious oscillations; the maximum error is confined within 5 dB, and the simulation at the back-scattering region is exact. The numerical results are generated on an equal increment grid of

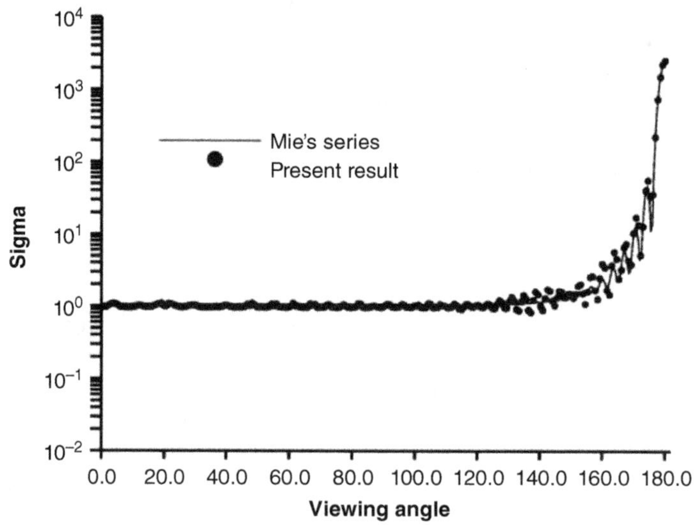

FIGURE 5.31 Comparing computational Bistatic RCS of PEC sphere with Mie's series, Ka = 50.0, VV polarity.

all independent variables (r, θ, φ); however, the cell volume of the finite-volume computation still clusters in the polar regions of the spherical coordinate. Additional computational resource saving is realizable by using an optimal mesh system to further alleviate the overly restrictive computational stability condition and advances further into a higher wave number regime. Today, the speedup of modern concurrent computers increases almost exponentially to expand the predictable frequency range of incident wave approaching the lower optical regime. The increased computational capability is essential to remove the up limit of CEM for practical engineering applications.

The concurrent computational capability and the general curvilinear coordinate transformation open the possibility for CEM of complex electrical scattering. As an illustration, the scattered electromagnetic field from a reentry vehicle is solved by the Maxwell equations cast on nonorthogonal curvilinear coordinate. The X24C-10D reentry vehicle is 74 feet 10 inches long with a wingspan of 24 feet and 2 inches and a height of 20 feet 7 inches. The basic configuration is a radically swept delta wing with a five-fin after-body, a central, and two middle fines, and two strakes. The mesh generation for a body-conformal system employs a 181 streamwise variable geometry cross-sectional planes, each defined by 59×162 mesh points. The interface between two media is easily defined on the coordinate surfaces. On the coordinate plane, all discretized nodes on the surface are precisely prescribed without the need for an interpolating procedure. The outward normal of the surface that is essential for the boundary conditions of the Maxwell equations can be computed by $n = \nabla s/||\nabla s||$. In the transformed space, computations are performed on a uniform mesh space, but the corresponding physical space can be highly clustered to enhance the numerical resolution.

In the multiple cross-sectional planes, the transformed coordinates are constructed by a single-block grid topology; the near-field mesh adjacent to the solid surface is generated by a hyperbolic grid generator that preserves the orthogonal property of the coordinate and extends to far field by a transfinite technique. The two mesh systems are merged by the Poisson average technique. All solid surfaces of the X-24c-10D are mapped onto the parametric surface in the transformed space (ξ, η, ζ), defined by $\eta = 0$. The number of cells between the surface and the far field boundary is 59, and the circumference of the cross-sectional pane is described by 162 nodes for the half-plane bisected by the plane of symmetry of the vehicle. The electromagnetic excitation is introduced by a harmonic TE incident wave along the vehicle axis of symmetry and along any direction the wavelength is resolved by at least 18 cells. The fringe pattern of the scattered electromagnetic wave on the X-24C-10D is presented in Figure 5.32 for a characteristic length-to-wavelength ratio $L/\lambda = 9.2$. A salient feature of the scattered field is brought out by the surface curvature: the smaller the radius of the surface curvature, the broader the diffraction pattern. It is the classic behavior of a specular backscatter. The numerical simulation exhibits highly concentrated contours at the line intersection between the upper and lower vehicle surfaces (chine) of the forebody and the leading edges of strakes and fins. It is interesting to note that the reflection and diffraction patterns are similar to the shock wave structure encountered in the computational aerodynamic simulation.

The time-domain computational procedure using the structured, over-set grid generation process has applied to a scattering simulation of a much more complex

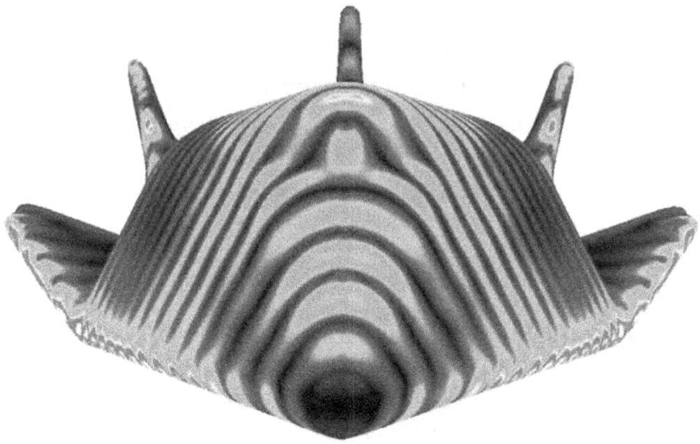

FIGURE 5.32 Electromagnetic wave fringe over on space vehicle X-24C-10D.

full-scale B-1B aircraft (Blake and Bulter, 1997; Blake and Shang, 1998). This particular simulation is intended to demonstrate the range of practical application of time-domain methods and to seek an alternative grid topology for the discretized field approximation. Two separate multiblock grids for the fore- and aft- portion of the B1-B are constructed according to the distinct geometric features of the aircraft. A widely used grid generation package from the computational fluid dynamic community is adopted for the process. The two major multiblock grid systems are overlapped to allow the information exchange between the two systems during the data processing. The level of effort required to generate this overall grid system is substantial from the time of the available CAD (computer-aided design) model of the aircraft until a workable grid is completed. The over-set grid system was used to produce the electromagnetic scattering field at 100 MHz consisting of a total of 711 grid blocks with 383 and 328 blocks distributed between the fore and aft grid systems, respectively. The fringe pattern on the B1-B aircraft is displayed in Figure 5.33. A total of 34.5 million grid cells were used, with 23.5 million cells concentrated in describing the forebody of this aircraft including the complex engine nacelles with the air-breathing inlet. The aftbody configuration is equally challenging with the rudder and elevator assembly. The numerical simulation requires then the unprecedented problem size of more than 192,000,000 unknowns. On a 258-node SGI Origin 2000 system, which is a first-generation parallel computer, the data-processing rate is estimated to be about 24.32 Gflops. This speedup benchmark in data-processing rate by parallel computing is easily surpassed by the new generation of multicomputer.

Meanwhile, progress in grid generation is shifting to the unstructured grid approach that represents the most general grid connectivity. Basically, the discretized field points are constructed by a group of contiguous tetrahedrons from the solid surface of a complex electric objective extending to far field. This coordinate transforming process requires an explicit connectivity of adjacent grid points. The identical need also exists for the mapping of a numerical procedure to multicomputers by the domain decomposition approach. The explicit connectivity of adjacent grid cells or data blocks for grid generation is also optimal for load balancing and limiting inter

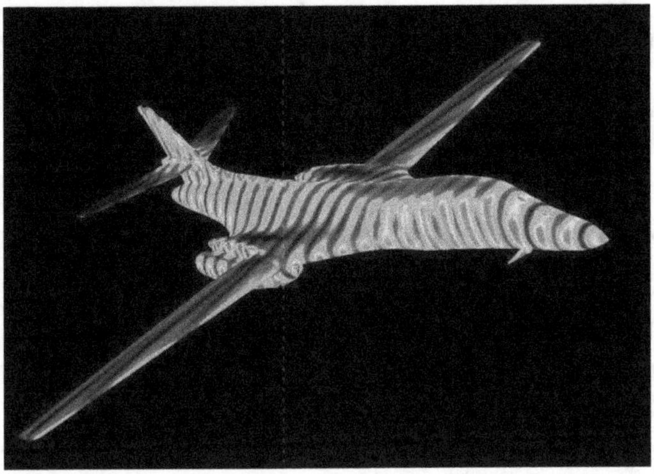

FIGURE 5.33 Electromagnetic wave fringe pattern on B-1B aircraft.

processor communication for concurrent computing to become the most attractive approach for simulating the complex scatter configuration.

The progress on concurrent computational technology is driven by the scientific and technological needs as the interdisciplinary simulation capability is required for advances design and evaluation processes, as well as the increasingly demand of physical fidelity by computational simulations. The rate of data processing has reached the Peta flops level (10^{15} floating operation points per second) by the concurrent operation of more than 28,000 multicomputers. The achieved computational capacity can only be attained together with improved data communication and data management paradigms. Based on the established trend, concurrent computational capability is increased by one thousand folds per decade. The enhanced computing efficiency becomes a telling factor for progress and maturation for computational simulation disciplinary.

As the technology is advancing, the increasing demands for more detailed physical phenomena are also elevating. CEM-aerodynamics occupies a unique position in that the governing equations built on fundamental laws are based on a sustained period of phenomenological observations and fully substantiated by the quantum physics. As an independent discipline, it does not have known theoretical limits, and its physical fidelity of simulation is completely dependent on the atomic or molecular states of a working gas medium for interdisciplinary applications.

REFERENCES

Balanis, C.A., *Antenna Theory*, 3rd ed., John Wiley & Sons, Hoboken, NJ, 2005.

Blake, D.C., and Bulter, T.A., Domain decomposition strategies for solving the Maxwell equations on distributed parallel architectures, *J. Appl. Comp. Electromagn.*, vol. 12, 1997, pp. 111–119.

Blake, D.C., and Shang, J.S., A procedure for rapidly prediction of electromagnetic scattering from complex object, *AIAA* 98–2925, 1998.

Bowman, J.J., Senior, T.B.A., and Uslenghi, P.L.E., *Electromagnetic and Acoustic Scattering by Simple Shapes*, Hemisphere Pub. Corp., New York, 1987.

Crispin, J.W., and Siegel, K.M., *Methods of Radar Cross-section Analysis*, Academic Press, New York, 1968.

Fedrick, R.A., Blevins, J.A., and Coleman, H.W., Investigation of microwave attenuation measurements in a laboratory-scale motor plume, *J. Spacecraft Rockets*, vol. 32, no. 5, 1995, pp. 923–925.

Elliott, R.A., *Electromagnetics*, Ch. 5. McGraw-Hill, New York, 1966.

Golant V.E., Methods of Diagnostics Based on the Interaction of Electromagnetic Radiation with a Plasma, *Basic Plasma Physics I*, editors Galeev, A. and Sudan, R., North Holland Publishing Co., Amsterdam, 1983, pp. 630–680.

Gropp, W., Lusk, E., and Skjellum, Using MPI: *Portable Parallel Programming with the Messaging-Passing Interface*, MIT Press, Cambridge, MA, 1994.

Harrington, R.R., *Time-Harmonic Electromagnetic Fields*, McGraw-Hill, New York, 1961.

Heald, M.A., and Wharton, C.B., *Plasma Diagnostics with Microwaves*, John Wiley & Sons, New York, 1965.

Jiao, D., Jim, J., and Shang, J.S., Characteristic-based finite-volume time-domain method for scattering by coating objects, *Electromagnetics*, vol. 20, 2000, pp. 1-257–268.

Jiao, D., Jin, J., and Shang, J.S., Characteristic-based finite-volume time-domain method for scattering by coated objects, *Electromagnetics*, vol. 20, 2000b, pp. 1257–1268.

Jiao, D., Jin, J., and Shang, J.S., Characteristic-based time-domain method for antenna analysis, *Radio Sci.*, vol. 36, no. 1, 2001, pp. 1–8.

Jordan, E.C., *Electromagnetic Waves and Radiating System*, Prentice-Hall Inc., NJ, 1960.

Katz, D., Piket-May, M., Taflove, A., and Umashankar, K., FDTD analysis of electromagnetic wave radiation from systems containing horn antennas, *IEEE Trans. Antenna Propagat.*, vol. 39, no. 8, 1991, pp. 1203–1212.

Koelbel, C.H., *The High Performance FORTRAN Handbook*, MIT Press, Cambridge, MA, 1994.

Krause, J.D., *Electromagnetics*, 1st ed., McGraw-Hill, New York, 1953.

Lochte-Holtgreven, W., (ed.) *Plasma Diagnostics*, North Holland Publishing Co., Amsterdam, 1968.

Luebbers, R.J., Kunz, K.S., Schneider, M., and Funsberger, F., A finite-difference time-domain near zone transformation, *IEEE Trans. Antennas Propag.*, vol. 39, no. 4, 1991, pp. 429–433.

Meuth, H., and Sevillano, E., Microwave diagnostics, vol. 1, Edited by Auciello, O. & Flamm, D., Academic Press, 1989, pp. 239–311.

Mitchner, M., and Kruger, C.H., *Partially Ionized Gases*, John Wiley & Sons, New York, 1973.

Pan, Y., Shang, J.S., and Guo, M., A scalable HPF implementation of a finite-volume computational electromagnetics application on a CRAY T3E parallel system, concurrency and computation, *Pract. Exp.*, vol. 15, no. 6, 2003, pp. 607–621.

Shang, J.S., Wagner, M., Pan, Y., and Blake, D.C., Strategies for adopting FVTD on multicomputers, *Comput. Sci. Eng.*, vol. 2, no. 1, 2000, pp. 10–21.

Shang, J.S., Ganguly, B., Umstattd, R., Hayes, J., Arman, M., and Bletzinger, P., Developing a facility for magnetoaerodynamic experiments, *J. Aircraft*, vol. 17, 2000, pp. 1065–1072.

Shang, J.S., Simulating microwave radiation of pyramidal horn antenna for plasma diagnostics, *Commun. Comput. Phys.*, vol. 1, no. 4, 2006a, pp. 677–700.

Shang, J.S., Simulating plasma microwave diagnostics, *J. Sci. Comput.*, vol. 28, September 2006b, pp. 507–532.

Shang, J.S., Time-domain electromagnetic scattering simulations on multicomputers, *J. Comput. Phys.*, vol. 128, 1996, pp. 381–390.

Smoot, L.D., and Underwood, D.L., Prediction of microwave attenuation characteristics of rocket exhausts, *J. Space Rockets*, vol. 3, no. 3, 1966, pp. 302–309.

Stratton, J.A., and Chu L.J., Diffraction theory of electromagnetic waves, *Phys. Rev.*, vol. 56, no. 1, 1939, pp. 99–107.

Swanson, D.G., *Plasma Waves*, Academic Press, Boston, 1989, pp. 77–117.

White, M., and Shang, J.S., Three-dimensional simulation of microstrip patch antenna radiation, *AIAA* 2000-2674, Denver, June 19–22, 2000.

Yee, K.S., Numerical solution of initial boundary value problem involving Maxwell equation in isotropic media, *IEEE Trans. Antennas Propagat.*, vol. 40, 1966, pp. 302–307.

6

COMPUTATIONAL FLUID DYNAMICS

INTRODUCTION

In most applications, electromagnetic effects enter electromagnetic–aerodynamic interactions by bringing additional force and energy conversion mechanisms through an extra physical mechanism, but the relative magnitude of electromagnetic effects appears only on a perturbation level. A measuring index of the relative magnitude between electromagnetics and inertia of fluid motion is the Stuart number $N_s = \sigma B^2 l/\rho u$, which has a much lower value than unity in most engineering applications (Sutton et al., 1965; Shang, 2001). To achieve a significant influence to change flow field behavior, electromagnetic effects must be amplified by mutual interactions to produce a significant contribution (Shang, 2002, 2009). Therefore, it is necessary to understand unique features of fluid dynamics phenomena and the simulating capability of computational fluid dynamics (CFD) for enhancing multidisciplinary applications. For this reason, topics of the following discussions explore opportunities for fluid dynamic phenomena that can be modified by electromagnetic perturbations through viscous–inviscid interactions, self-sustained oscillations, and vortical dynamics.

The first known numerical solution of the Navier–Stokes equations is often attributed to A. Thom, who published an article in 1933 titled, "The Flow Past Circular Cylinder at Low Speeds" in the Proceedings of the Royal Society of London (Thom, 1933). However, the most coherent development of CFD was initialed in the now Los Alamos National Laboratory of the United States in the 1940s. This

Computational Electromagnetic-Aerodynamics, First Edition. Joseph J.S. Shang.

systematic effort was driven by the war-time need for advanced weapon development and from this period of path-finding research. A substantial amount of fundamental knowledge of CFD was generated by von Neumann, Harlow, Hirt, Fromm, and others (Harlow, 1964; Von Neumann, 1944). Especially, Von Neumann developed the critical numerical stability criterion for finite-difference approximation and addressed the issue of resolving a hydrodynamic shock as early as 1944; his more accessible article in the literature was published 6 years later. These pioneering works were performed to meet an urgent engineering need that set a pattern of science and technology development for practical applications.

Harlow and his colleagues developed the particle-in-cell method based on a combined Lagrangian–Eulerian description of fluid motion. The particle-in-cell method has shown how well it is suited for studying time-dependent and multidimensional fluid motions. The effectiveness of this method is demonstrated through applications to shock interaction, supersonic wakes, and hypersonic flows over a sharp leading edge. This method is still in use today for magneto-hydrodynamic research (Shang, 2009). At the same time, Hirt made a lasting contribution to turbulent transport properties simulation by highlighting the numerical challenges and providing guidance for future research. All these accomplishments were accompanied by exhaustive proof in order to illustrate the validity of approximations and numerical accuracy. This rigorous process actually established the standards for all future CFD algorithm and numerical procedure research. The detailed derivation of formulations and extensive discussions on computational accuracy and limitations became the accepted standard in CFD research. During that time, all valuable information was largely unknown to the aerospace science community.

Prior to the 1970s, design and analysis tools in the aerodynamic community were primarily based on linearized potential flow theory, method of characteristics, and boundary-layer analyses (Shang, 2009). These tools were effective in skillful hands for commercial aircraft design, components of supersonic vehicle evaluation, and thermal management of reentry configuration. However, in the integrated aerodynamic performance of high-speed aerospace vehicles, the shortcoming of these modeling and simulation tools becomes glaring.

An outstanding example of technical challenge is the evidence of a seriously damaged ramjet pylon and lower lifting surface of the hypersonic test aircraft X-15 by viscous–inviscid interaction through a bow shock impingement (Korkegi, 1971). Numerous attempts have been made to treat numerically the viscous–inviscid interaction by coupling an external flow field to boundary-layer equations with methods of asymptotic matching. This approach to viscous–inviscid interaction is mathematically elegant and rigorous, but subject to severe limitations for practical application. A critical mechanism that usually triggers a strong inviscid–viscous interaction is boundary-layer separation; it is the consequence of a small adverse pressure gradient encountered by the wall-bound shear layer. It is also realized that much greater computational capability for the compressible Navier–Stokes equation must be developed for any further practical progress.

Chapman and Mark of NASA Ames Research Center are instrumental in sustaining CFD development to aerodynamic applications for the next four decades

(Chapman, 1979). Their vision for CFD development was crystal clear, and their most important contribution beyond science and technology is a tradition of unselfish knowledge sharing and mutual respect in the CFD community. A remarkable lesson in the management of a cutting-edge technology development is by making a clear distinction between science and technology.

For the past 40 years, increased efficiencies in numerical algorithms have been developed on the basis of eigenvalue and eigenvector analyses of governing equations. Through flux vector splitting techniques, one of the challenging issues for resolving the piecewise discontinuous solution, such as the shock wave jump or the flame front in combustion, has been treated satisfactorily by the approximate Riemann formulation. A wide range of coordinate transformation techniques, computational grid generations, and different grid topologies have also been devised to accommodate complex aerodynamic configurations. In that, the new concepts of the overset grid and multi-grid-block construction have developed, which opened new avenues for simulations of previously unsolvable aerodynamic performance problems. Equally important, an array of multicomputers, efficient data transfer from memory units, and data management paradigms were provided substantially improvements to the computational capability. Scientific achievements in CFD are vast and impressive; however, some are beyond the scope of the present discussion. Therefore, only subjects pertinent to electromagnetic–aerodynamics interaction are addressed.

6.1 GOVERNING EQUATIONS

The microscopic motion of gas particles can be completely described by the detailed collision process that is the basic approach of the kinetic theory of gas. This formulation describes the dynamic trajectories of each individual particle from known initial conditions that is a cornerstone of Newtonian mechanics. The fundamental integro-differential Boltzmann equation that describes the rate of change with respect to position and time by the velocity distribution function is based on the general concept of statistics. In the kinetics theory of gases, the position and the velocity coordinates are independent of each other to become six independent variables. Therefore, the solution of the Boltzmann equation is probabilistic rather than deterministic.

In the continuum regime, the Knudson number $\lambda/L \simeq M/\sqrt{\text{Re}}$ (λ is the free-mean-path between collisions) is less than unity (Hayes and Probstein, 1959; Vincenti et al., 1965). A direct link is established between the Boltzmann and equation of transfer by Enskog's successive approximation for velocity distribution functions. As a consequence, the equation of gas dynamics has an upper limit in the rarified gas regime, and the second or high-order approximation to the velocity distribution function is identified by phenomena of nonuniform states (Chapman et al., 1964). The macroscopic conservation formulation leading to the Navier–Stokes equation must include some irreversible processes such as molecular diffusion, dissipation, and thermal conductivity; for this reason, the differential system is not closed. Additional auxiliary relations are needed to include the global characteristics of collision processes through statistical means—the transport properties, equations of state, and chemical kinetics for an inhomogeneous gas mixture.

The fundamental principle of Newtonian mechanics is the conservation of mass, which means that the exchange of mass and energy is not considered. The species conservation equations of the multicomponents chemically reacting system can be given as

$$\partial \rho c_i / \partial t + \nabla \cdot (\rho u c_i + \rho u_i c_i) = dc_i / dt \tag{6.1a}$$

where c_i denotes the concentration of species i by mass fraction, $c_i = \rho_i/\rho$. For an n number chemical species system, only $(n - 1)$ species conservation equations are required. Any one of the n species equations may be replaced by the global equation of continuity for the mixture. In practice, all species conservation equations are usually computed and verified by the balance of the global continuity equation. This equation can be obtained by summing over all species conservation equations by noting

$$\Sigma \rho c_i u_i = 0 \tag{6.1b}$$

and

$$\Sigma dc_i / dt = 0 \tag{6.1c}$$

From which the global mass conservation law yields

$$\frac{\partial \rho}{\partial t} + \nabla \cdot (\rho u) = 0 \tag{6.1d}$$

The conservation of momentum equation is

$$\frac{\partial \rho u}{\partial t} + \nabla \cdot (\rho u u - \tau) = \rho \Sigma f_i \tag{6.2a}$$

The conservation of momentum equation is the only vector equation in the system, and it is Newton's second law of motion. The nonlinear convective transport of momentum is represented by a dyadic, which is a major component of the so-called inviscid terms. In fact, it is the source of turbulence by interactions of multiple scales vortices of the entire flow field (Bradshaw, 1994). The shear stress term is a second rank tensor described as

$$\tau = (-p + \lambda \nabla \cdot u)\bar{\bar{I}} + \mu def(u) \tag{6.2b}$$

Here, λ and μ are the bulk and molecular viscosity and $\bar{\bar{I}}$ is the identity matrix. The last term of the stress tensor is referred to as the deformation tensor,

$$def(u) = \nabla u + (\nabla u)^T \tag{6.2c}$$

The transpose of an operator for the gradient u, $(\nabla u)^T$ is simply by replaced the rows by columns in the matrix component of ∇u.

The deformation tensor has an important fluid mechanical interpretation in that diagonal derivatives represent the longitudinal strain while the off-diagonal derivatives represent the angular rotation of any line element in space. As the consequences, the viscous flow at the solid–fluid interface boundary produces both shear deformation and rotation. The inviscid terms produce only the expansion and compression by the hydrodynamic pressure term.

The conservation of energy equation is expressed as

$$\frac{\partial \rho e}{\partial t} + \nabla \cdot (\rho e u + q + u \cdot \tau) + Q = \rho \Sigma c_i f_i (u + u_i) \tag{6.3a}$$

The heat transfer term for a chemically reacting system is often given in a directly useful form as

$$q = -\kappa \nabla T + \rho \Sigma c_i h_i u_i + q_{\text{rad}} \tag{6.3b}$$

This equation corresponds to the Fourier law for conductive heat transfer, the convective heat transfer by individual species velocities, and the radiative energy exchanges across the control volume, respectively. The heat of formation for the reacting gas mixture is usually given as $Q = (dc_i/dt)\Delta h_i^o$. In here, e denotes the total internal energy and Δh_i^o is the standard heat of formation for species i.

The governing equations for a compressible medium are more complex and significantly different from the original incompressible Navier–Stokes formulation derived by M. Navier in 1827 and incorporated with G. G. Stokes' law of friction or the relationship between stresses and rates of strain in 1845 (Navier, 1827; Stokes, 1845). However, the governing equations are established on the identical foundation of Newtonian mechanics through gas kinetic theory. By inspecting equations (6.1a), (6.2a), and (6.3a), there are $(n + 5)$ scalar equations but $(n + 10)$ dependent variables (u, v, w, e, p, T, ρ, κ, μ, λ, c_n), so the equation system is not closed. The closure of the equation system is achieved by prescribing auxiliary relationships for transport properties, an equation of state, and chemical kinetics. The final description of the partial differential equations system can be completed only by imposing all appropriate initial and boundary conditions.

The time-dependent, nonlinear Navier–Stokes equations in the presented form are recognized as the incomplete parabolic partial differential type (Shang, 1985). However, solutions for shock wave capturing are by splitting the inviscid and viscous terms of the governing equations into hyperbolic and elliptic types and solving them separately but as a total sum of the split fluxes.

In the literature, the Navier–Stokes equations are most frequently written in a flux vector form in Cartesian coordinates (Shang, 2002, 2009). This coordinate system is always adopted as the fundamental frame of reference for a successive coordinate transformation to describe a particular flow field structure around any configuration. Thus

$$\frac{\partial U}{\partial t} + \frac{\partial F}{\partial x} + \frac{\partial G}{\partial y} + \frac{\partial H}{\partial z} = R \tag{6.4}$$

The dependent variable U is easily identifiable as ρ, ρu, and ρe. The flux vectors F, G, and H include components of vector and tensor components of the divergence operators. The inhomogeneous terms of the governing equation are designated as R on the right-hand side of equation (6.4). Thence

$$U = U(\rho, \rho u, \rho v, \rho w, \rho e)^T \tag{6.5}$$

$$F = \begin{bmatrix} \rho u \\ \rho u^2 - \tau_{xx} \\ \rho uv - \tau_{xy} \\ \rho uw - \tau_{xz} \\ \rho eu - \kappa(\partial T/\partial x) - (u\tau_{xx} + v\tau_{xy} + w\tau_{xz}) \end{bmatrix} \tag{6.6a}$$

$$G = \begin{bmatrix} \rho v \\ \rho uv - \tau_{yx} \\ \rho v^2 - \tau_{yy} \\ \rho vw - \tau_{yz} \\ \rho ev - \kappa(\partial T/\partial y) - (u\tau_{yx} + v\tau_{yy} + w\tau_{yz}) \end{bmatrix} \tag{6.6b}$$

$$H = \begin{bmatrix} \rho w \\ \rho uw - \tau_{zx} \\ \rho vw - \tau_{zy} \\ \rho w^2 - \tau_{zz} \\ \rho ew - \kappa(\partial T/\partial z) - (u\tau_{zx} + v\tau_{zy} + w\tau_{zz}) \end{bmatrix} \tag{6.6c}$$

The radiative heat transfer has been dropped from the equation (6.3b), which is customarily treated as a heat sink or source term in applications. The elements of the shear stress tensor on the Cartesian frame are

$$\begin{aligned}
\tau_{xx} &= (2/3)\mu[2(\partial u/\partial x) - (\partial v/\partial y) - (\partial w/\partial z)] \\
\tau_{yy} &= (2/3)\mu[2(\partial v/\partial y) - (\partial u/\partial x) - (\partial w/\partial z)] \\
\tau_{xx} &= (2/3)\mu[2(\partial w/\partial z) - (\partial u/\partial x) - (\partial v/\partial y)] \\
\tau_{xy} &= \mu[(\partial u/\partial y) + (\partial v/\partial x)] = \tau_{yx} \\
\tau_{xz} &= \mu[(\partial w/\partial x) + (\partial u/\partial z)] = \tau_{zx} \\
\tau_{yz} &= \mu[(\partial v/\partial z) + (\partial w/\partial y)] = \tau_{yx}
\end{aligned} \tag{6.7}$$

For most practical applications, Cartesian coordinates are rarely adequate in describing the geometric configuration. If the governing equations are solving on this fundamental Cartesian frame, an extensive interpolation procedure becomes necessary for imposing boundary conditions on a nonrectangular configuration. It is also very difficult to implement a systematic grid space clustering immediately adjacent to a curvilinear surface for high numerical resolution. Thus, a generalized coordinate mapping is introduced in the following form:

$$\begin{aligned} x &= x(\xi, \eta, \zeta) \\ y &= y(\xi, \eta, \zeta) \\ z &= z(\xi, \eta, \zeta) \end{aligned} \tag{6.8}$$

By means of the chain rule of differentiation, the governing equation, equation (6.4) acquires the following expression:

$$\begin{aligned} \frac{\partial U}{\partial t} &+ (\xi_x, \xi_y, \xi_z)(\partial F/\partial \xi, \partial G/\partial \xi, \partial H/\partial \xi)^T + (\eta_x, \eta_y, \eta_z)(\partial F/\partial \eta, \partial G/\partial \eta, \partial H/\partial \eta)^T \\ &+ (\zeta_x, \zeta_y, \zeta_z)(\partial F/\partial \zeta, \partial G/\partial \zeta, \partial H/\partial \zeta)^T = R \end{aligned} \tag{6.9}$$

Again through metric identities of coordinate transformation, the equation in chain-rule conservation form can be rewritten in the strong conservation form,

$$\begin{aligned} \frac{\partial (U/J_c)}{\partial t} &+ \frac{\partial}{\partial \xi}[(\xi_x F + \xi_y G + \xi_z H)/J_c] + \frac{\partial}{\partial \eta}[(\eta_x F + \eta_y G + \eta_z H)/J_c] \\ &+ \frac{\partial}{\partial \zeta}[(\zeta_x F + \zeta_y G + \zeta_z H)/J_c] = R \end{aligned} \tag{6.10}$$

The strong conservation form has a remarkable telescopic effect, namely the internal cancellation of differencing approximation within the control volume is total for a given differencing approximation. The relative merits of the strong versus the chain-rule conservation form have been recognized over the years in computational simulations. The best practice is mostly by adopting the strong conservation form for simplicity and is almost used exclusively in CFD applications (Shang, 2004).

The physical fidelity and limitation of CFD have been established for more than 60 years over millions of practical applications (Shang, 2009). Since the governing equations are derived from the kinetic theory of a diluted gas, any phenomena that can be described by this theory shall be replicated accurately. In principle, the multi-space-and-time-scales turbulent fluid motion belongs to this category, but it remains unsolvable and is constrained only by insufficient numerical resolution. An inherent exception from this observation is that when the physical phenomenon takes place at the molecular and atomic scales. The physics are beyond the realm of kinetic theory but must be governed by the quantum physics. These circumstances are common occurrences for electromagnetic–aerodynamics interactions, and the quantum physical phenomenon must be modeled.

It has been long realized that applying plasma actuator to control gas dynamic phenomena is most effective at a threshold of flow instability or flow bifurcation (Shang et al., 2014). The reason is that the electromagnetic effects are mostly appeared as a small perturbation to flow fields. At the critical point of flow bifurcation, a small disturbance always leads to a greater change in dynamics of gas. The bifurcation of a dynamic system is best described as the transition between different dynamic states. It always involves changes of at least by one controlling parameter at junction of stability boundaries. Another definition is given as replacing an unstable flow by a new stable equilibrium flow at a critical value of a parameter (Tobak and Chapman, 1985; Seydel, 1988). Therefore, bifurcation has roots in the instability of both temporal and spatial flow field dynamics and is always initiated by a small perturbation. For example, the famous von Karman street by shedding vortices from the trailing edge of a blunt body is simply a manifestation of mode change from convective to global instability. The richness of spatial and temporal developments in separated flow topology, self-sustained flow oscillation, and laminar–turbulent transition indicates a unique dependence on the Reynolds number. Nonlinear dynamic phenomena in aerodynamics also take place at the outer edge of performance envelope of flight vehicles. Any improvement in performance needs to focus on removing limitations by introducing additional mechanisms such as electromagnetic effects.

Bifurcations encountered in spike-tipped nose buzzing, asymmetric vortical formation and shedding, flow separation, compressor surge, stalls of rotating blade rows and lifting surface, vortex breakdown, aileron buzz, and transition to turbulence. These phenomena are aerodynamic performance limiters. For engineering applications in an open or closed system, the critical parameter for bifurcation is usually known from experimental observations. For an example, the limiting cycle of translational instability is known to be controlled by the ratio of sonic speed and the convective velocity of the vortex within an embedded separated shear layer. Some difficulty may reside in identifying a controlling parameters and determining the critical point precisely of demarcation that separates subcritical and supercritical states. A thorough and better understanding of fundamental fluid dynamics behavior is always required for practical applications. For this reason, the viscous–inviscid interaction, including flow separation, self-sustain oscillation, and vortex dynamics, is accentuated in the following discussions.

6.2 VISCOUS–INVISCID INTERACTIONS

A viscous–inviscid interaction arises when a shear layer and an oncoming external stream infringe onto each other and the flow field must undergo a significant adjustment to accommodate and transmit the disturbance (Dolling, 2001). The phenomena are exemplified as vortex breakdown, boundary-layer separation, and vortex or shock wave impingements. Classic viscous–inviscid interactions without flow separation are noted on leading edges of an aerodynamic shape from vorticity and Mach wave interactions. The vorticity interaction is initiated by a curved bow shock over a blunt body. If the vorticity in the shock layer is sufficiently strong, the boundary-layer structure is affected by the external vorticity. This is often referred to as vorticity interaction or the streamline interaction (Hayes and Probstein, 1959). Since the coupling between the momentum and energy equations is weak, the effect of viscous–inviscid interaction to heat transfer rate shall not be as large as on the surface shear. However, when the vorticity interaction becomes very intense, the analysis has to be carried out by solving the Navier–Stokes equations.

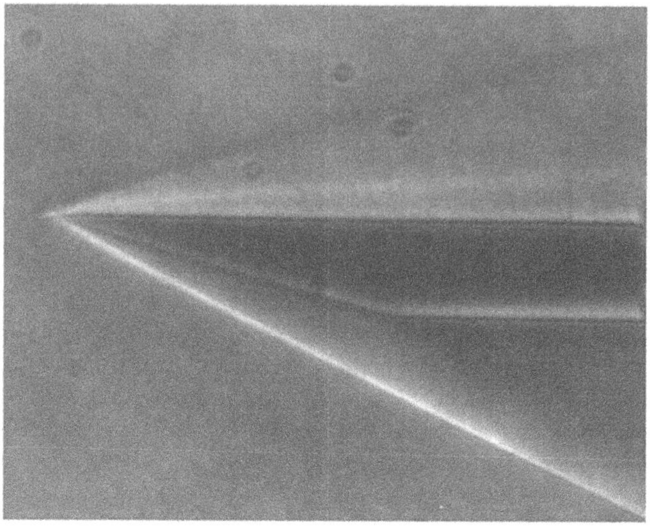

FIGURE 6.1 Schlieren photograph of leading edge shock over the flat plate of a wedge at zero incident, $M_\infty = 5.15$.

The second type and the most noticeable viscous–inviscid interaction is induced by an outward displacement of the shear layer in supersonic flow known as the pressure or Mach wave interaction. The pressure interaction occurs at the sharp leading edge of a solid surface that is another unique phenomenon in hypersonic flow involving Mach waves through flow deflection and elevated surface pressure. In hypersonic flow, the boundary layer is no longer negligible to the oncoming free stream. In fact, according to the boundary-layer theory, the growth rate is singular at the sharp leading edge. Regardless, the growth rate of the boundary-layer displacement thickness is significant at the leading edge. This physical phenomenon has been routinely observed in hypersonic flow over a sharp leading edge even if the surface is immersed at a zero incidence to the oncoming stream. In Figure 6.1, the leading-edge shock structure over a flat plate of a half wedge at zero angle of attack is clearly displayed by a schlieren photograph. This Mach wave interaction is negligible in low supersonic or subsonic flows. The photographed flow field is generated by a sharp leading edge wedge at a Mach number of 5.15 and a Reynolds number of 1.61×10^6/m.

The pressure interaction is clearly demonstrated by numerical simulations by solving the compressible Navier–Stokes equations. At the sharp leading edge, the sudden surface pressure jump is induced by a positive and finite slope of a local finite displacement surface of the boundary layer. In Figure 6.2, two solutions by the compressible Navier–Stokes equations at Mach numbers of 5 and 10, and with a Reynolds number based on the running length of 1.5×10^5 are shown. The surface pressure distributions are obtained for an isothermal wall. The sharp rises of surface pressure following by a rapid expansion is the characteristic pattern of Mach wave interaction and the elevated leading-edge pressure level increases with the free stream Mach number.

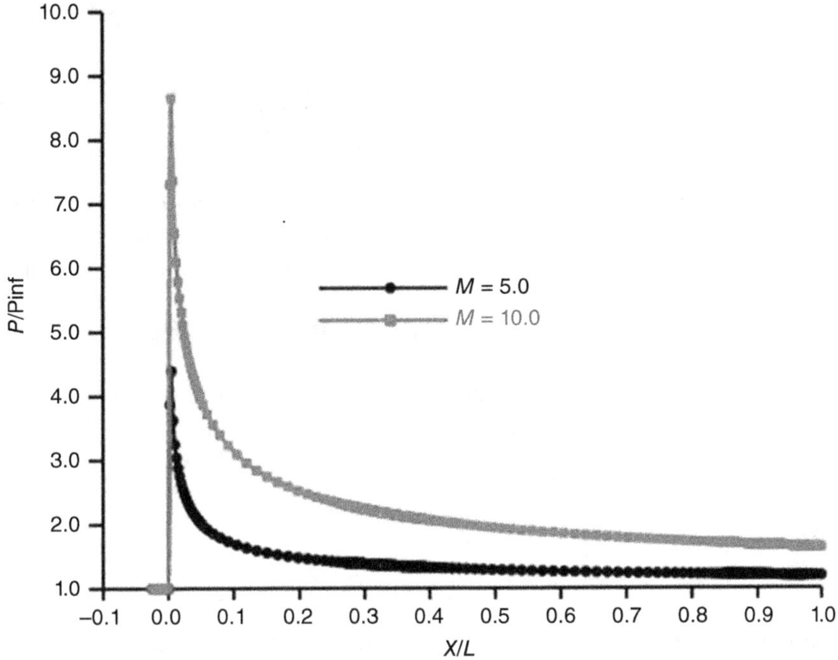

FIGURE 6.2 Pressure interaction at hypersonic sharp leading edge, $Re = 1.51 \times 10^5$.

The hypersonic Mach wave interaction is induced by the longitudinal curvature of a boundary-layer displacement thickness. According to classic boundary-layer theory, the equivalent body to an inviscid flow is the original body inclination plus the displacement thickness of an attached boundary layer. The local hypersonic similarity parameter is then

$$\kappa = M_\infty \left(\alpha_b + \frac{d\delta^*}{dx} \right) \tag{6.11}$$

For the case of a flat plate, the boundary-layer thickness can be scaled as

$$\delta^2 \sim \left(\frac{\mu_e x}{\rho_e u_\infty} \right) \sim \left(\frac{\mu_e}{\mu_\infty} \right) \left(\frac{\rho_\infty}{\rho_e} \right) \left(\frac{\mu_\infty x}{\rho_\infty u_\infty} \right) \tag{6.12}$$

Using a linear viscosity-temperature relation, $\mu_\infty/\mu = C_\infty T/T_\infty$, the density ratio across the boundary layer is the reciprocal of the temperature ratio that is proportional *to* $(\gamma - 1)M^2{}_\infty/2$, yielding

$$\frac{\delta^*}{x} \sim \frac{\delta}{x} \sim \frac{\gamma - 1}{2} \frac{M_\infty^2 \sqrt{C_\infty}}{\sqrt{\mathrm{Re}_x}} \tag{6.13}$$

At the leading edge of a sharp flat plate at zero incidence, the hypersonic similarity parameter becomes

$$\kappa = M_\infty \left(\frac{\delta}{x}\right) = \frac{M_\infty^3 \sqrt{C_\infty}}{\sqrt{\mathrm{Re}_x}} = \overline{\chi} \tag{6.14}$$

The induced surface pressure of the plate is obtained by substituting the hypersonic similarity parameter, $\overline{\chi}$ into the tangent wedge approximation (a simplified oblique shock equation), and expanding the surface to free stream pressure ratio according to the large and small values of $\overline{\chi}$. The induced pressure is further classified into the strong and weak interaction according to whether the value of $\overline{\chi}$ is greater than or smaller than the value of 3.

The strong and weak pressure interaction distributions at the leading edge are expressible, respectively, as (Hayes and Probstein, 1959)

$$\frac{p_w}{p_\infty} = 0.514\overline{\chi} + 0.759$$
$$\frac{p_w}{p_\infty} = 1 + 0.31 + 0.05\overline{\chi}^2 \tag{6.15}$$

The pressure interaction generates a streamwise distribution which is inversely proportional to the square root of the distance from leading edge. For weak pressure interaction, flow deflection is small and the interaction appears as a small perturbation and the boundary layer still grows like $x^{1/2}$. However, when the interaction parameter is sufficiently large, the pressure interaction through a feedback loop alters the structure of flow field and the boundary layer grows like $x^{3/4}$. It is an important illustration that a small local perturbation to the surface shear layer growth can substantially modify the flow field structure through viscous–inviscid interaction. The comparison with experimental data by Bertram (1957) is given in Figure 6.3 for flows over an insulated flat plate with a Prandtl number of 0.725 and the specific heats ratio of 1.4. It is remarkable how well the results by asymptotic analysis agree with experimental measurement. However, the detailed interacting phenomenon can be achieved only by solving the Navier–Stokes equations.

A large group of strong viscous–inviscid interactions, including flow separation, has been conducted over the past 70 years via experimental observations and computational simulations (Chapman et al., 1958; Dolling, 2001). The majority of these phenomena are focused on the most complex shock–boundary-layer interaction, because they impact strongly on the increased surface shear stress and catastrophic heat load. The results of viscous–inviscid interaction is also extremely remarkable that drastic different flow structures are induced by boundary-layer separation triggered by a small streamwise adverse pressure gradient. It becomes a critical issue for flow control across the entire spectrum of fluid dynamics. To describe inviscid–viscous interactions by solving the Navier–Stokes equations is one of the cornerstones of fluid dynamic research; it is also a challenging building block in computational simulations. Therefore, only the shock impingement, compression ramp, and corner flow are discussed at here.

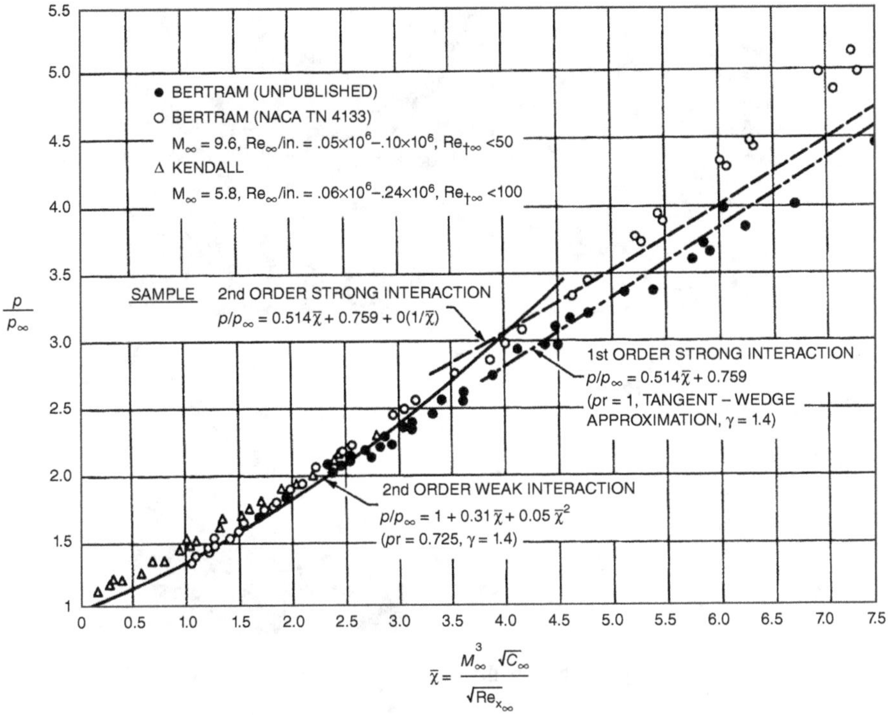

FIGURE 6.3 Verified Mach wave interaction theory with experimental data. From Bertram, H., NACA TN 4133, 1957.

A benchmark for this type of viscous-inviscid interaction is the compression ramp that is also the generic configuration encountered in high-speed aerodynamic applications. A side-by-side presentation of numerical simulation and experimental observation for turbulent flow over a 25° compression ramp is depicted in Figure 6.4. The incoming flow has a free stream Mach number of 2.96 and a Reynolds number of 10^7 (Shang and Hankey, 1975). The solution of the numerical simulation utilizes a relaxation eddy viscosity model that is based on the observation that the Reynolds shear stress after an abrupt perturbation will convey along a streamline exponentially to approaching a new state as an extra-rate-of-strain correction. The interacting turbulent flow simulation using this turbulent closure compares well with the experimental observation by an interferogram. Upstream of the corner, a clearly defined boundary layer can be discerned. The normal velocity component in the boundary layer first reveals an abrupt increase in magnitude ahead the ramp and then exhibits a rapid pressure rise in the wall region as the corner is approached. Adjacent to the separation point, the coalescence of compression waves into a shock is evident. The penetration of the induced shock wave reaches the shear dominated inner region. Extensive separation is located around the corner; the boundary-layer growth increases rapidly but still reveals a clearly identifiable outer edge. Further downstream of the corner,

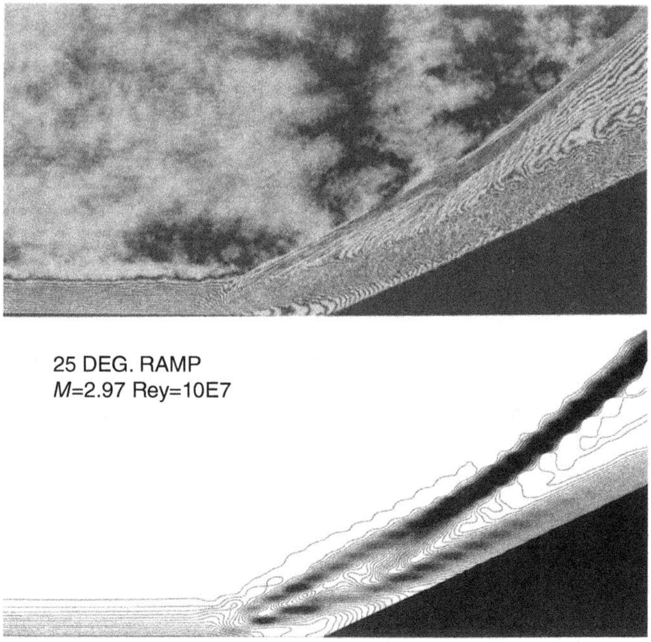

FIGURE 6.4 Computational result and interferogram for flow over a 25 ramp.

a second coalescing shock wave at the reattachment becomes discernible. The reattaching wave eventually intersects and merges with the leading separation shock and produces a distinctive overall change in the shock wave system. The outer edge of the shear layer exhibits a highly irregular pattern associated with low amplitude but high-frequency oscillations. The numerical result duplicates the essential features of the experimental observation.

For a two-dimensional computational simulation at a Mach number of 2.96, the point of separation is defined by the zero surface shear. For all the investigated cases from 15° to 25°, the incipient point of separation and the predictions of the separated flow domain are generally confined within the experimental uncertainty. An important verification from the computational simulation is that the local maximum heating is recorded at the reattachment of the separated shear layer. This particular behavior at the reattachment location is unique to all viscous–inviscid interaction involving flow separation.

From a pure analytic viewpoint, there is a class of inviscid–viscous interactions frequently referred to as free-shear interaction and all have similar flow field topology including flow separation. The compression ramp and shock wave impingement on a boundary-layer problem belong to this category (Shang et al., 1976). The two flow fields possess nearly the identical general features in all aspects. However, a small difference also exists, namely the expansion region embedded between the separation and the reattachment shocks of the impingement is substantially larger

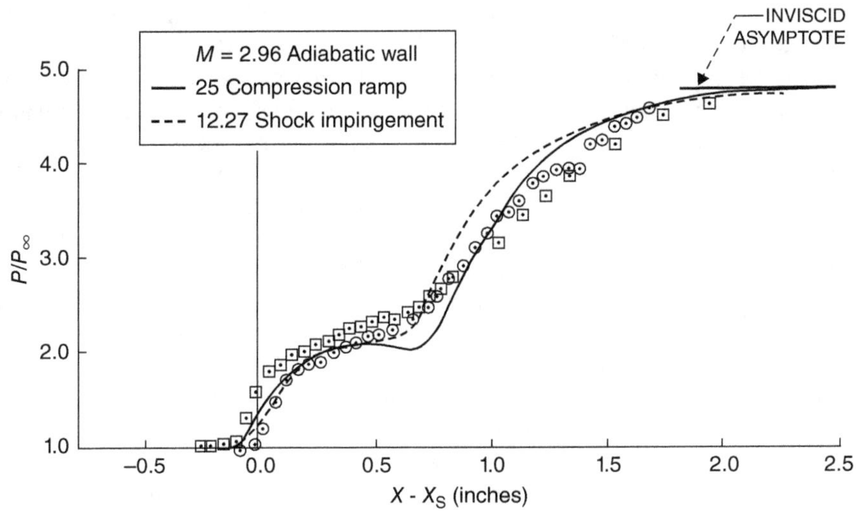

FIGURE 6.5 Comparison of surface pressure distribution of compression ramp (25°) and shock wave impingement (12.27°) with data.

and more obvious than the compression ramp. In this sense, the shock wave system produced by an incident oblique shock may be considered to be a variant to the compression ramp configuration.

The surface pressure distribution and region of flow separation of a 25° ramp and the shock impingement interaction that is generated by a 12.27° wedge are presented together with experimental data in Figure 6.5. The shock impingement data are collected for a Reynolds number of 1.2×10^7, which is slightly greater than that for the compression ramp of 1.0×10^7 but the Mach number is identical for both cases. The maximum deviation between the data and the calculated value is about 15% and occurs near the reattachment point. However, the overall agreement in pressure distribution with experimental data is observed. A common feature of the viscous–inviscid interaction is the upstream pressure propagation in the supersonic flow. The adverse pressure gradient is actually transmitting against the external flow direction in the subsonic portion of the boundary inner layer to induce the flow separation. Another important feature shared between two viscous–inviscid interactions is the dual-minimum distribution of the skin-friction coefficient in separated flow region that associated with the pressure plateau upstream to the downstream inviscid pressure asymptotes. It is a classic behavior of the upper and lower branch solutions of the boundary layer theory associated with the pressure gradient.

The hypersonic equivalent principle or the law of plane cross section reveals that the streamwise variation is less strenuous in comparison with the cross-flow plane (Il'yushin, 1956). In fact, this equivalence is valid for the entire flow field to second-order accuracy of the small flow deflection angle in comparison with the characteristic length dimension. This approximation is based on two premises; first, the maximum transverse dimension of the body is much smaller than the body length, $\tau = \delta/l \ll 1.0$. Second, the hypersonic similarity parameter $M_\infty \tau \geq 1.0$ is the upper limit of hypersonic small perturbation. These small-disturbance criteria are met by the corner flow and demonstrated by computational simulations. The level of perturbation to create a modification to flow field structure is achievable by electromagnetic effects.

An axial corner is formed by the intersection of a 15° wedge and a flat plate. The free stream Mach number is 12.5 and the Reynolds number based on the axial length is 1.21×10^7. By taking advantage of this salient feature of the three-dimensional corner flow field structure, a relatively sparse three-dimensional grid system ($8 \times 32 \times 36$) with a highly clustered frustum at the apex is adopted (Shang and Hankey, 1977). The free stream Mach number is 12.5 and the Reynolds number based on the axial length is 1.21×10^7/m. Laminar flow is found to exist throughout the investigated domain. Experimental investigation reveals a rather complex intersecting shock wave system with slip surfaces, and the surface oil film flow indicates extensive flow separation. Moreover, the experimental observation also recognizes an approximately conical flow from surface pressure measurement and shock wave structure formation. In other words, the inviscid dominant flow field is nearly invariant with respect to a generating ray that originates from the apex of the corner, regardless of a different scaling law for the growth of the viscous region.

The comparison of the surface oil flow pattern with the calculated surface shear stress components is depicted in Figure 6.6 (West and Korkegi, 1972). The magnitude of the shearing force decays rapidly toward the outer region differing by an order of magnitude at the corner. The convergent lines are easily detected and, according to the topology of the limiting streamline, the converged surface shear line is a necessary

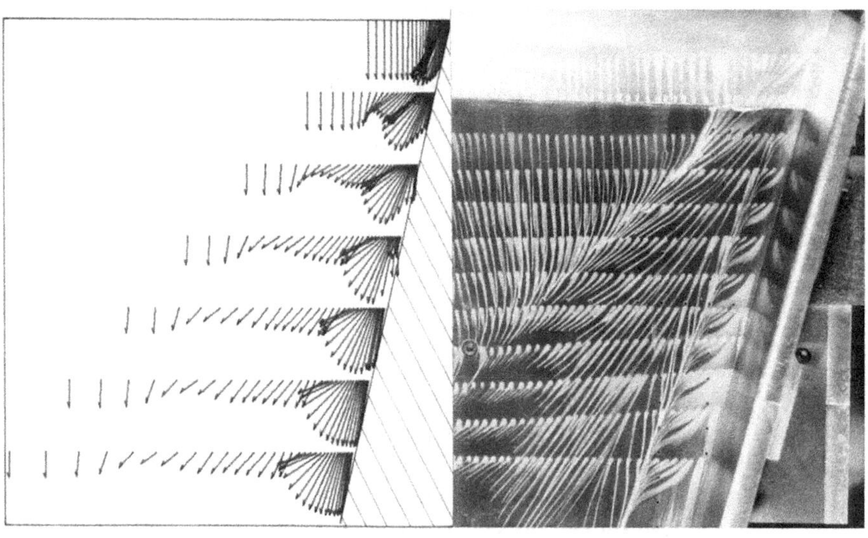

FIGURE 6.6 Comparison of the surface shear stress component with oil flow pattern, three-dimensional corner by intersecting 17° wedge and plat, $M = 12.5$, $\mathrm{Re} = 1.21 \times 10^7$.

condition for flow separation (Tobak and Peake, 1982). The outer convergent line corresponds to the wedge shock impingement at an angle of 44° with respect to the free stream, while the inner convergent line has a deflection angle of 22°. These lines vary only slightly from the leading edge as they develop downstream. The divergent lines are more difficult to detect than the convergent lines. The innermost divergent line can be recognized to be deflected about 17° from the free stream, while the outer divergent line is oriented at 27° as revealed by the oil film image. The numerical result of the surface shear duplicates the outer and inner convergent lines but fails to capture the secondary vortex footprint. In spite of the fact that the numerical resolution has resolution shortcomings, the outstanding features of the experimental observation have been reproduced by the computational simulation.

As a further verification, experimental surface pressure measurements on the flat plate along with numerical results were compared (Shang and Hankey, 1977, 1979). The agreement between the data and calculation was good, in spite of the sparse grid-density resolution. In general, the computational simulations either by the coarse and finer mesh systems are confined within measurement uncertainty. An interesting flow structure within the asymmetric three-dimensional corner is that the projected velocity vector exhibits a sink-like topology. In the cross flow plane, the velocity components are flowing toward the center of the vortex by forming a reverse flow zone above the plate surface. The surface pressure distribution in the cross-flow thus now can be viewed to be aligned with the cross-flow orientation. The trough-like pressure variation outward from the impingement shock that is generated by the wedge has been identified from experimental observation as the indication of the embedded flow separation. A better definition of the surface pressure behavior prior to the strong compression toward the corner is captured by the finer mesh-spacing calculation. It needs to be borne in mind that the three-dimensional flow separation projected on the cross-flow plane really is the result of a small perturbation in the streamwise direction.

An intricate wave pattern by shock-wave–boundary-layer interaction is further delineated by a detailed Pitot pressure survey given by Figure 6.7. The shock structures generated by a wedge perpendicular to a flat plate surface or by two symmetrical wedges are similar. For the corner of the intersection of a plate and wedge, the coalesced shock wave over the plate due to the growth of the boundary layer also forms over the plate and is perpendicular to the wedge shock. For the double-wedge intersecting corner at 9.48°, the oblique shocks have identical strength and yield a symmetrical shock formation with respect to the bisector of the corner. The projected intersection of these two shocks is embedded in the corner region where the local flow is subsonic. As a consequence, in the inner corner region, the shock waves must be bridged by two triplet points to terminate these shocks before entering the local subsonic region (Shang and Hankey, 1979). The shock intersection is deflected away from the corner at the triple point, the shock perpendicular to the wedge surface eventually forms a lambda shock normal to the plate surface. A mirror-like shock structure is also formed over the other wedge surface. Of cause, a simpler and unsymmetrical shock formation results over the intersection of the wedge-plate corner. The triplet shock structures are presented for both laminar

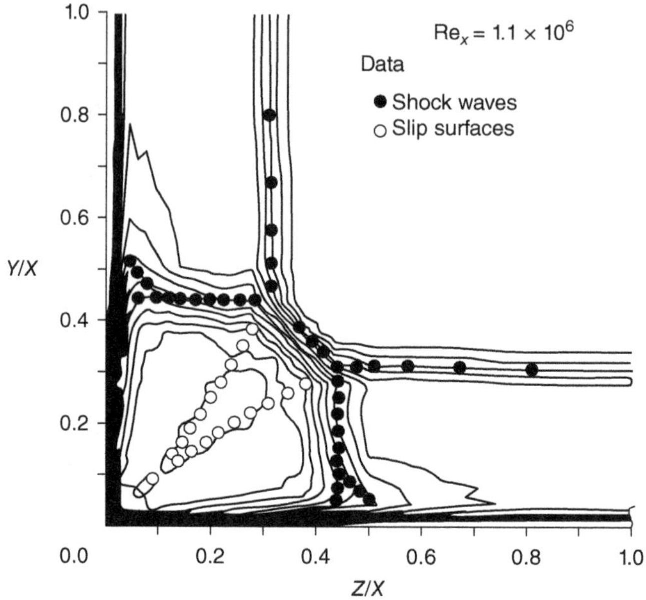

FIGURE 6.7 Triplet shock structure in the corner region.

($Re_x = 0.27 \times 10^6$/m) and turbulent ($Re_x = 1.1 \times 10^6$/m) flows. Verification of the computational simulations is accomplished by comparing with experimental measurements on the location of the shock waves, triplet points, and slip surface. The numerical results also show a reduced terminating embedded lambda shock decreases from laminar to turbulent flow.

The most striking feature of the asymmetrical corner flow field is the formation of a completely embedded supersonic cross-flow and a high-energy stream within the primary separation zone by expanding rapidly around the triple point. The cross-flow Mach number based on the magnitude of the conical velocity component is presented in Figure 6.8. This graph defines the subsonic region of the conical cross-flow plane.

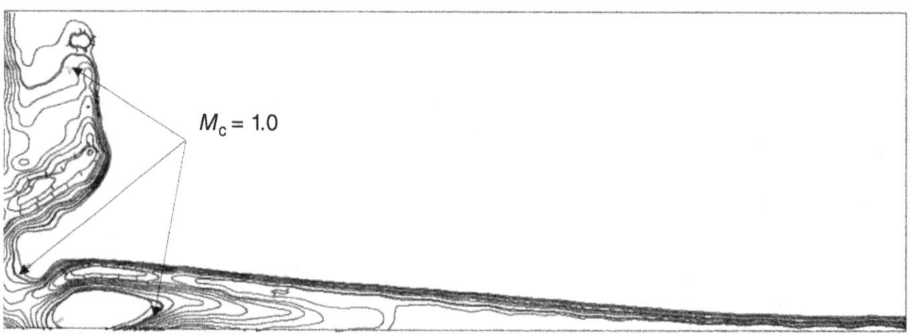

FIGURE 6.8 Cross-flow Mach number contour of an asymmetric corner.

For the asymmetrical corner, the flow region inboard of the triple point is subsonic, thereby negating the possibility of a pair of triple points like that of the symmetric corner. The cross-flow Mach number immediately adjacent to the corner even reaches a maximum values of 1.56 after the expansion at the location 1.0 mm above the plate and 5.28 mm from the corner. The embedded high-energy jet-like structure impinges and scavenges the wedge surface to generate a very high heat-transfer rate away from the corner. The heating rate is an order of magnitude greater than that on the surfaces away from the corner. Unfortunately, there are no accompanied data to verify the computational results. Nevertheless, this interacting flow phenomenon may have provided an explanation to the serious thermal damage on the ramjet pylon X-15 (Korkegi, 1971).

Research on three-dimensional shock-wave–turbulent-boundary-layer interactions has been sustained for further understanding. Noticeable progress has been made on the shock-boundary-layer interaction within symmetric or asymmetric double-fin geometries (Gaitonde and Shang., 1998). Therefore, the shock wave always intersects the boundary layer. The most severe challenge for this type of viscous–inviscid interaction is always associated with strong adverse pressure gradient, which leads to flow separation. Under this circumstance, the flow field constitutes complex vortical structures that impinge and reflect from the sidewalls and eventually realigning to exit (Gaitonde et al., 1999). For most computational simulations, two primary deficiencies emerge: sufficient numerical resolution for the unsteady motion, and the other formidable and irresolvable issue is the turbulent closure (Babinsky and Harvey, 2011). A major observation from investigations of a three-dimensional interaction is that the vorticity field consists of an inner core and an outer annulus. The separation space between the distinct vortices is the incoming boundary layer. Experimental observations of increasing interaction strength and the Pitot pressure profile show a pinching effect by the entrainment flow from side walls.

The evolution of vorticity reveals that by increasing the interaction strength, there is a corresponding increase in the longitudinal component of vorticity. Strong viscous–inviscid interactions are always initiated by a small perturbation started by flow separation, and the wall-bounded shear layer is triggered by a relatively small adverse pressure gradient. Thus, a modulation of the onset perturbation can profoundly alter the chain of events in viscous–inviscid interactions. All insights may become key elements when a new flow controls mechanism by introducing an electromagnetic force or energy.

6.3 SELF-SUSTAINED OSCILLATIONS

The fundamental process of self-sustained oscillation in a fluid is linked to the inherently unstable, amplifiable free shear layer fluctuation and a feedback mechanism by forming a closed loop of communication (Hankey and Shang, 1980). A resonant situation arises when a forcing function excites the shear layer in the frequency range where amplification is possible. The disturbance grows until it reaches a limiting cycle by viscous dissipation. This large group of unsteady fluid dynamic phenomena exists in very wide circumstances from open cavity oscillations, inlet buzzing, spike-tipped resonance, and periodic pulsation in counterflowing jet injection (Shang et al., 1982, 1985, 2001, 2002, 2003). The highly organized oscillations by selective disturbance amplification are often encountered in shear-layer impingement. The feedback loop is usually closed by the upstream pressure propagation through the subsonic portion of a flow field. The wave propagation is controlled by acoustic, but the characteristic wave speeds in a partially ionized gas or highly excited plasma may elevate to approach the speed of light. It begs the question whether some of the detrimental oscillations can be alleviated or completely suppressed by plasma flow control. Therefore, it needs to better understand and to correctly identify the control mechanisms of self-sustained oscillatory motion in fluid motions.

Rayleigh was the first to show that velocity profiles containing an inflection point are unstable for inviscid incompressible flows (Rayleigh, 1880). More recently, it has been found that the instability of a shear layer exists only at low frequency $f\delta/U_\infty < 1/4\pi$, where δ denotes the shear-layer thickness. From the analysis of a small perturbation induced by a low amplitude traveling wave, the stability equation of the linearized compressible shear layer can be obtained as (Lees and Lin, 1946; Michalke, 1964),

$$a^2\{[(u-c)\phi_y - u_y\phi]/[a^2 - (u-c)^2]\}_y = \alpha^2(u-c)\phi \tag{6.16}$$

where c is the complex propagation speed, α denotes the dimensionless wave number, and ϕ characterizes the amplitude of perturbation velocity. The eigenvalue of the stability equation can be calculated to determine the propagation velocity for a given velocity profile. From this information, the existence of a self-sustained oscillation can be determined. In a case study by assuming a shear-layer velocity profile, the amplification is found to be a function of wave number and Mach number. When a hyperbolic tangent velocity profile is adopted for an open cavity oscillation, instability occurs only for wave numbers less than unity, which means that resonance will not take place for a short cavity ($L < 2\pi\delta$), and the instability vanishes for Mach number greater than 2.5.

From observations on open cavities, the fluctuating separated shear layer causes periodic mass addition and expulsion from the cavity. At the rear attachment point of the free shear layer, a stagnation point is created on the rear bulkhead of the cavity that increases the local cavity pressure. As a consequence, the shear layer bulges outward and ejects mass from the cavity. The high pressure pulse then moves forward in the cavity and trailing by an oblique shock in the external stream. The pressure wave reflects at the forward bulkhead and doubles the pressure in the cavity but the flow external to the cavity is not reflected, thus the local high pressure across the shear layer deflects the oncoming external stream. A part of fluid will be entrained into the cavity. The entire process repeats itself in a periodic cycle.

The experimental data show that the rearward traveling wave is moving at the speed of sound in the cavity. The forward traveling wave is noted to be about half of the free stream value. The propagation velocity ratio is determined experimentally and found to be around $0.5 < \kappa < 0.6$. The frequencies of cavity oscillation have been given by Rossiter's formula (Rossiter, 1964):

$$f = mu_{\infty} / l(M + k^{-1}) \tag{6.17}$$

where m is the mode number and l is the cavity length. The external Mach number is denoted by M and κ is the propagation velocity ratio with respect to the free stream speed. This simple equation has been used successfully for describing cavity resonant frequencies, but there is no satisfactory prediction method for determining the disturbance intensity of different modes, m.

A first Navier–Stokes solution for a two-dimensional cavity was carried out at a Mach number of 1.5, for a cavity with a length of 91.44 cm and a depth of 40.64 cm, which yields the *L*/*D* ratio of 2.35 (Hankey and Shang, 1980; Shang and Hankey, 1984). The incoming boundary-layer thickness was prescribed at 2.54 cm and at the unit Reynolds number of 1.28×10^6/m. The numerical results were generated on a (78 × 52) rectangular mesh with grid spacing clustering along the opening of the cavity. The flow field within the confined cavity is subsonic, except perhaps in the region adjacent to the cavity opening. The orderly development of the free shear layer originates from the forward bulkhead and reattaches as a turbulent boundary layer downstream at the rear bulkhead. A recirculation flow is developed in the cavity with a much lower magnitude than the external stream, and the center of the recirculation is located near the upper corner of the rear bulkhead. A quantitative comparison of the calculated mean velocity profiles immediately upstream and downstream of the cavity with experiment indicated a difference of 7% with experimental measurements. The calculated time-average surface pressure distribution in the cavity also attained a similar accuracy with the experimental data.

From the time evolution of the static pressure at a fixed location in the cavity ($x/L = 0.66$ and $y/L = 0.96$), a persistent oscillation is recorded over nine characteristic periods ($t = 1.82 \times 10^{-2}$ s). Since the pressure oscillation over the cavity is composed of several frequencies of different amplitudes, it is a compound wave system and the discrete frequencies of modes of oscillation are difficult to discern. The basic information such as the relative phase angle and amplitude of each distinctive wave requires a spectral analysis. The numerical result is obtained by treating the pressure wave frequency to be commensurable (Shang and Hankey, 1984). Spectral analysis reveals four distinctive discrete frequencies of 154, 308, 462, and 616 Hz recognized as the first, second, third, and fourth modes of the oscillatory pressure disturbance. Figure 6.9 shows that the higher modes of oscillation decay rapidly and the numerical result dissipates even faster to show the insufficient wavenumber resolution. Nevertheless, an agreement with experimental data is reached to show a dominant second mode of the pressure oscillation.

A large number of subsequent computational simulations for the oscillating cavity flow have been performed, including three-dimensional simulations with a sophisticated turbulence closure and high-resolution LES (large eddy simulation). Additional details in flow field structure and better understanding have also been recovered, but the essential physical of the oscillating cavity remains unaltered (Shang, 2004).

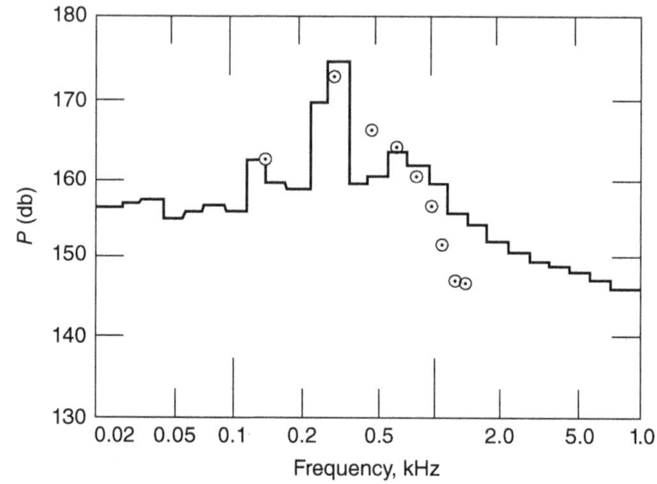

FIGURE 6.9 Comparison of spectral results, $M = 1.5$, $\text{Re} = 2.6 \times 10^7$, $l/D = 2.25$.

The occurrence of buzz on spike-tipped reentry configuration has been studied since the 1970s (Maull, 1962; Finley, 1966). It is also a class of self-sustained oscillation generated by shear-layer impingement. The spike-tipped configuration consists of a protruding spike from a blunt forebody. The bow shocks generating over the spike impinges on the following afterbody triggers flow separation. The bow shock formation oscillates between an oblique conical envelope extending from the spike to a normal shock and then collapses over the entire configuration, and the cyclic pattern will repeat persistently. The amplification of perturbations in the shear layer is still the consequence of the hydrodynamic instability from an inflection point in the velocity profile. Flow separation is not a necessary condition for the shear-layer instability but rather its existence is ensured by an inflection point in the velocity profile. The oscillating motion appears to occur in translational or rotational or a combination of both to be a phenomenon that can occur in axisymmetric flows.

A composite presentation of oscillating flow fields over a spike-tipped configuration is depicted by Figure 6.10 (Shang and Hankey, 1984). An identical temporal sequence of shock oscillations is captured by schlieren photographs and density contours of a computational simulation at the same flow conditions, $M = 3.0$, and a nominal Reynolds number of 7.87×10^6/m. The Reynolds number based on the spike length (38.1 cm) is 2.75×10^5. The grid-point density of the truncated three-dimensional computational simulation is sparse for the body conformal $(61 \times 42 \times 4)$ grid, and processed on four axial planes at 3° apart to reinforce the symmetric flow field condition. The numerical results actually display more information than that of the schlieren photographs. The computational simulations capture the complex shock structure over the intersecting shocks over the blunt forebody even including the shear-layer separation and reattachment, and the slip streams emerging from the shock interaction.

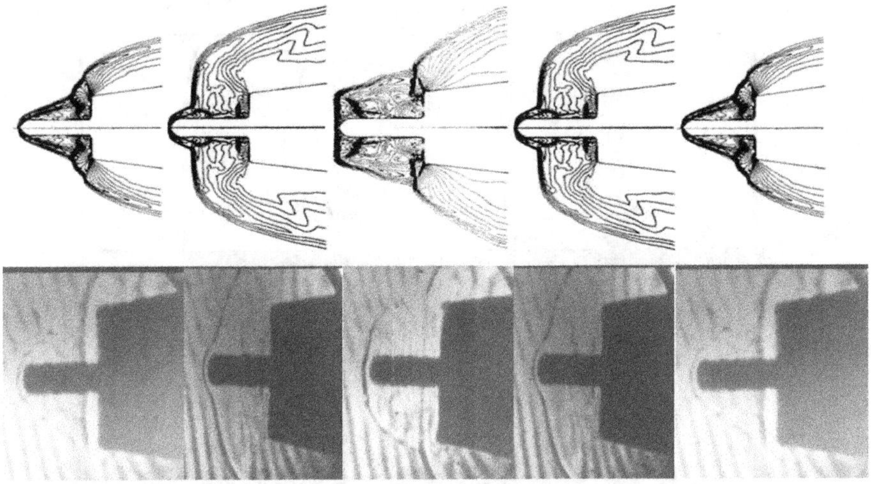

FIGURE 6.10 Flow oscillations over spike-tipped body, $M = 3.0$, Re = 7.87 × 106/m, $l = 38.1$ cm.

The RMS pressure data at the shoulder of the forebody reaches a maximum value for a spike length of 44.5 cm and decreases gradually as the spike is shortened. The computational simulations duplicate the experimental observation when the oscillation ceased for a short spike length of 12.7 cm. At the other extreme of a long spike length, the RMS value also drops rapidly after reaching a maximum. The cutoff frequency is related to the ratio of spike length to the afterbody diameter, l/d and local Mach number downstream of the bow shock.

For a spike length of 38.1 cm, the periodic cycle lasts around 0.0020 s or by advancing the numerical solution in 2000 time steps. The oscillatory solution is sampled over four periods and the simulation result maintains its amplitude and wave form; no noticeable numerical damping or decay is recognized. The self-sustained oscillation is evident and the magnitude of the pressure pulsation is about 17 times of the free stream value. A comparison between an oscillograph and the computed results is given in Figure 6.11. The first four modes of the discrete frequency are predicted within 7% of the experimental measurements. The discrete frequencies by a spectral analysis are 2.665, 5.331, 7.997, and 10.66 kHz. The discrete frequencies of the higher modes of oscillation are commensurate with the principal mode. The amplitude of the principal mode (2.665 kHz) is dominant over the rest of the higher buzz modes. In spite of insufficient computational accuracy and a lack of the accurate turbulence closure, the agreement between the experimental data and computational results still reflect the dominant phenomenon of viscous–inviscid interaction.

There are also numerous unanswered questions in the supersonic spike-tipped buzzing. For example, hysteresis in the measured surface pressure and oscillatory frequencies by increasing and decreasing the spike length is not fully understood. A convincing explanation of a cessation of buzzing at the limits of the short and long spike configurations is also not forthcoming according to the classic mechanics.

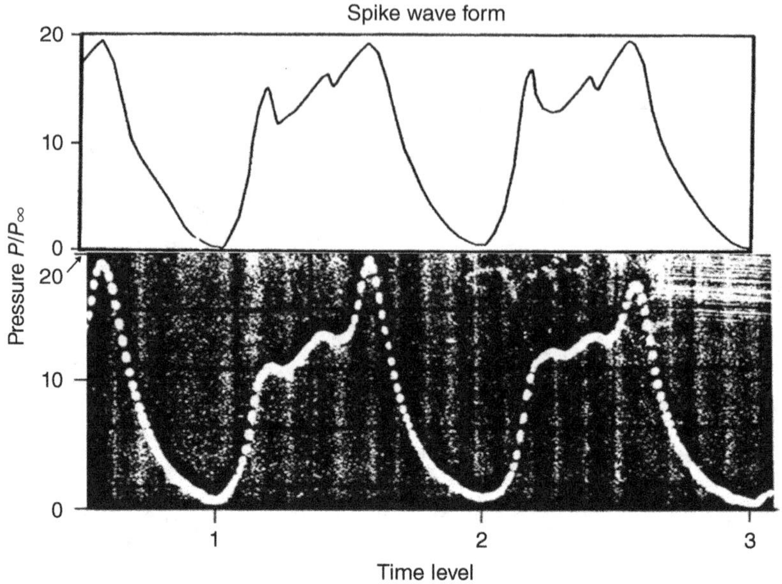

FIGURE 6.11 Wave form of spike-tipped oscillation, $M = 3.0$, $Re = 2.75 \times 105$, $l = 38.1$ cm.

These flow bifurcations are not completely determined by a set of known initial conditions but are lumped into chaotic and nonlinear fluid dynamics behavior.

According to the shear–layer stability theory and experimental observation of flow oscillations over a spike-tipped blunt body, buzzing will not take place either at a high supersonic Mach number or beyond a limited range of spike length to the after-body diameter, $0.4 \leq l/d \leq 0.88$. In the hypersonic domain, an oscillatory motion has been detected by experiment in the range of $0.25 \leq l/d \leq 2.5$. In both flow regimes, the oscillatory frequencies are from 50 to 600 Hz. These results are substantiated by a series of studies by experimental and computational simulations at a free stream Mach number of 5.80 and spike length-to-diameter ratios of 3.84, 12.0, 16.0, and 20.0. For the ratios of spike length-to-diameter tested, the fluctuating pressure amplitude normalized by the free stream value is 0.006 at 32.05 kHz. For all practical purposes, the flow field can be considered as steady. Both schlieren photographs in multiple images in long exposures and the computed streamline patterns of Figure 6.12 support the assertion.

All experimental observations were recorded from a blow-down, open jet wind tunnel with the installation of a weakly ionized air by a direct current discharge (Shang et al., 2001, 2003). The spike serves as the cathode for the DC glow discharge and a copper ring grounded to the metal frame of the tunnel as the anode. The DC discharge is rated at 8 kW with a 10 kV output source at a breakdown voltage of 4 kV. In contrast, the air flow of the tunnel must be maintained a flow rate of 0.77 kg/s by a power input of 8.07×10^3 kW. It is reasonably assured that the energy input from the DC discharge is negligible, and the ionized air concentration is even less and appears as a trace amount.

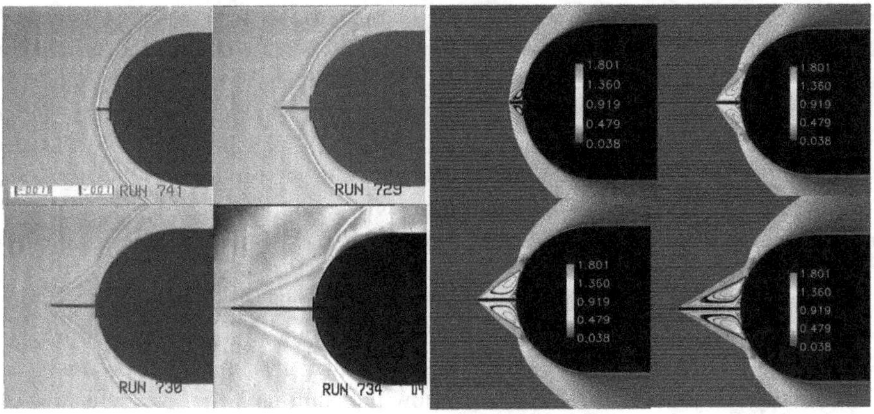

FIGURE 6.12 Spike-tipped blunt body in hypersonic flow, $l/d = 3.84$, 12.0, 16.0, and 20.0.

For this reason, only solutions for the compressible Navier–Stokes equations are used for this investigation. The validation of numerical simulation is derived from the comparison of the computed and experimental schlieren observations. In Figure 6.13, a composite schlieren image and a computational simulation are presented for the model with a spike length of 25.4 mm. The shock wave emanating from the spike and the hemisphere afterbody are generated under identically flow conditions. The

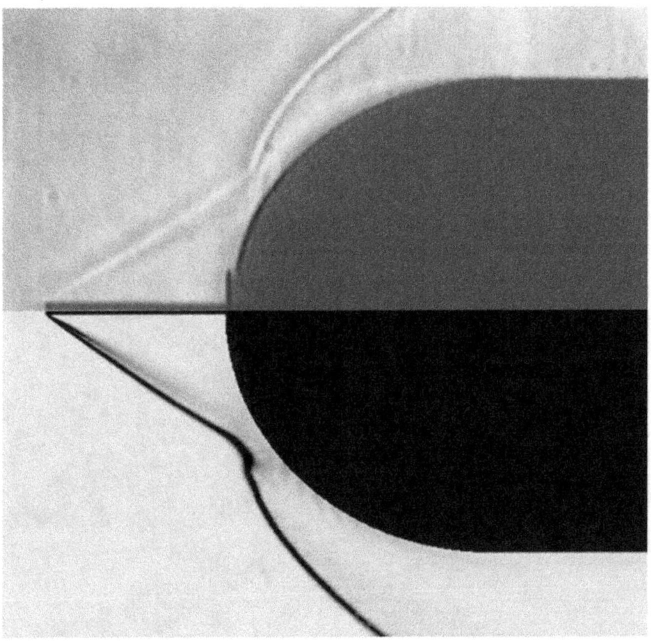

FIGURE 6.13 Comparison of schlieren image with photograph.

maximum number of cells using in the finite volume scheme is 57,657. The finest cell spacing adjacent to the solid surface is maintained by the law-of-the-wall variable on the order of magnitude of unity, $2.08 \leq y^+ \leq 3.24$ with an unstructured grid system. The interaction of the attached shear layer and the slip stream originating from the shock intersection is clearly duplicated, and the overall agreement of the observed image and computational simulation is excellent. The most important finding is that a small perturbation by electromagnetic forces for controlling an unstable flow is ineffective. The electromagnetic–aerodynamic interaction must be reinforced by an amplification mechanism.

Another well-known self-sustained oscillation is the counterflow jet injection from a blunt body. The mechanism of unsteady motion is due to shock wave bifurcation (Shang et al., 2002, 2003). The counterflow jet issuing from a blunt body in supersonic flow actually has two distinct states. At a low injection pressure, the shock system interacts with the jet replacing the upstream bow shock; the modified shocks in the subcritical state are appeared by a series of unsteady blunt conical formations known as LPM (long penetration mode). At higher injection pressures, the displaced shock retracts from the conical to a bow configuration and returns to a steady supercritical state. This is known as the SPM (short penetration mode) as shown in Figure 6.14. In essence, the flow field consists of the counterflow jet issuing from the stagnation point of the blunt body and induces a fluctuating and complex shock waves system. The jet expands into a lower stagnation pressure bow shock envelope and then terminates by a Mach disk, and reverses its direction toward downstream as a free shear layer. A

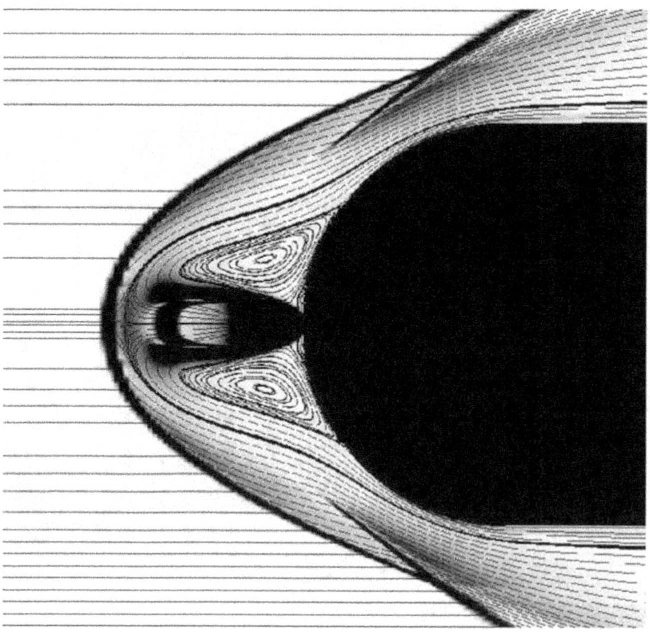

FIGURE 6.14 Unstable triple shock structure by counterflow jet injection.

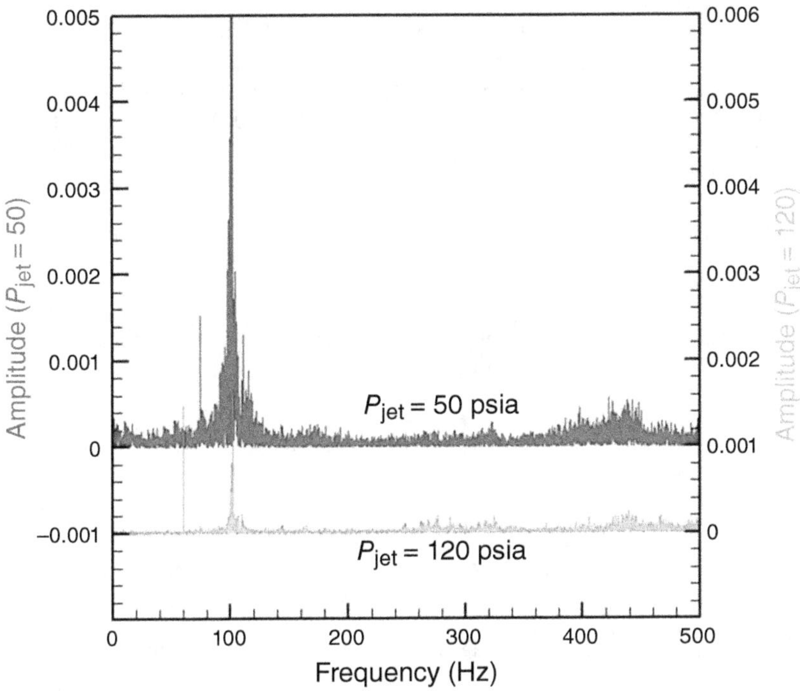

FIGURE 6.15 Spectral data of sub- and supercritical states of counterflow jet.

part of the shear layer is entrained to form a toroidal recirculation zone beneath the free shear layer. As the shear layer reattaches on the blunt body, it induces a series of compression waves that coalesce into a reattached ring shock. The counterflow jet interaction replaces the single bow shock over the blunt body by a triple-shock structure.

The spectrum substantiating the counterflow/shock bifurcation is displayed in Figure 6.15. The data are collected for an injecting nozzle with a throat diameter of 2.4 mm. The tunnel is operated at a stagnation pressure of 689.47 kPa. Only two typical data over the entire bifurcating range are included. One yields the subcritical state of the injecting stagnation pressure at P_j = 50 psia (344.74 kPa) in which the oscillatory motion is sustained. In the subcritical state, the pulsations cover a frequency spectrum up to 500 Hz. Two dominant discrete frequencies at 100 and 440 Hz are revealed. The oscillatory shock structures persist over the entire subcritical state. In high-speed drag reduction applications, this pulsating nonunique subcritical state is called the LPM that is mentioned above. Beyond the critical value of the bifurcation point, the flow field is nearly steady, or like that of the condition for which the counterflow jet is shut off. The extended conical shock structure is referred to as SPM, with the pulsation amplitude merely around 10 dB. The recording of the spectral data for the supercritical state at P_j = 120 psia (827.70 kPa) is superimposed in this figure.

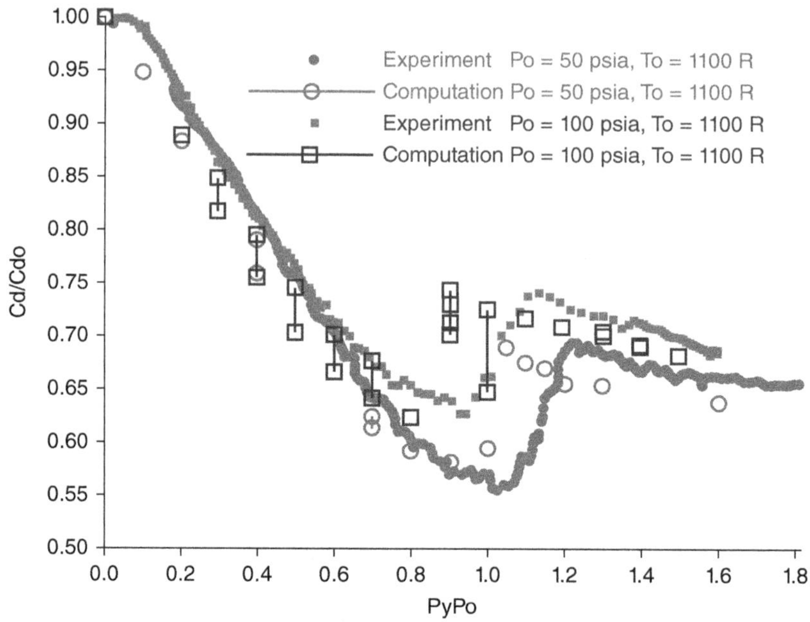

FIGURE 6.16 Data and computed drag coefficients over the entire injecting pressure range.

The bifurcated shock wave structure, in fact, leads to a substantial reduction of the wave drag. The calculated drag coefficients are depicted in Figure 6.16, which duplicate the experimental condition by an injecting nozzle with a diameter of 2.44 mm at the tunnel stagnation pressure of 432.4 and 684.8 kPa, and a stagnation temperature of 611 K. Both the three-dimensional and axisymmetric computations are carried out to determine whether the bifurcation is uniquely associated with three-dimensional motion. In other words, both computational simulations, either the three-dimensional or the axisymmetric computations, reproduce the identical shock bifurcation. Therefore, it is not necessary to have an additional transverse degree of freedom for a change of dynamic states in the longitudinal orientation. Numerical results also correctly recover the oscillatory flow structures for all subcritical states. All the calculated results in the subcritical states indicate a fluctuating band that is absent substantially reduced in the supercritical states. Numerical results, however, consistently underpredict the drag over the entire range of the injecting pressure. The tunnel stagnation pressure of 432.5 kPa yields a greater discrepancy of 6.1% against the data, in contrast to 2.7% for the case where the stagnation pressure is raised to 684.8 kPa. However, both calculations that duplicate the testing conditions underpredict the critical bifurcation of p_j/p_o by approximately 11%. A reason is that the drag calculations by the Navier–Stokes equations do not include the base drag. Also, the error may be attributed to the turbulent closure. It is important to note that the injection through a plasma torch has been determined to be small by a direct comparison with an air jet.

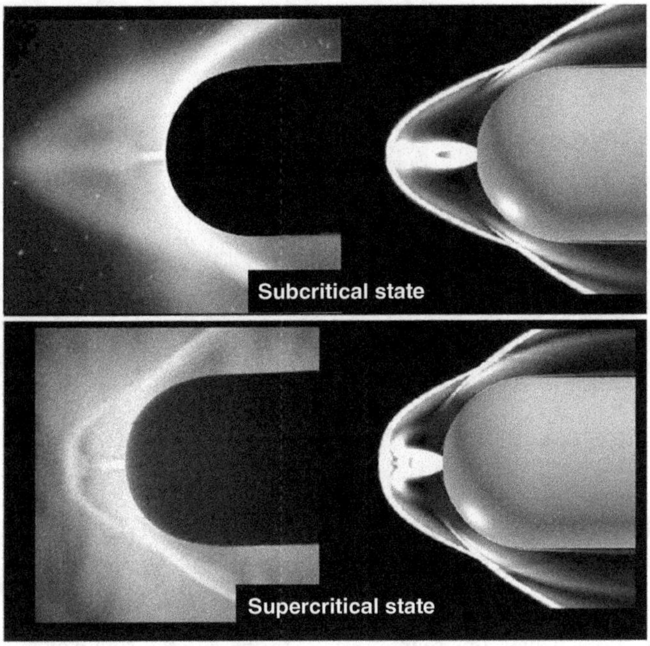

FIGURE 6.17 Schlieren and computational shock wave formations, $M = 5.28$, $p_o = 684.8$ kPa, $T_o = 611$ k, $480 < P_j < 965$ kPa.

In Figure 6.17, the bow shock bifurcation over the blunt body generated by the counterflow jet injection is described by the side-by-side presentation of schlieren photographs and the computational density contours at the subcritical and supercritical states. In the upper portion of the graphs, the fluctuating flow fields of the subcritical state LPM are produced by a counterflow jet through a supersonic nozzle operated at the stagnation pressures from 582 to 689 kPa. The schlieren photograph is recorded by a long exposure; thus, the pulsating shock wave structures at 100 Hz between the conical and collapsed triple shock formation appear to be blurred. The numerical simulation in this graph is an instantaneous result at the extreme location of the conical shock envelope. When the nozzle injection pressure elevates beyond the critical value at 689 kPa, flow oscillation ceases and the shock structure settles to the final and stable triple shock formation SPM, as shown in the lower half of the figure. The shock formation persists over the entire jet injection pressure from 689 to 965 kPa. Under this condition, both the schlieren photograph and computational results exhibit the irrefutable complex wave formations including the bow shock, Mach disk, and the reattaching ring shock.

The computational simulations also provide insight into the feedback mechanism of the self-sustaining oscillating fluid motion. As it mentioned earlier, any self-sustained oscillation flow field must be associated with an unstable, amplifiable free shear layer and a feedback communication link to reinforce the source of perturbation. For a supersonic shear layer, either in the closed domain of an inlet and cavity or

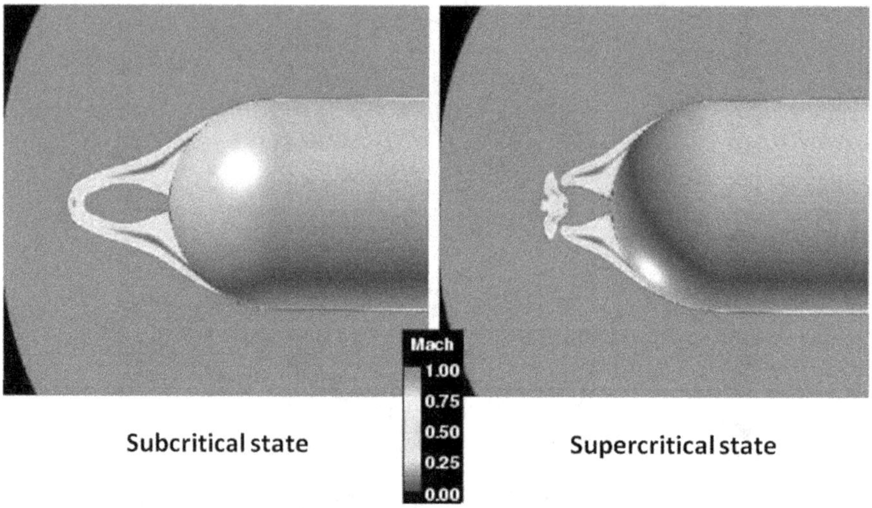

FIGURE 6.18 Mach number contours of the counterflow jet interaction.

enveloped by shock waves, upstream pressure propagation through an embedded subsonic domain is the necessary conduit. Mach number contours at the subcritical and supercritical states are given in Figure 6.18. For demonstration, all supersonic regions are presented in a uniform hue; only the contours below sonic speed are labeled. In the subcritical state, the impinging free shear layer above the injecting supersonic jet has an enclosed and continuous subsonic passage from the foremost location of the Mach disk and extends to the downstream reattachment and beyond. Thus, the sonic perturbation downstream can propagate upstream to close the feedback loop. In the supercritical state, the Mach disk that terminates the forward counterflow is detached from the reversed subsonic shear layer to permit the increase jet mass flow to expand continuously into the shock layer. The subtle change in the shear layer of the enveloping shock cuts off the feedback loop and the self-sustain flow oscillation ceases.

The opportunities for flow control using plasma actuator are obvious in self-sustain flow oscillations; a bistate of fluid dynamics is existed with distinctive characteristics. A small perturbation in either the magnitude or the signal propagating speed can lead to drastically different flow field structures. For example, boundary-layer separation is uniquely controlled by a streamwise adverse pressure gradient. The magnitude of this parameter is miniscule, which is within the effective range of the perturbation generating by electromagnetic forces. Equally important, the feedback mechanism of self-sustain oscillation through a small disturbance in a flow field is conveyed at the rate of sonic speed $c = \sqrt{(dp/d\rho)_s}$, which has a strong dependence to the ambient temperature. The feedback communication loop thus can be modified or even disrupted by volumetric Joule and electrode heating using a partially ionized gas. In fact, this type of disturbance can be introduced through transport properties that are modified by electromagnetic forces.

6.4 VORTICAL DYNAMICS

The vortex is a basic kinematic property of fluid motion and it is the necessary complimentary rotational component to the translation degree of freedom for fluid motion. The vortex is defined as a finite volume of rotational fluid bounded by irrotational fluid or a solid wall, a steep gradient around the kinematic structure is always presented (Saffman and Baker, 1979). In the macroscopic dimension, the interaction of vortical flow generated on a solid surface contributes significantly to aerodynamic lift. At the same time, vorticity may either enhance or degrade the aerodynamic performance. Vortex bifurcations at high incidence drastically affect aerodynamic behavior embracing asymmetric forebody vortex formation, vortex breakdown, and unsteady asymmetric force generation. As a consequence, a loss in control effectiveness combined with a cross flow alters completely the post-stall behavior for most lift generating surfaces. At the microscopic scale, vortices and its random interactions with the large-scale motion constitute turbulence (Bradshaw, 1994). Numerical resolution requirements for vortical dominant flow field are very stringent and for an accurate three-dimensional computational simulation are often beyond current capability. A formulation of the vorticity dynamic evaluation shall be very helpful for the basic understanding.

To derive vortical kinematics equation, only the continuity and the momentum equations are needed;

$$\frac{\partial \rho}{\partial t} + \nabla \cdot (\rho u) = 0$$
$$\frac{\partial \rho u}{\partial t} + \nabla \cdot (\rho u u + \tau) = 0 \tag{6.18a}$$

where the shear stress tensor is defined as

$$\tau = [p - (\lambda \nabla \cdot u)]\bar{\bar{I}} - \mu(\nabla u + \nabla u^*) \tag{6.18b}$$

By expanding the dyadic tensor of u and invoking the continuity condition, we have

$$\rho \frac{\partial u}{\partial t} + \rho u \cdot \nabla u + \nabla \cdot \tau = 0 \tag{6.19a}$$

From a vector identity o

$$\nabla(u \cdot u) \equiv 2[(u \cdot \nabla)u + u \times (\nabla \times u)] \tag{6.19b}$$

the divergence of τ, $\nabla \cdot \tau$, can be written into two equivalent expressions in terms of substantial derivatives;

$$\frac{Du}{Dt} = \frac{\partial u}{\partial t} + \frac{1}{2}\nabla(u \cdot u) - u \times (\nabla \times u) \tag{6.20}$$

However, only the preferred expression is given here:

$$\rho\frac{Du}{Dt} + \nabla(p - \lambda\nabla\cdot u) - \nabla\mu\cdot(\nabla u + \nabla u^{*}) + 2\mu\nabla(\nabla\cdot u) + \mu\times(\nabla\times u) = 0 \tag{6.21}$$

The vortical dynamic equation is achieved by dividing equation (6.21) by ρ and introduces the definition of vorticity as

$$\Omega = \nabla\times u \tag{6.22}$$

We have then

$$\begin{aligned}\frac{\partial u}{\partial t} + \frac{1}{2}\nabla(u\cdot u) - u\times\Omega + \nabla[(p-\lambda\nabla\cdot u)/\rho] - (p-\lambda\nabla\cdot u)\nabla(1/\rho)\\ -(1/\rho)\nabla\mu\cdot(\nabla u+\nabla u^{*}) - 2\nu\nabla(\nabla\cdot u) + \nu\nabla\times\Omega = 0\end{aligned} \tag{6.23a}$$

Instead of listing all the vector identities that have been applied for a detailed derivation, the term-by-term evaluation of equation (6.23a) after taking the curl operator is given as the following:

$$\begin{aligned}\nabla\times\partial u/\partial t &= \partial\Omega/\partial t\\ \nabla\times\frac{1}{2}\nabla(u\cdot u) &\equiv 0\\ \nabla\times(u\times\Omega) &\equiv (\Omega\cdot\nabla)u - \Omega(\nabla\cdot u) - (u\cdot\nabla)u\\ \nabla\times(p-\lambda\nabla\cdot u)\nabla(1/\rho) &= \nabla(p-\lambda\nabla\cdot u)\times\nabla(1/\rho)\\ \nabla\times[2\nu\nabla(\nabla\cdot u)] &\equiv 2\nabla\nu\times[\nabla(\nabla\cdot u)]\end{aligned} \tag{6.23b}$$

and finally

$$\nabla\times[\nu\nabla\times\Omega] = \nu\nabla\times\nabla\times\Omega + \nabla\nu\times(\nabla\times\Omega) \tag{6.23c}$$

Substituting the above identities to equation (6.23a), the vorticity dynamic equation acquires the following form:

$$\begin{aligned}\frac{\partial\Omega}{\partial t} + \Omega(\nabla\cdot u) + (u\cdot\nabla)\Omega - (\Omega\cdot\nabla)u + \nabla[p-\lambda(\nabla\cdot u)]\times\nabla\left(\frac{1}{\rho}\right)\\ -\nabla\times\left[\frac{\nabla\mu\cdot def(u)}{\rho}\right] - 2\nabla\nu[\nabla(\nabla\cdot u)] + \nabla\nu\times(\nabla\times\Omega) - \nu\nabla^{2}\Omega = 0\end{aligned} \tag{6.24}$$

The vorticity dynamic equation can be used to perform direct numerical simulation of turbulence. Some variations of equation (6.24) have been adopted for that purpose (Bhaganagar et al., 2002; Postl and Fasel, 2006). A quick verification of this equation is easily achieved by assuming that the flow medium is incompressible, so that the above equation reduces to the following classic form:

$$\frac{\partial \Omega}{\partial t} + (u \cdot \nabla)\Omega - (\Omega \cdot \nabla)u - \nu\nabla^2\Omega = 0 \tag{6.25}$$

At the same time, physical insights of vorticity dynamics can be gleaned from the above equation. First, the vorticity equation in incompressible medium is linear and homogeneous. Therefore, at any point in space if the vorticity vanishes, then the rate of change in time must vanish which yields the classic conclusion that vorticity can be neither created nor destroyed in the interior of a homogeneous fluid, and can be generated only at the solid boundaries. In contrast, in a compressible flow, when the vorticity is zero at a point in space, the rate of change in vorticity can still be generated by the following mechanisms such as gradient of density, dilatation or the normal component of stress tensor, gradient of viscosity, divergence of the velocity, and deformation tensor. Therefore, compressibility and the transport property of a flow medium have introduced additional complexity to vortical dynamics.

Second, the term $(\Omega \cdot \nabla)u$ of the vortical dynamics equation is commonly referred to as the vorticity stretching, which has no counterpart in equation of momentum. The effect can produce change to vorticity in time by a distinctive character only in a three-dimensional fluid motion (Chong et al., 1990; Tennekes and Lumley, 1972), in two-dimensional flows the vorticity stretching ceases identically. The vorticity stretching is a critical property of turbulence. For this reason, the description of vortex dominant turbulence is incomplete by treating it simply as a two-dimensional phenomenon. Finally, a frequently overlooked property of the vortex is that the divergence of vorticity is identically zero to make the vortex to be a divergentless vector.

To complete the description for compressible vortical motion, the conservation laws of mass, momentum, and internal energy are applied together with the second law of thermodynamics;

$$\begin{aligned} \frac{\partial \rho}{\partial t} + \nabla \cdot (\rho u) &= 0 \\ \frac{\partial \rho u}{\partial t} + \nabla \cdot (\rho u u + \tau) &= 0 \\ \frac{\partial \rho e}{\partial t} + \nabla \cdot (\rho e u + q + u \cdot \tau) &= 0 \end{aligned} \tag{6.26}$$

$$\frac{Ds}{Dt} = \frac{1}{T}\left[\frac{De}{Dt} + p\frac{D}{Dt}\left(\frac{1}{\rho}\right)\right] = \frac{1}{T}\left[\frac{Dh}{Dt} + \frac{1}{\rho}\frac{Dp}{Dt}\right] \tag{6.27}$$

A generalized Crocco's relationship for viscous flow relating the entropy, temperature, and vorticity for compressible flow becomes

$$T\nabla S + u \times \Omega = \frac{\partial u}{\partial t} + \nabla h_o + \frac{\nabla(\lambda \nabla \cdot u)}{\rho} + \frac{\nabla \mu \cdot def(u)}{\rho} + 2\nu\nabla(\nabla \cdot u) - \nu\nabla \times \Omega \tag{6.28}$$

It is noted that compressibility modifies the relationship between entropy and vorticity of the flow field through the bulk viscosity of the flow medium and the gradients of the transport properties.

Equation (6.28) reduces to the classic Crocco's theorem in incompressible medium.

$$T\nabla S + u \times \Omega = \frac{\partial u}{\partial t} + \nabla h_o \tag{6.29}$$

In summary, a fundamental behavior of an inviscid steady flow field for which if there exists a uniform total enthalpy and entropy condition, the vorticity must be identically zero.

Vortex dynamics presents the most intriguing and challenging structures in fluid dynamics; intrinsically, it always coexists with the swirling velocity component within combustors or high lifting generators (Tobak and Chapman, 1982; Tobak and Peake, 1982). Most of all, it is a truism to consider that the evolution of turbulence is the result of vortex dynamics, which will be deferred later in a special section of this chapter. Therefore, the following discussions are focused on vortex breakdown, asymmetric formation at high angle of attach over delta wing/slender body, vortical impingement to solid surface, as well as the horseshoe vortex formation ahead transversal injection and blunt fin.

One of unique features of vortex breakdown phenomenon is the creation of multiple stagnation points in the free stream at the breakdown onset location (Leibovich, 1984). In the past, the formations of breakdown are considered to consist of two distinct types; namely, the bulb and the spiral configuration. However, the difference in formations usually is a subset of each other or a continuation from the former to the latter. Nevertheless, vortex breakdown is a typical bifurcation in fluid dynamics that is always preceded by an upstream supercritical flow and followed by a subcritical wake. A bulb of recirculating and swirling fluid appears at the critical point. In the subcritical wake, the flow field is unsteady and often characterized by turbulent motion. In short, the pressure gradient along the vortex axis plays a controlling role in the initiation of breakdown. The structure of the instantaneous streamline patterns can be interpreted by critical point theory or phase space analysis. The vortex core constituted by two critical points or stagnation points that are associated with pairs of opposite spiral/saddle points after breakdown. Meanwhile, the instantaneous sectional topology indicates that the bubble is open, thereby permitting upstream propagation and allowing the flow within the bubble to exit downstream.

In spite of the complicated flow structure, a favorable comparison between experimental data and computational result has been attained for the laminar transient vortex breakdown over a delta wing at high incidence by a pitch-and-hold motion. The specific comparison of the measured and computed axial velocity profiles through the center of the vortex bubble is given in Figure 6.19 by Visbal (1994). The velocity profile carries a signature pattern of inflection points associating with fluid dynamic instability. For the first time, detailed information of the transient vortex breakdown is also validated with critical point theory. The validated physics by computational simulation would be a singular accomplishment if fluid dynamic research could attain the

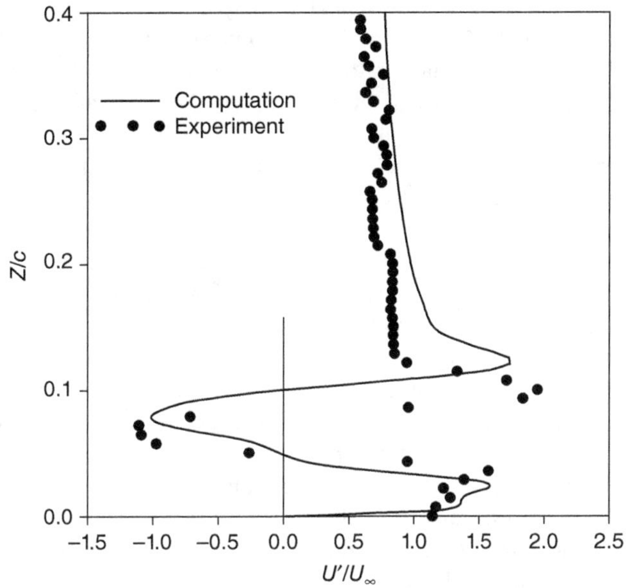

FIGURE 6.19 Axial velocity profile of bubble vortex breakdown.

same level of understanding and extend to turbulent flow or including the subcritical wake.

When a lifting surface or a slender body of revolution at incidence is moving with respect to the oncoming stream, a striking feature is the appearance of a pair of vortices formed by the rolling up of a vortex sheet (Stanbrook et al., 1964; Thomson and Morrison, 1971; Visbal and Shang, 1989; Degani and Tobak, 1992). For a highly swept delta wing, an adverse pressure gradient in the outboard region is generated by the wingtip expansion, leading to boundary-layer separation. For a slender body, the circumferential adverse pressure gradient is produced by flow realignment after it expands from windward to leeward. The separated flows generate a pair of counterrotating vortices over the leeward surfaces, and the resultant unsteady flow field patterns are well-known.

For a delta wing, the behaviors of the leading edge vortex have distinctive behaviors for subsonic, sonic, and supersonic leading edge flows. The distinguishing structure in supersonic flows does not exist for the slender body because the flow field is always enveloped by a bow shock. However, at an increasing incidence the counterrotating vortical structure becomes asymmetric and unsteady with zero sideslip (Rockwell, 1998; Ozen et al., 2012). The unsteady asymmetrical vortical formation unavoidably could induce a significant time-averaged side force greater than the normal force and triggers laminar–turbulent transition. This behavior not only creates an intrinsic pattern for a symmetrical shape in producing unsymmetrical flow pattern but also presents a serious flight control issue. The most undesirable feature of the unsymmetrical vortex formation is the unpredictability in the direction of the sudden exertion of a side force. The asymmetric vortex can be due to a small imperfection on the nose tip but shows poor repeatability. An artificial tip perturbation has been shown to be able to suppress the uncertain and arbitrary bistate behavior, but the position and size of an artificial perturbation must be predetermined. The outstanding feature of the asymmetric vortical formation reveals clearly that an extremely small perturbation to the vortex generation process at onset may produce drastically different consequence. A nonintrusive plasma actuator offers a very attractive option for the leading edge vortex flow control.

A vortex incident upon on a solid surface whether it is stationary or moving in rotational or pitching mode will distort the vortex and interact with the vorticity generated by the surface. The interactions always give rise to a substantial, unsteady force, acoustic noise, and vibrations on the impinged surfaces. However, it is possible to control the vortex formation by a phased excitation at very low amplitudes in the region of exponential amplification of a disturbance. It is also the most effective application by plasma-based control; the applications span a range from leading edge vortex breakdown, pitching and rotating wings, to flapping wings of biomimic motion. Vortices from different sources are easily distinguishable and the physics of the distortion and interaction have features in common to isolated vortices. For example, the complex unsteady vortical field of a pitching slender body of rotation, the formation of strong shear layer, and the roll-up into vortical sheet is similar to the lifting surface stall or the vortex shedding from a slender body at high incidence (Stanek and Visbal, 1993). In Figure 6.20, the numerical simulation is generated by pitch-and-hold motion of a slender configuration up to 20°. In the flow field around a slender body, the vortex–surface interaction produces multiple separation regions and the shedding of vortices from surface is clearly elicited. It is also detectable that the vortices have a finite convective speed that shifts the shed vortices trailing the body until the motion stops. At the final holding position, the unsteady and asymmetric vortices in the leeside of the body retain an identical pattern as the slender body at high incidence.

The last vortical structure of our discussion is the horseshoe vortex around an injection port or a protruding blunt fin over a plate. It is frequent in applications for supersonic combustion and flow control. The main feature of the vortex structure appears to be identical, but there are subtle differences in the formation of the surrounding vortex causing form entrainment (Shang et al., 1989a, 1989b). A side-by-side presentation of the vectors plot of the limiting surface streamlines for a transverse jet and a blunt fin is given in Figure 6.21. According to the limiting streamline theory for three-dimensional flow separation, the convergence of surface shear represents a necessary but not always sufficient condition for occurrence of flow separation. From the accompanied experimental measurements, the separated flows are irrefutably confirmed for both types of flows. There is a strong analogy between the two flow fields; the streamwise pressure gradient that induces three-dimensional flow separation around a source of disturbances is identical. However, a fundamental difference exists in the topology of the flow; the stagnation region downstream of the jet is absent, but this difference has been considered to be negligible to the upstream vortical structure. In fact, the counterclock rotating vortex entrained and reinforced by

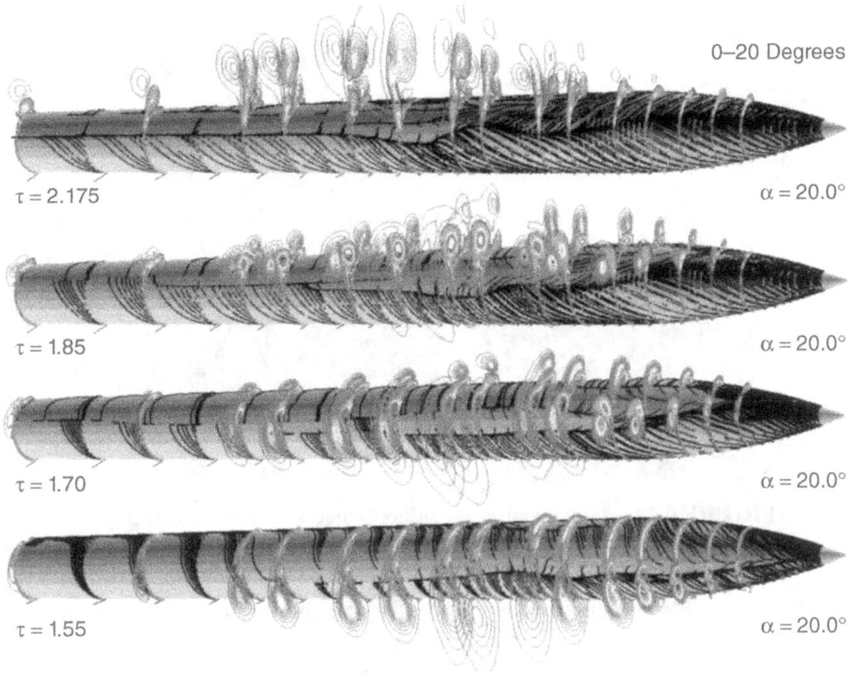

FIGURE 6.20 Time sequence of a pitch-and-hold sequence of vortex field around a slender body of revolution. From Stanek, M.J., and Visbal, M.R., *J. Aircraft*, vol. 30, 1993.

the transverse jet is no longer a secondary structure to present an excellent situation for flow control using electromagnetic–aerodynamic interactions.

Figure 6.22 depicts streamlines from a computational simulation of the hypersonic transverse jet displayed in the previous figure. The complex and detailed vortical structure is revealed in the symmetric plane. In addition, secondary counterrotating

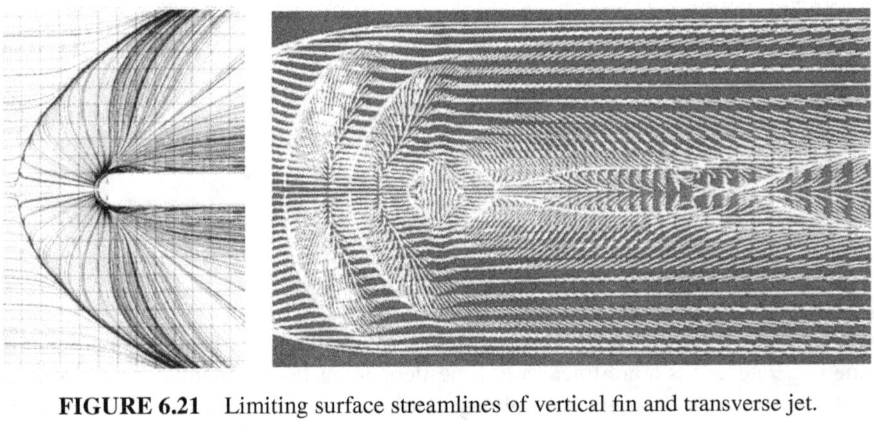

FIGURE 6.21 Limiting surface streamlines of vertical fin and transverse jet.

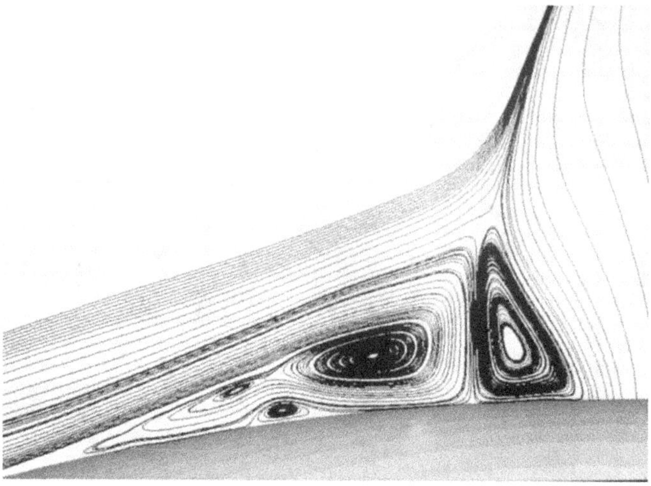

FIGURE 6.22 Vortical structure of horseshoe vortex of transverse jet.

vortices of the transverse jet are captured. The secondary counterclockwise vortex that attaches to the surface is pinched and elongated reverse flow region into two clockwise vortices. Upstream of the primary nodal point of attachment, all three vortices originated from the saddle point of separation in the plane of symmetry. The fluid immediately adjacent to the wall spirals into the primary clockwise vortex. The next layer of fluid similarly formed the forward-most secondary clockwise vortex system. The rest of the entrained fluid constitutes the secondary counterclockwise rotation vortex beneath the aforementioned two vortices and appears as the secondary separated flow region. Its footprint on the body surface is the second convergent and the first divergent line according to limiting streamline theory. The computational results are fully verified by experimental observations.

Vortex formation around transverse jet injection exerts a significant influence on the development of the injection and the mixing process downstream. Eventually, by reducing the loss of injection stagnation pressure, the penetration height of the transverse jet can be significantly modified for enhancing the combustion efficiency. This type of flow field depends strongly on the secondary vortical structures that are susceptible to small perturbations that can be generated easily by plasma actuator. Therefore, introducing electromagnetic forces to vortex dynamics for flow control is a promising research area.

6.5 LAMINAR–TURBULENT TRANSITION

Laminar–turbulent transition is one of the most challenging of fluid dynamics phenomena because it is a nonlinear and time-dependent fluid dynamic bifurcation and deeply rooted in hydrodynamic stability (Reshoko, 1976). The best understanding after a century of research shows that the transition over a plate surface starts with Tollmein–Schlicting wave incurred by hydrodynamic instability at high Reynolds number. The planar wave evolves into a spanwise vorticity and is followed by three-dimensional vortex breakdown (Tollmein, 1929). At high-speed regime, a second mode and higher harmonic instability is also presented in the linear instability growth regime. Downstream to the vortex breakdown, the Emmons spots appear (Emmons, 1951) and are followed immediately by fully turbulent flow. In this idealized but complex process, transition is susceptible to external disturbances such as free stream turbulence, noise, edge contamination, and surface irregularity including surface temperature variation. In addition, the impact by cross flow and surface curvature is especially pronounced in aircraft operations. The concave surface curvature is known to incur instability by a longitudinal vortex disturbance known as Taylor–Gortler vortices and makes the transitional boundary layer more complex (Gortler, 1954).

Based on the aforementioned observations, the receptivity theory by Morkovin becomes the mainstay in analyzing the transition phenomenon (Morkovin, 1969). The transition of the laminar boundary layer to turbulent is treated as the nonlinear response of a very complex stability problem. The disturbing spectrum is assumed to have the infinitesimal amplitude compared to laminar flow. This basic assumption of the normal mode analysis is the foundation in deriving the Orr–Sommerfeld equation. The stability equation is solved for its eigen solutions of the streamwise and spanwise wavenumber, growth rate, as well as the wave frequency and temporal growth rate (Malik, 1982). A typical and well-established thumb curve for the amplifiable disturbance within laminar boundary layer with/without cross-flow component actually can provide the critical Reynolds number for transition.

The temporal approach for studying the hydrodynamic instability has been achieved by systematically solving the Orr–Sommerfeld equation and became a powerful tool for laminar flow control. Cross-flow effects were first successfully calculated by Saric (Saric, 2007) to illustrate the impact to the stability property of a shear layer. They have depicted a direct comparison of the stability thumb curves by solving the Falkner–Skan equation for similar boundary layers nonparallel flows. The numerical results show that the nonparallel flow velocity component exerts a substantial influence to the stability property of the shear layer. It not only enlarges the amplifiable domain of instability but also lowers the onset of the critical Reynolds number based on the boundary-layer displacement thickness. The computational results are supported by a very convincing verification with experimental data. However, when a large amplitude disturbance is encountered, a laminar flow can lead to turbulent spots without amplification through the normal mode of the Tollmein–Schlicting wave. This possibility has been called high-intensity bypass by Morkovin and is not an eigenvalue problem.

Before any possibility for laminar flow control can be implemented, an understanding of basic characteristic of the flow field structure is mandatory. The turbulence is a complex, fluctuating, and disorderly fluid motion. Eddies or fluid packets of many sizes vary continuously from the entire shear-layer thickness down to the Kolmogorov length scale $(\nu^3/\varepsilon)^{1/4}$ (Bradshaw, 1994). The random fluid properties have a particular shape and each eddy has a specific continuous energy spectrum that vanishes at high wave number (small eddies). The turbulent motion can be described by vortical dynamics involving interactions between eddies and main flow. This observation can be easily appreciated by recognizing the second rank dyadic tensor in the conservation momentum equation, equation (6.2a). Therefore, it is logical to seek a micro-vortex generator/suppressor on the surface to create a favorable interaction to either suppressing the spreading helical vortices or diminishing their intensity.

The near-wall shear layer structure, whether it is a laminar or turbulent flow, is always associated with multiple vortices. Especially for the turbulent flow, the motion is inherently time dependent, three-dimensional, and rapidly changing in time and space. The energy and dynamics of the turbulent motion are also cascading from the large-scale eddies to the small-scale eddies and illustrated by Figure 6.23 (Chapman, 1979). In simulation of turbulent flow using either direct numerical simulation (Moin and Mahesh, 1998) by solving the Navier–Stokes equations or mostly the LES (Germano et al., 1991; Piomelli and Balaras, 2002), the solutions of the governing equations associated with irresolvable high wave numbers are filtered by low-pass numerical operators. The unresolved small-scale eddies that usually reside in near-wall region are taken into account through the modeling of the Subgrid scale

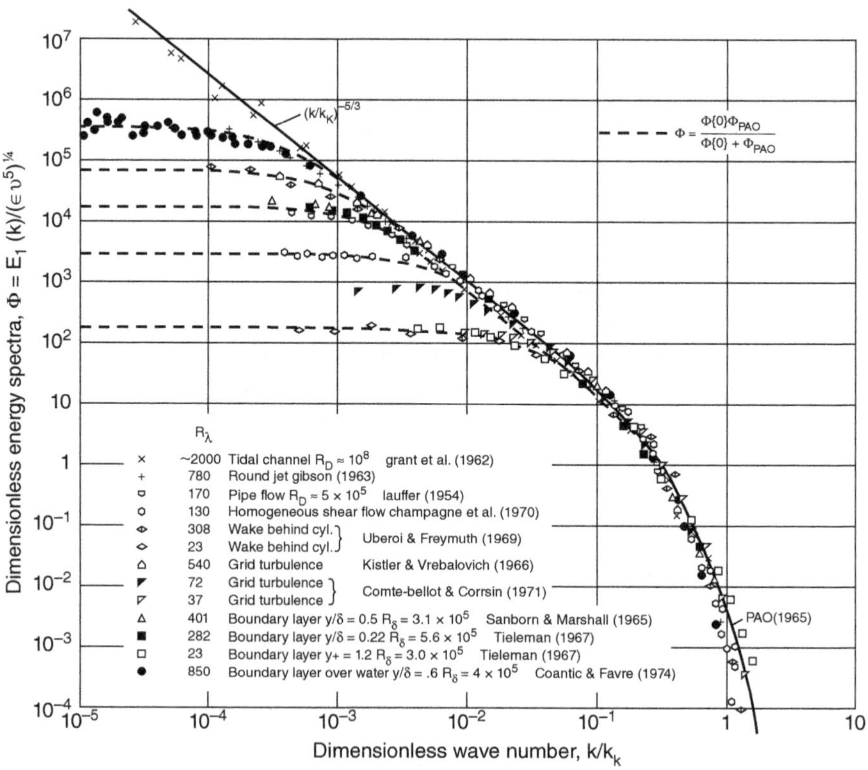

FIGURE 6.23 From Chapman, D.R., *AIAA J.*, vol. 17, 1979.

(SGS) modeling (Smagorinsky, 1963, De Stefano et al., 2005). The pioneering work by Smagorinsky in weather research accurately opens an avenue in linking the macroscopic to the microscopic turbulent dynamics for other to follow.

The physical understanding of turbulent flow is that large eddies that extract energy from the main flow are highly anisotropic. On the contrary, the small eddies dissipate energy and tend toward isotropy and are nearly universal in character. In Figure 6.23, the streamwise turbulent energy spectra for various flows are given. In this graph, the dimensionless energy spectra $\Phi = (\varepsilon/\nu^3)^{1/4}$ is presented versus the dimensionless wave number κ/κ_k, where κ_k is the Kolmogorov wave number $\kappa_k = (\varepsilon/\nu^3)^{1/4}$. A total of eight different types of turbulent flows are included in this figure. It is seen that the energy structure of large eddies change with both the Reynolds number and the type of flows, whereas the small eddies near the solid surface collapse on a simple curve. This convincing demonstration that the near wall, small energy dissipating eddies are clearly universal—it is independent of both Reynolds number and type of flows. This is the underlying principle for subgrid scale modeling for LES in computational simulations.

In LES formulation, the compressible Navier–Stokes equations are decomposed in temporal and spatial independent variables. The temporal scales separation is achieved by the Reynolds-average or the mass- or Favre-average for all ensemble-average of Navier–Stokes equations,

$$\bar{f}_R = \frac{1}{T}\int f(t)dt; \quad \tilde{f}_F = \frac{\int \rho f(t)dt}{\int \rho dt} \tag{6.30}$$

and the small spatial scale is eliminated by the low-pass filter;

$$\bar{f} = \int G(x - x')f'(x')dx' \tag{6.31}$$

The SGS models have also been generally categorized as deterministic and the stochastic models. The fundamental difference is rested in the fact that the stochastic formulation considers the interaction of the large- and the small-scale eddy interaction but not by the deterministic model. The stochastic model usually also consists of two components known as the coherent and incoherent subgrid-scale stress, but the later contributes little to the large-scale dynamics and the energy transfer. The energy-carrying large eddies are computed by LES, so it can be more accurate than the mass-average Navier–Stokes solution because the small-scale structures tend to be more isotropic and homogeneous than the large scale motion. As the consequence, the inner structure can be described by universal modeling.

In earlier research efforts by Mitchner and Kruger (Mitchner et al., 1973), the magnetic force lines were demonstrated to restraint the incompressible turbulent motion. Their observation warrants additional research, because recent research in surface plasma actuator has demonstrated that the weakly ionized gas can bring additional physical mechanisms to flow control through electromagnetic force. Indeed, Corke et al. (2007, 2010) showed that dielectric barrier discharge (DBD) can delay dynamic stall of an airfoil operating at low Reynolds number and at atmospheric pressure conditions. By applying a DBD over a solid surface, a periodic force at the AC frequency around 10 kHz generates a wall-jet-like electric wind up to 5–6 m/s at the outer edge of the plasma sheath that is usually around the thickness of the laminar sublayer or thinner. Surface plasma actuators have the attractive attributes that they are nonintrusive to the external flow field, easy to install, and can be actuated instantaneously. However, the power required is huge, because the gas-ionizing process is limited by electron impact with additional energy cascading to the vibrational excitation of air. The maximum ion number density of DBD and DCD (direct current discharge) is merely $10^{13}/cm^3$ in contrast to the neutral air molecules number density of $10^{19}/cm^3$ (Shang et al., 2009; Shang and Huang, 2010). The resultant ionization fraction exists in a trace amount. As the consequence, the electromagnetic force and energy are presented only at the perturbation level to the external flow field. For this reason, the DBD operation at the atmospheric condition is the most effective at the aerodynamic bifurcation such as the laminar–turbulent transition or dynamic stall at the very low Reynolds number environment.

REFERENCES

Babinsky, H., and Harvey, J.K., *Shock Wave–Boundary-Layer Interactions*, Cambridge University Press, Cambridge, 2011.

Bertram, M.H., Boundary-layer displacement effects in air at Mach number of 6.8 and 9.6, *NACA Tech. Notes No.*, vol. 4133, 1957.

Bhaganagar, K., Dempler, D., and Lumley, J.L., Direct numerical simulation of spatial transition to turbulence using fourth-order vorticity velocity, second-order vertical vorticity formulation, *JCP*. vol. 180, 2002, pp. 200–228.

Bradshaw, P., Turbulence: the chief outstanding difficulty of our subject, *Exp. Fluid*, vol. 16, 1994, pp. 203–216.

Chapman, D.R., Computational aerodynamics development and outlook, *AIAA J.*, vol. 17, 1979, pp. 1293–1313.

Chapman, D.R., Kuehn, D.M., and Larson, H.K., Investigation of separated flows in supersonic and subsonic streams with emphasis on the effect of Transition, *NACA Rept.*, vol. 1356, 1958.

Chapman, S., and Cowling, T.G., *The Mathematical Theory of Non-Uniform Gases*, 2nd ed., Cambridge University Press, 1964.

Chong, M.S., Perry, A.E., and Cantwell, B.J., A general classification of three-dimensional flow fields, *Phys. Fluids*, vol. 2, 1990, pp. 765–777.

Corke, T.C., Post, M.L., and Orlov, D.M., SDBD plasma enhanced aerodynamics: concepts, optimization and applications, *Prog. Aerosp. Sci.*, vol. 43, 2007, pp. 193–217.

Corke, T.C., Enloe, C.L., and Wilkinson, S.P., Plasma actuators for flow control, *Annu. Rev. Fluid Mech.*, vol. 42, 2010, pp. 505–525.

De Stefano, G., Goldstein, D.A., and Vasilyev, O.V., On the role of subgrid-scale coherent model in large eddy simulation, *JFM.*, vol. 525, 2005, pp. 263–274.

Degani, D., and Tobak, M., Experimental study of controlled tip disturbance effect on flow asymmetry, *Phys. Fluids*, vol. 4, 1992, pp. 2825–2832.

Dolling, D.S., Fifty years of shock/boundary interaction: what next, *AIAA J.*, vol. 39, 2001, pp. 1517–1531.

Emmons, H.W., The laminar–turbulent transition in boundary layer Part 1, *J. Aeronaut. Sci.*, vol. 18, 1951, pp. 490–498.

Finley, P.J., The flow of jet from a body opposing a supersonic free stream, *J. Fluid Mech.*, vol. 26, 1966, pp. 337–368.

Gaitonde, D., and Shang, J.S., Shock pattern of a triple-shock turbulent interaction, *AIAA J.*, vol. 36, no. 1, January 1998, pp. 113–115.

Gaitonde, D.V., Shang, J.S., Zheltovodov, A.A., and Maksimov, A.I., Investigation of 3-D turbulent interactions caused by asymmetric cross-shock configuration, *AIAA J.*, vol. 37, no. 12, December 1999, pp. 1602–1608.

Germano, M., Piomelli, U., Moin, P., and Cabot, W., A dynamic subgrid scale eddy viscosity model, *Phys. Fluids*, vol. A3, 1991, pp. 1760–1765.

Gortler, H., Nachr. Akad. Wiss. Gottingen, Math-Phys. Kl. I1a, 2, 1954 (NACA Tech. Memo. 1375).

Hankey, W.L., and Shang, J.S., Analyses of pressure oscillations in an open cavity, *AIAA J.*, vol. 18, no. 8, August 1980, pp. 892–898.

Harlow, F.H., The particle-in-cell computing method for fluid dynamics, *Method Comput. Phys.*, vol. 3, 1964, pp. 319.

Hayes, W.D., and Probstein, R.F., *Hypersonic Flow Theory*, Academic Press, New York, 1959.

Il'yushin, A.A., The law of plane section in the aerodynamics of high supersonic speeds, *PMM*, vol. 20, 1956, pp. 733–755.

Korkegi, R.H., Survey of viscous interaction associated with high Mach number flight, *AIAA J.*, vol. 9, 1971, pp. 771–784.

Lees, L., and Lin, C.C., Investigation of the stability of the laminar boundary layer in a compressible fluid, *NACA TN*, vol. 1115, 1946.

Leibovich, S., Vortex stability and breakdown: survey and extension, *AIAA J.*, vol. 22, 1984, pp. 1192–1206.

Malik, M.R., A black box compressible stability analysis code for transition prediction in three-dimensional boundary layer, 1982, NASA CR 165925.

Maull, D.J., Hypersonic flow over axially symmetric spiked bodies, *J. Fluid Mech.*, vol. 8, 1962, pp. 584–592.

Michalke, A., On the inviscid instability of the hyperbolic-tangent velocity profile, *J. Fluid Mech.*, vol. 19, 1964, pp. 543–556.

Mitchner, M., and Kruger, C.H., *Partially Ionized Gases*, John Wiley & Sons, New York, 1973.

Moin, P., and Mahesh, K., Direct numerical simulation; a tool in turbulent research, *Ann. Rev. Fluid Mech.*, vol. 30, 1998, pp. 539–578.

Morkovin, M.V., On the many faces of transition. In: C.S. Well, ed., *Viscous Drag Reduction*, Plenum, New York, 1969, and Air Force Flight Dynamics. Lab. Rep. AFFDL-TR-68-149.

Navier, M., Memoire sur les lois du mouvements des fluids, *Mem. De L'Acad. D. Sci.*, vol. 389, 1827.

Ozen, C.A., and Rockwell, D., Three-dimensional vortex structure on a rotating wing, *J. Fluid Mech.*, vol. 707, 2012, pp. 541–550.

Piomelli, U., and Balaras, E., Wall-layer models for large-eddy simulation, *Annu. Rev. Fluid Mech.*, vol. 34, 2002, pp. 349–374.

Postl, D., and Fasel, H.F., Direct numerical simulation of turbulent flow separation from a wall-mounted hump, *AIAA J.*, vol. 44, 2006, pp. 263–272.

Rayleigh, L., On the stability or instability of certain fluid motion, *Sci. Pap.*, vol. 1, 1880, pp. 474–484.

Reshoko, E., Boundary-layer stability and transition, *Ann. Rev. Fluid Mech.*, vol. 8, 1976, pp. 311–349.

Rockwell, D., Vortex–body interactions, *Annu. Rev. Fluid Mech.*, vol. 30, 1998, pp. 199–229.

Rossiter, J.E., Wind-tunnel experiment on the flow over rectangular cavities at subsonic and transonic speeds, *Br. A.R.C., R&M*, no. 3428, 1964.

Saffman, P.G., and Baker, G.R., Vortex interaction, *Ann. Rev. Fluid Mech.*, vol. 11, 1979, pp. 65–122.

Saric, W.S., Boundary-layer stability and transition, Springer handbook of experimental fluid mechanics, Springer-Verlag, Berlin-Heidelberg, 2007, pp. 886–896.

Seydel, R. *From Equilibrium to Chaos: Practical Bifurcation and Stability Analysis*, Elsevier Science Publication Co., New York, 1988.

Shang, J.S., An assessment of numerical solutions of the compressible Navier–Stokes equations, *J. Aircraft*, vol. 22, no. 5, 1985, pp. 353–370.

Shang, J.S., Computational fluid dynamics application to aerospace science, *The Aeronautical J.*, vol. 113, no. 1148, October 2009, pp. 619–632.

Shang, J.S., and Hankey, W.L., Numerical solution for supersonic turbulent flow over a compression ramp, *AIAA J.*, vol. 13, 1975, pp. 1368–1374.

Shang, J.S., Recent research in magneto-aerodynamics, *Prog. Aerosp. Sci.*, vol. 37, no. 1, 2001, pp. 1–20.

Shang, J.S., Plasma injection for hypersonic blunt–body drag reduction, *AIAA J.*, vol. 40, no. 6, June 2002, pp. 1178–1186.

Shang, J.S., Three decades of accomplishments in computational fluid dynamics, *Prog. Aerospace Sci.*, vol. 40, no. 3, 2004, pp. 173–197.

Shang, J.S., and Hankey, W.L., Numerical solution of the Navier–Stokes equations for a three-dimensional corner, *AIAA J.*, vol. 15, no. 11, November 1977, pp. 1575–1582.

Shang, J.S., and Hankey, W.L., Three-dimensional supersonic interacting turbulent flow along a corner, *AIAA J.*, vol. 17, no. 7, July 1979, pp. 706–713.

Shang, J.S., and Hankey, W.L., Numerical simulation of self-excited oscillations in fluid flows, Third Vol. *The Series on Recent Advances in Numerical Methods in Fluids; Computational Methods in Viscous Flows*, Editor W.G. Habashi, Pineridge Press, Swansen, UK, 1984, pp. 543–582.

Shang, J.S., and Huang, P.G., Modeling AC dielectric barrier discharge, *J. Appl. Phys.*, vol. 107, no. 11, 2010, pp. 1133021–1133027.

Shang, J.S., Hankey, W.L., and Law, C.H., Numerical simulation of shock wave-turbulent boundary layer interaction, *AIAA J.*, vol. 14, no. 10, October 1976, pp. 1451–1457.

Shang, J.S., Hankey, W.L., and Smith, R.E., Flow oscillations of spike-tipped bodies, *AIAA J.*, vol. 20, no. 1, January 1982, pp. 25–26.

Shang, J.S., Hayes, J., Menart, J., and Miller, J., Blunt body in hypersonic electromagnetic flow field, *J. Aircraft*, vol. 40, no. 2, 2003, pp. 314–322.

Shang, J.S., Hayes, J., Wurtzler, K., and Strang, W., Jet-spike bifurcation in high-speed flows, *AIAA J.*, vol. 39, no. 6, June 2001, pp. 1159–1165.

Shang, J.S., McMaster, D., and Golbitz, W., Supersonic, transverse jet from a rotating ogive cylinder in hypersonic flow, *J. Spacecraft Rockets*, vol. 26, 1989b, pp. 24–30.

Shang, J.S., McMaster, D., Scaggs, N., and Buck, M., Interaction of jet in hypersonic cross stream, *AIAA J.*, vol. 27, 1989a, pp. 323–329.

Shang, J.S., Menart, J.A., and Hayes, J.R., Spike-tipped blunt body in weakly ionized hypersonic stream, AIAA 2002-2167, Maui Hawaii, 2002.

Shang, J.S., Computational fluid dynamics application to aerospace science, *Aeronaut. J.*, vol. 113, no. 1148, October 2009, pp. 619–632.

Shang, J.S., and Huang, P.G., Surface plasma actuators modeling for flow control, progress. *In Aerospace Sci.*, vol. 67, 2014, pp. 29–50.

Smagorinsky, J.S., General circulation experiments with the primitive equations, *Mon. Weather Rev.*, vol. 29, 1963, pp. 99–164.

Stanbrook, A., and Squire, L.C., Possible types of flow at swept leading edges, *Aeronautical Q.*, vol. 15, 1964, pp. 72–82.

Stanek, M.J., and Visbal, M.R., Investigation of vortex development on a pitching slender body of revolution, *J. Aircraft*, vol. 30, 1993, pp. 711–718.

Stokes, G.G., On the theories of the internal friction of fluids in motion, *Trans. Cambr. Phli. Soc.*, vol. 8, 1845.

Sutton, G.W., and Sherman, A., *Engineering Magnetohydrodynamics*, McGraw-Hill, New York, 1965.

Tennekes, H., and Lumley, J. L., *A First Course in Turbulence*, MIT Press, Cambridge, MA, 1972.

Thom, A., The flow past circular cylinders at low speeds, *Proc. Royal Soc. Lond.*, vol. A141, 1933, pp. 651–666.

Thomson, K.D., and Morrison, D.F., The spacing, position and strength of vortices in the wake of slender cylindrical bodies at large incidence, *J. Fluid Mech.*, vol. 50, 1971, pp. 751–781.

Tobak, M., and Chapman, G.Y., Nonlinear problems in flight involving aerodynamic bifurcations, *AGARD CP*, vol. 386, May 1985, pp. 25: 1–15.

Tobak, M., and Peake, D.J., Topology of three-dimensional separated flows, *Ann. Rev. Fluid Mech.*, vol. 14, 1982, pp. 61–85.

Tollmein, W., Uber die Entestehung der Turbulenz, Nachr., Ges. Wiss. Gottingen, Math.-phys. Kl II, 1929. (Tollmien-Schlicting Wave, NACA technical Memo 609).

Vincenti, W.G., and Kruger, C.H., *Introduction to Physical Gas Dynamics*, John Wiley & Sons, New York, 1965.

Visbal, M.V., Onset of breakdown above a pitching delta wing, *AIAA J.*, vol. 32, 1994, pp. 1568–1575.

Visbal, M.R., and Shang, J.S., Investigation of the flow structure around rapidly pitching airfoil, *AIAA J.* vol. 27, no. 8, August 1989, pp. 1044–1051.

Von Neumann, J., Proposal and analysis of a numerical method for the treatment of hydrodynamical shock problems, *Nat. Def. Res. Com. Rept.* AM-551, March 1944.

West, J.E., and Korkegi, R.H., Supersonic interaction in the corner of intersecting wedges and high Reynolds numbers, *AIAA J.*, vol. 10, 1972, pp. 652–656.

7

COMPUTATIONAL ELECTROMAGNETIC-AERODYNAMICS

INTRODUCTION

Computational electromagnetic-aerodynamics (CEA) or computational electromagnetic fluid dynamics is an interdisciplinary science. The fundamental governing equations are the Boltzmann and Maxwell equations for describing the plasma dynamics in flow fields. The system of equations solves distribution functions of space and velocity in time that is a seven-dimensional formulation, and solutions to the governing equations are probabilistic. In cases the collision mean-free path is much greater than the characteristic scale of the studied phenomenon, the plasma can be approximated as collisionless and the Vlasov approximation is valid (Cheng and Knorr, 1976). The simplification is made by specific forms of the distribution functions for the rate of changes to achieve the Vlasov and Fokker–Planck equations. A widely adopted numerical approach for studying plasma dynamics is the particle-in-cell (PIC) method in a Lagrangian frame that is essentially a Monte Carlo method (Birdsall and Langdon, 1985). In order to close the governing equations system, the properties of transport of plasma are also needed for describing plasma medium with acceptable physical fidelity. By these approaches, the detailed descriptions of the chemical-physics properties are also venturing into a vast scientific unknown and uncertainty scientific discipline.

Within the framework of kinetic theory, the PIC simulation follows the particle dynamics through the Lagrangian formulation by tracking a large number of particles under the influence of the externally applied electromagnetic field. The integrated solution includes modification of the charged particle dynamics by the applied

Computational Electromagnetic-Aerodynamics, First Edition. Joseph J.S. Shang.

electromagnetic field. The distributions of the particles are determined by statistically means, and the accuracy of the solution is controlled by the statistic properties of the set of microscopic particles.

The Vlasov approach can also be conducted on the Eulerian frame. The numerical solutions of the particle distribution functions at each instant and fixed spatial location are sampled in phase space. The Maxwell equations are also satisfied by the distribution functions through the integration algorithm. This approach avoids statistical noise through the random selection of the distribution functions in phase space as mentioned in the PIC method. But the approach incurs a greater cost in computational resource.

At the another end of the plasma formulation spectrum that meets practical engineering needs is by the moments of the Boltzmann equation over velocity space for each constituent of the plasma. A further simplification is by the multifluid model, in which the electrons and all ionized chemical species are grouped into positively and negatively charged ions together with the neutral gas particles (Shumlak and Loverich, 2003). The closure of the equation system is achieved by taking moments of the Boltzmann equation in velocity space, which leads to the multifluid equations. The plasma is treated as an inhomogeneous, electrically conducting gas mixture. The concentrations of electrons, and positively and negatively charged chemical species, are determined by a system of continuity equations. Although all electromagnetic forces are included by the fundamental governing equations, the evolution of the electromagnetic fields through the distribution of the charge concentration and current density must also be coupled with Maxwell equations. In the multifluid model, the electrons and irons are treated as separated species but still need to preserve the plasma approximation of quasi neutrality. In fact, by decoupling the generalized Ohm's law, the common multifluid model is more general than the classic Magnetohydrodynamics (MHD) model in describing transport properties of plasma. However, the formulation subjects to the limitation by neglecting Hall effects and diamagnetic terms. In addition, by neglecting the electron mass in comparison with neutrals and ions as well as the displacement current, the resultant equations also are unable to describe the effect of Whistler wave resonance.

Nevertheless, all chemical–physical phenomena of interest are interlocking by the disciplines of fluid dynamics, electromagnetics, chemical kinetics, and quantum physical–chemical kinetics. The required physical length resolutions for numerical simulation are on the same order of magnitude of atomic and molecular scales, while the time scale is needed to be compatible up to the speed of light. Since only macroscopic information of the plasma system is sought after for most engineering applications, computational simulations rely heavily on physics-based modeling. For fluid dynamics or aerodynamics simulations, the conservation laws by the time-dependent, compressible Navier–Stokes equation are applicable in the continuum regime. All electromagnetic effects are introduced as source terms to the conservation laws. For this reason, the signal transmitting rate and communication of information are limited to the speed of sound. The vastly higher rate of dynamics at the speed of light is incorporated by a suitable time averaging process.

The electromagnetic phenomena are fully governed by the Maxwell equations in the time domain. Additional simplifications will be appropriately applied through the pertaining laws of the Maxwell equations that depend upon whether externally applied electric and magnetic fields are present. The most challenging components of CEA are still the nonequilibrium chemical kinetics and ionization processes for determining the composition and transport property of plasma. The state-of-the-art in these areas of science is completely governed by chemical kinetics and quantum physics models that are built entirely on empirical data to become the least rigorous portion of plasma dynamics modeling.

7.1 MULTIFLUID PLASMA MODEL

The multifluid model of plasma describes the coupling between the outstanding electromagnetic properties of plasma and fluid dynamics in the macroscopic scale. A basic approach of the multifluid model neglects the irreversible intermolecular interaction but focuses on the equilibrium solution to Boltzmann equations with constant coefficients (Birdsall and Langdon, 1985). By this assumption, the plasma is composed of multiple different fluids with different velocities and partial pressures as a nonhomogeneous medium. In most analytic works, this model is further simplified to the well-known two-fluid composition with electrons and positively charged ions (Shumlak and Loverich, 2003). The starting point of deriving the multifluid model for plasma dynamics is the kinetic theory by a combination of the Boltzmann and Maxwell equations:

$$\frac{\partial f_i}{\partial t} + u_i \frac{\partial f_i}{\partial r} + \frac{q_i}{m_i}(E + u_i \times B)\frac{\partial f_i}{\partial u} = \left(\frac{\partial f_i}{\partial t}\right)_{\text{collison}} \tag{7.1a}$$

The attractive attribute of the multifluid model is its ability to treat the individual dynamics of electrons and positively and negatively charged ions without the complication from the chemical kinetic model (Surzhikov et al., 2004). The process is parallel to the derivation of the Euler equations by taking the moments of the governing equation for the distribution function to eliminate the velocity space. The dependent variables of the resultant equations are the macroscopic property for mass, momentum, and energy.

The zero-order moment of the Boltzmann equation,

$$n(r,t) = \int f(r,v,t)dv \tag{7.1b}$$

yields the essentially the species continuity equation

$$\begin{aligned} \frac{\partial n_e}{\partial t} + \nabla \cdot (n_e u_e) &= \frac{\partial w_e}{\partial t} \\ \frac{\partial n_+}{\partial t} + \nabla \cdot (n_+ u_+) &= \frac{\partial w_+}{\partial t} \\ \frac{\partial n_-}{\partial t} + \nabla \cdot (n_- u_-) &= \frac{\partial w_-}{\partial t} \end{aligned} \tag{7.2a}$$

The first moment of the Boltzmann equations for each species,

$$\rho_i u_i(r,t) = \int vf(r,v,t)dv \tag{7.2b}$$

gives the momentum equation for the individual components

$$\begin{aligned}
n_e m_e \left[\frac{\partial u_e}{\partial t} + (u_e \cdot \nabla)u_e\right] &= -\nabla p_e - n_e e(E + u_e \times B) + R_e \\
n_+ m_+ \left[\frac{\partial u_+}{\partial t} + (u_+ \cdot \nabla)u_+\right] &= -\nabla p_+ + n_+ e(E + u_+ \times B) + R_+ \\
n_- m_- \left[\frac{\partial u_-}{\partial t} + (u_- \cdot \nabla)u_-\right] &= -\nabla p_- - n_- e(E + u_- \times B) + R_-
\end{aligned} \tag{7.3a}$$

The second moment of the Boltzmann equation,

$$\bar{\bar{\sigma}}(r,t) = \int (v-u)(v-u)f(r,v,t)dv \tag{7.3b}$$

yields the energy equation in thermodynamic variables in which temperatures are related to partial pressures of individual charged species:

$$\begin{aligned}
\frac{n_e}{\gamma - 1}\left[\frac{\partial T_e}{\partial t} + (u_e \cdot \nabla)T_e\right] &= p_e \nabla \cdot u_e \\
\frac{n_+}{\gamma - 1}\left[\frac{\partial T_+}{\partial t} + (u_+ \cdot \nabla)T_+\right] &= p_+ \nabla \cdot u_+ \\
\frac{n_-}{\gamma - 1}\left[\frac{\partial T_-}{\partial t} + (u_- \cdot \nabla)T_-\right] &= p_- \nabla \cdot u_=
\end{aligned} \tag{7.4a}$$

By defining the electric current density for each individual charged species $J_i = en_i u_i$, the momentum equation can be expressed as the partial current densities in divergence form:

$$\begin{aligned}
\frac{\partial J_+}{\partial t} + \nabla \cdot \left(\frac{j_+ J_+}{en_+} + \frac{ep_+ I}{m_+}\right) &= \frac{e^2 n_+ E}{m_+} + \left(\frac{e}{m_+}\right)(J_+ \times B + R_{ei}) \\
\frac{\partial J_-}{\partial t} + \nabla \cdot \left(\frac{j_- J_-}{en_-} + \frac{ep_- I}{m_-}\right) &= \frac{e^2 n_- E}{m_-} - \left(\frac{e}{m_-}\right)(J_- \times B + R_{ei}) \\
\frac{\partial J_e}{\partial t} + \nabla \cdot \left(\frac{j_e J_e}{en_e} + \frac{ep_e I}{m_e}\right) &= \frac{e^2 n_e E}{m_e} + \left(\frac{e}{m_e}\right)(J_e \times B + R_{ei})
\end{aligned} \tag{7.4b}$$

By combining the above equation with the corresponding energy equation, the resultant equation appears as the conservation of current densities in divergence form. By coupling the energy equation and the divergence equation of electric current density, equation (7.4b), Joule heating appears explicitly in the conservation of internal energy law for each species, namely,

$$
\begin{aligned}
\frac{\partial n_e}{\partial t} - \nabla \cdot \left[(e_e + p_e) \frac{J_e}{en_e} \right] &= J_e \cdot \left(E + \frac{\mathrm{R}_{ei}}{en_e} \right) \\
\frac{\partial n_+}{\partial t} + \nabla \cdot \left[(e_+ + p_+) \frac{J_+}{en_+} \right] &= J_+ \cdot \left(E + \frac{\mathrm{R}_{ei}}{en_+} \right) \\
\frac{\partial n_-}{\partial t} - \nabla \cdot \left[(e_- + p_-) \frac{J_-}{en_-} \right] &= J_- \cdot \left(E + \frac{\mathrm{R}_{ei}}{en_-} \right)
\end{aligned}
\tag{7.5}
$$

The multifluid model equations have often been simplified to consist of only the electrons and positively charged ions. The equations system establishes the relation among the species number densities, electric current densities, and the internal energy of each charged species. In this formulation, the species production/depletion, electrostatic force, Lorentz acceleration, and Joule heating appear as source terms that are results of electron–ion collision process. These source terms of the motion of plasma affect the evolution of the electromagnetic field. Thus the accurate solution of the multifluid plasma model also requires the Maxwell equations to accommodate the modifications to the electromagnetic field after a solution advanced in time. Only Faraday's induction law and Ampere's electric circuit law are required in this formulation:

$$
\begin{aligned}
\frac{\partial B}{\partial t} + \nabla \times E &= 0 \\
\varepsilon_o \frac{\partial E}{\partial t} - -\nabla \times \frac{B}{\mu_o} - e(n_+ u_+ - n_e u_e - n_- u_-) &= 0
\end{aligned}
\tag{7.6a}
$$

The two Gauss's divergence laws for electric field strength and magnetic flux density,

$$\nabla \cdot B = 0 \tag{7.6b}$$

and

$$\nabla \cdot E = e(n_+ - n_e - n_-)/\varepsilon_o \tag{7.6c}$$

are applied as constraints to eliminate nonphysical results.

A limitation of the multifluid model excludes the treatment of Hall effects and diamagnetic terms. Physically the model represents the individual motions of charged particles but still preserve the globally neutral property of the plasma. Meanwhile, these terms govern the ion current and the finite Larmor radius of the plasma components can be included. The Hall current has been added to the model equation that leads into the Whistler branch. In this branch, the dispersion relation shows that wave velocity increases with frequency, thereby reducing the envelope of computational stability.

In summary, the multifluid plasma model is more general than the classic MHD equation. Specifically, this model preserves the intrinsic quasi-neutral property of plasma by treating the charged particles separately. This feature actually is responsible for excluding the Hall and diamagnetic effects that are associated with the ion current and the finite Larmor radius of the plasma constituents.

7.2 GOVERNING EQUATIONS OF CEA

The rigorous and complex governing equations for CEA in the continuum regime consist of the compressible Navier–Stokes equations, the Maxwell equations for electromagnetics, the law of mass action including quantum chemical physics for nonequilibrium chemical reaction, and transport relations. However, from experimental evidence and order-of-the-magnitude analysis, the electromagnetic force and energy appear mostly as a perturbation to the inertia of the main flow field. In most plasma-based flow control environments, the magnetic Reynolds number $\mathrm{R_m} = ul/(\sigma\mu)^{-1}$, usually has a value much less than unity (Mitchner et al., 1973). This similarity parameter gives a measure of the relative importance of the inertia and electromagnetic force and is the product of the electrical conductivity σ, the magnetic permeability μ, a characteristic velocity, and a length scale of the under-study phenomenon. Under this low magnetic Reynolds number condition and by restricting the temporal scale of the dynamics to within the microwave spectrum, the induced magnetic flux density $\partial B/\partial t$ is negligible in comparison with the externally applied field. As the consequence, the Faraday induction law can be decoupled from Maxwell's equations (Shang, 2002b). This simplification focuses the study onto magneto-aerodynamic interaction with fluid dynamics rather than electromagnetic wave motion. In this formulation, the electrostatic force, the Lorentz acceleration, and the Joule heating appear as the perturbing source terms in the generalized Navier–Stokes equations. Therefore, the essential physics of aerodynamics–electromagnetics interaction for plasma flow control can be effectively approximated by a simplified governing equation system that is summarized as

$$\frac{\partial \rho_i}{\partial t} + \nabla \cdot [\rho_i(\vec{u} + \vec{u}_i)] = \frac{dw_i}{dt} \tag{7.7}$$

$$\frac{\partial \rho \vec{u}}{\partial t} + \nabla \cdot (\rho \vec{u}\vec{u} + p\vec{\vec{I}} - \vec{\vec{\tau}}) = \rho_e \vec{E} + (\vec{J} \times \vec{B}) \tag{7.8}$$

$$\frac{\partial \rho e}{\partial t} + \nabla \cdot [\rho e \vec{u} - \kappa \nabla T + \sum \rho_i u_i h_i + q_{rad} + \vec{u} \cdot p\vec{\vec{I}} + \vec{u} \cdot \vec{\vec{\tau}}] + Q_{vt} - Q_{et} = \vec{E} \cdot \vec{J}. \tag{7.9}$$

The vibrational energy conservation equations for polyatomic molecular species are

$$\frac{\partial \rho_i e_{iV}}{\partial t} + \nabla \cdot [\rho_i(\vec{u} + \vec{u}_i)e_{iV} + q_{iV})] = e_{iV}\frac{dw_i}{dt} + Q_{V,\Sigma}. \tag{7.10}$$

The electronic energy conservation equation has been traditionally given as

$$\begin{aligned}\frac{\partial \rho_i e_e}{\partial t} + \nabla \cdot [\rho_i(\vec{u} + \vec{u}_i)e_e + \vec{u} \cdot p_e\vec{\vec{I}} + q_e)] = e_e\frac{dw_i}{dt} + \vec{E} \cdot \vec{J} \\ + [\rho_e\vec{E} + (\vec{J} \times \vec{B})] \cdot (\vec{u} + \vec{u}_i) + Q_{e,\Sigma}.\end{aligned} \tag{7.11}$$

The definition of the internal energy is now given as

$$\rho e = \sum_{i \neq e} \rho_i \left(c_{V,i}T + \frac{\vec{u} \cdot \vec{u}}{2}\right) + \sum_{i \neq e} \rho_e e_{V,i} + \sum_{i \neq e} \rho_i h_i^o + \rho_e \left(c_{V,e}T_e + \frac{\vec{u}_e \cdot \vec{u}_e}{2}\right) \tag{7.12}$$

where h_i^o is the standard heat of formation for all reacting species.

It is important to note that in the governing equations (7.8), (7.9), and (7.11), the coupling between fluid motion and electromagnetic force/energy appears as the electrostatic force $\rho_e\vec{E}$, Lorentz acceleration $\vec{J} \times \vec{B}$, and Joule heating $\vec{E} \cdot \vec{J}$ (Shang et al., 2005, 2012; Moreau, 2007). The vector fields of $\vec{E}$ and $\vec{B}$ are the sum of the externally applied and induced electrical and magnetic field intensities. In the absence of an externally applied magnetic field, the charged particles are accelerated by the electrostatic force alone due to charge separation in free space. The electrostatic force is neglected by comparison with the Lorentz force in the classical MHD formulation that is consistent with the globally neutral property of the plasma except in the plasma sheaths, in cathode and anode layers of electrodes. In fact, the well-known, induced gas motion by dielectric barrier discharge (DBD) is the consequence of charge separation between electrodes resulting in the momentum transfer by collisions between ions and neutral molecules (Corke et al., 2007, 2010). On the contrary, direct current discharge (DCD) imparts thermal energy to the external stream through Joule heating mostly in the cathode layer in addition to the electrode heating. It is the basic mechanism of plasma actuator by thermal effect.

A vivid illustration of the Lorentz acceleration in altering the flow field structure is displayed in the classic experiment of Ziemer (1959). He convincingly demonstrated that an externally applied magnetic field can change the bow shock standoff distance by the magnetic pressure, $\vec{B} \cdot \vec{B}/2\mu$ in the shock layer. The electromagnetic–aerodynamic interaction with an externally applied magnetic field in partially ionized gas has been applied to hypersonic control surfaces and inlets, sharp-nose conical configurations, and for heat-flux mitigation at the stagnation region of blunt bodies (Shang et al., 2014).

A new approach has been adopted for evaluating the electric current density $\vec{J}$ directly from the number density of charged species in place of the simplified Ohm's law. The current density is calculated directly by the drift-diffusion theory for the partially ionized gas (Shang et al., 2011, 2012),

$$\begin{aligned}\vec{J} = e[&-d_+\nabla n_+ + d_e\nabla n_e + d_-\nabla n_- + (n_+\mu_+ + n_e\mu_e + n_-\mu_-)\vec{E} \\ &+[n_+\mu_+(u_+ \times B) + n_e\mu_e(u_e \times B) + n_-\mu_-(u_- \times B)]\end{aligned} \tag{7.13}$$

The transport property consists of the drift velocity and the diffusion of electrons and ions. The electron and ion mobilities (μ_e, μ_+, μ_+) are traditionally evaluated as functions of the reduced electric field intensity *E/P*, and the diffusion coefficients (d_e, d_+, d_-) of the charged particles that can be calculated by the Einstein formula. The diffusion and drift velocity for electrons and both positively and negatively charged ions are used to describe the individual random and forced motions of the charged particles.

Since a few collisions can fully excite the rotational degree of freedom of gas molecules, this internal mode is considered always in equilibrium with the translation mode. The net energy transfer between translational, vibration, electronic excitations and the sums in each degree of freedom that appear in equations (7.9), (7.10), and (7.11) are designated as Q_{Vt}, Q_{et}, $Q_{v,\Sigma}$, and $Q_{e,\Sigma}$ respectively. The energy-cascading processes for the partially ionized gas occur at the atomic and molecule scales that cannot be described by gas kinetic theory and chemical kinetics. Thus, these processes must be modeled through the quantum chemical and chemical–physical kinetics.

The quantum chemical–physical kinetics models adopted by the formulation must be based on some unique features of the ionization process. The thermodynamic conditions from thermal excitation are vastly different from electron impact ionization for which it occurs at the outer limit of the Maxwell–Boltzmann distribution. Energy transition is initiated from the averaged value for electron temperature of 3.9×10^4 K, which amounts to 2.5 eV to agree with classic estimations. Electrons in highly excited states reside within the cathode and anode layers (Shang et al., 2005). The typical temperatures of internal modes within a DCD are presented in Figure 7.1. The rotational and vibrational temperatures are acquired from emission spectroscopy measurements via the second positive group of nitrogen N_2 $C^3\Pi_u \rightarrow B^3\Pi_g$ at different applied electric currents. The rotational temperatures and the most likely translational temperature of the weakly ionized air are within the range from 70 and up to 200 K at a discharge current at 400 mA. The vibration temperature appears to saturate around the range 5000–7000 K. The electron temperatures are obtained by single and double Langmuir probes. The measure electron temperatures have a rather large data scatter at an ambient pressure around 10 torr, which indicate that the discharge has entered the abnormal glow discharge range.

The most common ionization process by thermal excitation in applications is through hypersonic shock compression. The extremely high kinetic energy carried by the oncoming free stream is converted by an enveloping shock in a few collisions to attain a high thermal state and decelerates to the stagnation condition at the foremost

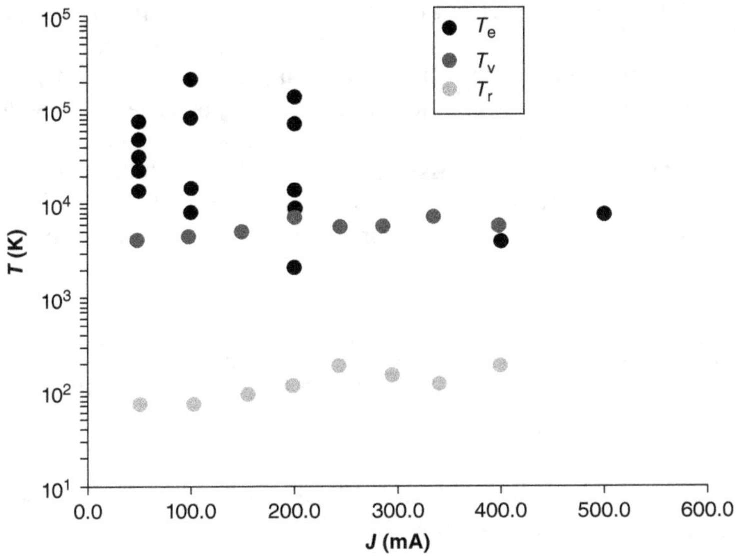

FIGURE 7.1 Temperatures of DCD.

point on the body surface (Shang et al., 2012a). A typical hypersonic flow structure is depicted in Figure 7.2. For a gas with a specific heat of $\gamma = 1.4$, the Rankine–Hugoniot relation indicates a compression approximately approaching 800 times and a temperature rising over the ambient condition by 120-fold. In the environment of the shock layer, the vibration energy level of the gas is far lower than that of translational

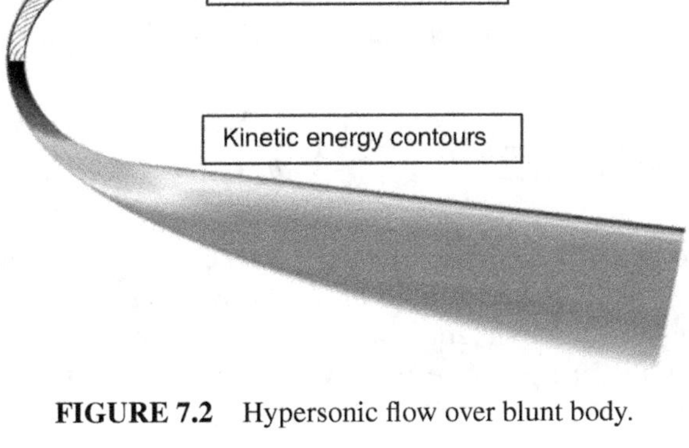

FIGURE 7.2 Hypersonic flow over blunt body.

mode immediately downstream to the enveloping bow shock. Furthermore, the energy transfer among the internal modes of excitation is dominated by the adjacent quantum levels, the so-called ladder-climbing process. Finally, the vibrational quantum states are mostly limited to the first few quanta above the ground level. Therefore, the anharmonic correction to the quantum vibrational mode is not necessary. From the principle of detailed balance, the classic result by Landau and Teller (1936) becomes the backbone for the quantum chemical–physical kinetics modeling.

The energy exchange between translational and vibrational modes is approximated by

$$Q_{t,v} = \rho\left[\frac{e_v^*(t) - e_v}{\tau}\right]; \quad e_v^*(t) = \frac{R\Theta_v}{\exp(\Theta_v/T) - 1}; \quad \tau = \frac{k_1 T^{5/6}\exp(k_2/T)^{1/3}}{p[1 - \exp(-\Theta_v/T)]} \tag{7.14a}$$

where Θ_v is the characteristic temperature of the vibrational excitation and τ is the relaxation time for vibration transition.

The energy transfer among the vibrational modes of different species is also modeled by the ladder-climbing relation via a single quantum level. However, the relaxation time scale is provided by the empirical correction of Millikan and White (1963),

$$Q_{v,v} = \rho\left[\frac{e_v^*(t) - e_v}{\tau_v}\right]; \qquad e_v^*(t) = \frac{R\Theta_v}{\exp(\Theta_v/T) - 1}$$

$$\tau_v = 1.16\times 10^{-3}\left(\frac{m_i m_j}{m_i + m_j}\right)^{1/2}\Theta_v^{4/3}\left[T^{-1/3} - 0.15\left(\frac{m_i m_j}{m_i + m_j}\right) - 18.42\right] \tag{7.14b}$$

The preferential dissociation model by Treanor and Marrone is also often adopted for the energy transfer between electronic and vibrational excitations (Treanor and Marrone, 1962). The basic idea is that the rate of molecular dissociation can occur preferentially from a higher vibrational quantum and is derived from the probability of dissociation from the excited vibrational states. Hence,

$$Q_{e,v} = \sum_i \rho_i\left[\frac{e_t - e_v}{\tau_v} - \frac{\bar{e}_{t,v} - e_v}{x_i}\left(\frac{dn_i}{dt}\right)_f + \frac{\bar{e}_{v,t} - e_v}{x_i}\left(\frac{dn_i}{dt}\right)_b\right] \tag{7.14c}$$

In the coupled vibration and dissociation (CVD) excitation model, n_i denotes the number density of the interacting molecules. The symbols $e_{t,v}$ and $e_{v,t}$ characterize the average energy lost from vibration in a dissociation and gained in recombination. The symbol e_v is the energy corresponding to the Boltzmann distribution of the vibrational excitation.

The energy transfer between elastic electron–ion collisions has been approximated by the formulation of the Coulomb logarithm and more recently estimated by the model of Chernyi et al (2004);

$$Q_{e,i} = 1.21 \times 10^{20} x_e x_i \frac{T - T_e}{T_e^{3/2}} \ln \Lambda \tag{7.14d}$$

$$Q_{e\mathbf{v}} = 2 \times 10^{-16} \sum_{v} n_e n_i \omega_{e,v} P_{1,0,v} + \left\{ \frac{e^{-(1.44\omega_{e,v}/Te)}}{1 - e^{-(1.44\omega_{e,v}/Te)}} - \frac{e^{-(1.44\omega_{e,v}/Tv)}}{1 - e^{-(1.44\omega_{e,v}/Tv)}} \right\} \tag{7.14e}$$

The total energy transfer in the vibrational and electronic modes is just the sum of each possible energy mechanisms, $Q_{v,\Sigma} = Q_{t,v} + Q_{v,v} + Q_{e,v}$ and $Q_{e,\Sigma} = Q_{e,v} + Q_{e,i}$. The internal mode energy exchanges given by equations (7.14a) through (7.14e) are just a typical implementation of the quantum chemical–physical models. Additional internal modes energy transfer or energy-cascading processes between internal modes can be added into the sum according to the specific circumstance. In fact, the kinetics of the species exchanges among the quantum states of vibration–translation, vibration–vibration, and vibration–dissociation can also be simulated through master equation for population density distribution (Josyula et al., 2011):

$$\begin{aligned}\frac{d\rho_{iv}}{dt} = \frac{1}{M} \Bigg\{ & \sum_{nv'} \left[k_{vt}\left(n_v' \to n_v\right) \rho_{v'}\rho - k_{vt}\left(n_v \to n_v'\right) \rho_v \rho\right] \\ & + \sum_{nw,nv',nw'} \left[k_{vv}\left(n_v', n_{w'} \to n_v, n_w\right) \rho_{v'}\rho_{w'} - k_{vv}\left(n_v, n_w \to n_v' n_w'\right) \rho_v \rho_w\right] \\ & + \left[\sum_{nv'} [k_{vd}(n_v \to \text{continuum})\rho_v \rho_i - k_{dv}(\text{continuum} \to n_v)\rho_{id}\rho_{id}]\right\}.\end{aligned} \tag{7.14f}$$

where n_v, n_v', n_w, and n_w' denote the quanta of vibrational excitation of the colliding partners. The rates of transitions and depletion are designated as k_{vt}, k_{vv}, and k_{vd} and can be evaluated by kinetic models or by the direct simulation Monte Carlo method for the random collision process. Unfortunately, an extensive computational resource requirement is the major weakness of this ionization formulation.

Figure 7.3 depicts the predicted translational, vibrational, and electronic temperature distributions along the stagnation streamline within the shock layer by the quantum chemical–physical models. All temperatures are calculated by assuming that the Boltzmann distribution holds for all internal modes of excitation to put all the results on an equal footing. The numerical results simulate the RAM CII sample return probe at an earth reentry speed of 7.62 km/s and at an altitude of 71.0 km. The numerical results exhibit correctly the relaxation behavior of the vibration degrees of freedom of molecular oxygen and nitrogen downstream the bow shock. The prolong

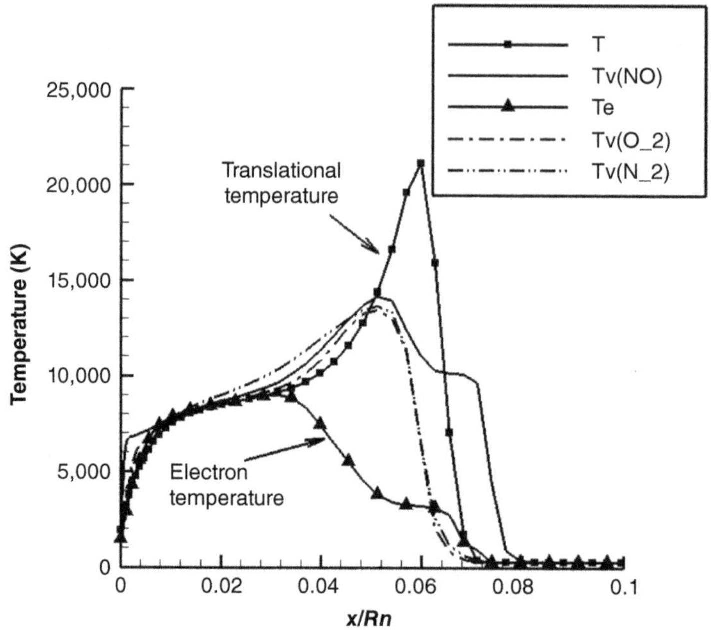

FIGURE 7.3 Temperatures distributions in shock layer reentry RAM C-II probe.

relaxation pattern for the electron temperature is also deemed physically realistic due to a greater number of collisions for the fully exited electronic excited mode than that for vibrational excitation. In all, the predicted peak translational temperature by the kinetic model reached a good agreement with results in literature.

The equations (7.10) and (7.11) from the quantum chemical-physics approximations to the solution of the master equation also have been successfully applied to earth reentry simulations from the RAMC-II probe to the fastest man-made sample return capsule Stardust. Figure 7.4 presents the nonequilibrium vibrational temperatures predicted by CVD model and master equation at the peak heating condition of Stardust at an altitude of 59.8 km, a velocity of 11.13 km/s, and an ambient temperature of 238.5 K. The difference in the predicted values that are obtained by assuming the Boltzmann distributions is around 12.5%. However, a greater discrepancy is detected for the predicted translational temperatures and the bow shock wave standoff distances. These discrepancies can be easily traced back to the computational accuracy in numerical resolution and the adopted numerical algorithm. The general agreement between published simulation results of the species concentration reflects the current status of nonequilibrium chemical kinetics models in practical applications, but with very limited verification from experimental observations.

A rare exception to the above status on the lack of experimental verification for prediction the ionized component is from the flight test data of the RAM-C-II as presented in Figure 7.5. The validation of the computational results is derived from

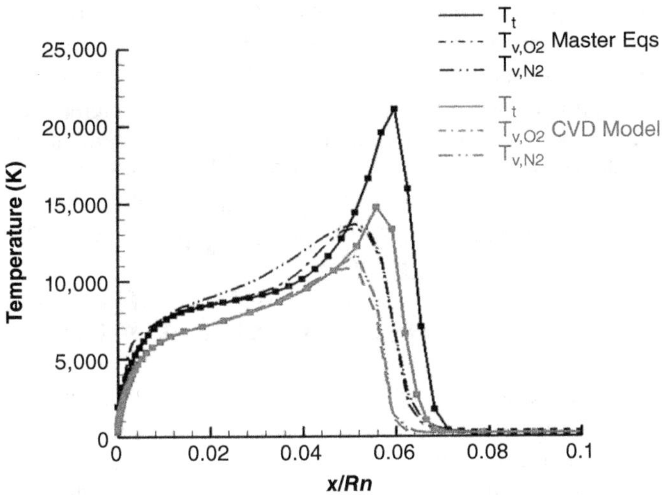

FIGURE 7.4 Compare vibration temperature by CVD model and master equation.

the comparison of electron number density with the data by Jones and Cross (1972). The major portion of their electron number density data is measured by a reflectometer and a checking point by an electrostatic probe at the distance around eight nose radii from the stagnation point. The data from the reflectometer represents the averaged

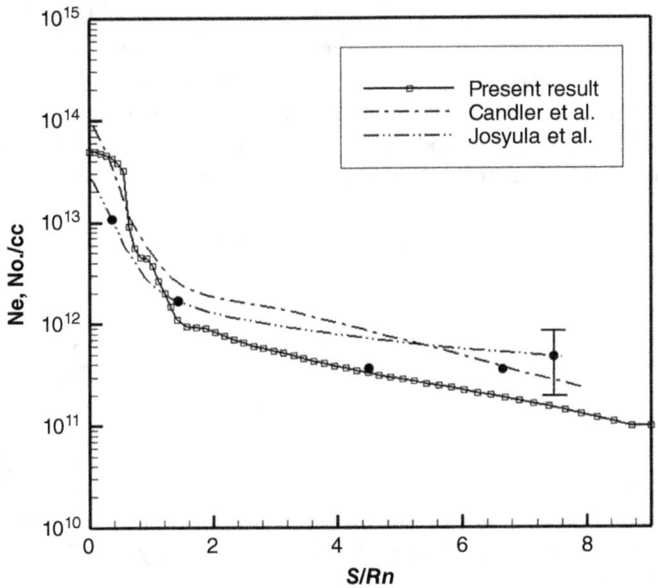

FIGURE 7.5 Verify electron number density with flight data of Radio attenuation measurements (RAM) C-II.

peak value at the sample location and the probe data is the time-averaged value among the innermost and outermost probes. The estimated error bar of the probe data covers a range of the peak-to-peak density fluctuation from 3.0×10^{11} to 1.2×10^{12} due to body motion. In spite of a noticeable difference in the predicted temperatures profiles in the shock layer by the computational simulations, the agreement with the flight data is reasonable and is comparable with earlier numerical results by Candler and MacCormack (1991) and Josyula et al. (2011). Indirectly, this comparison reflects the progress or the lack of it over nearly the past 30 years.

7.3 CHEMICAL KINETICS FOR THERMAL IONIZATION

In a closed system, the total mass of the reactants in a chemical reaction is an invariant. According to chemical kinetic theory, any complex chemical reaction can be broken down into multiple single-step, elementary reactions. The following equation indicates a relationship between the changes of number of moles of each species:

$$\Sigma v_i' N_i \rightleftarrows \Sigma v_i'' N_i \tag{7.15a}$$

The symbols v' and $v''$are referred to as the stoichiometric coefficients, and N_i is the species concentration in mole per unit volume. This fundamental law may be viewed as a statement of conservation of a total number of elementary atoms of reactants in the one-step reaction process.

The law of mass action establishes the foundation for modeling chemical reactions and the rate of species production and depletion in the species conservation law. This law is actually an empirical formulation confirmed by numerous experimental observations and states that the rate of generation and depletion of a chemical species is proportional to the products of the concentrations to a power equal to the corresponding stoichiometric coefficient, v_i. The forward rates are

$$R_{\mathrm{f},i} = k_{\mathrm{f},i}(T)\Pi\left(\frac{\rho_i}{M_i}\right)^{v_i'} \tag{7.15b}$$

Similarly, the rates of the backward chemical reaction are given as

$$R_{\mathrm{b},i} = k_{\mathrm{b},i}(T)\Pi\left(\frac{\rho_i}{M_i}\right)^{v_i''} \tag{7.15c}$$

The symbols $k_{\mathrm{f},i}$ and $k_{\mathrm{b},i}$ are the so-called reaction rate constants that are functions of the temperature and independent of the concentration of the reacting species. The net rate of change in the concentration of any chemical species is the combined results of the individual elementary, one-step reactions. In a complex overall reaction consisting of a number of elementary reactions, the rate of reaction of the system is often determined by the slowest rate of an elementary process. In this case, stoichiometric coefficients of the individual single-step reactions may not necessarily govern the overall reaction rate alone.

The following equation (7.15d) is described by the Arrhenius formula through the forward and backward chemical reactions.

$$\frac{dw_i}{dt} = M_i \sum_{j=1}^{J} \left(v''_{i,j} - v'_{i,j} \right) \left\{ k_{f,j} \prod_{k=1}^{N_{kf}} \left(\frac{\rho_k}{M_k} \right)^{v'_{k,j}} - k_{b,j} \prod_{k=1}^{N_{kb}} \left(\frac{\rho_k}{M_k} \right)^{v''_{k,j}} \right\} \tag{7.15d}$$

For a total number of chemical species of n, then $(n-1)$ concentration equations are required to be solved simultaneously. For the ionization of air by thermal excitation, the number of chemical species is usually more than 10 and involves more than 30 elementary reactions. For ionization by electron impact, the number of ionized species is even greater and the number of elementary reactions increases to hundreds. It is obvious that the required effort is tremendous and tedious. For most investigations, the forward and backward chemical reaction rates are adopted from a wide range of chemical kinetic models and experimental data.

For example, the high-temperature air mixture is widely considered to consist of 11 species: O, N, O_2, N_2, NO, O^+, N^+, $O_2{}^+$, $N_2{}^+$, NO^+, and e^- (Park, 2001). For a nonequilibrium chemical reaction, the production rate of chemical species and heat release are calculated by the rate constants for the forward and reverse chemical reactions. The energy conversion by chemical reactions becomes an increasingly important process in hypersonic flows. However, based on the accumulated experience in high temperature, hypersonic flow simulations, the chemical reactions, and aerodynamics can still be loosely coupled to achieve the most efficient computation. Two widely adopted nonequilibrium chemical reaction models are developed by Dunn and Kang (1973) and Park (1993). Park's model is widely adopted and modified simply because it has been frequently updated. In his model, 48 elementary reactions are included but the reverse reaction rates are not provided. The reverse reaction rate constants can be determined by the law of mass action, and the rate constant is determined by the Arrhenius equation.

From the law of mass action, the rate constant is determined by the Arrhenius equation,

$$k_{f/r,i} = A_{f/r,i} T^n \mathrm{Exp}(-E_{f/b,i}/kT) \tag{7.15e}$$

The symbol $E_{f/b,i}$ is referred to as the Arrhenius activation energy, and the exponential term is also known as the Boltzmann factor. The Arrhenius law simply states that only those reactants that possess energy greater than a certain potential barrier (activity energy) will lead to a reaction. The term $A_{f/r,i}T^n$ is called the collision frequency to reflect its roots in statistical quantum mechanics. The coefficient A is also known as the steric factor that depends on the orientation of the colliding species particles for reaction to occur. These reaction rate constants unfortunately are known to possess a large and varying degree of uncertainty.

The forward reaction rate is obtained from experimental observations and the reverse reaction rate can be determined by the Sharma approximation (Shrma, 2000); however, caution needs to be exercised in specifying the activity energy.

$$k_{f,i}/k_{b,i} \approx k_e T_a \tag{7.15f}$$

For the Stardust capsule at the earth reentry speed of 11.137 km/s, the capsule is protected by the phenolic–impregnated carbon ablator, PICA (Chen and Milo, 1999; Chen and Milos, 2005). Therefore, the high-temperature gas mixture in the shock layers and wakes of entering space vehicles contain optically active components such as CO_2, H_2O, CH_4, N_2, O_2, NO, N_2^+, C_2, CO, CN. The chemical reactions involves nine dissociative, eight exchange, and three electronic impact processes. The detailed rate constants of the chemical kinetic model of 18 species model by Olynick et al. (1999) are included in Table 7.1.

The interdisciplinary CEA equations (7.7) through (7.11) have been widely applied for nonequilibrium chemical reacting flows and are independent from the specific electron impact ionizing models (Olynick et al., 1999; Park, 2001; Golubovskii et al., 2002; Shang and Surzhikov, 2012). For nonequilibrium hypersonic flows, the initial values and boundary conditions have been established. The only implicit assumption

TABLE 7.1 Chemical reaction rate coefficients of 18-species nonequilibrium air model by Olynick et al. (1999)

		A_f, (cm³/mole)/s	n_f	E_f, K	A_r, (cm³/mole) /s	n_r	E_r, K
1	$N_2+M \leftrightarrow N+N+M$	0.70E+22	−1.6	0.113E+06	0.74E+17	−0.6	0.
2	$O_2+M \leftrightarrow O+O+M$	0.20E+22	−1.5	0.598E+06	0.267E+17	−0.5	0.
3	$NO+M \leftrightarrow N+O+M$	0.50E+16	0.0	0.755E+05	0.258E+12	1.0	0.
4	$C_2+M \leftrightarrow C+C+M$	0.37E+15	0.0	0.690E+05	0.117E+11	1.0	0.
5	$CO+M \leftrightarrow C+O+M$	0.23E+21	−1.0	0.129E+06	0.130E+16	0.0	0.
6	$CN+M \leftrightarrow C+N+M$	0.25E+15	0.0	0.710E+05	0.857E+10	1.0	0.
7	$CO_2+M \leftrightarrow CO+O+M$	0.69E+22	−1.5	0.633E+05	0.142E+17	−0.5	0.
8	$C_3+M \leftrightarrow C_2+C+M$	0.63E+17	−0.5	0.101E+06	0.523E+12	0.5	0.160E+05
9	$H_2+M \leftrightarrow H+H+M$	0.22E+15	0.0	0.483E+05	0.695E+10	1.0	0.
10	$NO+O \leftrightarrow O_2+N$	0.84E+13	0.0	0.195E+05	0.325E+14	0.0	0.348E+04
11	$N_2+O \leftrightarrow NO+N$	0.64E+18	−1.0	0.384E+05	0.132E+18	−1.0	0.467E+03
12	$CO+O \leftrightarrow O_2+C$	0.39E+14	−0.18	0.692E+05	0.164E+14	−0.18	0.
13	$CO_2+O \leftrightarrow O_2+CO$	0.21E+14	0.0	0.278E+05	0.323E+13	0.0	0.240E+05
14	$CO+C \leftrightarrow C_2+O$	0.20E+18	−1.0	0.580E+05	0.356E+17	−1.0	0.933E+03
15	$CO+N \leftrightarrow CN+O$	0.10E+15	0.0	0.386E+05	0.164E+14	0.0	0.
16	$N_2+C \leftrightarrow CN+N$	0.11E+15	−0.11	0.232E+05	0.340E+14	−0.11	0.
17	$CN+O \leftrightarrow NO+C$	0.16E+14	0.1	0.146E+05	0.106E+14	0.1	0.377E+03
18	$CN+C \leftrightarrow C_2+N$	0.50E+14	0.0	0.130E+05	0.543E+14	0.0	0.
19	$HCN+H \leftrightarrow CN+H_2$	0.80E+12	0.017	0.835E+05	0.564E+12	0.017	0.930E+05
20	$C+e- \leftrightarrow C^++e-+e-$	0.635E+16	0.0	0.138E+06	0.268E+15	1.0	0.403E+04
21	$N+e- \leftrightarrow N^++e-+e-$	0.508E+17	0.0	0.121E+06	0.667E+15	1.0	0.
22	$O+e- \leftrightarrow O^++e-+e-$	0.635E+16	0.0	0.106E+06	0.339E+15	1.0	0.

for the species conservation equations at the solid surface is either noncatalytic or fully catalytic. However, the mostly widely used condition is to simply let the chemical reaction control by the local thermodynamic condition identical to the interior domain. In other words, the species concentration in mass fraction $Y_i = \rho_i/\rho$ at the surface is assumed to be the same as the oncoming stream $Y_{i,b} = Y_i$. Sometimes, such a boundary condition is referred to as supercatalytic; technically, it is assumed that the solid surface is fully catalytic. The other limiting condition is the noncatalytic condition in which the gradient of the species concentration is set to zero, $(\vec{n} \cdot \nabla Y_i)_{bi} = 0$. For radiation heat transfer, the surface temperature is also to be calculated by the equilibrium radiative surface condition. Finally, on an ablating surface, the interface boundary conditions are much more complex than the initial values.

In describing the ablating surface, the majority of research efforts have been concentrated on the interface boundary conditions either from physical observations or by unique insights. By a rigorous approach, the interface boundary condition is derived from the governing equations on the surface of the control volume. These interface conditions can be summarized as follows:

For species conservation equations (Shang and Surzhikov, 2012):

$$\begin{aligned}[t] &[\rho_{i+}(u_+ + u_{i+} - u_{b+}) - \rho_{i-}(u_- + u_{i-} - u_{b-})] \cdot n_+ A \\ &= \frac{\lim}{v \to 0}\left[\iiint w_i dv - \frac{d}{dt}\iiint \rho_i dV\right] \to 0 \end{aligned} \tag{7.16a}$$

For momentum conservation equation:

$$\begin{aligned}[t] &[\rho_+ u_+(u_+ - u_{b+}) - \rho_- u_-(u_- - u_{b-}) - (\tilde{\sigma}_+ - \tilde{\sigma}_-] \cdot n_+ A \\ &= \frac{\lim}{v \to 0}\left[\iiint \sum_{i=1}^{n} \rho_i f_i dv - \frac{d}{dt}\iiint \rho u dV\right] \to 0 \end{aligned} \tag{7.16b}$$

For internal energy conservation equation:

$$\begin{aligned}[t] &[\rho_+ e_+(u_+ - u_{b+}) - \rho_- e_-(u_- - u_{b-}) + q_+ - q_- + u_+ \cdot \tilde{\sigma}_+ - u_- \cdot \tilde{\sigma}_-] \cdot n_+ A \\ &= \frac{\lim}{v \to 0}\left[\iiint q dV + \iiint \sum_{i=1}^{n} \rho_i f_i \cdot (u + u_i) dv - \frac{d}{dt}\iiint \rho e dV\right] \to 0 \end{aligned} \tag{7.16c}$$

In these equations, the subscript symbols + and − denote the variables evaluated above or beneath the ablating interface and thus indicate the required descriptions for the nonequilibrium gas emitting from the ablator. The velocity of the recessing surface is indicated as u_b, which is permitted to vary from point to point on the control surface. In this formulation, the ejection velocity of pyrolysis gas and the release vapor rate of the sublimated material on the ablating surface are required as input. Therefore, the details of gaseous motion through the porous ablating material are relegated to the research results of the ablative material. The heat transfer term q accommodates the heat by conduction, convection, as well as radiation transfer according to the energy-conservation equation (7.11). Finally, the normal stress component is designated as $\tilde{\sigma}$ and is the only tensoral stress component that can contribute to the work done by the gas media: the nonequilibrium gas, sublimating vapor, and pyrolysis gas of the ablating material.

The formulation of the interface boundary conditions is derived from the eigenvector structure of the governing equations. The present result intends to show that these interface boundary conditions for ablating surfaces are much more complex than that in the existing literature. Additional efforts are still required on how these conditions can be implemented and maintain computational stability and how spectral optical properties of boundary layer are changed by the ablative gases. At the present time, the implementations are not fully verified that the interface boundary condition can satisfy the well-posed criterion for numerical simulation. This issue remains open as a basic research objective of hypersonic ablation simulation.

Nevertheless, the interdisciplinary formulation has been successfully applied to a large group of interplanetary and low-earth reentry simulations. The species concentration in particle number density per cubic centimeter along the stagnation streamline of the reentering Stardust capsule is presented in Figure 7.6. The bow shock standoff distance is found at one centimeter upstream from the stagnation point of the capsule with a hemispherical nose radius of 22.9 cm, at a reentry speed of 11.137 km/s and an altitude of 59.77 km. In the mesosphere above the stratosphere, the air density is 2.34×10^{-4} kg/m^3 and the ambient temperature is 238 K. Nitrogen molecule has a number density around 6.0×10^{15}/cm^3 and oxygen molecule presents in the level of

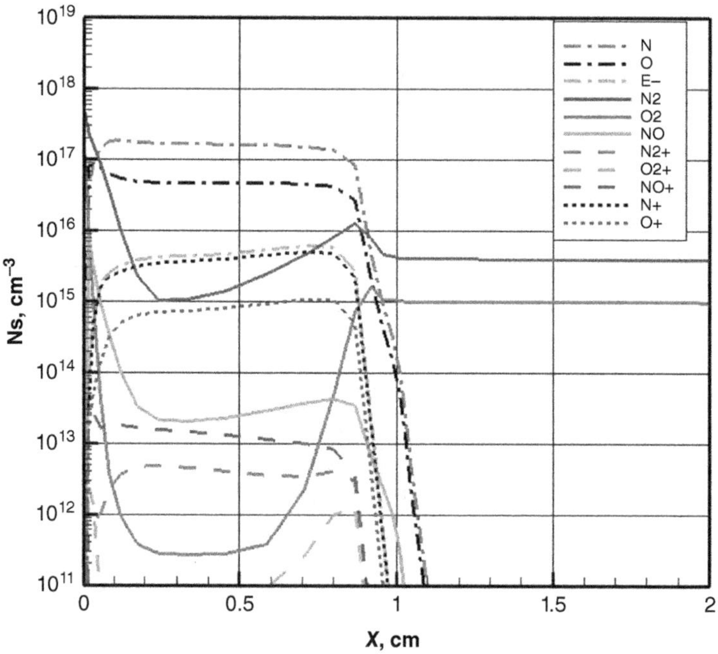

FIGURE 7.6 Species concentration of partially ionized air in the shock layer.

$1.0 \times 10^{15}/cm^3$. After the strong bow compression, the dissociated nitrogen atoms have the highest and a nearly uniform number density within the shock layer to attain a value of $1.0 \times 10^{17}/cm^3$. However, atomic nitrogen recombines at the catalytic surface and yields a high number density of $4.8 \times 10^{17}/cm^3$ for molecular nitrogen. The number of electrons increases rapidly downstream from the bow shock and reaches a maximum value of $7.0 \times 10^{15}/cm^3$ and then diminishes quickly toward the capsule surface. The profile of the positively charged nitrogen and oxygen atoms that are the dominant component of ions follows the behavior of the electrons closely. By summing all number density of the positive charged species, N^+, O^+, NO^+, $N_2{}^+$, and $O_2{}^+$, the global neutrality of a partially ionized plasma is unmistakably displayed. The chemical kinetic formulation indeed can model the fundamental property of ionization for plasma accurately.

Another physical insight of the relative influence of the internal degree of freedom on the chemical composition within the shock layer may also be gleamed from chemical kinetic modeling. Figure 7.7 depicts the species concentrations in molar fractions within the shock envelope of the Stardust probe at the peak heating condition at $t = 54$ s. From this presentation, the bow shock standoff distance of (1.3 cm and the profiles of all the species agree very well with the result of Olynick et al.). A high concentration of atomic nitrogen is observed in the stagnation region and extends to the outer edge of the heat shield. It becomes obvious that any chemical kinetic

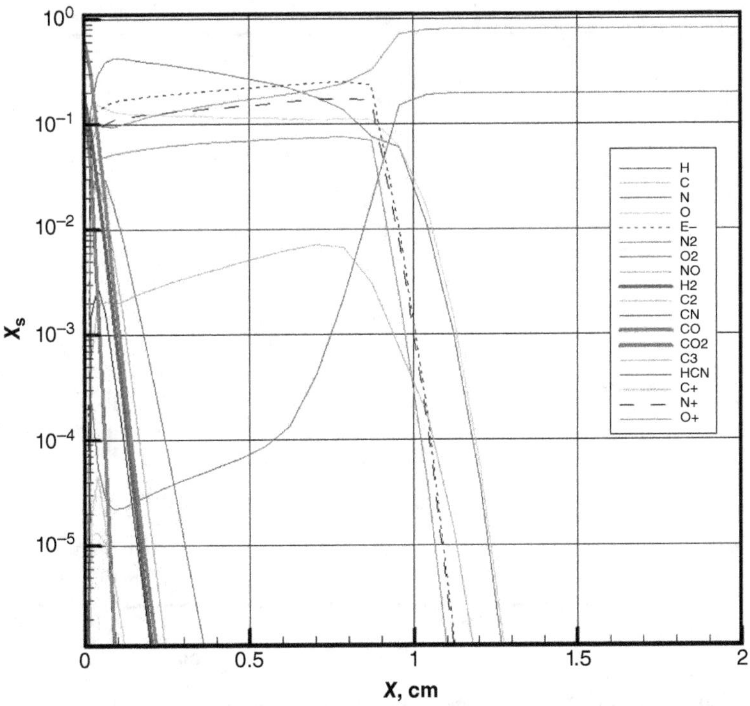

FIGURE 7.7 Species concentration in molar fractions, Stardust, $t = 54$ s.

model does not include nitridation and will underpredict the mass blowing rate in the stagnation region where the dissociated nitrogen has a significant concentration. Meanwhile, there is no significant presence of oxygen atoms adjacent to the ablator; the oxidation of carbon may have a limited effect on the ablation recession in the stagnation region. These important observations were first made by Chen and Milo in their study on the thermal response of spacecraft heat shield for ablation phenomena.

In early reentry simulations, the conservation equations of electronic excitation were routinely included. The processes of dissociation, ionization, electron–ion interaction, and energy cascading to and from other modes are important energy conversion and chemical kinetic mechanisms. The shortcoming of the state-of-the-art for the nonequilibrium, high-temperature chemical kinetic model for predictions of these phenomena is known to be woeful limited in physical fidelity. In the absence of an externally applied magnetic field, the electromagnetic force and energy of the globally neutral plasma are also negligible. Therefore, the value of including the electronic energy models into the governing equations is challenged. The recent works by Shang and Surzhikov have shown that the contribution by including the electronic energy equations is rather insignificant for conductive heat transfer, even at the fastest man-made entry speed of the Stardust sample return capsule (Table 7.2).

Figure 7.8 depicts the computational results of Stardust at an altitude of 59.79 km with a reentry speed of 11.14 km/s at the ambient temperature of 238.47 K. Two groups

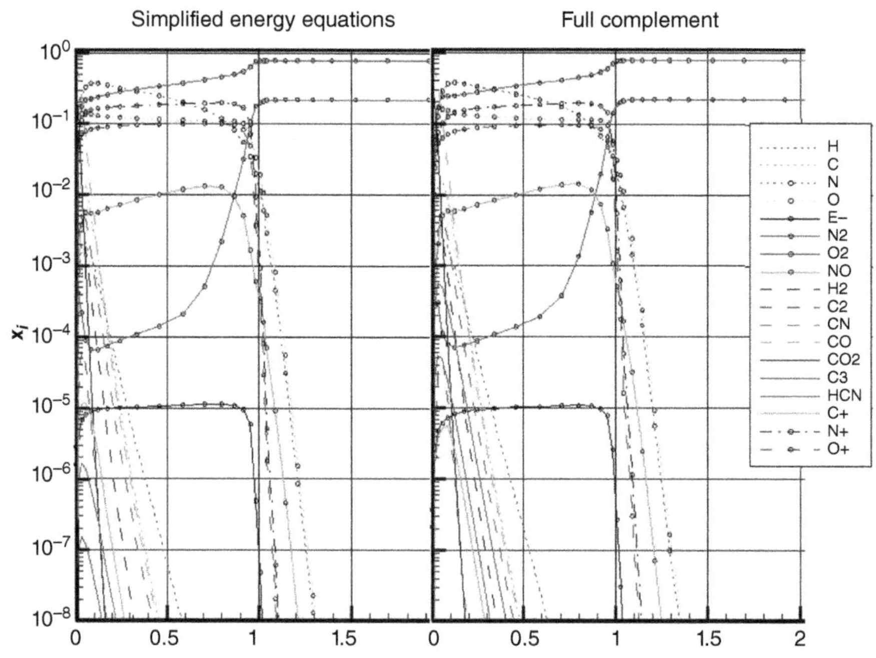

FIGURE 7.8 Comparison of species concentration along the stagnation line of Stardust capsule at reentry point at 54 s.

of species concentration profiles in molar fraction with and without the coupling of the electronic energy conservation equation are displayed side by side. The results by the traditional, full complement governing equations are included on the right-hand side of this figure. The chemical kinetic model is based on the formulation of Olynick et al. for a carbonaceous surface. An axisymmetric grid system is constructed by multiple blocks that have dimensions of (31 × 45), (31 × 27), (31 × 55), and (24 × 55). The grid system used was derived from the basic O and C type meshes. The outstanding feature of this multiblock construction is that the first grid block wraps around the corner of the fore- and after-body to provide a near-equal distance shock layer for shock capturing. The three-dimensional grid was generated simply by a coordinate rotation around the axis of the Stardust capsule. The two groups of the calculated species concentration are identical except for trace elements such as C_3, HCN, and C+ for the nonablating simulation. The magnitude of these species is less than 10^{-4} in molar fraction and the computational resource saving is 37%. This dilemma of requiring a huge amount of computational resource but producing negligible difference is made more evident by comparing the energy distributions among the various internal degrees of freedom of reacting chemical species that are generated with and without inclusion of the electronic energy conservation equation for the ionized components.

To further demonstrate the fact that the modeled energy transfer via the translation–electron and vibration–electron collisions, as well as the energy cascading from electronic to vibrational excitation, is not the dominant mechanism as the chemical kinetic processes at the simulated condition, the vibrational and electronic temperatures along the stagnation streamline of two different models are shown by Figure 7.9. The temperatures of different internal degrees of excitations are calculated by assuming the

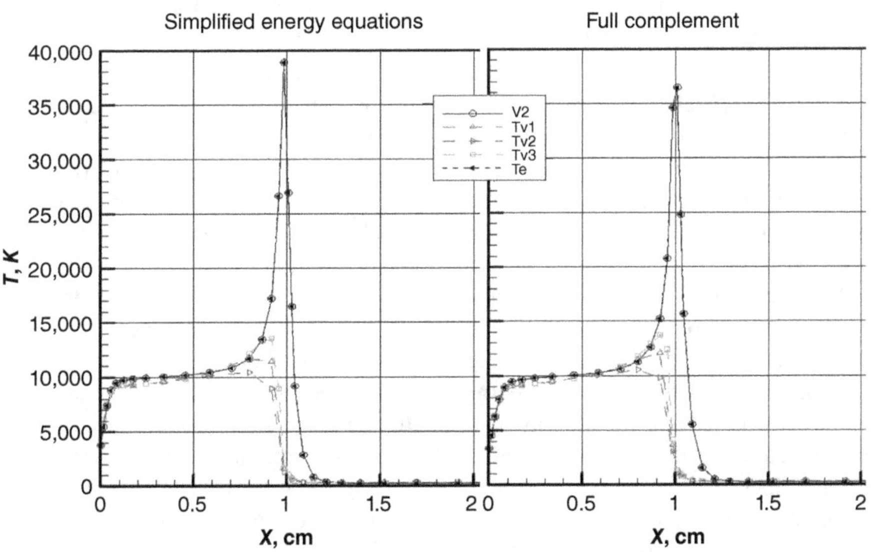

FIGURE 7.9 Comparing temperatures of two models of nonequilibrium chemical kinetics for Stardust simulation at reentry point of 54 s.

Boltzmann distributions that are still applicable. The numerical results of the full complement of model equations are shown on the right-hand side of the figure. First of all, the vibrational temperatures of O_2, N_2, and NO of two different models are essentially identical. The computed translational temperatures are bracketed between the values of 36,400 and 39,000 K to fall within the range of known numerical results. The lower translational temperature from the full complement of energy conservation equations correctly reflects the fact that a greater amount of energy is distributed to the electronic excitation. The difference in the predicted temperature is less than 7% without credible verification by either experimental observation or theoretical analysis. However, the slightly underpredicted population density of the ionized species has a significant impact on the radiative energy exchange. The shortage of a validating experimental data base becomes a glaring concern.

7.4 CHEMICAL KINETICS FOR ELECTRON IMPACT IONIZATION

The nonequilibrium chemical kinetics associated with DBD in the atmosphere is well known because it had been applied as an ozone generator for years. Elisson and Kogelshatz (1991) have identified plasma chemistry in microdischarge by 30 species through 143 elementary reactions. In a more recent work by Bogdanov et al. (2008), the chemical-physics kinetics of atmospheric plasma have been investigated by 576 chemical reactions involving vibrational excitations of nitrogen and oxygen, ozone, positive and negative ions, besides oxide nitrides. The complexity of the internal degrees of excitations includes molecular nitrogen and oxygen at few quanta above ground states; the atomic nitrogen, $(4S, 2D, 2P)$, oxygen $(^3P,^1S,^1D)$, the charged nitrogen molecules $(A^3\Sigma_u^+, B^3\Pi_g, A^1\Pi_g, C^3\Pi_u)$ and oxygen $(X^3\Sigma_u^+, A^1\Delta, B^1\Pi)$, ozone molecules O_3, as well as negatively charged ions $(N_2O^-, NO^-, O^-, O_2^-, O_3^-, \ldots)$ and positively charged ions $(N^+, O^+, NO^+, NO_2^+, \ldots)$, respectively.

In Figure 7.10, the ionization of air by electron collision for most process is seen to initiate and to approach the equilibrium state less than 40 ns. The highest species concentrations at the shortest time scale belong to the $O(^3P), N_2(A)$ and $O_2(A)$. Only the molecular compositions of O_3 and N_2O persist over the period more than one-tenth of 1 μs and beyond after the ignition. The other transient radicals such as $O(^1D), N(^2D)$, $O_2(o)$, and $O_2(b)$ have the life span less than 1 μs.

One may notice that most constituents of the partially ionized air are transient radicals. The chemical reaction is mainly initiated by the fast-moving electrons. The electrons collide with neutral molecules and then excited neutrals to higher energy levels but can replenish the lost energy from the electric field. For this reason, the reaction rate constants for a number of electron impact processes are described by the reduced electric field *E/N*, where *N* denotes the number density of the neutral particles. In fact, the Townsend similarity law for electron impact is characterized by the reduced electric field as *E/P*. For DBD, most charge carriers vanish after about 50 ns, and the chemical reactions proceed without interaction with the charge carriers. The ionization energy for these reactions varies widely from a few up to

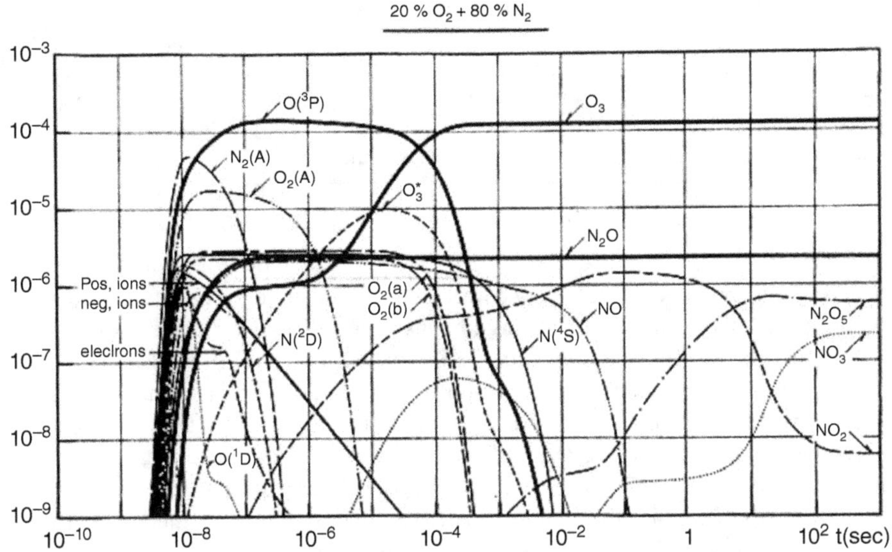

FIGURE 7.10 Species composition of DBD. From Eliasson, B., and Kogelschatz, U. *IEEE Trans. Plasma Sci.*, vol. 19, 1991. Used with permission from IEEE.

thousands of electron volts. Regardless, the reactions still can be presented in Arrhenius form:

$$k_i = k_{i,o} T^n \exp(-E_a/kT) \tag{7.17}$$

The rate constants for different reactions are given in terms of cm^3/s (two-body collision) or cm^6/s (three-body collision) and are usually given as a function of Vcm^2 and the temperatures of electrons and ions.

According to Elliason et al. (1991) the main chemical reactions by electronic impact fall into four groups, electron–molecule, atom–molecule, decomposition, and synthesis Tables (7.2) and (7.3).

TABLE 7.2 Ionization by electron–molecule and atom–molecule reactions

Electron–Molecular Reactions		Atom–Molecular Reaction	
Excitation	$e + A_2 \Leftrightarrow A_2^* + e$	Penning dissociation	$M^* + A_2 \Leftrightarrow 2A + M$
Dissociation	$e + A_2 \Leftrightarrow 2A + e$	Penning ionization	$M^* + A_2 \Leftrightarrow A_2^+ + M + e$
Attachment	$e + A_2 \Leftrightarrow A_2^-$	Charge transfer	$A^{\pm} + B \Leftrightarrow B^{\pm} + A$
Dissociative ionization	$e + A_2 \Leftrightarrow A^- + A$	Ion recombination	$A^- + B^+ \Leftrightarrow AB$
Recombination	$e + A_2^+ \Leftrightarrow A_2$	Neutral recombination	$A + B + M \Leftrightarrow AB + M$
Detachment	$e + A_2^- \Leftrightarrow A_2 + 2e$		

TABLE 7.3 Ionization by decomposition and synthesis

Decomposition		Synthesis	
Electronic	$e + AB \Leftrightarrow A + H + e$	Electronic	$e + A \Leftrightarrow A^* + e$ $A^* + B \Leftrightarrow AB$
Atomic	$A^* + B_2 \Leftrightarrow AB + B$	Atomic	$A + B \Leftrightarrow AB$

Bogdanov et al. also have correctly pointed out that the concentration of the negatively charged ions is relatively lower than the positively charged counterpart, but the effect of the negatively charged ions can still exert a critical directional demarcation for the electrostatic force during an AC cycle (Kim et al., 2007). This finding also shows that the O_2^- plays a dominant role in a DBD plasma actuator. Since DBD is a nonequilibrium transient discharge, the mean energy state of the electrons and heavy particles is considerably different. For this reason, the energy of charged particles can be effectively transferred to the internal states of molecule or atom.

In DBD applications, the charged particles interaction with the dielectrics is important. However, the photoemission is negligibly small for nitrogen because the energy of photons in the visible range is too small to cause the emission of electrons (Pancheshnyi et al., 2001). Raizer also has pointed out that photoemission cannot compete with electron impact ionization because the collision cross sections of molecular and atomic oxygen and nitrogen close to the emission threshold are rather high (10^{-16} cm^2) (Raizer, 1991). In gist, the partially ionized gas in DBD occupies only few quanta above the ground state that is capable of photoemission. This process, however, can supply the seeding electrons that are needed to start the avalanche in streamer propagation.

The other important processes in ionization are desorption of electrons and surface recombination. The electrons can be desorbed mostly by energy cascading to vibrational excitation or by depletion of the metastable molecule $N_2(A^3\Sigma_u^+)$. However, secondary emission of electrons from the surface during ionization is the dominant mechanism and leads to charge separation in the cathode layer. Meanwhile, surface charge accumulation also alters the electric potential within the discharge domain. In view of the relatively short duration of microdischarge versus the microsecond time scale of an AC cycle, the instantaneous surface recombination is a plausible and acceptable approximation. In this connection, one needs to be mindful of the fact that positive charge buildup on the dielectrics far downstream of the DBD has also been observed (Opaits et al., 2008). It is also noted that the surface potential coincides with the biased polarity and depletes slowly in duration of tens of minutes to degrade the performance of the DBD. Finally, and most important, the ionization of DBD is still a basic Townsend process (Howatson, 1975; Raizer, 1991). This observation is the foundation to all known formulation for ionization by electron collisions.

While the ionization process by electron impact is distinct from thermal excitation, the nonequilibrium ionized species concentrations must still satisfy the law of mass action. For this reason, ionization processes are treated as chemical reactions of molecules, atoms, and electrons or ions. Chemical kinetics models for low-temperature plasma generation span a very wide range of complexity, from the detailed description by Elisson and Kogelschatz (1991) and Bogdanoff et al. (2008) to simplified approximations by Solov'ev et al. (2008), Singh and Roy (2007), and Rafatov et al. (2012).

Singh et al. reduce the chemical kinetic model by omitting the metastable species along with N^{4+} and O^{4+} from these reactions that have the extremely high recombination rates. The model equations governing chemistry of discharge include the ionization, dissociation, and recombination of nitrogen, and the same for oxygen with an addition dissociative-attachment reaction. The system of equations is solved using a Galerkin variation formulation on the basis of a finite-element method. Numerical results show that the average force in DBD due to the negatively charged ions is small as compared with that of the positively charged ions, and the periodic electrostatic force increases slightly with the increase in the length of the grounded electrode beneath the dielectrics.

Golubovskii et al. (2002) simplify their selection of the elementary chemical reactions based on comparable reaction rates of species to the characteristic time scale on the order of 30 ns for DBD in atmospheric air. A total of 16 key reactions are considered, including four charge-exchange reactions in a chain of conversion. The main process of electron detachments from the negative ions is identified as

$$A^{+} + B^{-} + O_2 \rightleftarrows A + B + O_2 \tag{7.18a}$$

and

$$A^{+} + B^{-} + N_2 \rightleftarrows A + B + N_2 \tag{7.18b}$$

However, the characteristic detachment time is about 40 ns; thus, these processes do not make a significant contribution to the net number density of negatively charged ions. The main achievement of Golubovskii et al. (2002) work is that the numerical results predict the streamer length and charge number density that are in agreement with experimental data.

Rafatov et al. (2012) adopt an extended fluid model as a compromise between the strictly fluid and chemical kinetic model to analyze the DCD with seven elementary reactions for ionization from the elastic collision to the trapping radiation. The unique features of their chemical kinetic model are the further refined ionization process into the direct, stepwise, and penning ionizations and some of the rate constants are calculated from the local Boltzmann equation. In particular, the numerical examples even include the electron collision cross sections for ionization in argon as the functions of electron energy. This chemical kinetic model adequately describes most fundamental properties of glow discharge.

All these approaches are rational and the fidelity of numerical simulation to nonequilibrium chemical reaction is solely dependent on the empirical rate constants. For this reason, a validation process is difficult and remains incomplete. Therefore, the chemical rate constants are the weakest link of ionization models for both the high-temperature thermal excitation and electron collision mechanisms. In addition, the complete system of differential equations of species conservation is very stiff to become a formidable computational challenge.

7.5 TRANSPORT PROPERTIES

To be consistent with the kinetic model of the internal structure of a gas, transport properties of the gas mixture for thermal diffusion, molecular viscosity, and thermal conductivity need to be calculated from the Boltzmann equation by the Chapman–Enskog expansion. It is a landmark achievement of the kinetic theory of diluted gas mixtures by its ability to describe the transport property of any combination of gaseous mixture by the intermolecular potential functions. The required collision integrals and cross sections have been obtained by either the Lenard–Jones potential for nonpolar molecules or by a polarizability model for ion-neutral nonresonant collisions. The results from the kinetic theory for the molecular viscosity and thermal conductivity of individual species are used to generate a global property for a gas mixture. However, the additional forced diffusion mechanism induced by the charged particles in an electromagnetic field is dominant over the ordinary diffusion and which needs special attention.

Diffusion phenomenon in a high enthalpy gas mixture is dominant. From the kinetic theory of diluted gases, the diffusion coefficient is not affected by the presence of internal degrees of freedom in the molecules. However, in a chemically reacting system, the diffusion can dominate the mass and energy transfer processes. The net rate of production of chemical reactions will appear as a source term in the conservation equations for both homogeneous and heterogeneous reacting systems. Especially for the latter, chemical change takes place only in confined regions, such as on the interface boundary or shock wave. Diffusion becomes the major added complication for numerical simulations of nonequilibrium flow fields. In a diffusing gas mixture, the velocities of individual species can be significantly different from each other, particularly when the additional force diffusion is presented in electrically polarized species of an ionized gas. The species diffusion velocity v_i is defined by a relative velocity with respect to a local mass-averaged velocity, u of the organized motion:

$$u = \frac{\sum_i \rho_i v_i}{\sum_i \rho_i} \tag{7.19a}$$

where ρ_i is the density of species i. The relative diffusion velocity of a chemical species u_i is defined by the difference of the mass-average value and the diffusion velocity with respective to a stationary frame of reference.

$$u_i = v_i - u \tag{7.19b}$$

By definition, the sum of all the species mass diffusion velocities is identically zero

$$\sum_i \rho_i u_i \equiv 0 \tag{7.19c}$$

In a multicomponent chemically reacting system, the concentrations of various species are usually given as either the mass fraction

$$y_i = \rho_i / \rho \tag{7.19d}$$

or the molar fraction

$$x_i = (\rho_i / M_i) \Big/ \sum_i (\rho_i / M_i) \tag{7.19e}$$

Obviously, the fractions have the following properties: $\sum x_i = 1.0$ and $\sum y_i = 1.0$. The relationship between the two fractions is

$$x_i = \frac{(y_i / M_i)}{\sum (y_i / M_i)} \tag{7.19f}$$

The gas motion of a chemically reacting multicomponent medium is affected by the local concentration of different chemical species and the thermodynamic properties of the mixture. There are four distinct mechanisms of diffusion processes, ordinary (species concentration), thermal, pressure, and forced diffusions that exert substantial impact to the mass, momentum, and energy transfer. The species concentration, in general, is the primary diffusion mechanism for most gas mixtures. According to the Onsager reciprocal relationship of an irreversible thermodynamic process, if the temperature gradient gives rise to the diffusion velocity known as the Soret effect in mass transfer, then the concentration gradient must also produce a heat flux. This coupling effect is known as the Dufour energy flux that contributes to an additional amount to the total heat flux (Shang, 1991). However, the diffusion processes by the thermal and pressure gradients are usually negligible in comparison with species concentration and forced diffusion. The forced diffusion becomes significant only for the ionized gas in an electromagnetic field by the Coulomb force, Lorentz acceleration, and the shielding action between electron and ion known as the ambipolar diffusion.

In a homogeneous chemical reacting gas mixture, the energy fluxes are driven by the identical mechanisms for mass and momentum fluxes. In fact, there is a unique coupling between mass and energy transport. In case the species velocity differs from mass-averaged velocity of the organized gas mixture motion, the individual species will carry specific enthalpy equal to $\rho_i u_i h_i$ across the control volume. This additional energy flux is generated by the interdiffusion process. Therefore, the heat flux for a multicomponent system, including Joule effect, becomes

$$q = -k\nabla T + \sum \rho_i u_i h_i + J \cdot E \tag{7.20}$$

An extensive amount of effort has been devoted to describe the transport properties of high-temperature gases and simplified for practical applications. An outstanding and exhaustive review has been performed by Capitelli and his colleagues (Capitelli et al., 1996, 2000). The historical development and benchmark progress over the past 30 years are articulated and lay out a strategy for future improvement. The most recent approaches in basic research are directly applied to the formulation of kinetic gas theory through collision integrals and cross sections; see equations (1.61) through (1.64). The collision integral or transport cross section is obtained by performing three consecutive integrations: The first integral determines the classic deflection angle as the impact parameter. The second integral is an averaged process for the impact parameter to get the relevant cross section. Finally, an averaging process is conducted over the energy to produce the collision cross section as a function of temperature. A unique feature of this multiple integration process is the diminished dependence of the collision integrals on the potential functions. The final collision cross section is given as

$$\Omega^{(i,j)}(T) = \frac{4(i+1)}{(j+1)![2i+i-1(-1)^i]} \frac{1}{2kT} \int_0^\infty \left(\frac{\varepsilon}{kT}\right)^{j+1} Q^{(i)}(\varepsilon) \exp\left(-\frac{\varepsilon}{kT}\right) d\varepsilon \tag{7.21a}$$

where

$$Q^{(1)}(\varepsilon) = \frac{2\pi}{1-[1+(-1)^i]/2/(i+1)} \int_0^\infty [b(1-\cos^i\theta]db \tag{7.21b}$$

$$\theta(b,\varepsilon) = \pi - 2b \int_0^\infty \frac{1-b^2/r^2-\Phi(r)/\varepsilon}{r^2} dr \tag{7.21c}$$

In the above integration process, the Lenard–Jones potential

$$\Phi(r) = 4\varphi_o[(\sigma/r)^{12} - (\sigma/r)^6] \tag{7.21d}$$

can be used for all species up to a temperature of 2000 K. In the dissociation temperature range for nitrogen, an exponential repulsive form is adopted:

$$\Phi(r) = \varphi_o \exp(-r/\rho) \tag{7.21e}$$

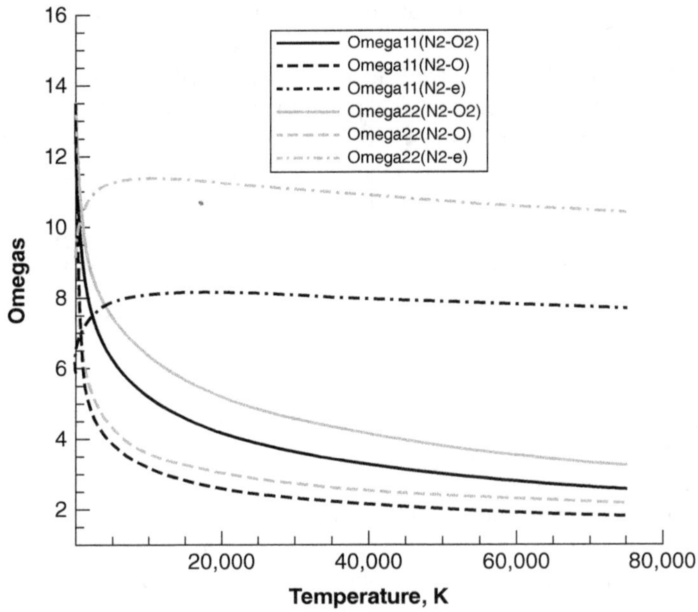

FIGURE 7.11 Collision integrals for transport properties.

For ion–ion and electron–electron interactions, the screened Coulomb potential is also included in the evaluation process:

$$\Phi(r) = (\varepsilon^2/r)\exp[-(r/d)] \tag{7.21f}$$

The results of kinetic theory for the viscosity and thermal conductivity of individual species are used to generate a global property for a gas mixture (Levin et al., 2004). One of the widely adopted formulas is known as Wilke's mixed rule, equations (1.65) and (1.66) (Hirschfelder et al., 1954; Bird et al., 1960). The collision integrals of very wide range of temperatures from 200 to 28,000 K of 11 species of air mixture are calculated and tabulated by Capitelli et al. (2000)

To illustrate the distinct characteristics of collision integrals, the values of molecule–molecule (N_2–O_2), molecule–atom (N_2–O), and molecule–electron (N_2–e) collisions are assembled over the temperature range from 300 to 75,000 K in Figure 7.11. The products of collision integrals and associated cross sections are scaled by a factor of π and, to cover the whole range of interparticle potential, are employed by a combination of the exponential repulsive, Lenard–Jones, and polarizability models with weight functions. It is easily recognized that the collision integrals $\Omega^{(1,1)}$ and $\Omega^{(2,2)}$ for molecular nitrogen–electron interaction stand out and have the greatest values than the rest. The collision integrals of the molecular nitrogen–oxygen atom have a smaller value than that of the molecular nitrogen and oxygen. The numerical results are in general agreement with the classic observation in that the ratios between $\Omega^{(2,2)}/\Omega^{(1,1)}$ yield a value around 1.1 to have range from 1.3519 (N_2–e), 1.2611 (N_2–O_2), to 1.196 (N_2–O).

The reminder issue for the forced diffusion will be addressed in the following chapter. Thus, only the pertinent physical phenomena are given in here. Forced diffusion of the transport property for the partially ionized gas emerges from the charged particle interaction with the electromagnetic field. The charge species velocity is obtainable from drift-diffusion theory that has been derived from the kinetic equations for electrons and positively and negatively charged ions. The total mass flux of charged particles in an electromagnetic field consists of the drift and diffusion components

$$\Gamma_i = \pm n_i \mu_i E \pm n_i \mu_i (u_i \times B) - d_i \nabla n_i \tag{7.22a}$$

In equation (7.22a), μ_i and d_i are characterized as the mobility and diffusion coefficient of charged particles. The plus sign is applied to the positively charged ions and the minus sign is associated with electrons and negatively charged ions. The electron motion is driving from the drift velocity by the Coulomb force and Lorentz acceleration as well as the diffusion generated by the concentration gradient of the charge number density. The proportionality constant μ_e is the mobility of the electron and is defined as $\mu_e = e/m_e \nu_e$, where ν_e is the averaged electron collision frequency for momentum transfer. However, the mean energy of the electrons depends on the electromagnetic field; the relationship between drift velocity and the mobility is not linear but has been chosen to give a qualitative estimate. At an electron temperature of a few electron volts, the electron mobility has a very good simple approximation as $\mu_e \approx \alpha_e/p$ with physical units of cm^2/s, and the pressure is measured in the unit of atmosphere. The widely adopted values of the coefficient α_e have a range from 3.75 to 4.4×10^5 for air plasma.

In a weak to moderate electromagnetic field, if there is limited charge transfer and the collision cross section is purely polarization induced, the drift velocity of the ions can also be approximated similarly as $\mu_i = e/m_i \nu_i \approx \alpha_i/p$. The physical dimension of the diffusion coefficient is cm^2/s. This approximation holds for either positively or negatively charged ions, and the coefficient α_i has the wider range of 1.45×10^2 to 6.55×10^2.

The diffusion coefficients can be calculated either by the collision cross section of a polarized interparticle potential or by a typical and dominant ion species. A simple and efficient evaluation can also be achieved by the Einstein formula between the charged particle mobilities and diffusion coefficients (Raizer, 1991; Surzhikov et al., 2004),

$$\begin{aligned} d_e &= \mu_e T_e \\ d_i &= \mu_i T_i \end{aligned} \tag{7.22b}$$

where the temperatures are measured in electron volts. Typical values used for electron impact ionization are provided in here for references: $\mu_e = 8.8 \times 10^4, d_e = 8.8 \times 10^4, \mu_i = 2.9 \times 10^2$, and $d_i = 7.5$.

The diffusion process in the presence of an externally applied magnetic field adds a gyrating component to the drift velocity. In the earlier discussion, the effect of the magnetic field contributes to the drift velocity is twofold. First, it is the component parallel to the gradient of the partial pressure gradient that transverse to the magnetic field is

$$d_{e\perp} = \frac{d_e}{1 + (\omega_e/\nu_e)^2} \tag{7.22c}$$

The second component is just a manifestation of the Hall Effect

$$d_{\mathrm{eH}} = \frac{(\omega_e/\nu_e)d_e}{1 + (\omega_e/\nu_e)^2} \tag{7.22d}$$

In these equations, the cyclotron frequency is given as $\omega_e = eB/m$. The drift across the field varies as $1/B$ known as Bohm diffusion but has not been predicted correctly for a strong magnetic field and high collision frequency. The diffusion of positively charged ions is affected similarly as that of electrons. The important fact is that the concentration gradient points in the same direction but Hall diffusion occurs in the opposite direction. Meanwhile, the Hall drift in the electric field moves in the same direction of the charged species. Therefore, the effect of magnetic field is best treated by the effective electric field in a transverse magnetic field.

A unique feature of diffusion in plasma will be discussed next. By virtue of the huge difference in mass between electrons and ions, the same electric field augments the drift velocity of ions and meanwhile retards that of the electrons. When the charged number density is sufficiently high, a considerable space charge is formed as a consequence of charge separation, and the polarized electric field impedes further breakdown of charge neutrality. The adjusted electric field restraints the faster-moving electrons within the Debye length and reinforces the mutual shielding to the same velocity by diffusion. This is the fundamental process of ambipolar diffusion. From the previous discussion in Chapter 1, the electron mobility is much greater than that of ion: $\mu_e \gg \mu_i$, the ambipolar diffusion in a pure electric field becomes

$$d_a = \frac{d_e \mu_i + d_i \mu_e}{\mu_e} \tag{7.23}$$

In the limiting case where the temperatures of electron and ion are equal in thermal ionization, then the ambipolar diffusion coefficient equals to twice the value of the ion diffusion $d_a = 2d_i$. At the other limit where the temperatures of ions are much lower than the electron in electron-impact ionization, the magnitude of ambipolar diffusion is the product of the ratio of mobilities of charged particles and diffusion coefficient of electron, $d_a = (\mu_i/\mu_e)d_e$.

Finally, additional and uncommon mechanisms such as the pressure, thermal, field force diffusion other than the electromagnetic field are included in the following discussion for completeness. Traditionally, the mass flux vector of diffusion has been approximated by the first-order spatial derivatives of the thermodynamic properties and mass-average velocity. Although the diffusive flux vectors are these originally derived for monatomic gases at low density, these have been applied to polyatomic gases with little error. The diffusive mass fluxes are given as the ordinary diffusive mass flux,

$$\rho_i u_i = \frac{c^2}{\rho RT} \sum_j M_i M_j D_{ij} \left[x_i \sum_{k \neq j} \frac{\partial G_i}{\partial x_k} \nabla x_k \right] \tag{7.24a}$$

In this equation, c, R, and T designate the speed of sound, the universal gas constant, and the temperature, respectively, where the Gibbs energy is characterized as G_i, which is defined by $G_i = E_i + PV - TS$.

The pressure diffusive mass flux is

$$\rho_i u_i = \frac{c^2}{\rho RT} \sum_j M_i M_j D_{i,j} \left[x_j M_j \left(\frac{v_j}{M_j} - \frac{1}{\rho} \right) \nabla p \right] \tag{7.24b}$$

The thermal diffusive mass flux can be approximated as

$$\rho_i u_i = -D_i^T \ln T \tag{7.24c}$$

and the forced diffusive mass flux then acquires the following form:

$$\rho_i u_i = -\frac{c^2}{\rho RT} \sum_j M_i M_j D_{i,j} \left[x_j M_j \left(F_j - \sum_k \frac{\rho_k}{\rho} F_k \right) \right] \tag{7.24d}$$

where D_{ij} and D_i^T are the binary and thermal diffusion coefficients, respectively. The Fj on the right-hand side of equation (7.24d) is the external body force other than the Coulomb force and Lorentz acceleration, such as the gravitation.

7.6 NUMERICAL ALGORITHMS

The types of the partial governing differential equations for CEA by thermal and electron impact are nearly identical. The distinctive physics are generated by the different initial values and boundary conditions. The physics-based model for electron-impact ionization or thermal excitation consists of mixed inhomogeneous hyperbolic and elliptic partial differential equations, equations (7.7) and (7.11); the generic form of these equations can always be expressed in the flux vector form:

$$\frac{\partial U}{\partial t} + \nabla \cdot F(U) = S \tag{7.25a}$$

in here, U is the dependent variable $(\rho_i, \rho\vec{u}, \rho_i e_{i,v}, \rho_i e_i, \rho e)$ and the function S contains all the inhomogeneous terms associated with the electromagnetic field. For most, the phenomena of interest, the solutions are always piece-wise continuous either across a shock wave or abridging the media interface. Since the interdisciplinary governing equation for CEA is basically the compressible Navier–Stokes equation modified by the source terms, the flux splitting technique is equally valid and applicable.

It will be a serious remiss if a terse description of numerical algorithms was not presented, because a major deficiency of all compressible flow numerical simulations is incurred by computational inaccuracy. For nonequilibrium chemical reacting flow simulations, numerical accuracy is controlled by the resolutions of strong shock waves and high gradient domains of rapid chemical reaction in the immediate postshock region and near the interface boundary. Both chemical physics phenomena take place in just a few numbers of collisions between particles; however, the chemical reaction is fundamentally a second order or a diffusive dominant process (Williams, 1971). Therefore, only strong shocks must be treated as a discontinuity in numerical analysis. The unique feature of weak solutions of nonlinear hyperbolic equation is recognized and successfully demonstrates the connection between solutions of wave systems and the general Riemann problem by Lax (1954). In 1959, Godunov first introduced a solution of the Riemann problem through flux computations for solving multidimensional compressible fluid dynamic problems (Godunov, 1959). Numerous approximations to the Riemann problem have been developed and it also has been pointed out by Roe that the most efficient solving scheme is achieved by exploiting the special feature of the system of equations (Roe, 1981).

The Rankine–Hugoniot jump condition across a shock wave is recoverable from the Euler equations. Therefore, the interdisciplinary governing equations are solved concurrently but separately by the inviscid and viscous terms. The governing equation, equation (7.25a), expressed in the split flux vector form in Cartesian coordinate can be given as

$$\frac{\partial U}{\partial t} + \frac{\partial (F_x + G_x)}{\partial x} + \frac{\partial (F_y + G_y)}{\partial y} + \frac{\partial (F_z + G_z)}{\partial z} = S, \tag{7.25b}$$

where the dependent variables, the flux vectors, and the source terms of the three-dimensional formulation are

$$U = (\rho, \rho u, \rho v, \rho w, \rho e)^T \tag{7.25c}$$

$$F_x(U) = \begin{bmatrix} \rho u \\ \rho u^2 + p \\ \rho uv \\ \rho uw \\ (\rho e + p)u \end{bmatrix}; \; G_x(u) = \begin{bmatrix} 0 \\ \tau_{xx} \\ \tau_{xy} \\ \tau_{xz} \\ u\tau_{xx} + v\tau_{xy} + w\tau_{xz} - q_x \end{bmatrix} \tag{7.25d}$$

$$F_y(U) = \begin{bmatrix} \rho v \\ \rho uv \\ \rho v^2 + p \\ \rho vw \\ (\rho e + p)v \end{bmatrix}; \; G_y(u) = \begin{bmatrix} 0 \\ \tau_{xy} \\ \tau_{yy} \\ \tau_{yz} \\ u\tau_{xy} + v\tau_{yy} + w\tau_{yz} - q_y \end{bmatrix} \tag{7.25e}$$

$$F_x(U) = \begin{bmatrix} \rho w \\ \rho uw \\ \rho vw \\ \rho w^2 + p \\ (\rho e + p)w \end{bmatrix}; \; G_x(u) = \begin{bmatrix} 0 \\ \tau_{xz} \\ \tau_{yz} \\ \tau_{zz} \\ u\tau_{xz} + v\tau_{yz} + w\tau_{zz} - q_z \end{bmatrix} \tag{7.25f}$$

$$S(U) = \begin{bmatrix} dw_i/dt \\ \rho_e E_x + J_y B_z - J_z B_y \\ \rho_e E_y + J_z B_x - J_x B_z \\ \rho_e E_z + J_x B_y - J_y B_x \\ E_x J_x + E_y J_y + E_z J_z \end{bmatrix} \tag{7.25g}$$

where the inviscid flux vector functions F_x, F_y, and F_z are identifiable by equations (7.25d) through (7.25f). The inviscid terms are further split across the shock to be treated as the approximated Riemann problem. The basic idea of flux vector splitting is to process data according to the direction of information propagation within the domain of dependence. Although split flux vectors are not unique, these vectors are unique functions of the eigenvectors and eigenvalues of the governing partial differential equation. The flux vectors are computed by the point value, including the metrics at the node if a general coordinate transform is applied from the Cartesian to any arbitrary body-oriented coordinates. The flux vector functions of the Euler equation have a remarkable property of a homogeneous function of degree 1 in the dependent variable U, which can also be written as the Jacobian of the derivative with respect to the dependent variables, namely (Steger and Warming, 1987)

$$F_x = \frac{\partial F_x}{\partial U} U = [A_x]U; \quad F_y = \frac{\partial F_y}{\partial U} U = [A_y]U; \quad F_z = \frac{\partial F_z}{\partial U} U = [A_z]U. \tag{7.26a}$$

This splitting flux vector formulation for solving hyperbolic partial differential equations not only ensures the well-posedness condition but also enhances the numerical stability for the numerical procedure (Anderson et al., 1984; Hirsch, 1990). Specifically, the flux vector functions, F_x, F_y, and F_z are split according to the sign of their respective eigenvalues. In this process, the Jacobian matrices of the flux vector function are first diagonalized by the similarity matrix and its left-hand inverse; S_i and S_i^{-1}, in which the eigenvectors constitute the elements of these matrices and the norms of these matrices are uniformly bounded. The split flux functions, according to the sign of the eigenvalues, are expressible as

$$\begin{aligned} F_x &= F_x^+ + F_x^- = (S_x\lambda_x^+S_x^{-1} + S_x\lambda_x^-S_x^{-1})U \\ F_y &= F_y^+ + F_y^- = (S_y\lambda_y^+S_y^{-1} + S_y\lambda_y^-S_y^{-1})U \\ F_z &= F_z^+ + F_z^- = (S_z\lambda_z^+S_z^{-1} + S_z\lambda_z^-S_z^{-1})U \end{aligned} \tag{7.26b}$$

The piecewise discontinuous solution is easily appreciated by examining the hyperbolic partial differential equation in flux vector form. The conservation laws of fluid dynamics in a divergence form require only the balance of the normal component of the flux vector on the control surface of an elementary control volume. The dependent variables can be discontinuous across the shock wave. Therefore, for all practical shock-capture schemes, it is critical for the grid topology to align with one of the coordinates with the shock wave, regardless of whether the structured or unstructured approach is adopted. As an example, along the outward normal direction, the typical numerical flux of the first-order Roe scheme can be adopted to balance the flux functions at the control surface

$$F_{i+1/2,j,k} = \frac{1}{2}\left[F_{i,i,k} + F_{i+1,j,k} - S_{i+1/2,j,k}\lambda_{i+1/2,j,k}S_{i+1/2,j,k}^{-1}(U_{i+1} - U_i)\right] \tag{7.27}$$

All the vector splitting formulations, at the best, are an excellent approximation to the Riemann problem. Numerical oscillations have always appeared in the calculated results at the location where the eigenvalues change sign as the consequence of Gibb's phenomenon. Most flux difference splitting algorithms applied to resolving a flow field including jump conditions must be implemented with numerical dissipation for suppression of numerical oscillation. Numerical dissipation is usually appended as the limiters.

Flux limiters are an imposed numerical procedure controlling the spatial derivative to reduce numerical oscillations for a piecewise continuous solutions in either a high gradient region or shock wave discontinuity. The shock wave in the continuum is formed just by a few molecular collisions with the physical dimension of order of mean-free path that is irresolvable by macroscopic conservation laws. The presence of shock compression in the computational domain is a necessary ingredient for thermal ionization; thus, flux splitting for the inviscid terms becomes essential (Anderson et al., 1984). The control parameter of a limiter is the ratio of successive gradients in the discretized space, $dr = (U_i - U_{i-1})/(U_{i+1} - U_i)$, it actually is a slope limiter to the numerical result across a discontinuity. Over 40 years, a very large group of limiters has been devised from the second-order TVD (total variation diminishing), minmod, symmetric TVD, Superbee, etc. All the flux-splitting schemes prove to be dissipative, and thoughtful questions have been raised when applying them to the Navier–Stokes equations. For most viscous flux reconstruction, limiters must satisfy

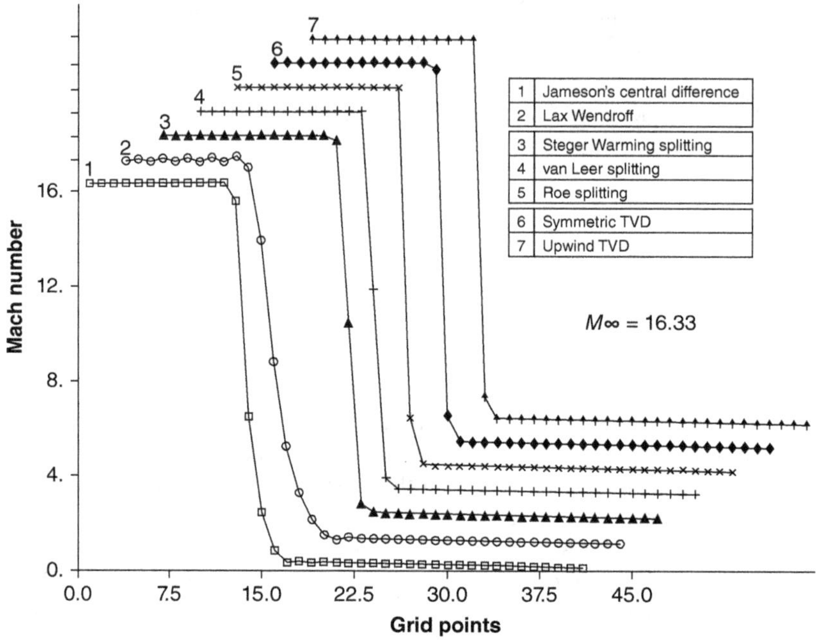

FIGURE 7.12 Shock-capturing schemes.

the discrete maximum principle and the linearity-preserving property in addition to the conservation law. In any event, the flux-splitting schemes are routinely and widely used in CFD to resolve shock waves in CFD applications. A collection of computed Mach number distributions across a normal shock at a free stream value of 16 is presented in Figure 7.12. It can be easily noted that when applying the second-order TVD or any numerical scheme across a piecewise discontinuous region, the solution will guarantee stability and lead to the capturing of a three-point shock (Shang, 1991). The first point in the shock is often located at the sonic point. These numerical results reflect the progress in shock-capturing technique that has been made over the years.

The widely used basic solving scheme is derived from the Beam–Warming algorithm. The gist of this solving scheme is the factored ADI but solved in a time-consecutive delta form (Anderson et al., 1984). To illustrate the fundamental character of this algorithm, only the inviscid terms of equation (7.25a) in two spatial dimensions are included,

$$\left[I + \frac{\theta \Delta t}{1+\xi}\left(\frac{\nabla_x A^+_{i,j}}{\Delta x} + \frac{\nabla_x A^-_{i,j}}{\Delta x} + \frac{\nabla_x B^+_{i,j}}{\Delta y} + \frac{\nabla_x B^-_{i,j}}{\Delta y}\right)^n\right] \Delta U^n_{i,j}$$

$$= -\left(\frac{\Delta t}{1+\xi}\right)\left(\delta^b_x F^+_{x(i,j)} + \delta^f_x F^-_{x(i,j)} + \delta^f_y F^+_{y(i,j)} + \delta^b_y F^-_{y(i,j)}\right)^n + \left(\frac{\xi}{1+\xi}\right)\Delta U^{n-1}_{i,j}$$

$$= R_{i,j} \tag{7.28a}$$

In this equation, Jacobian matrices are $A = \partial F_x/\partial U, F_x = AU$ and $B = \partial F_y/\partial U$, $F_y = BU$, and U for now is the conservative variables $U = U(\rho, \rho u, \rho e)$. The parameters θ *and* ξ determine the time-difference approximations for the trapezoidal formula $\theta = 1/2, \xi = 0$.

The left-hand side of equation (7.28a) can be factored into the product of two operators as

$$\left[I + \left(\frac{\theta\Delta t}{1+\xi}\right)\left(\frac{\nabla_x A^+_{i,j}}{\Delta x} + \frac{\nabla_x B^+_{i,j}}{\Delta y}\right)\right]\left[I + \left(\frac{\theta\Delta t}{1+\xi}\right)\left(\frac{\nabla_x A^-_{i,j}}{\Delta x} + \frac{\nabla_x B^-_{i,j}}{\Delta y}\right)\right] \Delta U^n_{i,j} = R_{i,j} \tag{7.28b}$$

In applications, this numerical scheme is always implemented by the sequence,

$$\begin{aligned} \left[I + \left(\frac{\theta\Delta t}{1+\xi}\right)\left(\frac{\nabla_x A^+_{i,j}}{\Delta x} + \frac{\nabla_x B^+_{i,j}}{\Delta y}\right)\right] \Delta U^*_{i,j} &= R_{i,j} \\ \left[I + \left(\frac{\theta\Delta t}{1+\xi}\right)\left(\frac{\nabla_x A^-_{i,j}}{\Delta x} + \frac{\nabla_x B^-_{i,j}}{\Delta y}\right)\right] \Delta U^n_{i,j} &= \Delta U^*_{i,j} \\ U^{n+1}_{i,j} &= U^n_{i,j} + \Delta U^n_{i,j} \end{aligned} \tag{7.28c}$$

When the spatial discretization is based on a semidiscrete finite-volume scheme, the upwind-biasing approximation is similarly applied to the convective and pressure terms. Central differencing is used for the shear stress and heat transfer terms. In the flux-difference splitting procedure, the flux vectors at the control surface are written as an exact solution to the approximate Riemann problem,

$$\begin{aligned} \nabla F_i = &\frac{1}{2}[F(U_L) + F(U_R) - |A|\,(U_R - U_L)_{i+1/2}] \\ &- \frac{1}{2}[F(U_L) + F(U_R) - |A|\,(U_R - U_L)_{i+1/2}] \end{aligned} \tag{7.29a}$$

where $\mathbf{U}_L$ and $\mathbf{U}_R$ are the interpolated values of dependent variables, ρ, ρu, ρv, ρw, and ρe on the interface of the control volume. **A** is the Jacobian matrix of the inviscid or convective term, $A = \partial F/\partial U$.

Time advancement is implicit for solving the flows. A slope limiter is also used to control the discontinuous pressure jumps at the shock front. Specifically, the minmod limiter is adopted for the present computations (Anderson et al., 1984):

$$(U_L)_{i+1/2} = U_i + 1/4[(1-k)\Delta_- + (1+k)\Delta_+]_i \tag{7.29b}$$

$$(U_R)_{i+1/2} = U_{i+1} - 1/4[(1-k)\Delta_+ + (1+k)\Delta_-]_{i+1} \tag{7.29c}$$

The one-sided difference operators are $\Delta_+ = U_{i+1} - U_i$ and $\Delta_- = U_i - U_{i-1}$, and the minmod operators are defined as

$$\Delta_- = \min \text{mod}[\Delta_-, (3-k)/(1-k)\Delta_+] \tag{7.29d}$$

$$\Delta_+ = \min \text{mod}[\Delta_+, (3-k)/(1-k)\Delta_-] \tag{7.29e}$$

For most investigations in CEA, the temporal accuracy is essential. Therefore, a greater stability bound enhancing by an implicit scheme for a larger time step is not necessary; instead, a successful iterative relaxation procedure is selected by the residual diminishing delta formulation in consecutive temporal evolution (Anderson et al., 1984). In essence, only the convective terms are retained on the left-hand side of the discretized equation and the complete governing equations in time is placed on the right-hand side by the approximation. The delta formulation of the discretized equation is used as shown in the following to linearize equation (7.25a):

$$\begin{aligned}\frac{\partial(\Delta U)}{\partial t} + \frac{\partial}{\partial x}(\Delta F_x) + \frac{\partial}{\partial x}(\Delta F_y) + \frac{\partial}{\partial x}(\Delta F_z) \\ = \frac{\partial}{\partial x}(\Delta G_x) + \frac{\partial}{\partial y}(\Delta G_y) + \frac{\partial}{\partial z}(\Delta G_z) + R^n\end{aligned}$$

where

$$\begin{aligned}R^n = \frac{\partial(\Delta U)^n}{\partial t}\frac{\partial}{\partial x}\left(F_x^n\right) - \frac{\partial}{\partial x}\left(F_y^n\right) - \frac{\partial}{\partial x}\left(F_z^n\right) + \frac{\partial}{\partial x}\left(G_x^n\right) \\ + \frac{\partial}{\partial y}\left(G_y^n\right) + \frac{\partial}{\partial z}\left(G_z^n\right) + S^n.\end{aligned} \tag{7.30a}$$

where

$$\Delta U = U^{n+1} - U^n, \Delta F_i = F^{n+1} - F^n \tag{7.30b}$$

and

$$\Delta G_i = G^{n+1} - G^n \tag{7.30c}$$

As a consequence, it achieves the optimized numerical procedure, the left-hand side of the discretized equation is required only to maintain the computational stability, and the final solution is determined by the error tolerance of the iterative procedure and the algorithm for solving the right-hand side of the equation. It should be noted that the right-hand side of equation (7.28a) is explicit and it can be evaluated using an intermediate solution with any high-order scheme provided it is numerically stable. In most studies, a central differencing scheme is adopted for the diffusive and source terms. A quadratic upwind differencing scheme is adopted for the convective terms.

In the above formulation, all diagonal terms of the discritization are moved to the left-hand side of the finite-difference approximation to maintain diagonal dominance and to enhance the computational stability. To improve temporal accuracy, a third-order Runge–Kutta scheme is applied to resolve the rapid variation of the electron impact ionization in the kHz range (Huang et al., 2011). By this formulation, the axiom of the optimal numerical simulation is fully realized in that the physical fidelity is controlled by the accurate right-hand side of the approximation. The left-hand side of the approximation is pure numerical and is constructed to maintain computational stability.

Another excellent algorithm for maintaining computational stability and yield sharp definition of a piecewise continuous numerical solution to the left-hand side of the governing equation is the Weighted essentially non-oscillatory (WENO) scheme (Balsara and Shu, 2000; Shu and Osher, 1989). The most recently improved WENO scheme operates on the same three substencils of three points by a convex combination of polynomials for the split flux computations (Borges et al., 2008). The fifth-order scheme is supported by an overall stencil of five points,

$$U^{+}(x) = \sum_{0}^{2} w_k U^k(x_{i+1/2}) \tag{7.31a}$$

The three-point polynomial $U^k(x_{i+1/2})$ generates the interpolated values on a stencil of $(x_{i+k-2}, x_{i+k-1}, x_{i+k})$ with $k = 0$, 1, 2, and the weights w_k are defined as

$$w_k = \frac{\alpha_k}{\sum_{0}^{2} \alpha_l}, \quad \alpha_k = \frac{d_k}{(\beta_k + \varepsilon)^p} \tag{7.31b}$$

In equation (7.50b), the d_k's are the ideal weight and the α_k's are referred to as the nonnormalized weights. The smoothness indicator or weight β_k is given as

$$\beta_k = \sum_{0}^{2} \Delta x^{2l-1} \int_{x-1/2}^{x+1/2} \left[\frac{d^l}{dx^l} U^k(x) \right]^2 dx \tag{7.31c}$$

The smoothness indicators measure the sum of normalized squares of the scaled L_2 norm for derivatives of the lower-order polynomials. For traditional WENO scheme, a nearly zero weight is assigned to the lower-order polynomial because the stencils contain the steep gradient domain. As a consequence, an essentially nonoscillatory solution is obtained at the discontinuities. In other words, the weight of the classic WENO scheme is used to suppress the lower-order polynomials within the total five-point stencil at the high gradient domain or shock. The innovation of the improved WENO also uses the same total number of stencils to generate a new smoothness indicator but set a high-order truncation error than the classic smoothness indicator

$$|\beta_0 - \beta_2| = \frac{13}{3} |U_i'' U_i'''| \Delta x^5 + O(\Delta x^6) \tag{7.31d}$$

The higher-order information incorporated into the weights definition improves the convergence order at the critical points of the smooth solution. Meanwhile, the algorithm decreases the dissipation near to discontinuities and retains the essentially nonoscillatory solution behavior. It is clear that the enhancement of the new scheme is derived from a large weight assigned to the discontinuous stencils, but not from their superior order of convergence at critical points.

In summary, most adopted numerical algorithms in CEA are shared with CFD, especially for thermally excited ionizations. These numerical procedures are built on either the classical ADI scheme for finite-difference approximation or the Gauss-Siedel relaxation for finite-volume approach. Both procedures rely heavily on the TVD concept to maintain numerical stability. A special need for accurate temporal resolution is enhanced by the delta formulation through the diminishing residual method for solving the system of nonlinear, inhomogeneous partial differential equations.

REFERENCES

Anderson, D.A., Tannehill, J.C., and Pletcher, R.H. 1984. *Computational Fluid Mechanics and Heat Transfer*. Hemisphere, New York, 1984.

Balsara, D., and Shu, C-.W., Monotonicity preserving weighted essentially non-oscillatory schemes with increasingly high order of accuracy, *J. Comp. Phys.*, vol. 160, 2000, pp. 405–452.

Birdsall, C.K., and Langdon, A.B., *Plasma Physics Via Computer Simulation*, McGraw-Hill, Singapore, 1985.

Bogdanov, E.A., Kudryavtsev, A.A., Kuranov, A.L., Kozlov, L.E., and Tkchenko, T.L., *2D Simulation of DBD Plasma Actuator in Air*, AIAA, 2008, pp. 1368–1377.

Borges, R., Carmona, M., Costa, B., and Don, W.S., An improved weighted essentially non-oscillatory scheme for hyperbolic conservation laws, *J. Comp. Phys.*, vol. 227, 2008, pp. 3191–3211.

Candler, G.V., and MacCormack, R.W., Computation of weakly ionized hypersonic flows in thermochemical nonequilibrium, *J. Thermophys.*, vol. 5, no. 3, 1991, pp. 266–272.

Capitelli, M., Celiberto, R., Gorse, C., and Giordano, D., Transport properties of high temperature air components: a review, *Plasma Chem. Plasma Proc.*, vol. 16, 1996, pp. 267S–302S.

Capitelli, M., Gorse, C., Longo, S., and Giordano, D., Collision integrals of high-temperature air species, *J. Thermophys. Heat Transfer.*, vol. 14, no. 2, 2000, pp. 259–268.

Chen, Y.K., and Milo, F.S., Ablation and thermal response program for spacecraft heatshield analysis, *J. Spacecraft Rockets*, vol. 36, no. 3, 1999, pp. 475–483.

Chen, Y.K., and Milos, F.S., Navier–Stokes solutions with finite rate ablation for planetary mission Earth reentries, *J. Spacecraft Rockets*, vol. 42 no. 6, 2005, pp. 961–970.

Cheng, C.Z., and Knorr, G., The integration of the Vlasov equation in configuration space, *J. Comp. Phys.*, vol. 22, 1976, pp. 330–351.

Chernyi, G.G., Losev, S.A., Macheret, S.O., and Potapkin, B.V., *Physical and Chemical Processes in Gas Dynamics: Physical and Chemical Kinetics and Thermodynamics of Gases and Plasmas. vol. 2. Progress in Astronautics and Aeronautics.* (Paul Zarchan, Ed.) vol. 197. 2004.

Corke, T.C., Enloe, C.L., and Wilkinson, S.P., Plasma actuators for flow control, *Annu. Rev. Fluid Mech.*, vol. 42, 2010, pp. 505–525.

Corke, T.C., Post, M.L., and Orlov, D.M., SDBD plasma enhanced aerodynamics: concepts, optimization and applications, *Prog. Aerosp. Sci.*, vol. 43, 2007, pp. 193–217.

Dunn, M.G., and Kang, S.W., *Theoretical and Experimental Studies of Reentry Plasma*, NSA CR 2232, April 1973.

Elisson, B., and Kogelschatz, U., Nonequilibrium volume plasma chemical processing, *IEEE Trans. Plasma Sci.*, vol. 19, 1991, pp. 1063–1077.

Godunov, S.K., Finite-difference method for numerical computation of discontinuous solution of equations of fluid dynamics, *Mat. Sb.*, vol. 47, 1959, pp. 271–306.

Golubovskii, Yu, B., Maiorov, V.A., Behnke, J., and Behnke, J.F., Influence of interaction between charged particles and dielectric surface over a homogeneous barrier discharge in nitrogen, *J. Phys. D: Appl. Phys.*, vol. 35, 2002, pp. 751–761.

Hirsch, C., *Numerical Computation of Internal and External Flows, vol. 2, Computational Methods for Inviscid and Viscous Flows*, Wiley, New York, 1990.

Hirschfelder, J.O., Curtiss, C.F., and Bird, B.B., *Molecular Theory of Gases and Liquids*, John Wiley & Sons, Inc., New York, 1954.

Huang, P.G., Shang, J.S., and Stanfield, S.A., Periodic electrodynamic field of dielectric barrier discharge, *AIAA J.* vol. 49, no. 1, January 2011, pp. 119–127.

Jones, L.J., and Cross, A.E., Electrostatic Probe Measurements of Plasma Parameters for Two Reentry Flight Experiments at 25,000 Feet per Second, NASA TN D 66-17, 1972.

Josyula, E., Bailey, W., and Suchyta, C., Dissociation modeling in hypersonic flows using state-to-state kinetics, *J. Thermophys. Heat Transfer.*, vol. 25, no. 1, 2011, pp. 34–47.

Kim, W., Do, H., Mungal, M.G., and Cappelli, M.A., On the role of oxygen in dielectric barrier discharge actuation of aerodynamic flows, *Appl. Phys. Lett.*, vol. 91, 2007, 181501.

Landau, L., and Teller, E., Zurtheorie der shallispersion, *Physik Z. Sowjetunion, B.*, vol. 10, 1936, p. 14.

Lax, P.D., Weak solutions of nonlinear hyperbolic equations and the numerical computations, *Commun. Pure Appl. Math.*, vol. 7, 1954, pp. 150–193.

Millikan, R.C., and White, D.R., Systematics of vibrational relaxation, *J. Chem. Phys.*, vol. 39, no. 12, 1963, pp. 3209–3213.

Mitchner, M., and Kruger, C.H., *Partially Ionized Gases*, John Wiley & Sons, New York, 1973.

Moreau, E., Airflow control by non-thermal plasma actuator, *J. Phys. D: Appl. Phys.*, vol. 40, 2007, pp. 605–636.

Olynick, D.R., Chen, Y.K., and Tauber, M.E., Aerodynamics of the Stardust Sample Return Capsule, *J. Spacecraft Rockets*, vol. 36, no. 3, 1999, pp. 442–462.

Opaits, D.F., Sheider, M.N., Miles, R.B., Likhanskii, A.V., and Macheret, S.O., Surface charge in dielectric barrier discharge plasma actuators, *Phys. Plasma.*, vol. 15, 2008, pp. 073505-1-5.

Pancheshnyi, S.V., Starikovkaia, S.M., and Starikovskii, A.Y., Role of photoionization processes in propagation of cathode-directed streamer, *J. Phys. D: Phys*, vol. 34, 2001, pp. 105–115.

Park, C., Review of chemical kinetics problems of future NASA missions, I earth entries, *J. Thermophys. Heat Transfer*, vol. 7 no. 3, 1993.

Park, C., Chemical-kinetic parameters of hypersonic earth entry, *J. Thermophys Heat Transfer.*, vol. 15, no. 1, 2001, pp. 76–90.

Rafatov, I., Bogdanov, E.A., and Kudryavtsev, A.A., On the accuracy and reliability of different fluid models of the direct current glow discharge, *Phys. Plasma*, vol. 19, 2012, pp. 033502-1-12.

Roe, P.L., Approximate Riemann solvers, parameter vectors and difference scheme, *J. Comput. Phys.*, vol. 43, 1981, pp. 357–372.

Sarma, G.S.R., Physico-chemical modeling in hypersonic flow simulation, *Prog. Aerosp. Sci.*, vol. 36, no. 3/4, 2000, pp. 281–349.

Shang, J.S., Shared knowledge in computational fluid dynamics, *Electromag. Magnetobiol.-Aerodyn., Prog. Aerosp. Sci.*, vol. 38, no. 6/7, 2002b, pp. 449–467.

Shang, J.S., and Surzhikov, S.T., Simulating Stardust earth reentry with radiation heat transfer, *J. Spacecraft Rockets*, vol. 48, no. 3, 2011, pp. 385–396.

Shang, J.S., and Surzhikov, S.T., Nonequilibrium radiative hypersonic flow simulation, *Prog. Aerosp. Sci.*, vol. 53, 2012, pp. 46–65.

Shang, J.S., Surzhikov, S.T., Kimmel, R., Gaitonde, D., Menart, J., and Hayes, J., Mechanisms of plasma actuators for hypersonic flow control, *Prog. Aerosp. Sci.*, vol. 41, 2005, pp. 642–668.

Shang, J.S., and Huang, P.G., Surface plasma actuators modeling for Flow control, *Prog. Aerosp. Sci.*, vol. 67, 2014, pp. 29–50.

Shu, C.W., and Osher, S., Implementation of essentially nonoscilatory shock capture scheme II, *J. Comp. Phys.*, vol. 83, 1989, pp. 32–78.

Shumlak, U., and Loverich, J., Approximate Riemann solver for the two-fluid plasma model, *J. Comp. Phys.*, vol. 187, 2003, pp. 620–638.

Singh, K.P., and Roy, S., Modeling plasma actuators with air chemistry for effective flow control, *J. Appl. Phys.*, vol. 101, 2007, pp. 123308-1–123308-8.

Solov'ev, V., Konchakov, A.M., Krivtsov, V.M., Aleksandrov, N.L., Numerical simulation of a surface barrier discharge in air, *Low-Temperature Plasma*, vol. 34, no. 7, 2008, pp. 594–608.

Steger, J. L., and Warming, R. F., Flux vector splitting of the inviscid gas dynamics equations with application to finite difference methods. *J. Comput. Phys.*, vol. 20, no. 2, 1987, pp. 263–293.

Surzhikov, S.T., and Shang, J.S., Coupled radiation-gasdynamic model for Stardust earth entry simulation, *J. Spacecraft Rockets*, vol. 49, no. 5, 2012, pp. 875–888.

Treanor, C.E., and Marrone, P.V., Effect of dissociation on the rate of vibrational relaxation, *Phys. Fluids.*, vol. 5, no. 9, 1962, pp. 1022–1026.

Williams, F.A., Theory of combustion in laminar flows, *Annu. Rev. Fluid Mech.*, vol. 3, 1971, pp. 171–188.

Ziemer, R.W., Experimental investigation in magneto-aerodynamics, *J. Am. Rocket Soc.*, vol. 29, 1959, pp. 642–647.

8

MODELING ELECTRON IMPACT IONIZATION

INTRODUCTION

Electron impact ionization is the mainstay for plasma-based flow control actuators because it actuates in less than microseconds and the electrode configuration can be made nonintrusive to oncoming stream. In addition, when the actuator is not needed, unlike most mechanical flow control devices, it will not be parasitic and degrade performance of aerospace vehicles. The ionizing energy for electron impact is derived from an externally applied electric field, and the electric field breakdown occurs when an applied voltage exceeds the ignition potential. The overall process is initiated by the secondary emission of electrons from the cathode and is followed by multiple primary avalanches. The energy transferred by colliding to excite another neutral species is continuously replenished by the electric field and maintains the discharge. The direct current discharge (DCD) is a self-sustaining process and appears as a glow discharge or an arc, depending on which gas species and the electric circuit conditions. In either case, the ionized gas becomes luminous.

The structure of the DCD is characterized by the positive column connecting the anode and cathode layers. The electric potential intensity drops by nearly two orders of magnitude from the electrode surface within the cathode fall. Four computational results are presented in Figure 8.1 at three different ambient pressures by two different computational simulations (Surzhikov et al., 2004; Shang et al., 2009). At the pressure level of 10 torr, the maximum normal electric field intensity on the cathode exceeds a value of 8.0×10^3, decreases to a value of 3.5×10^2 V/cm, and rises

Computational Electromagnetic-Aerodynamics, First Edition. Joseph J.S. Shang.

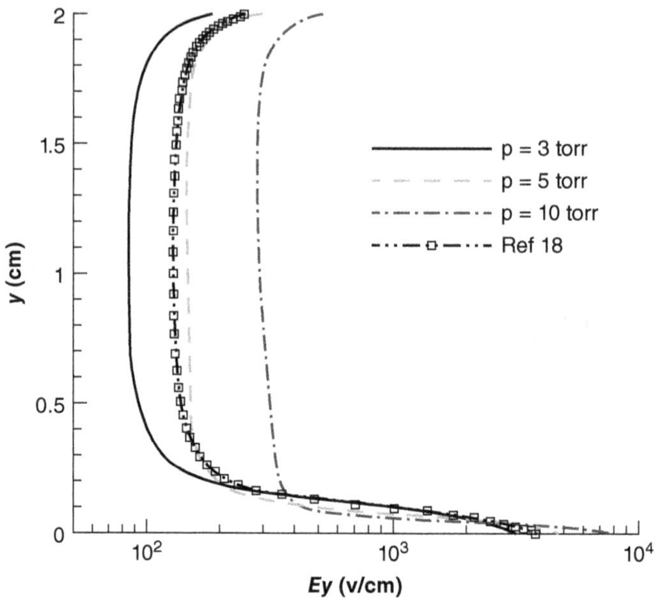

FIGURE 8.1 Surface normal electric field intensity, EMF = 2.0 kV, gap = 2.0 cm.

again to a moderate level on the anode. The cathode fall becomes steeper when the ambient pressure is elevated. An interesting observation has also been made obvious by numerical simulations: the thickness of the cathode layer increases when the secondary emission coefficient is lowered accordingly. This behavior reflects a reduction of ionization efficiency.

The steady DCD or the glow discharge has been observed as high as an ambient pressure of 50 torr in dry air. When the pressure approaches one atmospheric condition, the breakdown in a nonuniform field is strongly dependent on geometry of the electrodes and its polarity and is known as corona discharge. If the ignition is initiated from the cathode, it is the negative or anode-directed corona. Otherwise, it is known as the positive or cathode-directed corona discharge. Although the corona is basically a Townsend discharge, the structure of the discharge pattern depends on the polarity and relative dimensions of the electrodes. At a lower applied electric potential, the periodic Trichel pulse is presented and vanishes as the electric potential is increased. Beyond a critical value, a nearly steady state corona discharge is sustained until the spark breakdown takes place between the gap in the electrodes.

Electron impact ionization by an alternating current (AC) is not fundamentally different from a DC field. The essential supply of electrons is obtained mainly from secondary emission caused by the bombardment on the cathode by positively charged ions through the action of an electric current. At the atmospheric pressure, the discharge is dominated by the random streamer, filamentary, or microdischarge in time and space. The streamer, or the microdischarge, is a weakly ionized channel by the primary avalanche in a strong electric field (Raizer, 1991). The space charges induce an additional electric field potential that makes it different from the corona discharge. For the dielectric barrier discharge (DBD), one of the electrodes of dielectric barrier is encapsulated in dielectric film and the gap distance between electrodes is just a few millimeters apart. The dielectric serves two purposes: it limits the amount of charged transport by a single microdischarge, and it distributes the microdischarges over the entire electrode surface. A unique feature of the dielectric is that the displacement current can pass through an alternative electric potential source, and the required voltage is in the range from 5 to 20 kV in AC cycles in the microwave range. However, the most outstanding feature of DBD is the accumulated charges over the dielectrics that effectively reduce the electric field intensity and provide the self-limiting behavior from transition to spark.

Another salient feature of DBD is that the microdischarges have distinct characteristics within a complete AC cycle (Enloe et al., 2004, 2008). The microdischarges are initiated form electrodes as either the positive or negative corona discharges. The avalanche transforms discharge into streamer before it reaches the receiving electrode. The negative streamers propagate against the direction of electron drift velocity; thus, they tend to be more diffusive than the positive streamer. On the contrary, positive streamers require traveling a longer distance to reach the cathode. This distinct behavior has been captured by the high-speed, image-intensified photographs in each phase of the DBD by Enloe et al. Unfortunately, the current computational capability is unable to conduct a direct simulation for the multiple random temporal and spatial microdischarges (Figure 8.2).

The known electron concentration in earth reentry by thermal ionization is capped at the value of $10^{15}/cm^3$, which is about four orders of magnitude lower than the

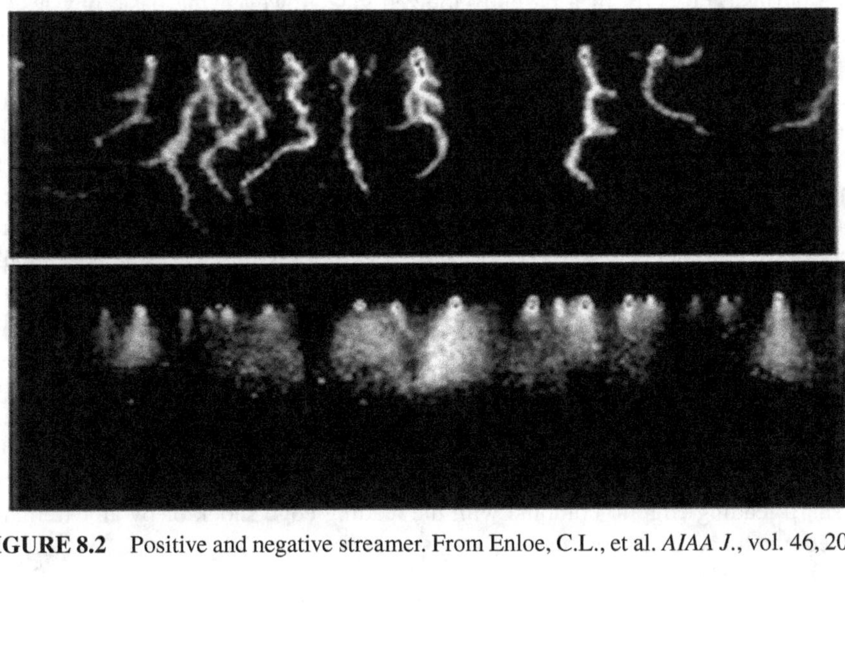

FIGURE 8.2 Positive and negative streamer. From Enloe, C.L., et al. *AIAA J.*, vol. 46, 2008.

natural species of $10^{19}/cm^3$. The electron number density by electron impact ionization is further lower than three orders of magnitude with the maximum value rarely exceeding $10^{12}/cm^3$ to appear as a tracing element in the weakly ionized gas (Shang et al., 2014). By solving the species conservation equation, the number densities of numerous transient radical species of positively and negatively charged ions are too minuscule for accurate determination, even if the reaction rates could be appropriately assessed. For computational electromagnetic-aerodynamics, all phenomena associated with electromagnetics are presented as a small perturbation that still could not be accurately accounted for by the known detailed chemical kinetic theory. In spite of that, the best physics-based, multifluid model for the electron impact ionization is presented in this chapter.

8.1 TRANSPORT PROPERTY VIA DRIFT-DIFFUSION THEORY

The transport properties of a partially ionized gas are paramount for the momentum transfer process, which is accentuated by the forced diffusion resulting from drift velocities of charged particles and Lorentz acceleration. The additional energy transfer mechanism by electromagnetics in the form of Joule heating is proportional to electric current density, which is also generated by individual species velocities of electrons and ions. Another complication incurred by the transport property of ionized gas is the communication blackout by a peculiar behavior of electromagnetic wave propagation in an electrically conducting medium. In plasma, a reflected incident microwave wave at the media interface and dissipated wave amplitude are tied directly to the charge number density or the plasma frequency, which also has roots in the intrinsic property of a partially ionized gas. However, the basic physics for transport properties of plasma is not describable by the kinetic theory of gas but must be included in analyzing the partially ionized plasma.

Using DCD for supersonic and hypersonic flow controls, both electrodes are exposed and flush mounted over the vehicle's surface, so the combined electrode and joule heating can produce a local thermal expansion over the electrodes. In turn, the elevated local temperature increases boundary-layer displacement thickness to displace the external stream further away from the solid surface. The ensuing viscous–inviscid interaction triggers a family of compressible waves and eventually coalesces into an oblique shock (Shang et al., 2005, 2008). Figure 8.3 displays a schlieren image over a sharp-leading edge of flat plate at zero incident in a Mach number 5.15 MHD channel. The leading-edge shock is a result of the classic Mach wave interaction at the sharp leading edge. The second shock wave downstream to the leading edge shock is induced by a side-by-side DCD electrode arrangement. The DCD is generated by an EMF of 1.2 kV with an electric current of 60 mA. The induced local pressure rise can be further elevated by placing the cathode closer to the sharp-leading edge to combine with the leading edge shock or by an externally applied transverse magnetic field.

FIGURE 8.3 Induced compression waves by DCD in supersonic stream, EMF = 1.2 kV, I = 50 mA.

The DBD creates a wall jet on top of overlapped electrodes of different materials through charge separation (Roth et al., 2000). The most significant effect of an encapsulated DBD electrode to the electromagnetic field is achieved by modifying the electric field intensity through surface charge accumulation. Under a discharging condition, the electrodes become a capacitor to diminish the electric field intensity across the electrodes. The quenching and replenishing rate of the surface discharge is dependent on applied alternating voltage across electrodes and electric permittivity of media. To a large degree, the rate is dependent on the material property of the dielectric. The electric current–voltage distribution of DBD is depicted in Figure 8.4 by an externally applied AC field at EMF = 4.0 kV and 5.0 kHz. The current–voltage relationship is a typical behavior of DBD over multiple AC cycles and displays a surge of conductive current component at the breakdown. The conductive current carries a very small amount, generally around a fraction of few mA, that is carried by the plasma between the exposed electrode and the encapsulated dielectric surface. The biased electric field intensity and the charge separation over the electrodes generate a periodic electrostatic force to move charged particles, which in turn transfer momentum by colliding with neutral particles. In fact, the momentum transfer between ions and neutral particles leads to the so-called electrical wind of the DBD. It is therefore logical to model the partially ionized air plasma using multiple-fluid particle approximations.

The modeling of weakly partially ionized gas actually consists of two parts: the transport property and the ionization process for the partially ionized plasma. The physics for electron impact ionization is well known through the luminary contributions by Townsend; the ionization is evolving from the electrons collision, secondary emission, and avalanche (Howatson, 1975; Raizer, 1991). The secondary emission from the cathode arises as the controlling mechanism for all electron impact ionization. A high concentration of positive charges over an electrode creates an

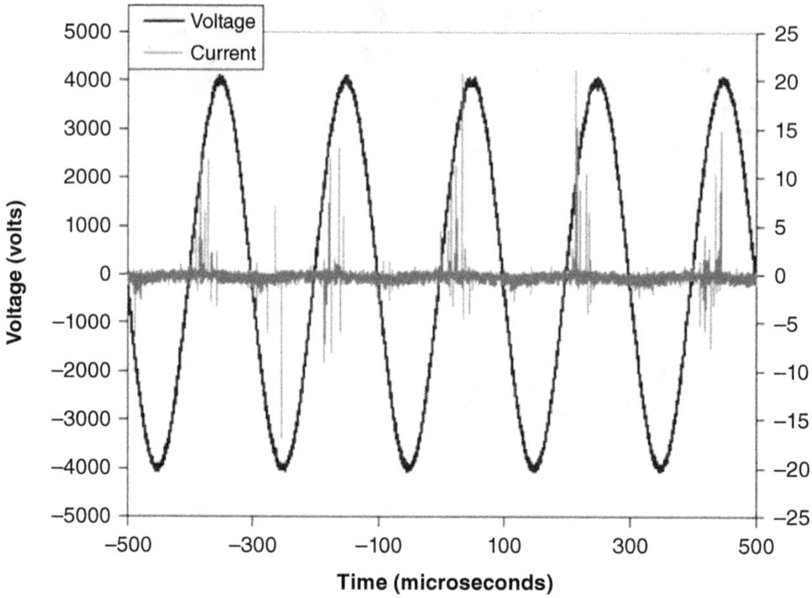

FIGURE 8.4 Current–voltage relationship of DBD EMF = 4.0 kV, f = 5.0 kHz.

extremely high electrical field potential, which is comparable to the magnitude in a nuclei. The electrons move toward the anode by free electrons tunneling through the electric barrier, followed by avalanche over the cathode. The physical process of the glow discharge has long been referred to as the Townsend mechanism. The transport properties of a charged carried gas mixture must include the force diffusion to an electrically conducting medium by an externally applied electromagnetic field, collectively known as the ordinary and bipolar diffusions.

In theory, all transport properties of an inhomogeneous, nonequilibrium chemically reacting gas mixture can be determined by the gas kinetic theory via the intermolecular potential functions. In practice, the predictive error band for an ionized species has a range from 16.3% to 24.2% (Capetilli et al., 2000; Levin et al., 2004). Furthermore, the theory is inapplicable to model a quantum transition, which is a common thermodynamic state of air plasma. Therefore, for the gas discharge physics, the modeling and formulation are based on the drift-diffusion theory for the transport property that was established in the late 1980s. The model for DCD flow control is first applied by Surzhikov and Shang (Surzhikov et al., 2004) and extends the discharge model to include an applied transverse magnetic field. A series of computational simulations using this model has been verified and validated by experimental observations conducted in a hypersonic plasma channel.

The forced diffusion mechanisms of plasma arise from the externally applied electromagnetic field, which is intrinsically different from the ordinary diffusion by the gradient of species concentration (Graves and Jensen, 1986; El-Bahy, 2005). The forced terms can be evaluated from the momentum equations of the multifluid model based on kinetic theory (Raizer and Surzhikov, 1990; Surzhikov et al., 2004; Surzhikov, 2011). The most general plasma model is constituted by four fluid components: natural particles, electrons, positively charged ions, and negatively charged ions. Under the influence of an electromagnetic field and the collisions mechanism with all fluid components, the motion of all charged species can be given as

$$m_e n_e \frac{\partial u_e}{\partial t} + m_e n_e (u_e \cdot \nabla) u_e = -\nabla p_e - \tau_e - e n_e (E + u_e \times B) - m_e n_e \nu_{en} (u_e - u_n) - m_e n_e \nu_{e+} (u_e - u_+) - m_e n_e \nu_{e-} (u_e - u_-) \quad (8.1a)$$

$$m_+ n_+ \frac{\partial u_+}{\partial t} + m_+ n_+ (u_+ \cdot \nabla) u_+ = -\nabla p_+ - \tau_+ + e n_+ (E + u_+ \times B) - m_+ n_+ \nu_{+n} (u_+ - u_n) - m_+ n_e \nu_{+e} (u_+ - u_e) - m_+ n_+ \nu_{+-} (u_+ - u_-) \quad (8.1b)$$

$$m_- n_- \frac{\partial u_-}{\partial t} + m_- n_- (u_- \cdot \nabla) u_- = -\nabla p_- - \tau_- - e n_- (E + u_- \times B) - m_- n_- \nu_{-n} (u_- - u_n) - m_- n_- \nu_{-+} (u_- - u_+) - m_- n_- \nu_{-e} (u_- - u_e) \quad (8.1c)$$

In these equations, the symbol τ_i denotes the shear stress tensor for each species and the subscripts are designated for different fluid species. The collision frequency ν_{ij} is the collision frequency between binary collisions of the four fluid species.

For a weakly ionized gas, the number densities of neutral particles are more than $10^{18}/\mathrm{cm}^3$, and the electrons and positively charged ions are around $10^{13}/\mathrm{cm}^3$. The negatively charged ions number density is usually three orders of magnitude lower than its positively charged counterpart to have a value of $10^{10}/\mathrm{cm}^3$ at the most. Thus, the number of neutral particles is much greater than the charged particles whether they are electrons or ions: $n_n \gg n_e, n_+, n_-$. Therefore, the viscous force of the charged particles becomes negligible. At the same time, the characteristic time intervals of the interparticle collision frequency are scaled by the speed of light. As a consequence, the inertia of the particle's organized motion is negligible

$$m_i n_i \partial u_i / \partial t + m_i n_i (u_i \cdot \nabla) u_i \simeq 0 \quad (8.1d)$$

From definition of partial pressure of each charged species, $p_i = n_i k T_i$, the three charged species of the approximated equation (8.1a) are expressible as

$$kT_e \nabla n_e + e n_e (E + u_e \times B) + m_e n_e \nu_{en} (u_e - u_n) + m_e n_e \nu_{e+} (u_e - u_+) + m_e n_e \nu_{e-} (u_e - u_-) = 0 \quad (8.2a)$$

$$kT_+ \nabla n_+ + e n_+ (E + u_+ \times B) + m_+ n_+ \nu_{+n} (u_+ - u_n) + m_+ n_+ \nu_{+e} (u_+ - u_e) + m_+ n_+ \nu_{+-} (u_+ - u_-) = 0 \quad (8.2b)$$

$$kT_- \nabla n_- + e n_- (E + u_- \times B) + m_- n_- \nu_{-n} (u_- - u_n) + m_- n_- \nu_{-+} (u_- - u_+) + m_- n_- \nu_{-e} (u_- - u_e) = 0 \quad (8.2c)$$

The huge disparity between velocities of neutrals and charged particles makes the velocities' differences among charged particles negligible. Again the fact that the collision frequency between charged particles and neutral is orders of magnitude greater than the collision frequencies between charged particles: $\nu_{en} \gg \nu_{e+}, \nu_{e-}$, then equation (8.2a) yields

$$n_e u_e = -\left(\frac{kT_e}{m_e \nu_{en}}\right)\nabla n_e - \left(\frac{e}{m_e \nu_{en}}\right) n_e E - \left(\frac{e}{m_e \nu_{en}}\right) n_e (u_e \times B) \quad (8.3a)$$

It is recognized immediately that

$$\mu_e = e/m_e \nu_{en} \quad \text{and} \quad d_e = kT_e/m_e \nu_{en} \quad (8.3b)$$

are the well-known electron mobility μ_e and the thermal diffusion coefficient d_e, respectively.

By the same argument that $\nu_{+n} \gg \nu_{+e}, \nu_{+-}$, and even if the electron velocity is much greater than the ion velocity, it still makes the velocity of organized motion of neutral particles negligible, and we have

$$n_+ u_+ = -\left(\frac{kT_+}{m_+ \nu_{+n}}\right)\nabla n_+ - \left(\frac{e}{m_+ \nu_{+n}}\right) n_+ E - \left(\frac{e}{m_+ \nu_{+n}}\right) n_+ (u_+ \times B) \quad (8.3c)$$

and

$$n_- u_- = -\left(\frac{kT_-}{m_- \nu_{-n}}\right)\nabla n_e - \left(\frac{e}{m_- \nu_{-n}}\right) n_e E - \left(\frac{e}{m_- \nu_{-n}}\right) n_- (u_- \times B) \quad (8.3d)$$

Again the ion mobility and diffusion coefficient are recognized as

$$\mu_\pm = e/m_\pm \nu_{\pm n} \quad \text{and} \quad d_\pm = kT_\pm/m_\pm \nu_{\pm n} \quad (8.3e)$$

For all practical processes, the masses of the positively and negatively charged particles are identical.

The individual charged species velocities from equations (8.3a), (8.3c), and (8.3d) can be summarized as the following (Raizer and Surzhikov, 1990; Surzhikov et al., 2004):

$$\begin{aligned} n_e u_e &= -d_e \nabla n_e - \mu_e n_e E - \mu_e n_e (u_e \times B) \\ n_+ u_+ &= -d_+ \nabla n_+ + \mu_+ n_+ E - \mu_+ n_+ (u_+ \times B) \\ n_- u_- &= -d_- \nabla n_- - \mu_- n_- E - \mu_- n_- (u_- \times B) \end{aligned} \quad (8.4)$$

These equations actually define diffusion velocities of an inhomogeneous, electrically conducting gas mixture of the multifluid model. The leading terms in equation (8.4) describe the ordinary diffusion due to local concentration of electrons, positively charged ions, and negatively charged ions. The middle terms are the force diffusion by drift motion for an electrically charged species in an applied electric field. The last terms are also the force diffusion by the Lorentz acceleration.

Once the species velocities are determined, the species conservation equation becomes a simplified approximation to the continuity equation, equation (7.1), for plasma. The diffusion-drift model is equally applicable to all weakly ionized gas from thermal to electron impact ionizations, including glow, corona, and DBDs. This formulation has been adopted by most recent numerical simulations (Shang et al., 2010). One of the weaknesses of this approximation is that the detailed distinctions between metastable species are no longer recoverable by this approximation. Instead, the weakly ionized gas is treated as a medium that consists of three charged and one neutral species. The species conservation equations for the three distinct charged particles are

$$\frac{\partial n_e}{\partial t} - \nabla \cdot [d_e \nabla n_e + n_e \mu_e (E + u_e \times B)] = \frac{dw_e}{dt} \tag{8.5a}$$

$$\frac{\partial n_+}{\partial t} - \nabla \cdot [d_+ \nabla n_+ - n_+ \mu_+ (E + u_+ \times B)] = \frac{dw_+}{dt} \tag{8.5b}$$

$$\frac{\partial n_-}{\partial t} - \nabla \cdot [d_- \nabla n_- + n_- \mu_- (E + u_- \times B) = \frac{dw_-}{dt} \tag{8.5c}$$

where the terms in the right-hand side of the equations, dw_e/dt, dw_+/dt, and dw_-/dt, designate production and depletion rates of the considered ionized species. These rates either have been determined by some chemical kinetic models or are simplified approximations based on known ionization processes. In most formulations, the electron, positively charged ion, and negatively charged ion number density fluxes are designated by symbols Γ_e, Γ_+, and Γ_-. These notations are used by following derivations:

$$\begin{aligned} \Gamma_e &= -d_e \nabla n_e - n_e \mu_e (E + u_e \times B) \\ \Gamma_+ &= -d_+ \nabla n_+ + n_+ \mu_+ (E + u_+ \times B) \\ \Gamma_- &= -d_- \nabla n_- - n_- \mu_- (E + u_- \times B). \end{aligned} \tag{8.6}$$

The important connection between the coefficients of diffusion d_i and the mobility μ_i of drift velocity is given by the Einstein relationship: the relative magnitude is a function of the characteristic electron energy to the elementary electric charge $|E|/P$ (Surzhikov et al., 2004):

$$\begin{aligned} \mu_e &= \frac{e}{m_e \nu_{en}} \approx \mu_e(|E|/P);\ d_e = \frac{kT_e}{e}\mu_e \\ \mu_\pm &= \frac{e}{m_\pm \nu_{\pm n}} \approx \mu_\pm(|E|/P);\ d_\pm = \frac{kT_\pm}{e}\mu_\pm \end{aligned} \tag{8.7}$$

The drift velocity is given as a function of the reduced electrical field $|E|/P$ with a physical dimension of V/cm-torr, which is the work done by field transform of energy into the energy of molecular vibration. In the range of $|E|/P$ ratios up to 50, the electron drift velocity in ionized air has a typical value of around 10^7 cm/s. The drift velocity of ions is much slower; from equation (8.7), the drift velocity of ions is estimated to be 10^5 cm/s due to the large differences between masses of electrons and ions. Therefore, when considerable numbers of free space charges are formed by charged separation, the polarized electric field between charges restrains the motion of electrons from that of ions to appear as the ambipolar diffusion. The unique diffusion phenomenon of plasma is also known as a function of the ratio of electron energy and ambient thermodynamic states. Estimated magnitudes of these unique transport properties have been discussed in Chapter 1 (Section 1.12).

The diffusion coefficients of charged particles have an explicit dependence to the characteristic temperatures of each species. Large differences in diffusion coefficients between electrons and ions can be anticipated in nonequilibrium plasma generated by electron impact, because in most partial ionized gases, ions usually remained in the thermodynamic state of their surrounding environment.

The drift-diffusion model has been very successfully applied for plasma-based flow control actuators. In Figure 8.5, a group of graphs of electron and ion number densities is displayed for three different ambient pressures of 3, 5, and 10 torr at a constant temperature of 300 K. The contours of electrons and ions consistently exhibit

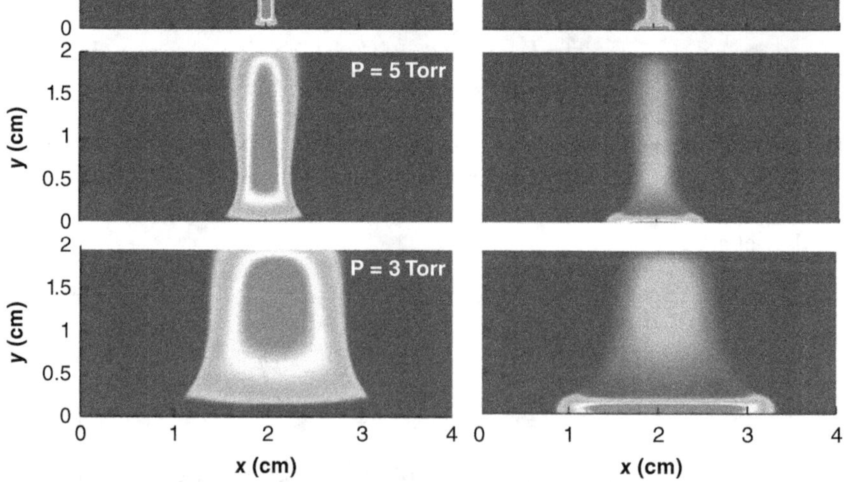

FIGURE 8.5 Electron and ion number density of DCD between parallel electrodes, EMF = 2.0 kV.

the constricted positive column as the ambient pressure is increased. The thinning cathode layers at the bottom of the configuration also clearly displayed the effect of the ambient pressure, or rather the density between the electrode gaps. The cathode layer thickness decreases as the ambient pressure is increased, and the width of the cathode layer also shrinks from 2.66 to 0.58 cm, corresponding to the pressure rise from 3 to 10 torr (Shang et al. 2009).

8.2 DRIFT-DIFFUSION THEORY IN TRANSVERSE MAGNETIC FIELD

A general formulation for the drift-diffusion model including an externally applied magnetic field is very complex because of gyrations of charged particles associated with the magnetic field polarity. For the purpose of a compact analytic formulation, it is convenient to resolve the velocity of charged particles into components parallel and perpendicular to the direction of an externally applied magnetic field (Surzhikov et al., 2004, 2014). In practical applications, a transversal magnetic field is the most effective for flow control. Therefore, a degenerated three-dimensional formulation in (x,y) plane for charged particle flux densities, including all the electromagnetic effects, is still viable by restricting the magnetic flux density B on the z coordinate.

The orientation of a conductive electric current is defined by the drift motion of charged particles; see equation (1.5). In the absence of an applied magnetic field, the number density flux of charge species has a one-to-one correspondence to the component electric field intensity in their respective coordinates. Under the influence of a transverse magnetic field, the charged particles flux number density will depend on all components of the electric field. The interaction of an externally applied magnetic field modifies charge particle velocities driven by the Coulomb force. A component of charged velocities is generated perpendicular to both the electric current and the magnetic field $F = J \times B$. According to Ohm law, equation (1.38), the induced velocity component is actually aligned with $E \times B$. This cross product involves all x and y components of electric field intensity and the magnetic flux density along the z coordinate. The electron velocity under the influence of a transverse magnetic field thus has two components, $u_e = u_e(u_{e,x}, u_{e,y})$. The detailed description of charged particles velocities is provided by collision dynamics, and the velocities of charged particles are described by equation (8.3a) through (8.3c).

$$
\begin{aligned}
n_e u_{e,x} &= \frac{1}{1+b_e^2}\left(-d_e\frac{\partial n_e}{dx} - \mu_e n_e E_x\right) - \frac{b_e}{1+b_e^2}\left(-d_e\frac{\partial n_e}{dy} - \mu_e n_e E_y\right) \\
n_e u_{e,y} &= \frac{1}{1+b_e^2}\left(-d_e\frac{\partial n_e}{dy} - \mu_e n_e E_y\right) + \frac{b_e}{1+b_e^2}\left(-d_e\frac{\partial n_e}{dx} - \mu_e n_e E_x\right)
\end{aligned}
\tag{8.8a}
$$

Effective electric field strength can be introduced to include all components of the electric field intensity, including the magnetic flux density as a Hall parameter b_e to simplify the formulation

$$
\begin{aligned}
E_{e,x} &= (b_e E_y - E_x)/(1+b_e^2) \\
E_{e,y} &= -(b_e E_x + E_y)/(1+b_e^2)
\end{aligned}
\tag{8.8b}
$$

The resultant components of the electron flux density are similar to that in absence of an externally applied magnetic field

$$\Gamma_{e,x} = -\mu_e n_e E_{e,x} - \frac{1}{1+b_e^2} d_e \frac{\partial n_e}{\partial x} + \frac{b_e}{1+b_e^2} d_e \frac{\partial n_e}{\partial y} \tag{8.8c}$$

$$\Gamma_{e,y} = -\mu_e n_e E_{e,y} - \frac{1}{1+b_e^2} d_e \frac{\partial n_e}{\partial y} - \frac{b_e}{1+b_e^2} d_e \frac{\partial n_e}{\partial x} \tag{8.8d}$$

For the positively charged ions, the effective electric field strength with the added Hall parameter is defined as

$$\begin{aligned} E_{+,x} &= (b_+ E_y - E_x)/(1+b_+^2) \\ E_{+,y} &= -(b_+ E_x + E_y)/(1+b_+^2) \end{aligned} \tag{8.9a}$$

It yields

$$\Gamma_{+,x} = \mu_+ n_+ E_{+,x} - \frac{1}{1+b_+^2} d_+ \frac{\partial n_+}{\partial x} - \frac{b_+}{1+b_+^2} d_+ \frac{\partial n_+}{\partial y} \tag{8.9b}$$

$$\Gamma_{+,y} = \mu_+ n_+ E_{+,y} - \frac{1}{1+b_+^2} d_+ \frac{\partial n_+}{\partial y} + \frac{b_+}{1+b_+^2} d_+ \frac{\partial n_+}{\partial x} \tag{8.9c}$$

Finally, by setting the effective electric field strength, including the contribution by the transverse magnetic field B, it follows that

$$\begin{aligned} E_{-,x} &= (b_- E_y - E_x)/(1+b_-^2) \\ E_{-,y} &= -(b_- E_x + E_y)/(1+b_-^2) \end{aligned} \tag{8.10a}$$

The components of the negatively charged ions are

$$\Gamma_{-,x} = -\mu_- n_- E_{-,x} - \frac{1}{1+b_-^2} d_- \frac{\partial n_-}{\partial x} + \frac{b_-}{1+b_-^2} d_- \frac{\partial n_-}{\partial y} \tag{8.10b}$$

$$\Gamma_{-,y} = -\mu_- n_- E_{-,y} - \frac{1}{1+b_-^2} d_- \frac{\partial n_-}{\partial y} - \frac{b_-}{1+b_-^2} d_- \frac{\partial n_-}{\partial x} \tag{8.10c}$$

In these equations, b_e, b_+, and b_- are the Hall parameters for electrons, positively charged ions and negatively charged ions, which are directly related to the Larmor frequencies of electrons and the ions, $b_e = \mu_e B_z/c = \omega_e/\nu_e, b_+ = \mu_+ B_z/c = \omega_+/\nu_+$, and $b_- = \mu_- B_z/c = \omega_-/\nu_-$. Here, ν_e, ν_- , and ν_+ are the averaged electron and ion collision frequencies.

The most general formulation for a multifluid model by drift-diffusion theory under an applied transverse magnetic field is actually a three-dimensional problem, but the charged particle dynamics are restricted in the two-dimensional plane perpendicular to the applied magnetic field. The governing equation system is closed by including the Poisson equation of plasma dynamics to provide a compatible electric field potential. The system of equations is casting on Cartesian coordinate to appear as

$$\frac{\partial n_e}{\partial t} + \frac{\partial}{\partial x}\left(-\mu_e n_e E_{e,x} - \frac{1}{1+b_e^2} d_e \frac{\partial n_e}{\partial x} + \frac{b_e}{1+b_e^2} d_e \frac{\partial n_e}{\partial y}\right)$$
$$+ \frac{\partial}{\partial y}\left(-\mu_e n_e E_{e,y} - \frac{1}{1+b_e^2} d_e \frac{\partial n_e}{\partial y} - \frac{b_e}{1+b_e^2} d_e \frac{\partial n_e}{\partial x}\right) = \left(\frac{dw}{dt}\right)_e \quad (8.11a)$$

$$\frac{\partial n_+}{\partial t} + \frac{\partial}{\partial x}\left(\mu_+ n_+ E_{+,x} - \frac{1}{1+b_+^2} d_+ \frac{\partial n_+}{\partial x} - \frac{b_+}{1+b_+^2} d_+ \frac{\partial n_+}{\partial y}\right)$$
$$+ \frac{\partial}{\partial y}\left(\mu_+ n_+ E_{+,y} - \frac{1}{1+b_+^2} d_+ \frac{\partial n_+}{\partial x} + \frac{b_+}{1+b_+^2} d_+ \frac{\partial n_+}{\partial y}\right) = \left(\frac{dw}{dt}\right)_+ \quad (8.11b)$$

$$\frac{\partial n_-}{\partial t} + \frac{\partial}{\partial x}\left(-\mu_- n_- E_{-,x} - \frac{1}{1+b_-^2} d_- \frac{\partial n_-}{\partial x} + \frac{b_-}{1+b_-^2} d_- \frac{\partial n_-}{\partial y}\right)$$
$$+ \frac{\partial}{\partial y}\left(-\mu_- n_- E_{-,x} - \frac{1}{1+b_-^2} d_- \frac{\partial n_-}{\partial x} - \frac{b_-}{1+b_-^2} d_- \frac{\partial n_-}{\partial y}\right) = \left(\frac{dw}{dt}\right)_- \quad (8.11c)$$

$$\frac{\partial^2 \phi}{\partial x^2} + \frac{\partial^2 \phi}{\partial y^2} = \frac{e}{\varepsilon}(n_e + n_- - n_+) \quad (8.11d)$$

where the source terms $(dw/dt)_i$ in equations (8.11a) through (8.11c), provide the generation and deprecation rates of ionized species of electrons, positively charged ions, and negatively charged ions.

The equations (8.11a) through (8.11d) have been validated by successfully simulating plasma hypersonic flow controls in the presence of an externally applied transverse magnetic field at a sharp leading edge of a control surface and in a virtual variable area inlet cowl. The numerical results have been verified by experimental data generated by a hypersonic plasma channel. On the sharp-leading edge, the experimental data and numerical results of the surface pressure reveal a substantially amplified viscous–inviscid interaction by a transverse magnetic field through the Lorentz force. The computational simulations also compare well with Pitot pressure surveys in the cross-sectional planes and surface pressure distributions within a square cross-sectional inlet.

The left-hand side column of Figure 8.6 shows typical experimental behaviors of DCD under the influence of an externally applied magnetic field. The DCD is generated in a uniform and constant transverse magnetic field aligned with the z coordinate of positive and negative polarity at 0.05 T (Shang et al., 2005, 2008; Kimmel et al., 2006). The top row of the graph depicts the condition where the magnetic field is absent and the glow is concentrated mostly over the cathode. The second row of the graph shows that the concentration of charge density is propelled

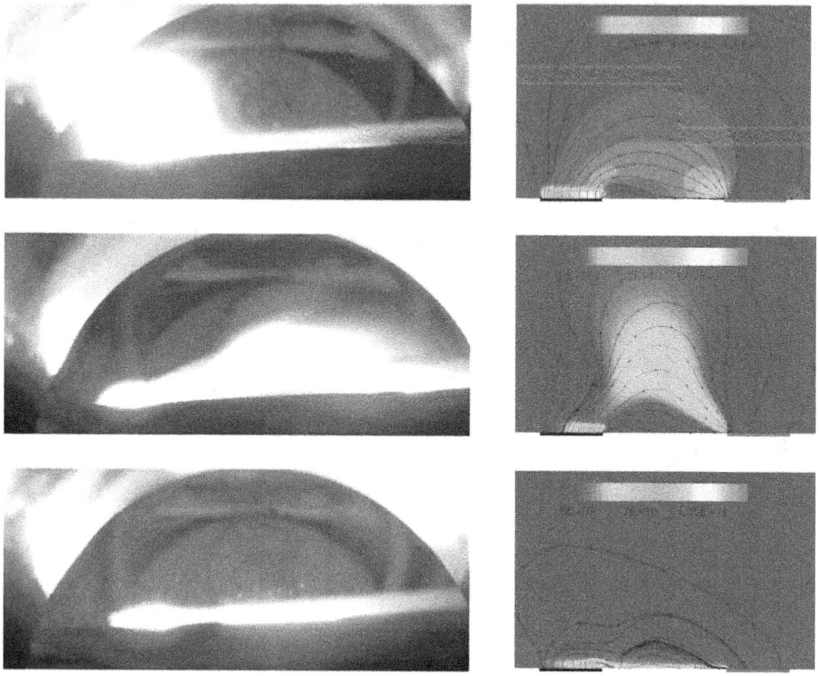

FIGURE 8.6 DCD in transverse magnetic field, $B_z = 0.00$, $B_z = -0.05$, and $B_z = +0.05$ T.

upward between electrodes by the magnetic field of a negative polarity, $B_z = -0.05$ T. When the polarity is reversed to $B_z = +0.05$ T, the ion is now suppressed toward the electrode and the dielectric surface between electrodes in the bottom row of the graph. The movement of electrons is nearly identical to that of the ion and is governed by the fundamental parameter of Debye length. In the right-hand column of this figure, computational results with a uniform magnetic field having the same magnitude of $B_z = \pm 0.05$ T are presented. The corresponding Hall parameters of electrons and ions under the simulation conditions are ± 0.44 and ± 0.00144, respectively (Shang et al., 2008). The numerical simulations based on the multifluid plasma model are in agreement with experimental observation, in that the Lorentz force effect to charge species is duplicated. In addition, the computational simulations also indicate that depending on the polarity of an applied magnetic field, an increased applied voltage is needed to sustain a fixed discharge electric current for plasma generation. This finding is among the first few observations that reveal the influence of the magnetic field to the ionization process via electron impact.

8.3 BOUNDARY CONDITIONS ON ELECTRODES

In deriving the species concentration equation for the multifluid model, the Maxwell equation is not explicitly included but the sum of equations (8.5a) through (8.5c) fully satisfies the charge conservation equation. This equation is obtained by taking the divergence of the generalized Ampere's electric circuit law

$$\partial(\nabla \cdot D)/\partial t - \nabla \cdot (\nabla \times H) + \nabla \cdot J = 0 \tag{8.12a}$$

and by virtue of a vector identity the second term vanishes. Invoking Gauss's law for the relation $\nabla \cdot D = \rho_e$ between the divergent of the electric displacement $D = \varepsilon E$ and the electric current density, yields

$$\frac{\partial \rho_e}{\partial t} + \nabla \cdot J = 0 \tag{8.12b}$$

In this equation, the electric charge density and the electric current density are defined as

$$\begin{aligned} \rho_e &= e(n_+ - n_- - n_e) \\ J &= e(\Gamma_+ - \Gamma_- - \Gamma_e) \end{aligned} \tag{8.12c}$$

The electron-impact plasma model is formally closed by the Poison equation of plasma dynamics for a compatible electric field that satisfies Gauss's law for the electric displacement, $\nabla \cdot (\varepsilon E) = \rho$, and assumes that the electric field intensity can be derived from the electrical potential. Specifically, the electric field intensity is equal to the negative gradient of the electric potential, $E = -\nabla\varphi$. The resultant well-known Poisson equation of plasma dynamics, equation (1.30), appears as

$$\nabla^2 \varphi = \frac{e}{\varepsilon}(n_e + n_- - n_+) \tag{8.13}$$

According to the theory of electromagnetics, the electric field across the interface of media such as a dielectric and plasma must satisfy the following conditions:

$$\vec{n} \times (E_{\mathrm{d}} - E_{\mathrm{p}}) = 0 \tag{8.14}$$

$$\vec{n} \cdot (D_{\mathrm{d}} - D_{\mathrm{p}}) = \rho_{e,s} \tag{8.15}$$

In these equations, subscripts p and d designate the variables residing either in plasma medium or on the dielectrics. The local surface charge density is evaluated by integrating the net charged number density over an infinitesimal normal distance, Δ over the control surface, and is defined as $\rho_{e,s} = e \int (n_+ - n_e - n_-) d\Delta$, which has a physical dimension of coulombs/cm^2.

Equation (8.14) simply states that the tangential electrical field strength is continuous across the media interface (Kong, 1986). It can be shown on the Cartesian frame: the rate of change for the electric potential in z, $\partial\varphi/\partial z$, must be identical across the interface along the x coordinate. Similarly, the rate of change of electrical potential in x, $\partial\varphi/\partial x$, must be equal along the z coordinate across the media interface. This requirement is automatically satisfied by the two-dimensional formulation (Shang et al., 2010, 2011).

The normal component of the electric displacement D across different media is discontinuous and must be balanced by the net surface charge on the interface by emission, desorption, and accumulation. In fact, the condition defined by equation (8.15) is independent of all chemical-physics processes on the medium interface but is required for all electromagnetic fields. This equation is further developed by introducing the electric potential for the partially ionized plasma, $E = -\nabla\varphi$, to become

$$\varepsilon_{\mathrm{p}}\frac{\partial\varphi_{\mathrm{p}}}{\partial n} - \varepsilon_{\mathrm{d}}\frac{\partial\varphi_{\mathrm{d}}}{\partial n} = \rho_{e,s}. \tag{8.16}$$

In this equation, ε_{p} and ε_{d} denote the relative electric permittivity of plasma and dielectrics, respectively. For the weakly ionized gas, the relative value of ε_{p} is around unity, and the ε_{d} can have a magnitude up to a value of 7.0 from polystyrene to rubber.

On the dielectric barrier, the surface diffusion and desorption can be modeled by the collisions of excited molecules and atoms using their collision frequency and the binding energy. The charge accumulation on the surface is the remainder from the instantaneous recombination of the charge particles with the active surface elements, after satisfying the imposed boundary condition for the species concentration. In other words, the surface is considered to be a perfect catalytic. For the two-dimensional formulation, equation (8.16) acquires the following form:

$$\frac{\partial\varphi_p}{\partial n} = \frac{\varepsilon_d}{\varepsilon_p}\frac{\partial\varphi_d}{\partial n} + \frac{e}{\varepsilon_p}\int (n_+ - n_e - n_-)d\Delta \tag{8.17}$$

Equation (8.17) is the critical boundary condition to be imposed on the dielectrics. All physics meaningful boundary conditions on the media interface have been demonstrated to be robust and accurately described the behavior of weakly ionized gas generated by electron collision.

On the cathode of DCD, the electric field potential vanishes on the metallic surface, and all negatively charged ions are repelled. Meanwhile, the number density of the positively charged ions reaches the cathode unaltered. The number density of electrons is derivable from the outward normal components of number flux density on the cathode. The secondary electron emission is a result of the Penning tunneling through very strong electrical field intensity in the cathode layer generated by positive-ion accumulation immediately adjacent to the electrode surface. The quantification of the electron emission from the cathode by positively charged ions is described by an emission coefficient γ, which has a well-known value range from 0.01 to

0.1, depending on the electrode material. The boundary conditions for DCD on the cathode are therefore

$$\begin{aligned} &\varphi = 0 \\ &n_- = 0 \\ &\vec{n} \cdot \nabla n_+ = 0 \\ &\vec{n} \cdot \Gamma_e = -\gamma \vec{n} \cdot \Gamma_+ \end{aligned} \tag{8.18}$$

On the anode, all the positively charged ions are repelled. The negatively charged ions and electron number densities are unaltered when the anode surface is reached, and this physical phenomenon is easily described by the no-change boundary condition. The electric potential on the anode is the balance between the electromotive force (EMF) and the loss along the external electric circuit:

$$\begin{aligned} &\varphi = EMF - IR \\ &\vec{n} \cdot \nabla n_- = 0 \\ &n_+ = 0 \\ &\vec{n} \cdot \nabla n_e = 0 \end{aligned} \tag{8.19}$$

where the symbol I is the total electrical current and R is the resistance of a closed electric circuit after electric breakdown including the discharging column.

The boundary conditions on the far field boundaries are uniformly imposed by the vanishing outward normal gradient approximation to describe the vanishing electrical field potential:

$$\begin{aligned} &\vec{n} \cdot \nabla \varphi = 0 \\ &\vec{n} \cdot \nabla n_- = 0 \\ &\vec{n} \cdot \nabla n_+ = 0 \\ &\vec{n} \cdot \nabla n_e = 0 \end{aligned} \tag{8.20}$$

These far field boundary conditions are equally applicable to all electron impact ionization, which is associated with a very rapidly decayed electrical field potential from electrodes. The placement of the far field of a partially ionized gas is easily satisfied, because the highly concentrated electric field intensity is always clustered around electrodes.

In DBD operation, the distinguishing discharge structures during an AC cycle are frequently referred to as the forward stroke or negative going and the back stroke or positive going. In fact, the discharge is exclusively governed by the polarity of electrodes. When the electrical potential is positive, $\varphi(t) > 0$ on the exposed electrode, it performs as the anode on which the electrical potential is dictated by the externally applied AC source with a frequency of ϖ and the voltage lost in a closed electric circuit (Figure 8.7). The resistance R in the circuit is the total value of the power supply unit and the discharging column. Electrons propagate very rapidly from the

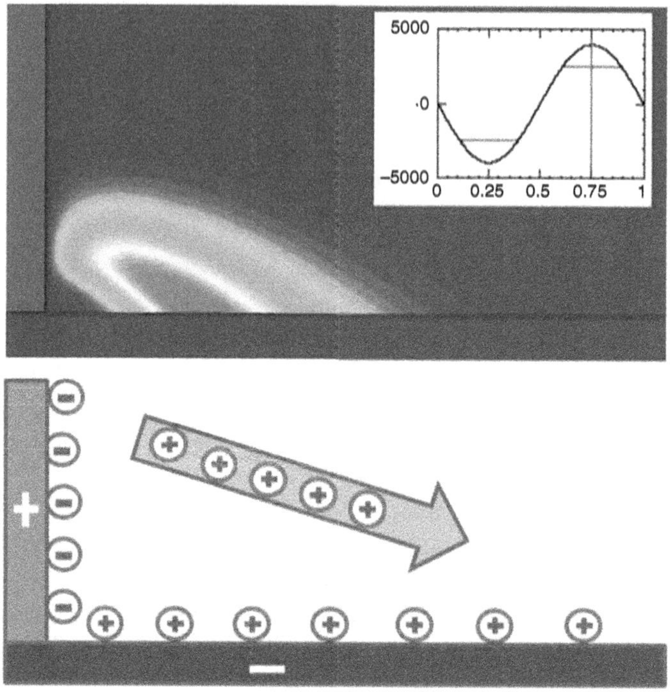

FIGURE 8.7 $\varphi(t) > 0$ on exposed electrode.

dielectric surface toward the exposed metallic electrode. All positively charged ions are repulsed and propagate toward the encapsulated dielectric. The imposed boundary conditions on the anode or the exposed electrode are

$$\begin{aligned} &\varphi(t) = EMF \sin(\varpi t) - R \int J dv \\ &n_+ = 0 \\ &\vec{n} \cdot \nabla n_e = 0 \text{ and } \vec{n} \cdot \nabla n_- = 0 \end{aligned} \tag{8.21a}$$

The *EMF* is the external voltage applied to a DBD electric circuit. It shall be recognized that the electric field potential across the exposed and the dielectric barrier after the breakdown is determined by the external circuit law $EMF = \varphi + R \int J dv$. The electric potential is thus a reduced value from the EMF of the applied AC field. The electric current density of the discharge is simply

$$J = e(\Gamma_+ - \Gamma_e - \Gamma_-) \tag{8.21b}$$

At this same instant, the dielectric barrier surface acts as the cathode on which the secondary electron emission is induced by an excessive accumulation of positively charged ions over the dielectric barrier to reduce the effective difference further in electric potential. At this AC phase, the imposed boundary conditions for the electrical potential and charged particle number densities on the dielectric or cathode shall be

$$\vec{n} \cdot (\nabla\varphi)_p = \vec{n} \cdot \left(\frac{\varepsilon_d}{\varepsilon_p}\right)(\nabla\varphi)_d + \frac{e}{\varepsilon_p}\int (n_+ - n_e - n_-)d\Delta \tag{8.22a}$$

$$\vec{n} \cdot \nabla n_+ = 0 \text{ and } n_- = 0 \tag{8.22b}$$

$$\vec{\Gamma}_e = -\gamma_d \vec{\Gamma}_+ \tag{8.22c}$$

During $\varphi(t) > 0$ AC cycle, equation (8.22a) describes the diminishing electrical field intensity across the electrodes through the surface charge accumulation and the difference between electric permittivity of the media. It is also the fundamental mechanism of the self-limiting characteristic in preventing the DBD transition into spark. Again, equation (8.22c) enforces the electron secondary emission condition for Townsend's discharge.

The roles of the exposed electrode and the dielectric barrier surface reverse, when the AC field switches polarity to a negative electric potential $\varphi(t) < 0$. The exposed electrode now performs as the cathode, and the secondary emission that is created by the positive-ion accumulation now occurs over the exposed metallic surface (Figure 8.8). Except the coefficient of the emission on the metal surface, γ_m is

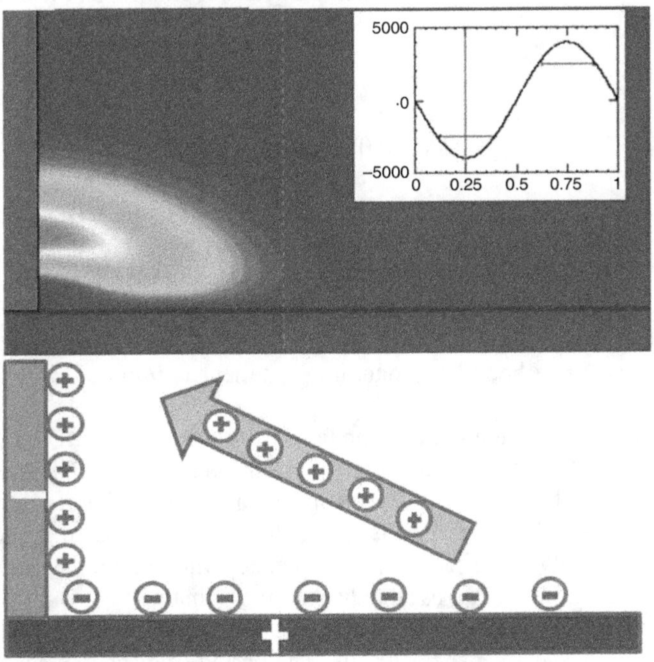

FIGURE 8.8 $\varphi(t) < 0$ on exposed electrode.

different from that of the dielectrics. On the cathode or the exposed metallic electrode, all the electrons and negatively charged ions are repelled completely from the cathode and propagate toward the dielectric surface. According to theory, the electrons shall move much more quickly than the negatively charged ions and the drift velocities of charged particles range from 10^5 to 10^7 cm/s. But in a small charge separation region of a few fractions of 1 mm in a DBD, the difference is hardly discernible by computational simulation.

The electric potential on the metallic surface is again the effective externally imposed electric field by satisfying the external circuit law. At an infinitesimal distance σ away from the cathode, the clustered positively charged ions further reduce the potential across the jump condition over the media interface. The boundary conditions over the exposed electrode or the cathode are

$$\begin{aligned} &\varphi(t) = EMF\sin(\omega t) - R\int Jdv \\ &\vec{n}\cdot\nabla n_+ = 0\ n_e \text{ and } n_- = 0 \\ &\Gamma_e = -\gamma_m\Gamma_+. \end{aligned} \tag{8.23}$$

At the same instance, the dielectric encapsulated electrode now acts as the anode. If the encapsulated electrode is not grounded, then the electric potential over the dielectric surface is exactly the same as that of the exposed electrode when $\varphi(t) > 0$, namely $\varphi = EMF - R\int Jdv$. However, the electrode beneath the dielectric is often grounded in DBD operations. All charged particles recombine on the dielectrics and maintain a neutralized state:

$$\begin{aligned} &\varphi(t) = 0 \\ &n_+ = 0 \\ &\vec{n}\cdot\nabla n_e = 0 \text{ and } \vec{n}\cdot\nabla n_- = 0 \end{aligned} \tag{8.24}$$

To complete the boundary conditions specification for the DBD simulation, the vanishing gradient conditions are imposed uniformly for all four dependent variables on the far field boundary of the computational domain; see equation (8.20). For the multiple length- and time-scale phenomena, the ionization and recombination of charge particles must keep pace with the time scale of the alternative electric field. Therefore, no separate boundary condition treatment is imposed during switching potential of an AC cycle.

It may be noticed that the imposed boundary conditions for DBD on the exposed electrode and the dielectric surface are fully reinforced for the DBD self-limiting mechanisms. These boundary conditions specifications must be synchronized with the alternating polarities of electrodes. The differences in the imposed boundary conditions appear only in the secondary emission coefficients on the cathode and the effective electrical potential across the discharging domain.

The DBD discharge is a periodic dynamic event. At AC frequency of 10 kHz, the characteristic time scale is a short 200 μs. The period of discharge starts only after

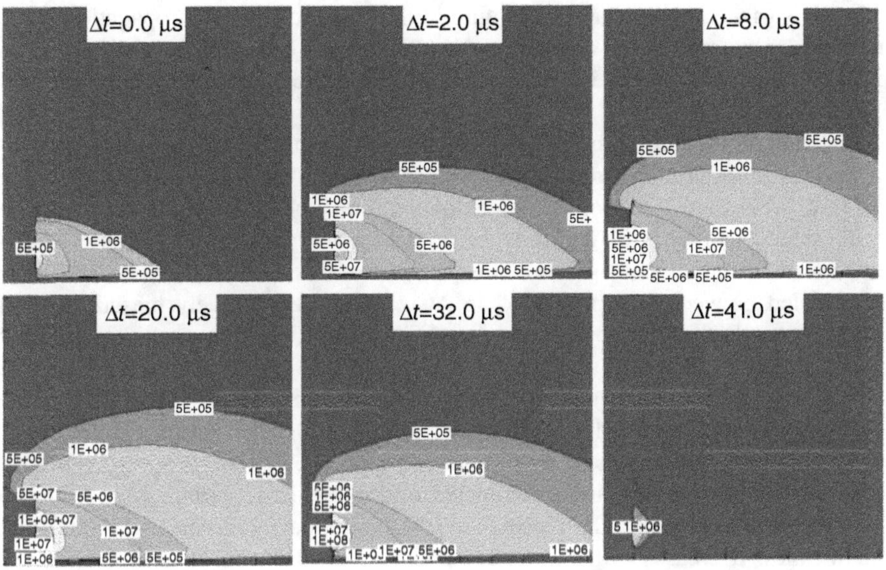

FIGURE 8.9 Electrons generation and propagation in a DBD cycle, $\varphi = +4.0$ kV.

the externally applied electric field voltage exceeds the value of ionization potential: therefore, meaningful computational results are presented at several instances after the discharge breakdown until the discharge ceased. Typical sequences of charged particles generation and propagation via number density contours are presented in the next three figures.

In Figure 8.9, in a negatively biased voltage phase of an AC cycle, the discharge lasts for the duration of 41 μs and possesses a physical domain of 0.150 mm and 0.425 mm in height and length. The discharge domain is relatively small in comparison with exposed and encapsulated electrodes with a length of 10 mm but is spilt over the vertically recessed dielectrics of 0.127 mm. Immediately after breakdown, electrons propagate away from the exposed electrode over the dielectric surface, which now acts as the anode. The discharge domain expands continuously until the externally applied electrical potential falls beneath the breakdown threshold. The concentration of electrons is lower than that of ions in the discharge domain, by virtue of greater mobility of electrons. The number density of electrons attains a lower value of $10^{10}/\text{cm}^3$ immediately adjacent to the dielectrics. The similar progression of the electron propagation occurs in the positively biased voltage phase. The discharge period is slightly shorter because of different values of electric permittivity of a dielectric and metal. The electron distribution over the dielectrics exhibits a more pronounced diffusion pattern than the counterpart that associates with a positive polarity, but the overall structures of electron movements are similar between the two AC phases. For the same reason, the discharge domain is more compact over the dielectric surface. An outstanding feature of the charged particle propagation is the high speed movement for all charged species. There shall be a substantial difference in speed between electron and ion motions, but it cannot be discerned by the numerical simulations.

Since the momentum transfer between colliding positively charged ions and neutral species is the principal mechanism for the so-called electrical wind by applying DBD for flow control, the ions propagation and movement of a complete AC cycle are presented in Figure 8.10. From the computational simulation, significant differences in formations and sizes of discharge domains of positively charged ions are revealed between two phases of an AC cycle. The positively biased exposed electrode expels the ions toward the dielectric surface, and the positively charged ions propagate rapidly and sustain their expansion over nearly the entire discharge period. In the negative phase of DBD discharge, the dielectric now performs as the anode; all positively charged ions are repulsed from it. The electric field is also vastly different from the arrangement when the encapsulated electrode is not grounded. A large number of electrons are accumulated on the dielectric surface. According to the computational results, the plasma sheath is extremely thin, around 0.082 mm. The thickness decreases with a lower value of Townsend's secondary emission coefficient. The propagation of the ions is clustered closer to the vertical surface of the exposed electrode, which leads to a small collision cross section between the ions and neutrals. In the inner most electrode juncture, the number density of ions dominates over that of electrons to have maximum values of $7.82 \times 10^{10}/cm^3$ and $6.92 \times 10^{10}/cm^3$, respectively. A large disparity of discharge domains of positively charged ions in an AC cycle may lead to an asymmetric momentum transfer collision cross section and a biased electric wind that are always observed for the DBD applications.

8.4 QUANTUM CHEMICAL KINETICS

An enormous amount of energy is needed to generate a localized volumetric plasma that must have a sufficient charged particle number density for strong electromagnetic–aerodynamics interactions. For example, the ionization potential is 34 eV for electron beam, 65.7 eV for DCD, and 81 eV per ion–electron pair for discharge at the radio frequency (Raizer, 1991). The ionization potential always underestimates the energy requirement in applications, because the nonequilibrium energy cascades to vibration excitation, recombination, and attachment processes. In the electron impact processes for ionization, the positive and negative charged ions still retain their ambient conditions. For this reason, the partially ionized gas is often identified as the low-temperature plasma with a charge number density generally limited to a magnitude of $10^{13}/cm^3$. As a weakly partially ionized plasma, the electromagnetic force usually exerts only a small perturbation to the mainstream flow, and the thermodynamic behavior is significantly different from the plasma generated by thermal excitation. Therefore, the plasma actuator for flow control is the most

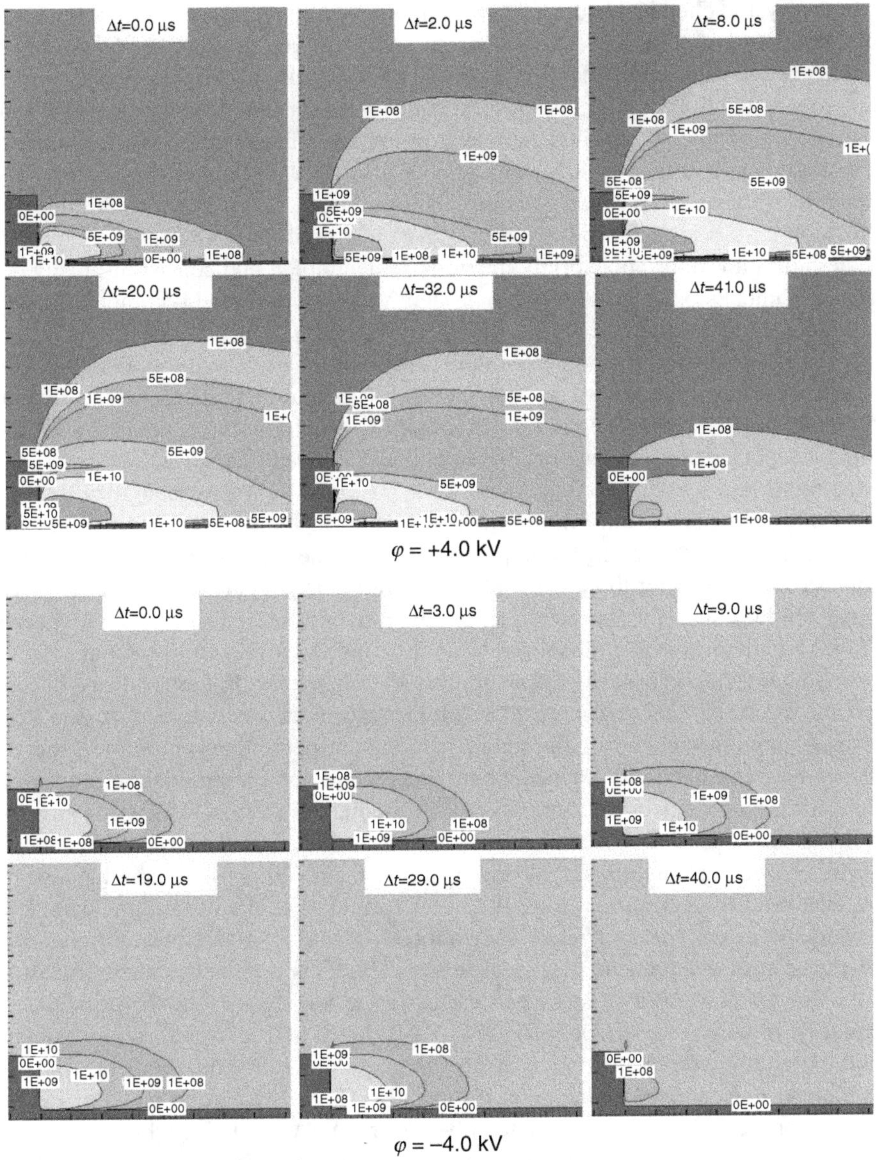

FIGURE 8.10 Ions generation and propagation in an AC cycle. $\varphi = +4.0\,\text{kV}$; $\varphi = -4.0\,\text{kV}$.

effective at flow bifurcations such as the onset of dynamic stall, laminar–turbulent transition, and vortical separation (Shang et al., 2014). However, the electromagnetic effect can also be amplified by an externally applied magnetic field or by inviscid–viscous interaction at the leading edge of hypersonic control surfaces (Macheret et al., 2001; Shang et al., 2005).

In theory, the chemical species of partially ionized air are numerous and the reaction processes are complex, but its concentration still can be calculated by the law of mass action. There are many groups of nonequilibrium reactions in the ionization process: the impact dissociations, exchange reactions, the associative ionization, dissociative recombination, and the electron-impact ionization. In fact, the impact dissociations are further divided into heavier particle and electron impact dissociations. In spite of these types of complications, successful models of the thermal ionization have been achieved. Especially for electron impact ionization, numerous species concentrations appear only in a minuscule amount and consist of mostly the transient radicals to make the equation system very stiff and rend it computationally unstable.

An alternative approach in determining the net rates of generation and depletion of individual ionized species can be achieved by quantum chemical-physics ionization modeling. This approach is also simplified by grouping the charge species into global categories of electrons, positively charged ions and negatively charged ions as the multifluid model. The justification for this approach is to recognize that the charged species appear only in the tracing amount in molar or mass fractions and are dominated by a few radical components. In fact, the ionization by electron collisions occurs at the outer limit of the Maxwell–Boltzmann distribution where the gas energy is lower than the ionization potential (Raizer, 1991). The dominant species of DBD are the metastable molecules $N_2(A^3\Sigma_u^+)$ and $O_2(b^1\Sigma_d^+)$ (Ellison et al., 1991). The main ionized processes of a nonequilibrium volumetric discharge consist of four types of reactions: electron/molecular, atomic/molecular, decomposition, and synthesis. At any instance during discharge, the ionized species concentration is the net balance between ionization, detachment, attachment, and recombination processes. The ionization model by quantum chemical kinetics is therefore focused on these mechanisms.

The classic formulation for electronic impact ionization is the similar law by Townsend and is an empirical formula. In the formulation, the ionization coefficient α, which measures the number of ionization by electron impact per unit distance, is a function of the reduced electrical field $p/|E|$. This quotient is also a measure of the energy gain by a charged particle between collisions from the principle of similarity. In fact, the coefficient $\alpha = A\exp(-Bp/\,|E|)$ of Townsend's similarity law holds extremely well in comparison with a large group of experimental data, in both the ionization frequency and degree of ionization. For discharge in air, $A = 15$ and $B = 365$ for the E/P range from 100 to 800 (V/torr cm). At a relatively higher value of E/P, an accuracy improvement may be needed but is not essential (Surzhikov et al., 2004). The coefficients of ionization for air can be summarized as

$$\left(\frac{\alpha}{p}\right) = 15\exp\left[-\frac{365}{(|E|\,/P)}\right],\quad 1/\text{cm}\times\text{torr} \tag{8.25}$$

The depletion of ionized species is accomplished by the recombination and attachment processes. The dissociative recombination is the fastest mechanism of the bulk recombination of a weakly ionized gas and is a simple binary chemical reaction. The decay rate with time of plasma is often given as

$$\left(\frac{dn_e}{dt}\right)_i = -\beta n_+ n_e \tag{8.26}$$

and typically coefficient β is assigned a value of 2×10^{-7} cm^3/s, and the characteristic decay time scale is less than 10^{-3} s.

Another main mechanism of charge neutralization is the ion–ion recombination process. In the low pressure environment, the recombination takes place through binary collision and the reaction is similar to charge transfer. The reaction rate constants have been estimated around the values of 10^{-10} cm^3/s for two-body collisions. At a moderate pressure, the process proceeds through triple collisions, and the reaction rates are much lower, at a value about 10^{-25} cm^6/s (Raizer, 1991). Again the rate of ion–ion recombination can be approximated as by collision kinetics:

$$\left(\frac{dn_i}{dt}\right)_r = -\beta_i n_+ n_- \tag{8.27a}$$

For an example, the recombination rate constants between O_2^- and O_4^+, as well as NO^+ and NO_2^-, are on the orders of magnitude of 10^{-25} and 10^{-26} cm^6/s. The maximum value of the ion–ion recombination at the one atmosphere has a value of 10^{-6} cm^3/s.

The electron attachment and detachment are the formation and depletion processes of negatively charged ions in partially ionized air. The chemical reaction rates are uncertain and have a wide range from 10^{-8} cm^3/s to 10^{-30} cm^6/s, and especially for the triple-collision reactions, strong electronic temperature dependence is noted (Gibalov and Pietsch, 2004). For the strongly bound molecules, a sufficiently high energy is required for the dissociative attachment of an electron. The electron attachment is a main mechanism for removing the electron from the negatively charged ions. The lost electron number density can be given as

$$\left(\frac{dn_e}{dt}\right)_a = -\nu_a n_e \tag{8.27b}$$

The attachment frequency of electrons in dry air at one atmosphere condition is around 10^8/s.

The coefficients of the detachment of electrons κ_d in partially ionized gas at room temperature have the value of 10^{-10} cm^3/s. The value for the metastable molecules $N_2(A^3\Sigma_u^+)$ and $O_2(b^1\Sigma_d^+)$ in air are unknown. But by an indirect estimate, the discharges are characterized by a value of 10^{-14} cm^3/s. It is also interesting to note that weakly ionized plasma is sustainable in negatively charged gas at a lower value of *E*/*p* than that by a short pulse discharge.

From the above brief discussion, the plasma generation and depletion processes are modeled by the quantum chemical physics for the multifluid model (Shang et al., 2014):

$$\begin{aligned}
&\frac{\partial n_e}{\partial t} - \nabla \cdot (n_e \mu_e E + d_e \nabla n_e) = \alpha(|E|)|\Gamma_e| - \beta n_+ n_e - \nu_a n_e + \kappa_d n_n n_- \\
&\frac{\partial n_+}{\partial t} + \nabla \cdot (n_+ \mu_+ E - d_+ \nabla n_+) = \alpha(|E|)|\Gamma_e| - \beta n_+ n_e - \beta_i n_+ n_- \\
&\frac{\partial n_-}{\partial t} - \nabla \cdot (n_- \mu_- E + d_- \nabla n_-) = \nu_a n_e - \kappa_d n_n n_- - \beta_i n_+ n_-
\end{aligned} \tag{8.28}$$

In fact, the sum of these equations satisfies fully the charge conservation law; the rate of change in charge number density must be balanced by the gradient of the electric current density, which is derived from the Maxwell equations; see equation (8.11). Again in the above approximation, the number density of neutral particles of air n_n is treated as a constant to have a value of $2.69 \times 10^{19}/\text{cm}^3$. The attachment frequency ν_a is formally defined as $\nu_a = (\alpha_a/p)u_{e,\text{drift}} \cdot p$. Here, a commonly adopted value is $(\alpha_a/p) = 0.005/\text{cm} \times \text{torr}$, and $u_{e,\text{drift}}$ is the drift velocity of an electron. From the drift-diffusion theory, the velocity can be consistently evaluated as $u_{e,\text{drift}} = \mu_e |E|$.

The coefficient of ion–ion recombination β_i is estimated from the chemical kinetics of ionized and molecular nitrogen and oxygen to have a value of $\beta_i = 1.6 \times 10^{-7}\,\text{cm}^3/\text{s}$. The detachment coefficient is estimated to have a value range of $10^{-14} < \kappa_d < 8.6 \times 10^{-10}\,\text{cm}^3/\text{s}$. This coefficient is important for the negatively charged ion number density computation, because it is the principal mechanism that governs the depletion of this species. Fortunately, its value can be verified by the discharged electrical current from experimental data.

The equation (8.28) of electron collision ionization is actually replaced by the energy conservation equations of vibrational and electronic internal excitations, equations (7.10) and (7.11), as well as the law of mass action equations (7.15b) and (7.15c), for chemical kinetics modeling. The electron attachment mechanism has been modeled to be proportional to Townsend's ionization process for a nonphysical but computational stability concern. In this regard, some investigators adopt the similar basic collision formulation for the attachment and recombination mechanisms of the ionization process with empirically determined coefficients (Beouf et al., 2005; Solov'ev et al., 2008). Others also model the negative-ion ionization by splitting a portion of the electron generation as the electron attachment process (Unfer and Boeuf., 2009). On the contrary, the ionization model used by Likhanskii et al. (2008) is nearly identical to that by Surzhikov et al. (2004), Huang et al. (2011), and Shang et al. (2014).

The final and the most important verification of the chemical-physics ionizing model is presented by a direct comparison with a measured conductive current during the discharge (Shang et al., 2014). The ionization model is built on Townsend's discharge mechanism together with the bulk and ion–ion recombination, as well as the electron attachment and detachment. If the ionization model by the chemical-physics kinetics is not able to describe a reasonable discharge composition, the

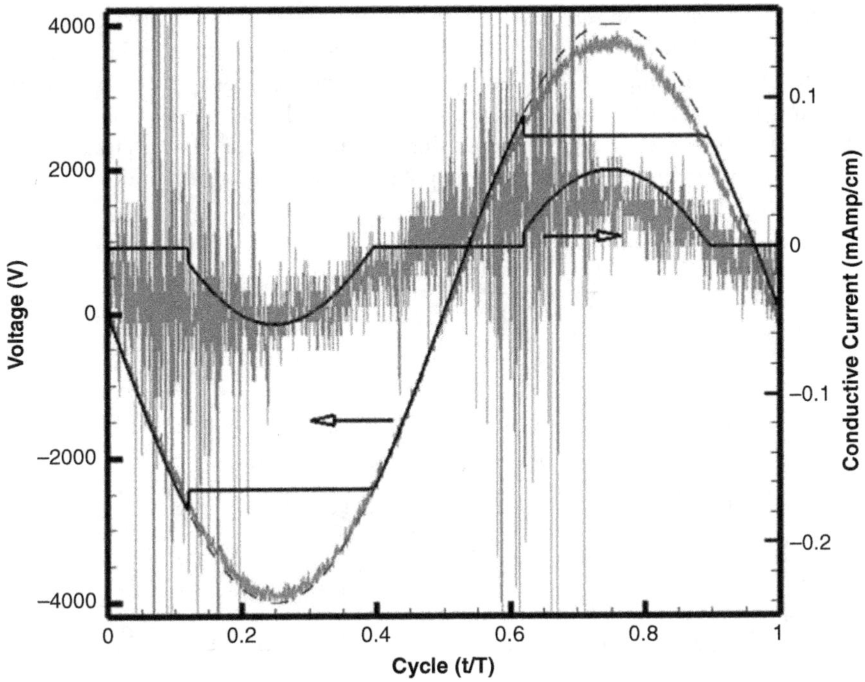

FIGURE 8.11 Comparison of DBD conductive current with experiment, EMF = 4 kV, f = 5 kHz.

calculated conductive current of a DBD will be clearly erroneous. The experimental data is collected at an AC cycle of 5 kHz and an electrical potential of 4.0 kV. In Figure 8.11, the calculated conductive current is designated by a black line that lies within the scattering traces of the experimental data to reflect the correct magnitude and duration of the discharge. Equally important, the effects of different electrical permittivity and secondary emission coefficients on metallic electrode and dielectric surface are also accurately captured.

In theory, the above ionizing models can be further improved by expanding the experimental data base or by an *ab initio* approach to the quantum chemical-physics. In this regard, the state-of-the-art status in low-temperature ionization modeling is similar to that of chemical-physics models using the approximated chemical reaction rates, but it has a decided advantage in substantially reducing the computational resource required for simulations.

8.5 NUMERICAL ALGORITHMS

The types of the partial governing differential equations for thermal excitation, equations (7.7) through (7.11), and electron impact, equations (8.5) and (8.13), are nearly identical, but distinctive physics are now generated by the different initial values and boundary conditions. The physics-based model for electron-impact ionization consists of a mixed inhomogeneous hyperbolic and elliptic partial differential equations, equations (8.11a) through (8.11d) and (8.28). The generic form of these equations can always be expressed in the flux vector form by splitting the flux into the convective and diffusive terms. A two-dimensional formulation of a two-ionized species multifluid model includes an externally applied transverse magnetic field and can be cast in the following form:

$$\frac{\partial U}{\partial t} + \frac{\partial F_x(U)}{\partial x} + \frac{\partial F_y(U)}{\partial y} = \frac{\partial G_x(U)}{\partial x} + \frac{\partial G_y(U)}{\partial y} + S \tag{8.29a}$$

where

$$U = \begin{bmatrix} \phi \\ n_e \\ n_+ \end{bmatrix}, F_x(U) = \begin{bmatrix} 0 \\ -\mu_e n_e E_{e,x} \\ \mu_+ n_+ E_{+,x} \end{bmatrix}, F_y(U) = \begin{bmatrix} 0 \\ -\mu_e n_e E_{e,y} \\ \mu_+ n_+ E_{+,y} \end{bmatrix},$$

$$G_x(U) = \begin{bmatrix} \dfrac{\partial \phi}{\partial x} \\ \dfrac{d_e}{1+\beta_e^2}\dfrac{\partial n_e}{\partial x} - \dfrac{\beta_e d_e}{1+\beta_e^2}\dfrac{\partial n_e}{\partial y} \\ \dfrac{d_+}{1+\beta_+^2}\dfrac{\partial n_+}{\partial x} + \dfrac{\beta_+ d_+}{1+\beta_+^2}\dfrac{\partial n_+}{\partial y} \end{bmatrix}, G_y(U) = \begin{bmatrix} \dfrac{\partial \phi}{\partial y} \\ \dfrac{d_e}{1+\beta_e^2}\dfrac{\partial n_e}{\partial y} + \dfrac{\beta_e d_e}{1+\beta_e^2}\dfrac{\partial n_e}{\partial x} \\ \dfrac{d_+}{1+\beta_+^2}\dfrac{\partial n_+}{\partial y} - \dfrac{\beta_+ d_+}{1+\beta_+^2}\dfrac{\partial n_+}{\partial x} \end{bmatrix}, \tag{8.29b}$$

$$S = \begin{bmatrix} -\dfrac{e}{\varepsilon}(n_e - n_+) \\ \alpha\left|\Gamma_e\right| - \beta n_e n_+ \\ \alpha\left|\Gamma_e\right| - \beta n_e n_+ \end{bmatrix}, \left|\Gamma_e\right| = \sqrt{\left(\mu_e n_e \frac{\partial \phi}{\partial x} - d_e \frac{\partial n_e}{\partial x}\right)^2 + \left(\mu_e n_e \frac{\partial \phi}{\partial y} - d_e \frac{\partial n_e}{\partial y}\right)^2}$$

The eigenvalues of the inhomogeneous partial differential equations system span a wide range to make it very stiff for computation. A number of innovative measures must be introduced to ensure stability and accuracy of the numerical simulation. In earlier efforts, the entire equations system was solved by a straightforward Penta-diagonal implicit scheme, which possesses a narrow convergent band. A rigorous and actively maintained time evolution procedure is needed to attain a converged solution. An alternative exponential-base-function Sharfetter–Grummel algorithm has also been applied; therefore, a set of stringent initial values must be specified to meet a converged tolerance within a reasonable time (Scharfetter et al., 1969). An alternative numerical algorithm becomes a pacing item for computational electromagnetic-aerodynamics.

One more step in this approach is taken by the Delta formulation via the iteratively diminishing residual approach (Huang et al., 2011). In fact, this algorithm is considered to be the ideal numerical procedure for time-dependent problems. The delta formulation is set the dependent variable in temporal difference, $\Delta U = U^n - U^{n-1}$, where the superscript n is the current iteration level and $n-1$ is the previous iteration level. The governing equation in flux vector form acquires the expression,

$$\frac{\partial \Delta U}{\partial t} + \nabla \cdot [F(U^n) - F(U^{n-1})] = S - \left[\frac{\partial U}{\partial t} + \nabla \cdot F(u) + \nabla \cdot G(u)\right]^{n-1} \tag{8.29c}$$

The delta forms of the flux vectors are adopted and obtained by using the linearized equation (8.29a):

$$\begin{aligned}
\Delta F_x &= F_x^{n+1} - F_x^n \approx A_x^n \Delta U,\\
\Delta F_y &= F_y^{n+1} - F_y^n \approx A_y^n \Delta U,\\
\Delta G_x &= G_x^{n+1} - G_x^n \approx B_1^n \frac{\partial \Delta U}{\partial x} + B_2^n \frac{\partial \Delta U}{\partial y}, \text{ and}\\
\Delta G_y &= G_y^{n+1} - G_y^n \approx B_1^n \frac{\partial \Delta U}{\partial y} - B_2^n \frac{\partial \Delta U}{\partial x}
\end{aligned} \tag{8.30a}$$

where n is the iteration level and

$$\begin{gathered}
A_x = \frac{\partial F_x}{\partial U} = \Lambda \left[0, -\mu_e E_{e,x}, \mu_+ E_{+,x}\right], A_y = \frac{\partial F_y}{\partial U} = \Lambda \left[0, -\mu_e E_{e,y}, \mu_+ E_{+,y}\right],\\
B_1 = \Lambda \left[1, \frac{d_e}{1+\beta_e^2}, \frac{d_+}{1+\beta_+^2}\right], \text{and } B_2 = \Lambda \left[0, -\frac{\beta_e d_e}{1+\beta_e^2}, \frac{\beta_+ d_+}{1+\beta_+^2}\right]
\end{gathered} \tag{8.30b}$$

It should be noted that in the present formulation, the equations are purposely decoupled to render the off-diagonal elements as zero, and thus these equations can be solved in a sequential manner.

By dropping the time-dependent term and substituting equation (8.30a) into equation (8.29a), the result can be written as

$$\begin{gathered}
\frac{\partial}{\partial x}(\Delta F_x) + \frac{\partial}{\partial x}(\Delta F_y) = \frac{\partial}{\partial x}(\Delta G_x) + \frac{\partial}{\partial y}(\Delta G_y) + R^n\\
\text{where } R^n = -\frac{\partial}{\partial x}\left(F_x^n\right) - \frac{\partial}{\partial x}\left(F_y^n\right) + \frac{\partial}{\partial x}\left(G_x^n\right) + \frac{\partial}{\partial y}\left(G_y^n\right) + S^n
\end{gathered} \tag{8.31a}$$

It should be noted that the right-hand side of equation (8.31a) is explicit, and it can be solved using the intermediate solution with any high-order scheme, provided that it is numerically stable. In the formulation, a central differencing scheme is applied to the diffusive and source terms and a quadratic upwind differencing scheme is adopted for the convective terms. The detailed treatments are discussed in the remaining section.

The discretization of the second-order diffusive terms is done by the central differencing scheme. For example,

$$\frac{\partial}{\partial x}\left(G_x^n\right) \approx \frac{G^n_{x_{i+\frac{1}{2}j}} - G^n_{x_{i-\frac{1}{2}j}}}{x_{i+\frac{1}{2}} - x_{i-\frac{1}{2}}} \tag{8.31b}$$

where

$$\begin{aligned} G^n_{x_{i+\frac{1}{2}j}} &= B_1\left(\frac{\partial U}{\partial x}\right)^n_{i+\frac{1}{2}j} + B_2\left(\frac{\partial U}{\partial y}\right)^n_{i+\frac{1}{2}j} \\ &= B_1\frac{U^n_{i+1,j} - U^n_{i,j}}{x_{i+1} - x_i} + B_2\frac{\left(U^n_{i,j+1} + U^n_{i+1,j+1}\right)/2 - \left(U^n_{i,j-1} + U^n_{i+1,j-1}\right)/2}{y_{j+1} - y_{j-1}}. \end{aligned} \tag{8.31c}$$

By contrast, the convective flux is discretized by a high-order upwind differencing scheme as an approximate Riemann problem

$$\frac{\partial}{\partial x}\left(F_x^n\right) \approx \frac{F^n_{x_{i+\frac{1}{2}j}} - F^n_{x_{i-\frac{1}{2}j}}}{x_{i+\frac{1}{2}} - x_{i-\frac{1}{2}}} \tag{8.31d}$$

and

$$F^n_{x_{i+\frac{1}{2}j}} = \frac{1}{2}A^n_{x_{i+\frac{1}{2}j}}\left(U^n_{R_{i+\frac{1}{2}j}} + U^n_{L_{i+\frac{1}{2}j}}\right) - \frac{1}{2}\left|A^n_{x_{i+\frac{1}{2}j}}\right|\left(U^n_{R_{i+\frac{1}{2}j}} - U^n_{L_{i+\frac{1}{2}j}}\right). \tag{8.31e}$$

The values of $U_{R_{i+\frac{1}{2}j}}$ and $U_{L_{i+\frac{1}{2}j}}$ are obtained by a high-order upwind Lagrange interpolation from the values of U at the nodal point values, $U_{i,j}$'s, to the interface location, $(x_{i+1/2}, y_j)$. The subscripts R and L represent the upwind interpolation from right and left hand sides, respectively. In the following development, a quadratic upwind differencing scheme is used; for example, $U_{R_{i+\frac{1}{2}j}}$ is obtained by the interpolation of a quadratic polynomial constructed by $U_{i,j}$, $U_{i+1,j}$ and $U_{i+2,j}$ at (x_i, y_j), (x_{i+1}, y_j), and (x_{i+2}, y_j) locations, respectively, and $U_{L_{i+\frac{1}{2}j}}$ is evaluated by the interpolation of a quadratic polynomial constructed by $U_{i-1,j}$, $U_{i,j}$ and $U_{i+1,j}$ at (x_{i-1}, y_j), (x_i, y_j) and (x_{i+1}, y_j) locations, respectively.

Finally, the ionization rate, as it appears in S, is evaluated explicitly by the central differencing scheme approximated at the nodal point (x_i, y_j). While the right-hand side of equation (8.31a) is evaluated explicitly, the left-hand side is solved implicitly. First, the central differencing scheme identical to that shown in equations (8.31b) and (8.31c) is used to discretize the diffusive terms of ΔU. It should be noted that the cross diffusion terms appearing in equation (8.31a) were purposely dropped to simplify the solution matrix coefficients. As will be shown later, any modification to the left-hand side should not affect the final solution as long as numerical convergence can be reached, because the exact discretized governing equation is always satisfied fully on the right-hand side of the formulation. Second, the convective terms are approximated by the first-order upwind differencing scheme to ensure numerical stability (Yee, 1986). For example, equation

$$\frac{\partial}{\partial x}\left(\Delta F_x\right) \approx \frac{\Delta F_{x_{i+\frac{1}{2},j}} - \Delta F_{x_{i-\frac{1}{2},j}}}{x_{i+\frac{1}{2}} - x_{i-\frac{1}{2}}} \tag{8.32a}$$

where

$$\Delta \mathrm{F}_{x_{i+\frac{1}{2},j}} = \frac{1}{2}A^n_{x_{i+\frac{1}{2},j}}(\Delta U_{i+1,j} + \Delta U_{i,j}) - \frac{1}{2}\left|A^n_{x_{i+\frac{1}{2},j}}\right|(\Delta U_{i+1,j} - \Delta U_{i,j}) \tag{8.32b}$$

By substituting diffusive and convective approximations into equation (8.31a), the final discretized equation can be written as

$$a_{i,j}\Delta U_{i,j} = a_{i+1,j}\Delta U_{i+1,j} + a_{i-1,j}\Delta U_{i-1,j} + a_{i,j+1}\Delta U_{i,j+1} + a_{i,j-1}\Delta U_{i,j-1} + R^n \tag{8.33a}$$

where

$$a_{i+1,j} = \frac{\frac{1}{2}\left(\left|A^n_{x_{i+\frac{1}{2},j}}\right| - A^n_{x_{i+\frac{1}{2},j}}\right)}{x_{i+\frac{1}{2}} - x_{i-\frac{1}{2}}} + \frac{B_1}{(x_{i+1} - x_i)\left(x_{i+\frac{1}{2}} - x_{i-\frac{1}{2}}\right)} \tag{8.33b}$$

$$a_{i-1,j} = \frac{\frac{1}{2}\left(\left|A^n_{x_{i-\frac{1}{2},j}}\right| + A^n_{x_{i-\frac{1}{2},j}}\right)}{x_{i+\frac{1}{2}} - x_{i-\frac{1}{2}}} + \frac{B_1}{(x_i - x_{i-1})\left(x_{i+\frac{1}{2}} - x_{i-\frac{1}{2}}\right)} \tag{8.33c}$$

$$a_{i,j+1} = \frac{\frac{1}{2}\left(\left|A^n_{y_{i,j+\frac{1}{2}}}\right| - A^n_{y_{i,j+\frac{1}{2}}}\right)}{y_{j+\frac{1}{2}} - y_{j-\frac{1}{2}}} + \frac{B_1}{(y_{j+1} - y_j)\left(y_{j+\frac{1}{2}} - y_{j-\frac{1}{2}}\right)} \tag{8.33d}$$

$$a_{i,j-1} = \frac{\frac{1}{2}\left(\left|A^n_{y_{i,j-\frac{1}{2}}}\right| + A^n_{y_{i,j-\frac{1}{2}}}\right)}{y_{j+\frac{1}{2}} - y_{j-\frac{1}{2}}} + \frac{B_1}{(y_j - y_{j-1})\left(y_{j+\frac{1}{2}} - y_{j-\frac{1}{2}}\right)} \tag{8.33e}$$

$$a_{i,j} = a_{i+1,j} + a_{i-1,j} + a_{i,j+1} + a_{i,j-1} + q \tag{8.33f}$$

$$q = \frac{A^n_{x_{i+\frac{1}{2},j}}}{x_{i+\frac{1}{2}} - x_{i-\frac{1}{2}}} - \frac{A^n_{x_{i-\frac{1}{2},j}}}{x_{i+\frac{1}{2}} - x_{i-\frac{1}{2}}} + \frac{A^n_{y_{i,j+\frac{1}{2}}}}{y_{j+\frac{1}{2}} - y_{j-\frac{1}{2}}} - \frac{A^n_{y_{i,j-\frac{1}{2}}}}{y_{j+\frac{1}{2}} - y_{j-\frac{1}{2}}} \tag{8.33g}$$

where R^n is the right-hand side defined in equation (8.31a).

As the solution converges to satisfy equation (8.29a), or $R = 0$ in equation (8.31a), the solution of $\Delta U's$ will return a zero-value field. This implies that the exact form of the matrix coefficients a's is not that relevant to the final solution and hence the main consideration here is to modify a's such that a numerically stable approach to the final solution is ensured. Here, the following measures are recommended to modify a's in order to enhance numerical stability. First, the contribution of q in $a_{i,j}$ is dropped to maintain the diagonal dominance of the matrix coefficients. It should be noted that by so doing, the coefficients in equation (8.33b) through (8.33g) are all positive, and the dropping of q can ensure an unconditional stable system for the linearized equations. Second, to further enhance numerical stability of the equations, the positive contribution of source terms is recovered to the diagonal contribution as follows:

$$a_{i,j} = \begin{bmatrix} a^{\phi}_{i,j} \\ a^{n_e}_{i,j} + \beta n_+ \\ a^{n_+}_{i,j} + \beta n_e \end{bmatrix}. \tag{8.34a}$$

Finally, the definition of A's and B_1 in equation (8.30b) is the following:

$$A_x = \Lambda\left[0, \quad -\mu_e\tilde{E}_{e,x}, \quad \mu_e\tilde{E}_{+,x}\right], A_y = \Lambda\left[0, \quad -\mu_e\tilde{E}_{e,y}, \quad \mu_e\tilde{E}_{+,y}\right], \text{ and}$$

$$B_1 = \Lambda\left[1, \frac{d_e}{1+\beta_e^2}, \frac{d_e}{1+\beta_+^2}\right]. \tag{8.34b}$$

It should be noted that we purposely replaced the ion mobility and diffusivity coefficients, μ and d, by the corresponding electron counterparts. This replacement may seem odd because the transport coefficients for the electron are at least two orders of magnitude larger than those of the ion, while the mass of the electron particles is at least 2000 times less than that of the ion particles. However, one may find that both the electrons' and the ions' transport equations share the same source terms, which is solely in proportion to the magnitude of the electron flux density,

$|\vec{\Gamma}_e|$. As a result, this stability measure can be viewed as a numerical underrelaxation to force the convergence rate of the ion equation to synchronize to that of the electron equation.

The solution of equation (8.31a) can be performed by a successive line relaxation method in the x- and y-directions, sequentially:

$$a_{i,j}\Delta U_{i,j}^{**} + a_{i+1,j}\Delta U_{i+1,j}^{**} + a_{i-1,j}\Delta U_{i-1,j}^{**} = a_{i,j+1}\Delta U_{i,j+1}^{*} + a_{i,j-1}\Delta U_{i,j-1}^{*} + R^n,$$
$$a_{i,j}\Delta U_{i,j}^{***} + a_{i+1,j}\Delta U_{i,j+1}^{***} + a_{i,j-1}\Delta U_{i,j-1}^{***} = a_{i+1,j}\Delta U_{i+1,j}^{**} + a_{i-1,j}\Delta U_{i-1,j}^{**} + R^n \tag{8.35}$$

For the present iterative solution procedure, the convergence criterion is established on the balance of the electric current density on the cathode and anode. This is a physical measure for convergence because in order to satisfy the overall conservation of the current flow, the local conservation must be satisfied first. Since this physically meaningful property varies a great deal from different electrode configurations, the duplication of at least five digits of its mantissa is enforced.

8.6 INNOVATIVE NUMERICAL PROCEDURES

The most recent approach implements the total variation diminishing (TVD) discretization by an implicit time accurate marching scheme coupled with a multigrid method, and the implicit boundary conditions on electrodes (Shang et al., 2010, 2011, 2014). The numerical procedure attempts to establish a computationally stable and efficient method for simulating the surface DBD plasma actuator. First, the two-dimensional charged species conservation law, equation (8.28), is discretized over the control volume as shown in the following for the variable n_+:

$$\begin{aligned}\frac{\partial n_+}{\partial t} &= -\left[\frac{(n_+E_x)_{(i+1/2,j)} - (n_+E_x)_{(i-1/2,j)}}{\Delta x_{(i,j)}} + \frac{(n_+E_y)_{(i,j+1/2)} - (n_+E_y)_{(i,j-1/2)}}{\Delta y_{(i,j)}}\right] \\ &\quad + \frac{1}{\Delta x_{(i,j)}}\left[\left(d_+\frac{\partial n_+}{\partial x}\right)_{(i+1/2,j)} - \left(d_+\frac{\partial n_+}{\partial x}\right)_{(i-1/2,j)}\right] \\ &\quad + \frac{1}{\Delta y_{(i,j)}}\left[\left(d_+\frac{\partial n_+}{\partial y}\right)_{(i,j+1/2)} - \left(d_+\frac{\partial n_+}{\partial y}\right)_{(i,j-1/2)}\right] + s_+ - sn_+\end{aligned} \tag{8.36a}$$

In this formulation, the diffusive terms are treated with the central differencing and the source terms are separated in two parts; the term S_i does not explicitly contain the pertaining dependent variable n_i and the nonlinear term S is a product of collision process with other charged species. The typical convective flux on the control surface $(n_+E_x)_{(i+1/2,j)}$ of equation (8.36a) can be expressed as

$$
\begin{aligned}
(n_+E_x)_{(i+1/2,j)} = {} & (E_x)_{(i+1/2,j)} \left[(n_+)_{(i,j)} + (n_+)_{(i+1,j)}\right]/2 + \left|(E_x)_{(i+1/2,j)}\right| \\
& \left[(\Delta\tilde{n}_+)_{(i,j)} + (\Delta\tilde{n}_+)_{(i+1,j)}\right]/4 \\
& - \left|(E_x)_{(i+1/2,j)} + \left|(E_x)_{(i+1/2,j)}/2\right| \left[(\Delta\tilde{n}_+)_{(i+1,j)} - (\Delta\tilde{n}_+)_{(i,j)}\right] / \right. \\
& \left. \left[(n_+)_{(i+1,j)} - (n_+)_{(i,j)}\right]\right| \left[(n_+)_{(i+1,j)} - (n_+)_{(i,j)}\right]/2
\end{aligned} \tag{8.36b}
$$

The TVD limiter by Limiter module (MINMOD) scheme is adopted as (Yee, 1986)

$$\Delta\tilde{n}_{(i,j)} = \tilde{s}_{(i,j)} \max[0, \min(|\Delta n_{(i+1,j)}|, \tilde{s}_{(i,j)}\Delta n_{(i-1,j)}] \tag{8.36c}$$

where $\Delta n_{(i+1,j)} = n_{(i+1,j)} - n_{(i,j)}$ and $\tilde{s}_{(i,j)} = sign(\Delta n_{(i+1,j)})$.

The discretized charged species conservation equations are expressible by a simple algebraic equation of the following form:

$$
\begin{aligned}
\frac{\partial n_{+,(i,j)}}{\partial t} = {} & a_{E,(i,j)}n_{+(i+1,j)} + a_{W,(i,j)}n_{+(i-1,j)} + a_{N,(i,j+1)}n_{+(i,j+1)} + a_{S,(i,j)}n_{+(i,j-1)} \\
& - \left(a_{P,(i,j)} + s_{(i,j)}\right) n_{+(i,j)} + s_{+,(i,j)}
\end{aligned} \tag{8.36d}
$$

In this equation, the subscripts E, W, N, and S denote discretized grid-point locations of the multigrid structure beyond the current evaluated locations (Brandt, 1977).

One more step taken in this approach is to use the Delta formulation via the iteratively diminishing residual approach. By setting $\Delta n_+ = n_+^n - n_+^{n-1}$, where n is the current iteration level and n- is the previous iteration level, the discretized equations system can be summarized as

$$
\begin{aligned}
\frac{\partial \Delta n_{+,(i,j)}}{\partial t} = {} & a_{E,(i,j)}\Delta n_{+(i+1,j)} + a_{W,(i,j)}\Delta n_{+(i-1,j)} + a_{N,(i,j+1)}\Delta n_{+(i,j+1)} + a_{S,(i,j)}\Delta n_{+(i,j-1)} \\
& - \left(a_{P,(i,j)} + s_{(i,j)}\right) \Delta n_{+(i,j)} + RHS^*_{+,(i,j)}
\end{aligned} \tag{8.36e}
$$

In this equation, the notation $RHS^*_{+,(i,j)}$ is an abbreviation of the right-hand side of equation (8.36a), according to the diminishing residual delta formulation, and is evaluated at the previous iterative level $n - 1$.

A third-order TVD type of Runge–Kutta algorithm for time advancement is applied by the consecutive implicit operators. It is actually the complete time-dependent governing equation, according to the diminishing residual delta formulation to be evaluated at the previous iterative level $n - 1$; $[\partial U/\partial t + \nabla \cdot F(u) = S]^{n-1}$. The left-hand side of the discretized equation consists mainly of the nonlinear convective term; thus, the only requirement is to maintain computational stability until the residual of the discretized equation has reached a predetermined magnitude. In general, the preferred numerical procedures are the TVD and the weighted essentially nonoscillatory schemes (Shu et al., 1989; Balsara et al., 2000; Borges et al., 2008). For the TVD approach, all scalar components of the vectors can be discretized using the central differencing scheme. In the Delta formulation, the equation yields

$$(a\Delta U)_{(i,j)} = (a\Delta U)_{(i+1,j)} + (a\Delta U)_{(i-1,j)} + (a\Delta U)_{(i,j+1)} + (a\Delta U)_{(i,j-1)} + RES^{*}_{(i,j)} \tag{8.37}$$

The TVD discretization is implemented by an implicit time accurate marching scheme coupled with a multigrid method and the implicit boundary conditions on interface. The numerical procedure establishes a computationally stable and efficient method for simulating the surface DBD plasma actuator. Here, RES is the residual value of equation (8.13), namely the Poisson equation of plasma dynamics. Both equations (8.36e) and (8.37) require a very efficient implicit solver. The Strong Implicit Procedure (SIP) or the Implicit Lower and Upper decomposition scheme by Stone (1968), or a modified version of it, MSIP (Schneider and Zedan, 1981), is adopted. The algorithm is very efficient for solving a sparse matrix equation system; $[A]U = R$. Here, $[A]$ is the 5-point stencil matrix whose coefficients are made up from $a's$ shown in equations (8.36a) and (8.37). By adding a complementary matrix $[C]$ to the $[A]$, the resulting matrix *[M]* can be solved implicitly using the lower and upper matrix decomposition

$$[M] = [A] + [C] \tag{8.38a}$$

The identical implicit, iterative solving procedure is then

$$[M]U^{n} = [C]U^{n-1} + R \tag{8.38b}$$

Stone further introduced a partial cancellation term in both sides of the equation (8.38b) to speed up the rate of convergence.

For the surface discharge phenomenon, a steep gradient always occurs near the corner of the exposed electrodes and dielectric surface; thus, a high nonuniform mesh is generally recommended to solve the equations. Because of the large number of highly stretching grid points used to cover the whole computational domain, the multigrid method offers the best solving option. The multigrid method renders the error of a rapidly varying Fourier component to be first eliminated by the fine mesh system. The slowly decaying error associated with the spectral radius becomes a smooth function. Through this implementation, an amazingly convergent acceleration is observed, but the most important improvement to the numerical simulation is the computational stability achieved by the combination of semi-infinte programming (SIP) (or MSIP; Schneider and Zedan, 1981), the implicit third-order Runge–Kutta scheme, and TVD spatial discretization, which contributes greatly to the accuracy of the numerical solution.

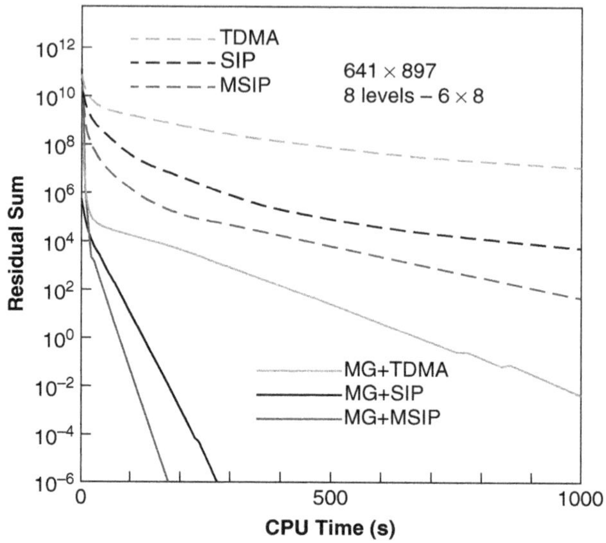

FIGURE 8.12 Convergence acceleration for Poisson equation of plasma dynamics.

In Figure 8.12, a comparative study of the implicit solving schemes for the Poisson equation of plasma dynamics is depicted on a highly stretched (641×891) grid. The remarkable convergent acceleration from the conventional tridiagonal line relaxation time division multiple access (TDMA) to SIP and finally the combination of SIP and the multigrid procedure is clearly illustrated. The acceleration in residual reduction by the SIP algorithm alone exceeds four orders of magnitude. The rate of convergence is further vastly improved by the combined application of the SIP and multigrid techniques: although the computational stability is not easily quantified, its impact is impressive. Through these superior computational attributes, the grid independent solutions are truly realizable.

Equation (8.13) shows that a force vector must follow the direction of the electric field vector; it is useful to study how the electric field intensity is modified by the separated free-space charges, namely the net balance of the charged number density. If one neglects effects of the separated charge density, the solution of the Poisson equation of plasma dynamics is dictated only by the instantaneous normal electric potential field generated by an externally applied AC field. It is actually a solution to the Laplace equation for electric potential as shown by the subsets (a-1) and (b-1) in Figure 8.13. The contour and vector plot in the subset (a-1) are the dimensionless electric potential φ and electric field $\vec{E}$, respectively. The contours of the magnitude of the electric field are displayed in the subset (b-1). The subsets (a-2), (b-2) and (a-3) and (b-3) in this figure display the same dimensionless quantities when the EMF of a 5-kHz AC cycles are in its maximum and minimum values of 8.0 kV. From this figure, the separated charge number density has only a small effect to the solution of the Poisson equation of plasma dynamics. The instantaneous force field is essentially acting along an undisturbed externally applied electric field potential, with alternating directions when the polarity on the exposed electrode changes sign.

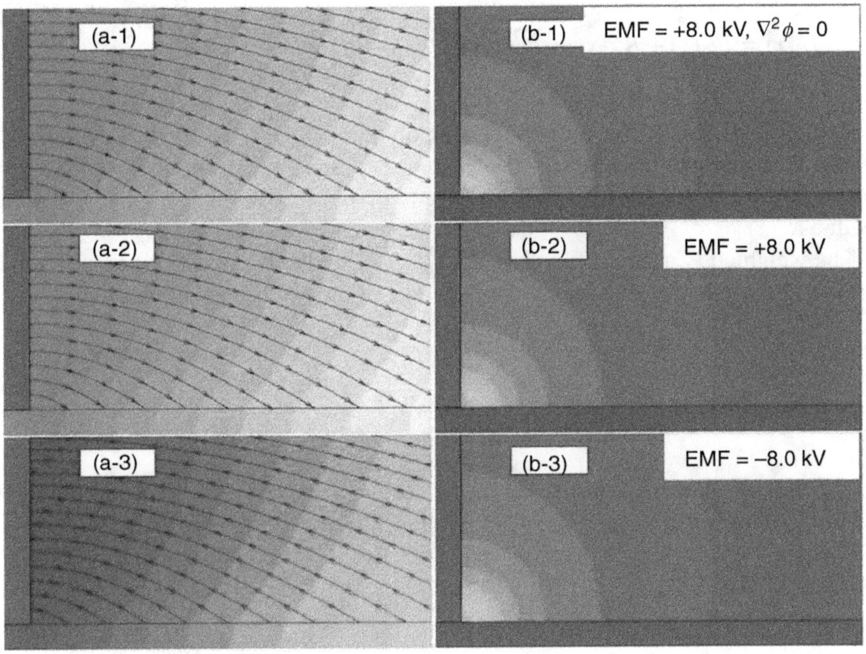

FIGURE 8.13 Solutions of Poisson equation of electromagnetics.

The high-resolution solution of the delta formulation via interactively diminishing residue is preordained by the convergent tolerance, and the accuracy of the left-hand side of the discretized equation is ultimately controlled by the high-resolution algorithm. Two numerical procedures appear very attractive. One is the polynomial grid refinement, which is derived from the Gauss quadrature formulation (Gautsche, 2004). The entire computational domain is divided into global grid blocks, and the local high resolution is achieved within each grid block on the unequal-spacing roots of an orthogonal polynomial. The early effort by Korpriva et al. exemplifies this new approach in achieving the high local numerical resolution (Korpriva and Kolias, 1996; Shang et al., 2010). The local grid refinement approach has also been extended for solving the conservation laws on unstructured grids. For discontinuity capturing, an artificial dissipative term is still required within the subgrid domain for suppressing numerical oscillations, and as a consequence it degrades numerical accuracy.

In applying the polynomial refinement for high local numerical resolution, the first step is to approximate the dependent variable, $U(x)$, in the global grid by any orthogonal polynomials through the Gauss quadrature formula via an independent variable transformation $x = [2z - (a+b)]/(a+b)$. The integral limits a and b characterize the interval between control surfaces within a global grid:

$$\int_a^b U(\eta)d\eta = \frac{2}{n(n-1)}[U(a) + U(b)] + \sum_{i=1}^{n-1} L_i U_i(\eta) + R_n \qquad (8.39a)$$

$$U(x) = P_n(x) + R_n(x) = \sum_{i=1}^{n} L_i(x)U(x_i) + \left[\prod_{i=1}^{n}(x - x_i)\right] \frac{U^{n+1}(\xi)}{(n+1)!} \tag{8.39b}$$

The approximation by an orthogonal polynomial $P_n(x)$ within the global grid is able to generate a spectral-like resolution, and the numerical accuracy is uniquely determined by the chosen degrees of the ploynomial, as shown by the residue $R_n(x)$.

The weighted (Cardinal) function, $L_i(x)$, given by the Newton divided-difference formula, is defined on all unequal spaced roots of the orthogonal polynomial,

$$L_i(x) = \prod_{\substack{j=0 \\ j \neq i}}^{n} \frac{(x - x_j)}{(x_i - x_j)} \tag{8.39c}$$

The derivative of any approximation functions for either spatial or temporal independent variable can be computed by simply differentiating the Cardinal function $L_i(x)$.

$$\frac{dF(x_i)}{dx} = \sum_{n} \left[\frac{dL_i(x)}{dx}\right] f(x_n) = \sum_{i=1}^{l} \frac{\sum\limits_{q=1,q\neq i}^{l} \left[\sum\limits_{m=1,m\neq i,m\neq q}^{l} (x_p - x_t)\right]}{\prod\limits_{m=1,m\neq i}^{l} (x_i - x_m)} F(x_n). \tag{8.40a}$$

It is immediately recognized that the above formulation provides the recursive relationship for the first-order derivative computation through the Gauss quadrature and can be expressed by an $n \times n$ matrix as follows (Shang et al., 2010):

$$\frac{dF(x_i)}{dx} = \begin{bmatrix} a_{1,1}(x_1) & a_{1,2}(x_2) & a_{1,3}(x_3) & .. & a_{1,n-1}(x_{n-1}) & a_{1,n}(x_n) \\ a_{2,1}(x_1) & a_{2,2}(x_2) & a_{2,3}(x_3) & .. & a_{2,n-1}(x_{n-1}) & a_{2,n}(x_n) \\ .. & .. & .. & .. & .. & .. \\ a_{n-1,1}(x_1) & a_{n-1,2}(x_2) & a_{n-1,3}(x_3) & .. & a_{n-1,n-1}(x_{n-1}) & a_{n-1,n}(x_n) \\ a_{n,1}(x_1) & a_{n,2}(x_2) & a_{n,2}(x_3) & .. & a_{n,n-1}(x_{n-1}) & a_{n,n}(x_n) \end{bmatrix} \times \begin{bmatrix} F(x_1) \\ F(x_2) \\ . \\ F(x_{n-1}) \\ F(x_n) \end{bmatrix} \tag{8.40b}$$

This approximation is restricted to generate the first derivative of the approximated function, because all hypogeometric differential equations that are associated with the orthogonal polynomials are second order. When attempts are made to generate the second derivative from the weighted function, the eigenvalue structure often leads to oscillatory behavior. For this reason, the second spatial derivative in the present analysis is obtained by consecutive approximations using the calculated first derivative.

A unique feature of the local polynomial refinement approach is that it is independent from the global mesh system. There is no need to reconstruct the overall grid system for local refinement to capture the fine-structure features: but just increasing the degrees of polynomials within the grid block is sufficient (Shang et al., 2010; Barnes, 2011). The numerical procedure is equally applicable to the temporal advancement of a dynamic problem. In some cases, it may even be possible to evaluate integrals in which the integrand has discontinuity within the integral interval. The singularity can be relegated to the weighting function. To apply the approximate Riemann formulation on the control surface, the Gauss–Lobatto quadrature offers a distinct advantage. By this formulation, the integral limits of the Gauss quadrature become two of the root points and the quadrature now involves the weighted sum of $n+1$ functional values.

In essence, the above equation gives the recursive relationship for the first derivative computation. A unique behavior of the recursive formula is that the calculated derivative depends on all discretized points within the global grid. In fact, all high-resolution schemes strive to achieve a spectral-like accuracy by employing all discretized points in an array to mimic the Gauss-quadrature formulation. It would be interesting to understand how a steep gradient is resolved by the Gauss quadrature formula as it is shown in Figure 8.14. The sudden change of the dependent variable is prescribed near the coordinate origin by a jump of a dependent variable from the ambient condition of 300 to the coordinate origin of 1500. In the present numerical experiment, a decreasing integral interval from 0.02 to 0.000625 cm was specified. A 50° Legendre polynomial was used for all the calculations by the Gauss–Lobatto formulation. At the greatest integral interval of 0.02 cm, the numerical resolution is insufficient even for the Gauss–Lobatto quadrature, which is reflected by oscillation in the numerical result along the higher-temperature limit. The steep gradient is captured by a few roots near the lower-temperature boundary value. As the interval is reduced, an increasing number of roots are distributed to connect the jump across the different boundary values. At the finest integral interval, nearly all roots of the Legendre polynomial are distributed over the vertical slope. This behavior is identical whether increasing the degrees of the orthogonal polynomial or decreasing the integral interval of the numerical jump.

Figure 8.15 depicts the state-of-the-art, high-resolution numerical algorithms applying to the one-dimensional model of shock-density wave interaction (Barnes, 2011). The numerical result is conducted at the lower limit of a hypersonic Mach number of 3. By using only a fourth-degree Gauss–Lobatto polynomial, the numerical result is capable of reproducing the benchmark solution, the fine scale density oscillation of the shock-entropy wave interaction. However, a diffusion switch has to be activated at appropriate cells at each time step. It is clear even from the most recent innovations that the fundamental challenge remains for hypersonic flow simulation: Namely, all the high numerical resolution schemes still incur unavoidable degradation at the discontinuity of the solution space.

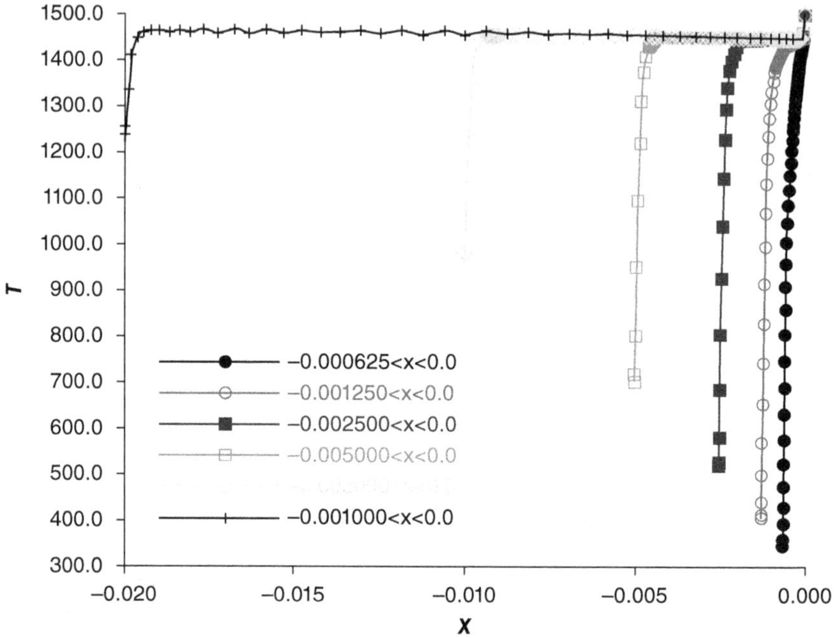

FIGURE 8.14 Gauss–Lobatto quadrature across a finite jump.

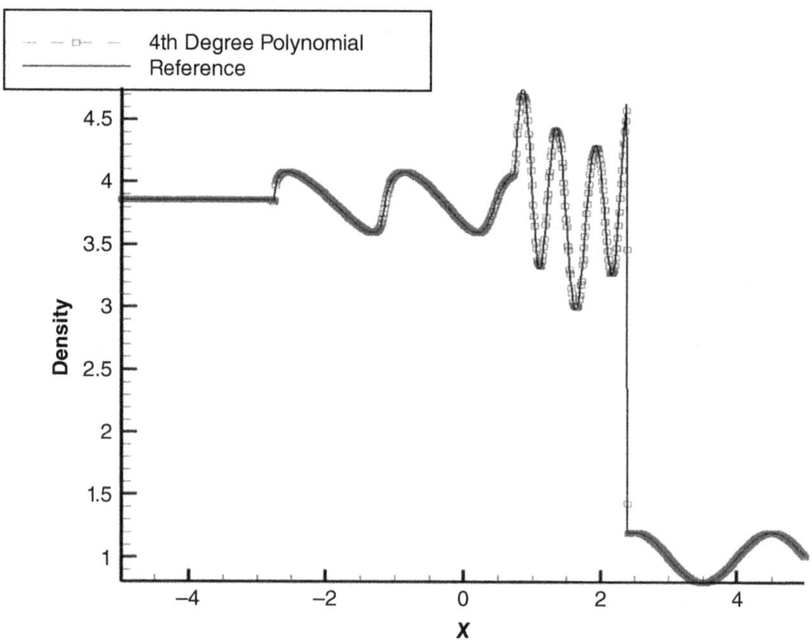

FIGURE 8.15 Local polynomial refinement solution to the shock acoustic wave.

Another high-resolution algorithm that is suitable for solving the left-hand side of the discretized equation is the compacting differencing approximation. Lele has pointed out using the formal order of accuracy of truncation errors exclusively to select an algorithm for time-dependent simulation is an oversimplification. The desired feature of a numerical scheme may be derived from optimization in Fourier space rather than by focusing on the lowest possible truncation error. In light of this line of reasoning, compact differencing is a viable method by which to achieve high resolution. The basic approach provides an algorithm that has a small stencil dimension and yet maintains a lower level of dispersive and dissipative error than the conventional numerical schemes. The basic idea of the compact difference approximation is based on Hermite's generalization of the Taylor series.

If the solving procedure is limited to the six-order accuracy, the compact differencing formulation is reducible to a simple tridiagonal formulation for evaluating the first and second derivatives evaluating of the coefficients of equation (4.32) of Chapter 4:

$$\alpha\left(\frac{dU}{dx}\right)_{i-1}+\left(\frac{dU}{dx}\right)_{i-1}+\alpha\left(\frac{dU}{dx}\right)_{i-1}=a\frac{U_{i+1}-U_{i-1}}{2\Delta x} \tag{8.37a}$$

$$\alpha\left(\frac{d^2U}{dx^2}\right)_{i-1}+\left(\frac{d^2U}{dx^2}\right)_{i-1}+\alpha\left(\frac{d^2U}{dx^2}\right)_{i-1}=a\frac{U_{i+1}-2U_i+U_{i-1}}{\Delta x^2} \tag{8.37b}$$

The basic formulation of compact differencing is an implicit procedure for evaluating the derivatives of the dependent variables. The stencil structure for most compact-difference formulas is spatially central and requires additional derivative values at boundaries. The spatially central scheme is inapplicable to the immediately adjacent mesh point at a boundary. An additional transitional numerical scheme from the boundary to the interior domain is required. The transitional boundary scheme requires not only transmitting data from the boundary but also preserving the stability and accuracy of the interior domain for global resolution. This unique feature of compact-difference schemes also is one of the sources of spurious high-frequency oscillations.

However, this requirement is alleviated by the tridiagonal formulation: instead, the derivatives at the boundary are provided by a one-sided derivative as the boundary condition for the matrix inversion procedure. For the fourth-order approximation, these formulations have been given in equations (4.49). Nevertheless, the slight difference in the leading truncation term between the spatially central and one-sided approximation can trigger the time-delay instability. This shortcoming has been successfully overcome by the low-pass filter in suppressing the irresolvable Fourier components (Gaitonde et al., 1999).

For simulating DCDs, there are three high-gradient regions that require high numerical resolution for simulating DCD, and they are located immediately adjacent to electrodes. Two of these regions are within the cathode and anode layers immediately over the electrodes. The gradient of electric potential in the cathode layer is particularly pronounced in the cathode fall, indicated by the drop of the electric potential frequently constituting half of the total voltage across the gap of the electrodes. The most severe numerical challenge, however, resides at the sharp edges of the electrode. At this location, the discharge behaves more like a corona, where the electric field distribution is highly nonuniform. The local electric field intensity distribution also exhibits a singular behavior on both the cathode and the anode. For numerical simulations, the highly clustering mesh space without numerical smoothing is recommended, and the grid point distribution needs to cluster closely to the sharp edge of the electrode. Specifically, the artificial dissipation or the slope limiter commonly associated with the approximate Riemann method for treating a piece-wise discontinuity shall be avoided to introduce excessive artificial diffusion.

Figure 8.16 presents a numerical resolution study of the discharge arrangement between two side-by-side electrodes. The discharge is maintained at an *EMF* of 2.0 kV in the electrical circuit under the ambient pressure of 5 torr. The discharge breakdown occurs within the gap of 2 cm. Two grid systems of (160 × 120) and (120 × 60) are used, and the finest mesh spacing used in the coordinate normal to electrode is $\Delta y_{min} = 1.67 \times 10^{-3}$ cm. In this numerical simulation, the cathode is placed on the left and has a separation distance of 1.0 cm from the anode. Both

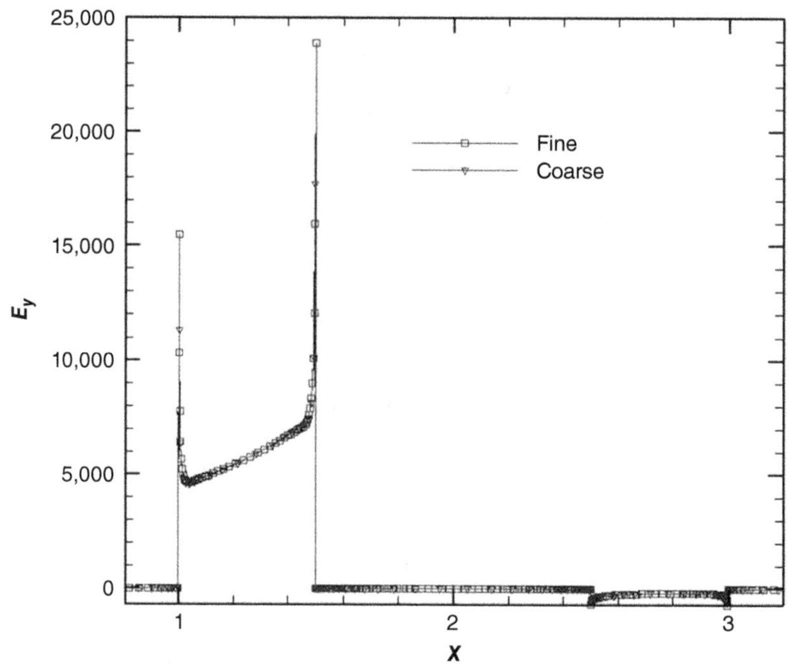

FIGURE 8.16 Mesh refinement near the edge of electrodes, EMF = 2.0 kV, gap distance of 2.0 cm.

electrodes have the width of 0.5 cm and are subjected to an EMF of 2.0 kV and under an ambient pressure of 5 torr. The extremely high gradient regions exist at the edge of electrodes abridging the perfect electrically conducting and the dielectric surfaces of a side-by-side discharge. A grid density clustering is required locally to resolve the sharp jump while maintaining the computational stability. The computed normal components of the electric field intensity from the coarse and fine mesh systems essentially duplicate each other, except the peak values on the edges of the cathode. The discrepancy between two numerical results over the electrodes is less than 0.1%. It is important to note that the numerical oscillations near the piecewise discontinuity are conspicuously absent and the numerical results by the two grid system are nearly identical. Therefore, it is ascertained that a grid independent numerical simulation for the diffuse DCD is achieved.

As a convincing illustration for the needed high-resolution procedure for plasma-based actuator simulations, the effects on an externally applied transverse magnetic field are displayed by the previous figure. The side-by-side electrode configuration is exactly the same as depicted in Figure 8.17, with the identical EMF but subject to two externally applied transverse magnetic fields of opposite polarities. First, the range number density flux vector of electrons in a DCD spans a range from 0 to 10^{+18}. The number density is peaked at the outer edge of the cathode and appears as a sharp

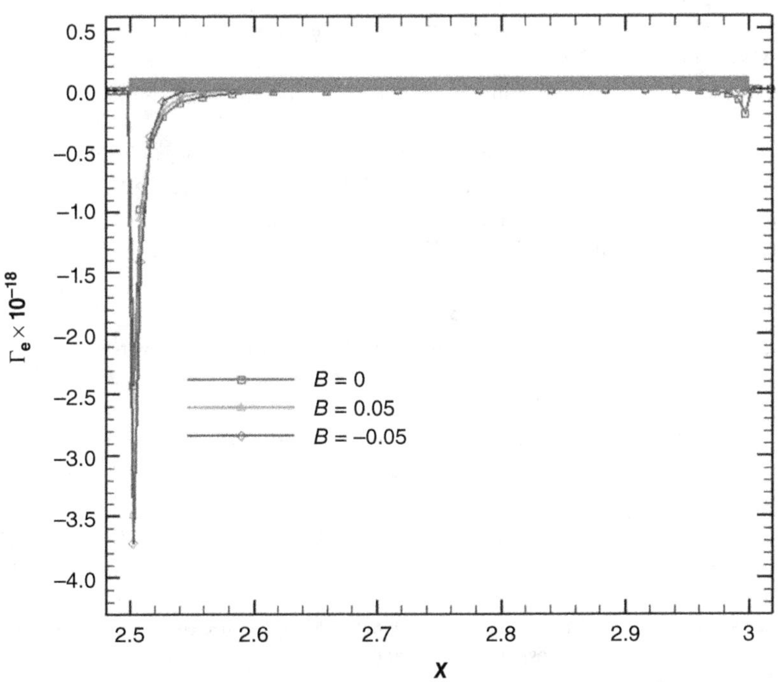

FIGURE 8.17 Number density flux under externally applied transverse magnetic field.

spike with the dimension less than 0.7% of the electrode surface. The edge effect is much less pronounced away from the anode. The externally applied magnetic flux density intensity exhibits a sudden jump especially when the negative magnetic field is applied, but both magnetic fields increase the higher number density of electrons at the outer edge of the electrode. The supporting experimental observation is obtainable only by the concentrated glow at the electrode edge; even quantification of the electric field intensity is unavailable, but the effect is consistent with the well-known sharp point discharging phenomenon.

REFERENCES

Balsara, D., and Shu, C.W., Monotonicity preserving weighted essentially non-oscillatory schemes with increasingly high order of accuracy, *J. Comp. Phys.*, vol. 160, 2000, pp. 405–452.

Barnes, C., *An Implicit High-Order Spectral Difference Method for the Compressible Navier–Stokes Equations Using Adaptive Polynomial Refinement*, Master Thesis, Wright State, September 2011.

Boeuf, J.P., and Pitchford, L.C., Electrohydrodynamic force and aerodynamic flow acceleration in surface dielectric barrier discharge, *J. Appl. Phys.*, vol. 97, 2005, pp. 103307-1-10.

Borges, R., Carmona, M., Costa, B., and Don, W.S., An improved weighted essentially non-oscillatory scheme for hyperbolic conservation laws, *J. Comp. Phys.*, vol. 227, 2008, pp. 3191–3211.

Brandt, A., Multilevel adaptive solutions to boundary value problem, *Math. Comp.*, vol. 31, 1977, pp. 333–390.

Capitelli, M., Gorse, C., Longo, S., and Giordano, D., Collision integrals of high-temperature air species, *J. Thermophys. Heat Transfer.*, vol. 14, no. 2, 2000, pp. 259–268.

Elisson, B., and Kogelschatz, U., Nonequilibrium volume plasma chemical processing, *IEEE Trans. Plasma Sci.*, vol. 19, 1991, pp. 1063–1077.

El-Bahy, M.M., A numerical modelling of microdischarge threshold in uniform electric fields, *J. Phys. D: Appl. Phys.*, vol. 38, 2005, pp. 103–112.

Enloe, C.L., Mcharg, M.G., and McLaughlin, T.E., Time-corrected force production measurements of the dielectric barrier discharge plasma aerodynamic actuator, *J. App. Phys.*, vol. 103, 2008, 073302.

Enloe, C.L., McLaughlin, T.J., Vandyken, R.D., Kachner, K.D., Jumper, E.J., Corke, T.C., Post, M., and Haddad, O., Mechanisms and responses of a single dielectric barrier plasma actuator: plasma morphology, *AIAA J.* vol. 42, 2004, pp. 595–604.

Gaitonde, D.V., Shang, J.S., and Young, J.L., Practical aspects of higher-order numerical schemes for wave propagation phenomena, *Int. J. Numer. Methods Eng.*, vol. 45, 1999, pp. 1849–1869.

Gautsche, W., *Orthogonal Polynomials, Computational and Approximation*, Science Publications, Oxford University Press, Oxford, 2004.

Gibalov, V.I., and Pietsch, G.J., Dynamics of electric barrier discharges in coplanar arrangement, *J. Phys. D: Appl. Phys.*, vol. 37, 2004, pp. 2082–2092.

Graves, D.B., and Jensen, K.E., A continuum model of DC and RF discharges, *IEEE Trans. Plasma Sci.*, vol. PS-14, no. 2, 1986, pp. 78–91.

Kimmel, R.L., Hayes, J.L., Menart, J.A., and Shang, J., Effect of magnetic fields on surface plasma discharges at Mach 5, *J. Spacecraft Rockets*, vol. 42, no. 6, 2006, pp. 1340–1346.

Korpriva, D., and Kolias, J., A conservative staggered-grid Chebyshev multidomain method for compressible flows, *J. Comp. Phys.*, vol. 125, 1996, pp. 244–261.

Levin, E., and Wright, M.J., Collision integrals for ion–neutral interactions of nitrogen and oxygen, *J. Thermophys. Heat Transfer*, vol. 18, no. 1, 2004, pp. 143–147.

Likhanskii, A.V., Shneider, M.N., Macheret, S.O., and Miles, R.B., Modeling of dielectric barrier discharge plasma actuator in air, *J. Appl. Phys.*, vol. 103, 2008, pp. 055505-1-13.

Macheret, S.O., Shneider, M.N., and Miles, R.B., Magnetohydrodynamic control of hypersonic flows and scramjet inlets using electron beam ionization, *AIAA J.*, vol. 40, 2001, pp. 74–81.

Raizer, Yu.P., and Surzhikov, S.T., Diffusion of charges along current and effective numerical method of eliminating of numerical dissipation at calculations of glow discharge, *High Temp.*, vol. 28, no. 3, 1990, pp. 324–328.

Roth, J.R., Sherman, D.M., and Wilkinson, S.P., Electrohydrodynamic flow control with a glow-discharge surface plasma, *AIAA J.*, vol. 37, 2000, pp. 1166–1172.

Scharfetter, D.L., and Gummel, H.K., Large signal analysis of a silicon read diode, *IEEE Trans. Electron. Dev.*, vol. 16, 1969, pp. 64–77.

Schneider, G.E., and Zedan, M., A modified strongly implicit procedure for the numerical solution of field problems, *Numer. Heat Transfer.*, vol. 4, 1981, pp. 1–19.

Shang, J.S., and Huang, P.G., A local resolution refinement algorithm using Gauss–Lobatto quadrature, The Sixth International Conference on Computational Fluid Dynamics, St. Petersburg, Russia, July 12–16, 2010, pp. 515–525.

Shang J.S., and Surzhikov, S.T., Magnetoaerodynamic actuator for hypersonic flow control, *AIAA J.*, vol. 43, no. 8, August 2005, pp. 1633–1643.

Shang, J.S., Kimmel, R., Menart, J., and Surzhikov, S.T., Hypersonic flow control using surface plasma actuator, *J. Propul. Power.*, vol. 24, no. 5, 2008, pp. 923–934.

Shang, J.S., Roveda, F., and Huang, P.G., Electrodynamic force of dielectric barrier discharge, *J. Appl. Phys.*, vol. 109, no. 11. 2011, pp. 113301-1-8.

Shang, J.S., and Huang, P.G., Surface plasma actuators modeling for flow control, progress. *In Aerospace Sci.*, vol. 67, 2014, pp. 29–50.

Shu, C.W., and Osher S., Implementation of essentially nonoscilatory shock capture scheme II, *J. Comp. Phys.*, vol. 83, 1989, pp. 32–78.

Solov'ev, V., Konchakov, A.M., Krivtsov, V.M., Aleksandrov, N.L., Numerical simulation of a surface barrier discharge in air, *Low-Temperature Plasma*, vol. 34, no. 7, 2008, pp. 594–608.

Stone, H.L., Iterative solution of implicit approximations of multidimensional partial differential equations, *SIAM J. Numer. Anal.*, vol. 5, no. 3, 1968, pp. 530–558.

Surzhikov, S.T., The drift-diffusion and quasi-neutral models of gas discharges for flow control problems, RTO-AVT-VKI Lecture Series, 2011.

Surzhikov, S.T., and Shang, J.S., Normal glow discharge in axial magnetic field, *Plasma Sour. Sci. Technol.*, vol. 23, 2014, pp. 054017, 1–8.

Surzhikov, S.T., and Shang, J.S., Two-component plasma model for two-dimensional glow discharge in magnetic field, *J. Comput. Phys.*, vol. 199, no. 2, September 2004, pp. 437–464.

Unfer, T., and Boeuf, J.P., Modeling of a nanosecond surface discharge actuator, *J. Phys. D: Appl. Phys.*, vol. 42, 2009, pp. 194017-1-12.

Yee, H.C., Linearized form of implicit TVD scheme for the multidimensional Euler and Navier–Stokes equations, *Comp. Math. Appl.*, vol. 12A, 1986, pp. 413–432.

9

JOULE-HEATING ACTUATORS

INTRODUCTION

In 1958, Resler and Sears first advocated the idea of incorporating electromagnetic force into aerodynamics for improving aerospace vehicle performance (Resler and Sears, 1958). Convincing experimental evidence was provided by Ziemer (1959) a year later. He demonstrated that stand-off distance of a bow shock wave over a hemispherical cylinder could be displaced upstream in ionized gas by an externally applied magnetic field. In fact, the magnetic pressure $B^2/2\mu_m$ constitutes an added pressure component that reduces the value of the gas density within the shock layer. To satisfy the continuity condition across the bow shock, the shock layer thickness increases and displaces the standoff distance outward. This is an irrefutable demonstration that the electromagnetic effect can alter ionized gas flow field for modifying aerodynamic performance of aerospace vehicles. A period of intensive basic research in magneto-aerodynamic ensued and ebbed for the next 30 years (Lieberman et al., 1994).

A rejuvenated interest in the middle 1990s to magneto-aerodynamic research was stimulated by the innovative A JAX concept vehicle from Russia (Gurijanov and Harsha, 1996). Many exciting innovations were put forth in areas of MHD propulsion, remote energy deposition for drag reduction, plasma actuator, radiation-driven hypersonic wind tunnel, sonic boom meditation, and enhanced plasma ignition and combustion stability (Bletzinger et al., 2005). A very promising innovation is the concept of an MHD-bypass scramjet engine (Fraishtadt et al., 1998). Extensive

Computational Electromagnetic-Aerodynamics, First Edition. Joseph J.S. Shang.

research, however, has revealed that additional and definitive evaluations of pertinent physics are necessary for modeling and analyzing this innovative concept to reach a conclusive assessment. From lessons learned, most recent research activities tend to refocus on basic and simpler aerodynamic–electromagnetic interactions.

It is logical to apply an electromagnetic force in for hypersonic flow modification, because, in a high-temperature and low-density environment, ionized species already exist in the gaseous medium. For this reason, surface plasma discharge has been widely researched as a hypersonic flow control mechanism (Shang et al., 2005). The electromagnetic effects generated by electron impact ionization can be summarized as Lorentz acceleration, electrostatic force, and plasma and electrode heating, all of which have a limited magnitude. The Lorentz acceleration is significant only in the presence of an externally applied magnetic field; because of the relative low current density, the resultant force is just a few kN/m^3. The electrostatic force is mostly detectable in the gap between electrodes where the space charge separation occurs. Under atmospheric conditions, a dielectric barrier discharge (DBD) generates an oscillatory force around one kN/m^3 (Beouf et al., 2007; Shang et al., 2010, 2011). In short, the electromagnetic forces by a plasma actuator are generally orders of magnitude lower than the aerodynamic inertia of typical hypersonic and transonic flows (Shang et al., 2005). The orientation of the electrostatic force of DCD and DBD in the absence of an applied magnetic field is mostly pointed downward or parallel to the electrodes. In a boundary layer, the downward force cannot be supported by shear stress and is transmitted directly to the solid surface. The gist is that electromagnetic effects under most known aerodynamic applications exert only a small perturbation to the mainstream; thus, it can be viable only for specific applications.

The thermal effects of DCD, corona, and arc discharge to fluid medium include plasma and electrode heating. The plasma heating consists mainly of Joule heating and stochastic heating; the latter however is nearly negligible except in an oscillating plasma sheath. The Joule heating concentrates within the cathode layer and is a rather small but effective spatial heat source to the surrounding gas stream. In the electron collision process for plasma generation, the electrode heating is an unavoidable consequence. This heat flux of electrode heating enters flow field by conductive process and generally has a magnitude around 10 W/m^2. In a laboratory environment, the Joule and electrode heating fluxes have a comparable order of magnitude but are still very much lower in comparison to energy content of hypersonic flows. And yet experimental and computational investigations have repeatedly and persistently shown perceptible effectiveness to shock structure (Bityurin et al., 1999), shock layer formation (Borghi et al., 2006), and electromagnetic–aerodynamic interactions (Shang et al., 2005; Cristofolini et al., 2008).

The renewed interest in plasma dynamics research is also based on observations that a heterogeneous discharge has a significant impact on the shock structure and wave propagation. Discharges have been applied to control the flow separation in supersonic flows and some interesting experimental results have been obtained for control of low-density hypersonic flows. Characterizations of electromagnetic–aerodynamic interactions for several basic configurations with an externally applied transverse magnetic field are supported by their substantiated experimental efforts.

The potential application is highlighted for adjusting the oblique shock system within an inlet at off-design operating conditions. On the contrary, a series of side-by-side computational and experimental investigations of DCDs was conducted in a hypersonic MHD flow channel (Shang et al., 2005). An externally applied transverse magnetic field has been included to enhance the flow control effectiveness. Meanwhile, DCD applications have been extended to axisymmetric configurations by Borghi et al. and Cristofolini et al. The transverse magnetic field also has been applied to mitigate stagnation point heat transfer in which the polarity of the externally applied transverse magnetic field is critical (Matsuda et al., 2008; Gulhan et al., 2009). An altered heat transfer rate has been detected, but undisputed and detailed descriptions of electromagnetic–aerodynamics interactions have been elusive from experimental measurements.

The interactions between aerodynamics and the surface electron impact discharge have been simulated by computational electromagnetic aerodynamics using a multifluid, drift-diffusion weakly ionized gas model. The investigated phenomena are mostly focused on DCDs over the sharp leading edges of a wedge and constant cross-sectional area, rectangular and cylindrical inlets. For these research activities, direct verifications with data or experimental observations are available. Some side-by-side experimental and computational studies are conducted with/without an externally applied magnetic field. A large number of numerical simulations were performed for hydrodynamic stability and the most challenging laminar–turbulent transition phenomena included in simplified ionized gas models. Many scientific findings are difficult to validate against the underlying physics for assessing physical fidelity. Thus, only computational results verifiable with available experimental observations are presented.

9.1 FEATURES OF DIRECT CURRENT DISCHARGE

The pioneering efforts in glow discharge research by Von Engel and Steenbeck (1932) and Raizer (1991) have established a solid foundation for physics of gas discharges. Detailed electrodynamic structure of a discharge column is clearly understood and has been experimentally verified. A uniform diffuse discharge between parallel electrodes without the edge effect is an idealized configuration. When the ambient pressure is lower than 1 torr, the discharge becomes so diffusive that the discharge column becomes almost a one-dimensional structure. When numerical simulations were conducted under a relatively high pressure condition, more than 1 torr, the discharge became constricted. The discharge column of an idealized discharge simulation is nonunique. In other words, the center of the discharge column can occur in any location along parallel electrodes. In the actual computation, the solution is allowed to move to its own most stable position, which is a function of initial conditions and preferential numerical sweeps. Despite the nonuniqueness of the discharge location, it should be mentioned that the electromagnetic structure of the constricted discharge is independent of the location of the discharge column.

For the constricted discharge of the classical electrode arrangement, there is no experimental measurement for validation; therefore, the verification for this group of computations is limited by comparing results with other numerical results. DCD simulations across a given electrode gap reflect a strong dependence on the ambient pressure and the applied voltage or eventually the *P*/*E* ratio to reach affinity with theory and previous calculations. Most computational results for this electrode configuration are performed at a constant electromotive force (EMF) of 2 kV in an external circuit with a constant resistance of 3 kilo-ohms. These computational simulations are presented in Figure 8.5.

The overall structure of a DCD discharge across parallel electrodes with finite dimensions is presented next. For this electrode arrangement, the electrode edge plays an interesting role. At ambient pressures more than 5 torr, the discharge is constricted and is biased toward the edges of the parallel electrodes. Validating experimental data is extremely difficult to obtain, because the Langmuir probe distorts the discharge column and a nonintrusive microwave absorption is limited by number density range and nonuniform three-dimensional discharge structure along the beam path. Therefore, the verification is restricted to comparison with visual observation and integrated electric current. The composite presentation of electron number density contours and electric field vector traces, together with a photograph, are given in Figure 9.1. The numerical results are generated on a (161 × 121) mesh with grid space clustering normal to and around the edges of the electrodes. The minimum grid spacing is 1.67×10^{-3} cm. The anode is placed at the upper border of the graph and cathode is located on the bottom boundary. As can be seen from this figure, the electric field is intensified at the sharp edges of the electrodes. The electron number density has a maximum value of $1.7 \times 10^{10}/\text{cm}^3$ at one edge of the anode. This value is 3.4 times greater than an infinite parallel plate electrode counterpart.

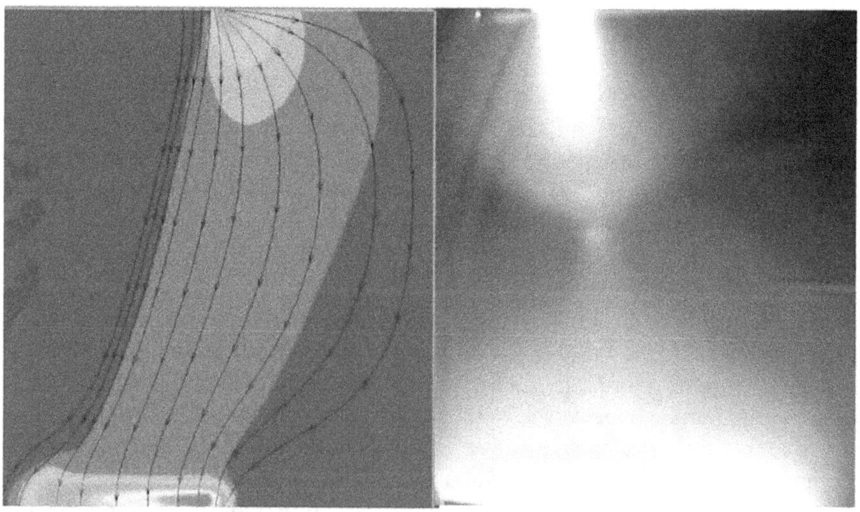

FIGURE 9.1 Constricted DCD between parallel electrodes of finite dimensions.

From this contour presentation, the cathode layer can be clearly discerned, even though the structure is no longer uniform. At the nearest edge between the anode and cathode, the anode layer is completely distorted by the stronger, local electric field. A similar discharge pattern was also observed in experiments, which show the DCD between two parallel plates, and each with the identical dimensions of 1.4 × 4.6 cm. The discharge is maintained at 840 V with a total discharging current of 100 mA. The ambient pressure of the test condition is around 1 torr, the gap distance between electrodes is 5.1 cm and the charge is constricted. In spite of the differences of the discharge environment, the numerical result bears a close resemblance to the photographed discharge pattern. In fact, the biased discharge pattern for flow control is often observed in the constrictive DCD between electrodes of finite dimensions.

A constricted DCD between electrodes of finite dimension actually has a bistable state. The discharge column biased toward anode placement and can occur at either edge. Meanwhile, a computational simulated discharge column can switch from one edge of the electrode to another, depending on the initial condition and the preferred order of the numerical sweep. Therefore, for a finite parallel electrodes arrangement, the constricted electromagnetic field has a bistable state and this bifurcation has also been observed in experiments. Despite the fact that switching from one configuration to another can be achieved by a small perturbation in the numerical analysis, both solutions are identically symmetric. More support evidence for the permissible bistable states of a constrict discharge between electrodes of finite dimension is presented in Figure 9.2. The discharge electric current vector traces and electric potential contours of the bifurcated DCD are depicted side by side to form a mirror image. High-gradient, electric field intensity regions of the numerical simulations are located on the right and left edges of the cathodes, respectively. Once the electric field strength on the cathode exceeds the breakdown threshold, electric current will emit from the cathode toward the anode. The discharge current density is biased toward the sharp edge of the cathode and the intensity elevates to an even greater level at

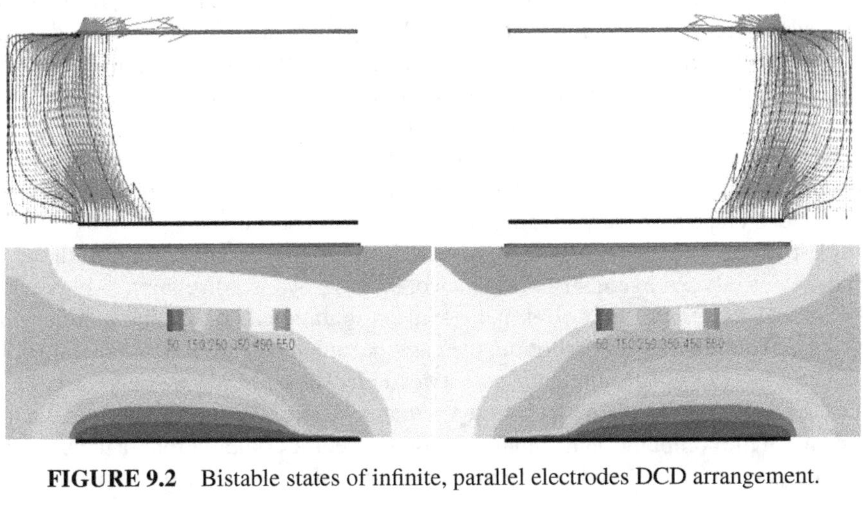

FIGURE 9.2 Bistable states of infinite, parallel electrodes DCD arrangement.

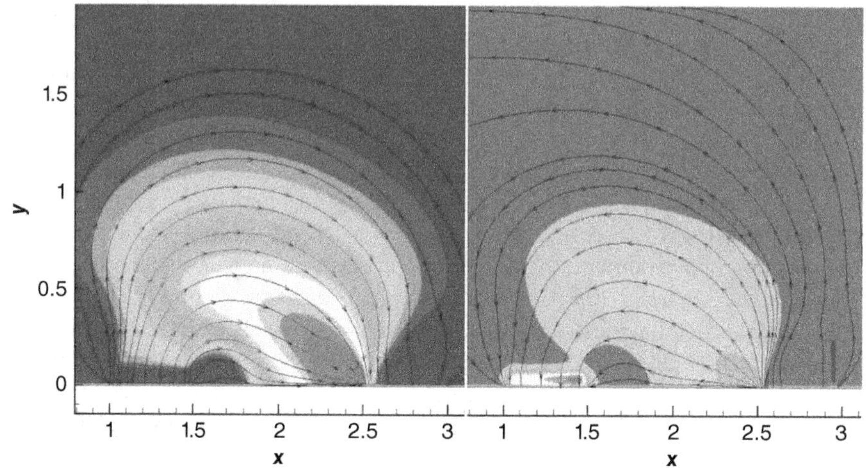

FIGURE 9.3 Ions (left) and electron (right) number density contours and electric field structure for side-by-side DCD, $\varphi = 439.4$ V, $I = 5.2$ mA, $P = 5.0$ torr.

the edge of the anode. This behavior is detected with the numerical simulations. To confirm this bifurcation phenomenon, multiple numerical results were carried out and sustained beyond an established convergence criterion. An identical numerical result of the bistable solutions has been repeatedly verified to eight-digit accuracy. Therefore, it is ascertained that the numerical simulation has captured the DCD bifurcation between parallel and finite-dimension electrodes.

The side-by-side electrodes placement (see Figure 9.3 is the most efficient configuration for high-speed flow control. First, this arrangement produces the least unwanted flow deflection by added pressure drag from the induced oblique shock waves. Second, the volumetric discharge generates a near-surface expansion to alter the effective shape of a flight vehicle at a designated location. In practice, the cathode is placed upstream of the anode, because the Joule heating is concentrated close to the cathode. The combined Joule and electrode heating increases the displacement thickness of a shear layer that creates an external flow deflection and leads to a stronger viscous–inviscid aerodynamic interaction. In fact, the magnitude of the pressure interaction is inversely proportional to the Reynolds number based on the distance measured from the leading edge. Therefore, a DCD is the most effective when placed close to a sharp leading edge of a control surface.

Side-by-side DC discharges are generated under similar conditions to those of parallel electrode arrangements; for instance, the side-by-side discharge in Figure 9.3 uses an EMF of 2.0 kV. The two-dimensional computational domain has a dimension of 5.0×5.0 cm. In this configuration, the cathode and anode have the identical width of 0.5 cm and the separation distance between electrodes is 1.0 cm. The cathode is placed on the left of the anode with edges being located at 1.0 and 1.5 cm from the left boundary of the computational domain. For these specific computations, a nonuniform 120×60 grid mesh was used, with a minimum grid spacing of 5×10^{-3}cm clustering near the horizontal edges of the electrode and with a minimum perpendicular spacing of 1.67×10^{-3} cm over the electrode surfaces. The result of a finer 250×120 nonuniform mesh was also obtained for the purpose of comparison. The agreement was very good and indicates computational simulations with a minimum numerical error. For an ambient pressure of 5 torr, the computed electrical potential across the electrodes was 437.37 and 441.75 V for the fine and coarse grid systems, respectively; the corresponding total current in the discharge field yields the value of 5.21 and 5.19 mA, respectively.

In Figure 9.3, the electric field intensity vector traces are superimposed over ion number density contours on the right-hand side of the composite graph. Similarly, the discharge current vector traces are appended to electron number density distribution to the left-hand side of the graph. The high concentration of charged particle number densities at the edge of electrodes reflects the shortest separation distance between electrodes and is clearly and correctly displayed in Figure 9.3. The maximum ion number density has a value of $5.76 \times 10^{10}/\text{cm}^3$ over the edge of the cathode and the maximum electron number density of the discharge field is 3.02×10^{10}/cc. The electrons are nearly absent in the cathode layer except the secondary emission from the electrode, which is easily observed. The average thickness of the cathode layer from the computational simulation is 0.125 cm, which is comparable to the classic one-dimensional result of 0.15 cm by Von Engel and Steenbeck (1932). It is also noted that the strong electrical field is perpendicular to the cathode. In flow control applications, the resultant electrostatic force is exerted downward toward the cathode, which cannot be supported by the shear stress to alter the flow structure but transmits directly onto the cathode. For the first time, the detailed DC discharging field over for a side-by-side electrode arrangement is duplicated by numerical simulation and is not contaminated by numerical artifact.

The normal components of the electric field intensity E_y under ambient pressure values of 1, 5, and 10 torr are depicted in Figure 9.4. Attention needs to focus on the behavior of the normal component of electric field strength E_y at the edges and over the cathode. The sharp jumps of the electric field intensity are anticipated and have been captured by numerical simulations. For a uniform diffuse discharge at an ambient pressure of 1 torr, the value of E_y is 4800 V/cm. At an ambient pressure of 5 torr, the field intensity over the entire cathode exceeds this threshold and the discharge has covered the entire cathode. At the highest ambient pressure investigated, a steep gradient exists over the cathode and the magnitude of E_y monotonously increases toward the edge adjacent to the anode. The electric field intensity exceeds a value of 4800 V/cm only over half the length of the cathode on which the discharge in the cathode layer becomes visible. High electric field intensity at the sharp electrode edge is well known and is characterized as the point discharge effect. It is noted for all cases considered; the normal component of the electric field intensity on the cathode overwhelmingly dominates that over the anode.

The Joule heat distribution over the side-by-side electrode arrangement is depicted in Figure 9.5. The numerical results are obtained by the inner product of the electric field intensity $E = -\nabla\phi$ and the electric current density $J = e(\Gamma_+ - \Gamma_e - \Gamma_-)$. The symbols Γ_i's represent the species charge number density fluxes of ionized species.

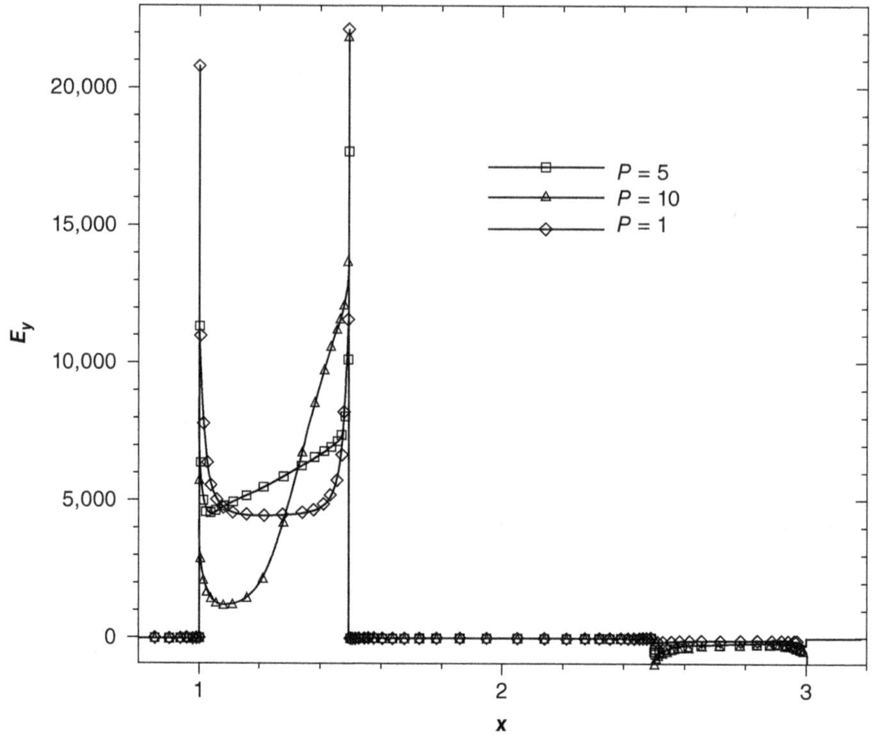

FIGURE 9.4 Normal component of electric field strength over pressure range of $1 < p < 5$ torr, EMF = 2.0 kV, Gap = 1.0 cm.

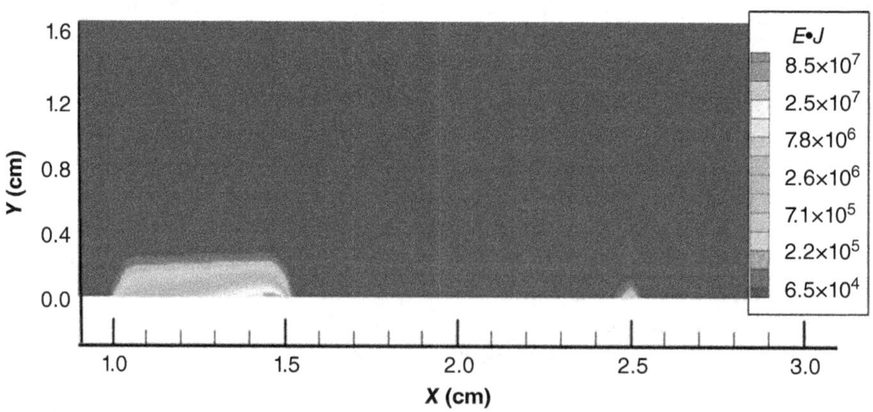

FIGURE 9.5 Joule heating over side-by-side DCD, $\varphi = 439.4$ V, $I = 5.2$ mA, $P = 5.0$ torr.

The volumetric heating occurs mainly within the cathode layer and with a significant concentration over the edges of cathode near anode (x =1.5 cm). A hot spot is also located immediately over the anode edge (x = 2.5 cm), but with much lower intensity. In flow control applications, Joule heating becomes an internal heating source to a flow field that effectively modifies the local density of high-speed flows. The numerical result has duplicated the physical phenomenon for which the Joule heating is pronounced over the edges of electrodes. For the simulation, the Joule heating released to the entire discharge field is 1.30 J/s/cm^2. This value is relatively small in comparison with the total amount of power input for the surface DC plasma generation of 21.52 J/s/cm^2, but the thermal energy is released as a heat source directly into the interior domain of the flow field. In flow control applications over a solid surface, this energy deposition is not by surface heat conduction, but by volumetric heating in space that is extremely efficient to alter the flow field structure. This additional energy source creates a local expansion to thicken the displacement thickness of an attached shear layer and to become the triggering mechanism for an oblique shock by the viscous–inviscid and electromagnetic–aerodynamics interactions. Although the energy released by a DC discharge appears as a small perturbation, it is amplified by the viscous–inviscid interaction to become a viable flow control mechanism in low-density flows.

9.2 VIRTUAL LEADING EDGE STRAKE

Flow modification by plasma can be understood easily in the light of a unique feature of hypersonic flows and the subsequent viscous–inviscid interaction (Hayes et al., 1959). In essence, the flow displacement by a thin boundary layer is no longer negligible under the high-speed flow condition. For flow over a blunt body, the viscous–inviscid interaction is characterized by the vorticity interaction within the shock layer, which is induced by the varying entropy from the curvatures of the enveloping shock. At a sharp leading edge, the outward boundary layer displacement triggers compression waves that coalesce into an oblique shock, which is the consequence of the classic Mach wave interaction. The near-surface DCD becomes an additional triggering mechanism to produce a suddenly increased displacement thickness through the heat sources in the shear layer and from the solid surface. The injected thermal flux heats the surrounding air stream and lowers the density within the shear layer, which leads to a sudden increase in the displacement thickness. This chain of events accentuates the pressure interaction for which the induced pressure has been successfully formulated and verified by experiments by a single interaction parameter, $\bar{\chi} = M^3\sqrt{C/\mathrm{Re}}$. This interaction phenomenon actually forms a close feedback loop between the growth rate of the displacement thickness and the induced streamwise pressure gradient. At the lower $\bar{\chi}$ value, the growth rate is unaltered from the laminar flow value of $\delta^* \sim x^{1/2}$ but changes to three-quarter power $\delta^* \sim x^{3/4}$when the parameter $\bar{\chi}$ exceeds the value of 3. Most importantly, the magnitude of the induced pressure increases according to the cubic power of the free stream Mach number to become a viable hypersonic flow control mechanism.

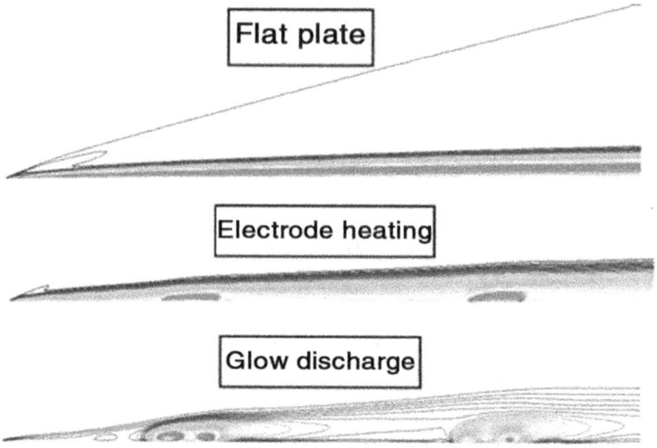

FIGURE 9.6 Temperature contours on flat plate with electrodes, $T_w = 330$ K, $T_{el} = 600$ k, $M = 5.15$.

The computed temperature contours of the electro-aerodynamic interaction are depicted in a composite graph. Figure 9.6 consists of computational results of a hypersonic boundary layer with and without electrode and Joule heating via DCD over the plate surface (Shang et al., 2005). The electrode conduction and Joule heating clearly introduce a significant perturbation to the structure of a hypersonic shear layer. The conductive heating exhibits the typical convective and diffusive pattern of the heat transfer process. The numerical result captures the distinguishing characteristic of an electric discharge at the sharp edges of the electrodes. In this computational simulation, the cathode is placed closer to the leading edge and the total electrical energy input is limited to 60 watts. It is clearly shown that intensive heat exchange takes place at the leading and trailing edges of the embedded electrodes. The rate of heat transfer is greater at the cathode than at the anode, which agrees with experimental observation. The effect of Joule heating is much more pronounced over the edges of the electrode rather than uniformly distributed over the electrodes like that of the electrode heating.

A dominant electrodynamic effect of the DCD for flow control without an externally applied magnetic field is manifested by Joule heating. This phenomenon is clearly illustrated in Figure 9.7 by the two static temperature profiles over the flat plate surface located midway between the electrodes. Both computations are generated at a free stream Mach number of 5.15 and a Reynolds number of 1.08×10^5. At this location, the surface temperature of the model has a constant value of 300 K. When the DCD is activated by an electric potential of 450 V, a total electric current in the external circuit is 50 mA. The Joule heating concentrates mostly around the cathode layer and acts as an internal heat source to the flow field. Therefore, the profile generated by the ignited DCD shows an elevated local temperature by Joule heating in the boundary sublayer close to the plate surface. (Artana et al., 2002; Borghi et al., 2006). The hottest point is not obtained at the plate surface because the plate absorbs heat from the plasma.

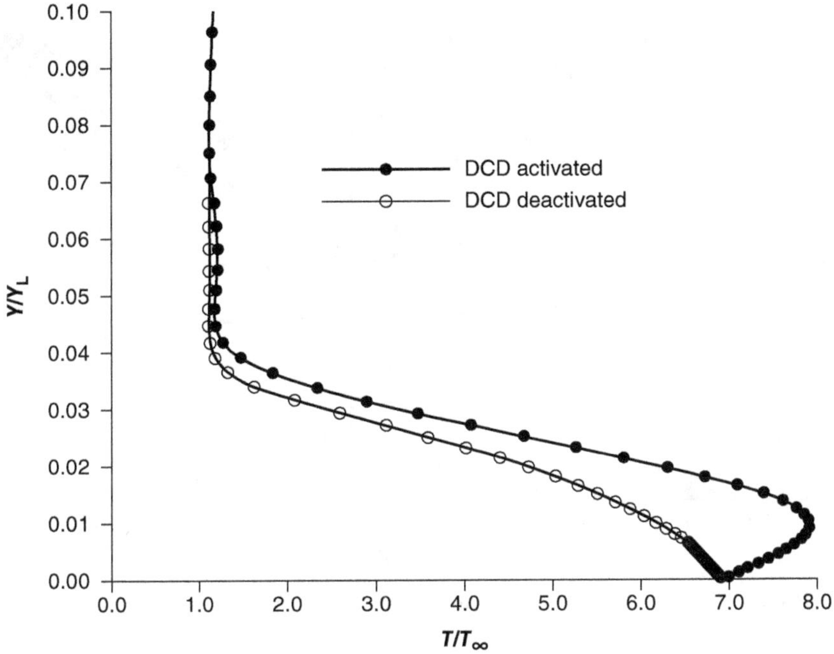

FIGURE 9.7 Comparison of temperature profiles with/without activated DCD.

Figure 9.8 displays an induced oblique shock generated by a simple DCD actuator on a sharp leading plate with a low power input of 164 watts in a Mach number 5.15 hypersonic flow (Shang et al., 2005). The second shock downstream to the leading edge shock by Mach wave interaction is much weaker but is still captured by the schlieren photograph and the density contours of the accompanying computational simulation. The electrode and Joule heating at the cathode creates a sudden outward growth of the boundary-layer displacement thickness that leads to a viscous–inviscid interaction. In turn, the induced family of surface compression waves coalesces into an oblique shock. This phenomenon is captured by both the schlieren photograph and computational simulation in a Mach 5 plasma channel. The family of compression waves initiated at the leading edge of the cathode produces an induced surface pressure rise that constitutes an effective hypersonic flow control technique. The bulk electric conductivity is estimated by assuming that the electric field intensity is proportional to the applied voltage across the gap of electrodes. The results indicate a range from 0.06 to 0.07 mho/m for conditions tested with applied voltages varying from 0.5 to 1.1 kV and at electric currents from 50 to 200 mA. It must be stressed that these values are effective for evaluated bulk conductivities, because the electric field is nonuniform, especially near the cathode. The electric conductivity varies rapidly. Langmuir probe measurements show that the electron number density at a given height above plate can vary by orders of magnitude between cathode and anode, and a similar variation is observed in the vertical direction between the cathode and the edge of the discharge

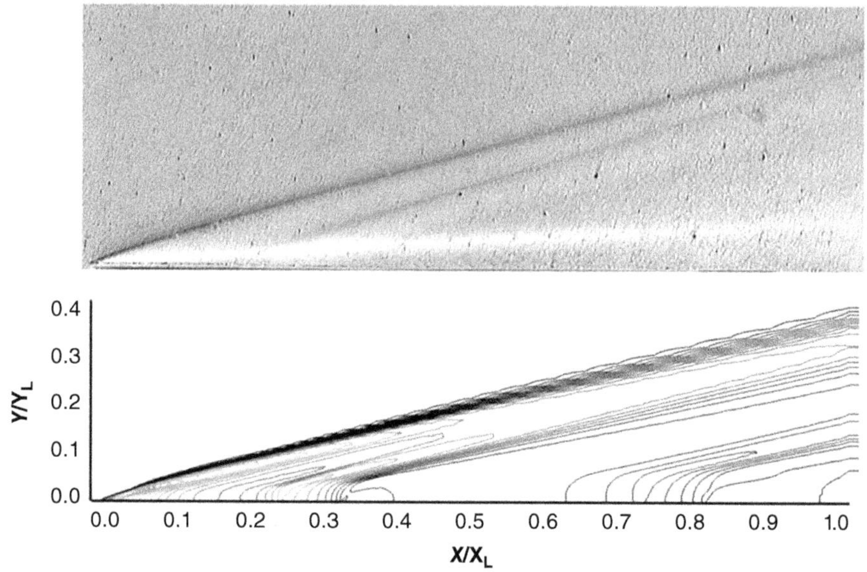

FIGURE 9.8 Oblique shock generated by DCD over a sharp leading edge plate, $M = 5.15$.

domain. Nevertheless, this is a bona fide aerodynamics–electromagnetic interaction (Rosa, 1968).

The quantification of an induced pressure rise and vastly different time scales between the electrode and volumetric Joule heating are illustrated in Figure 9.9 by single Pitot probe measurements. The heating in short-duration discharges is primarily Joule heating. During longer discharges, the electrode heating and some fluid-heating may arise from convective heat transfer from electrodes. The Pitot pressure probe is inserted into the electro-aerodynamic interacting flow field until a steady state is reached. It is seen that for a 40-s discharge, the Pitot pressure jumps about 14% immediately after the discharge is actuated. After this, approximately 150 s elapses before it decays to its undisturbed value. This slow decay indicates that heating of the electrode has occurred during the discharge. As the plasma activating period decreases, the recovery time also decreases. For a 5-s discharge period, the Pitot-pressure decay occurs immediately after the plasma is turned off. In this case, the electrode and plate have had little time to heat up during the discharge. The observed effect is due to Joule heating and the Pitot pressure recedes quickly after the discharge is extinguished. It must be noted that the plate and electrodes constitute a complex thermal system; the combined conduction and convective heat transfer mechanisms proceed at a much slower rate than volumetric Joule heating.

Effective simulations of electromagnetic–aerodynamic interactions have been demonstrated by integrating the multifluid partially ionized gas model into the compressible Navier–Stokes equations (Shang et al., 2005, 2014). Namely, the electromagnetic–aerodynamic governing equations are described by equations (7.8)

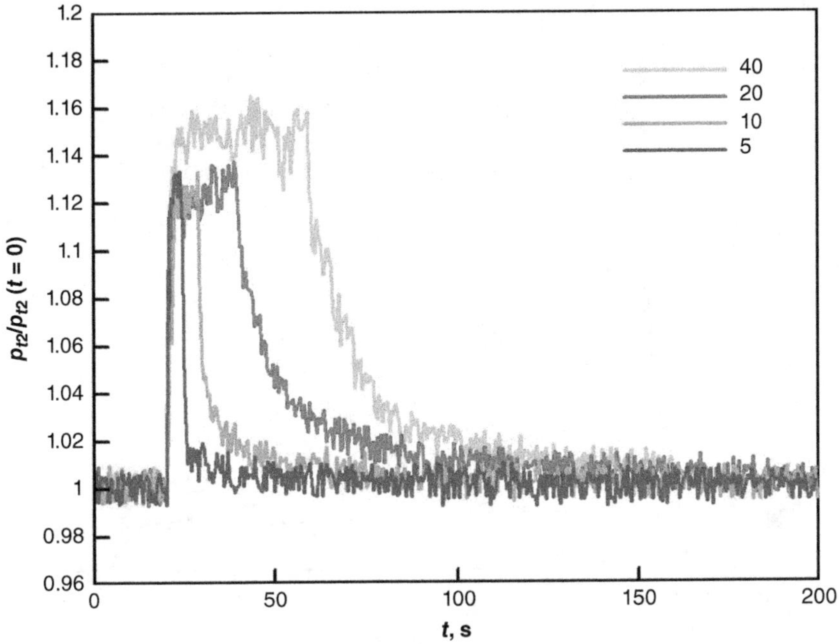

FIGURE 9.9 Surface pressure response to DCD actuation, $M = 5.15$, $\varphi = 440$ V, $p =$ 5.0 torr.

and (7.9) and replacing the species conservation equation (7.7) by the electron impact-ionized gas model. This physics-based model provides both concentration of charged species and the necessary forced diffusive effects. While the energy transfer between internal degrees of excitation becomes negligible for electron impact ionizations, which are embedded only in the cathode layer and by a trace amount. The compatible electric field intensity and electric current density in the drift-diffusion formulation are determined by the number density of charged species through the Poisson equation of plasma dynamics. The equations used to describe the plasma are

$$\begin{aligned}
&\frac{\partial n_e}{\partial t} - \nabla \cdot [d_e \nabla n_e + n_e \mu_e (E + u_e \times B)] = \alpha(E)|\Gamma_e| - \beta n_+ n_e \\
&\frac{\partial n_+}{\partial t} - \nabla \cdot [d_+ \nabla n_+ - n_+ \mu_+ (E + u_+ \times B)] = \alpha(E)|\Gamma_+| - \beta n_e n_+ \\
&\nabla^2 \varphi = \frac{e}{\varepsilon}(n_e - n_+)
\end{aligned} \tag{9.1}$$

$$\frac{\partial \rho u}{\partial t} + \nabla \cdot (\rho u u + pI - \tau) = \rho_e E + (J \times B) \tag{9.2}$$

$$\frac{\partial \rho e}{\partial t} + \nabla \cdot \left[\rho e u - \kappa \nabla T + \sum \rho_i u_i h_i + u \cdot pI + u \cdot \tau\right] = E \cdot J \tag{9.3}$$

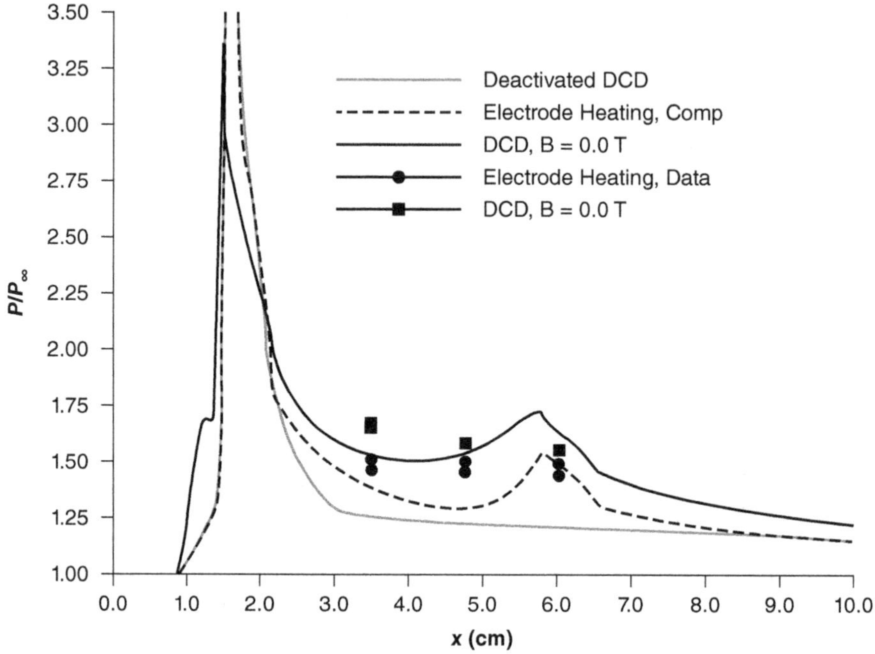

FIGURE 9.10 Induced pressure rise over plate, $M = 5.15$, $\varphi = 1.2$ kV, $I = 50$ mA.

The effectiveness of using DCD actuator as flow control implementation is demonstrated by numerical solutions of the these governing equations and verifications with experimental observations. The relative magnitude of an induced surface pressure over a sharp leading edge plate by electrode convection and volumetric Joule heating are confirmed by experimental and computational simulations. Experimental and numerical results are collected at a free stream Mach number of 5.15, and Reynolds number, based on a plate length, of 1.08×10^5. In Figure 9.10, solutions to the electromagnetic–aerodynamics equations using the drift-diffusion model compare reasonably well with measurements. The induced surface pressure plateau by an activated DCD is greater in magnitude and streamwise extent direction more than that by electrode convective heating alone. Induced surface pressure distributions are elevated over the plate in comparison with a deactivated DCD. The maximum relative value of the surface pressure rise over the deactivated counterpart is 33% at the tested conditions of p_∞=0.59 torr and T_∞=43 K, but this difference decays in a short distance downstream. The DCD indeed generates a bona fide electromagnetic–aerodynamic interaction that is not possible by electrode heating alone. It also demonstrates that the DCD can be applied as a virtual leading edge strake without any moving parts. At a very low level power input to the DCD of 60 watts, the higher surface pressure over the deactivated DCD condition produces a surface pressure ratio of p/p_∞ = 1.275 through the electromagnetic–aerodynamic interaction. The net result is an equivalent surface deflection of 1 degree at a free stream Mach number of 5.15. From

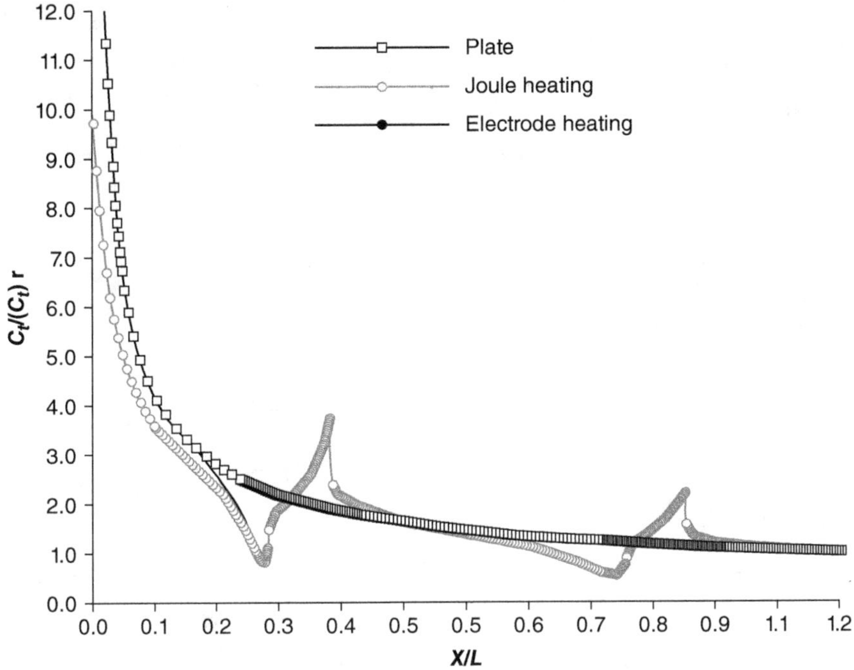

FIGURE 9.11 Effect of DCD on surface shear stress by a phenomenological model.

this result, it may be stated that the scaling of power required for the plasma actuator per electrode length is 18.90 watts/centimeter per degree of deflection at these flow conditions.

A hypersonic flow field in the leading edge region of a flat plate is characterized by a continuous and favorable pressure gradient due to the Mach wave or pressure interaction. The aerodynamic–electromagnetic interaction alters the expanding flow by multiple compression and expansion domains adjacent to the electrodes. Since the boundary-layer structure must respond to the streamwise pressure gradient, the corresponding surface shear force is displayed and normalized by using the far downstream value as a reference. The familiar inverse square-root decay with increasing distance from the leading edge is clearly shown for a laminar boundary layer. The surface shear stress in terms of skin-friction coefficient C_f is depicted in Figure 9.11, and modification to C_f reveals that the boundary-layer structure responds to the plasma flow control mechanism. The numerical results were generated by using a simplified phenomenological model in a preliminary assessment and have been completely verified by subsequent computational results (Shang et al., 2005). The numerical results show that the surface shear decreases upstream and increases downstream of the electrodes according to the local adverse and favorable surface pressure gradient distribution. The decreased and increased surface shears tend to negate each other, resulting in a zero net change in skin friction drag by the electromagnetic–aerodynamics interaction.

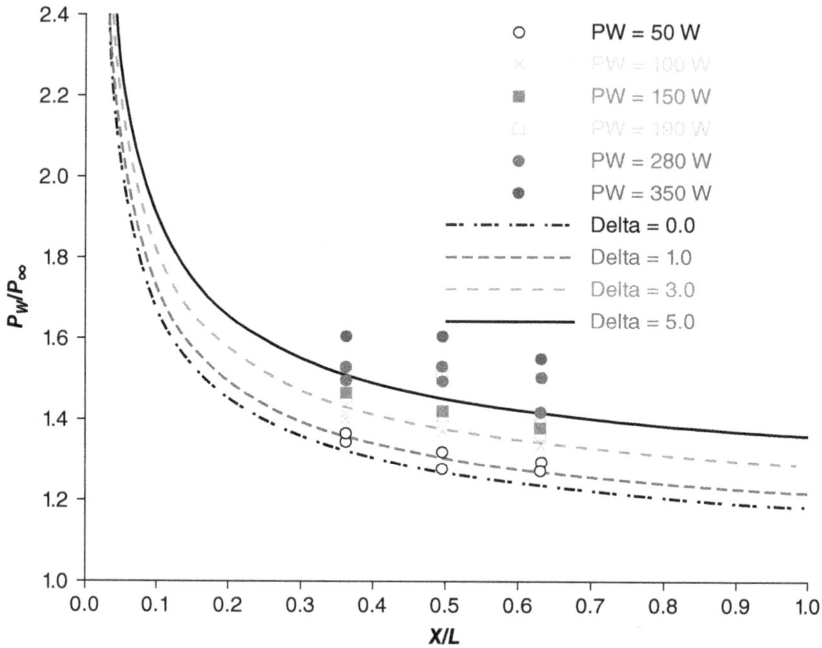

FIGURE 9.12 Power requirement for virtual leading edge strake.

The effectiveness of a DCD actuator as a virtual leading strake at a free stream Mach number of 5.15 is depicted in Figure 9.12. All computational simulations have been conducted by using the drift-diffusion plasma model and duplicated the identical experimental conditions. The DCD is maintained by an electric potential of 1.2 kV in an ambient pressure of 78.4 Pa (0.59 torr). The supporting experimental data by the MHD channel are collected from the wedge model with embedded electrodes with dimensions of 3.175 × 0.635 cm (length × width). The power supply to the electrode spans a range from 60 to 350 watts. The surface pressure data are plotted together with the calculated surface pressure distribution according to the classic Mach wave or pressure interaction theory. At a power input of 280 watts, the plasma actuator produces a flat plate stream deflection over the electrode that is equivalent to a deflection angle of 5°. Remember that the spanwise dimensions of the electrodes and the model are highly limited in the narrow MHD channel used to perform this work; the electrodes do not cover the entire span of the model to include span edge effects. The power scaling requirement for flow control is reduced slightly from 18.9 to 17.6 watts per degree per cm (Shang et al., 2005; Shang, 2008). At the maximum power input of 350 watts for DCD, the surface pressure distribution attains a streamline deflection over 7°. However, the plasma generation process is now operated at the outer limit in an abnormal discharge mode.

In essence, the combined computational and experimental investigations have demonstrated that an electromagnetic–aerodynamic interaction induced by a DCD over a flat surface in a hypersonic stream can be adopted as an effective flow control mechanism. The net result is a virtual leading edge strake over a control surface. The similar flow control arrangements have been implemented on two-dimensional and axisymmetric configurations to have varying degree of success (Borghi et al., 2006; Cristofolini et al., 2008). The effectiveness of a DCD applied as a virtual control surface for flow modification is critically dependent on the placement of the electrodes relative to a sharp leading edge to take full advantage of the Mach wave interaction. Next, increasing control effectiveness will be demonstrated by applying an external transverse magnetic field via using a Lorentz force to push charged particles away from electrodes. Momentum transfer by ion-neutral collisions further deflects the gas stream outward and creates compression waves over the control surface to amplify electromagnetic–aerodynamic interactions. However, uniform plasma generation by direct current discharges is restricted by an effective application in low ambient pressures or high-altitude environment.

9.3 MAGNETIC FIELD AMPLIFICATION

From electromagnetic theory, the magnetic intensity generates a gyrating field force and the orientation of a magnetic flux density, B is conventionally defined by the right-hand rule by its polarity. Electrical current at right angles to magnetic field lines, coupled with particle collisions, give phenomenal rise called the Hall effect. In an ionized gas, the resultant Lorentz force becomes an added mechanism that confines charged particles and enhances the collision with neutrals, thereby increasing the momentum and energy transfer. The Hall parameter is defined as a ratio between the gyro frequency and plasma collision frequency, $\beta = f_b/\nu = eB/m\nu$. As a consequence, the strength of an applied magnetic field can be a mechanism for increasing the intensity of the aerodynamic–electromagnetic interaction. A weak interaction is evident especially in a low-charged particle number density plasma generated by electron impact ionization. The intensity of the electromagnetic–aerodynamic interaction can be amplified by an externally applied magnetic field, which can be provided by an array of rare earth permanent magnets or by steady state current running through a solenoid.

To augment an applied magnetic field immediately adjacent to a solid surface, compact neodymium rare earth permanent magnets can be inserted onto a small testing model in the laboratory. The NdFeB magnet has a higher maximum strength than most other rare earth permanent magnets. These magnets come in disk or block shape with a thickness less than 3.2 mm and a pole area surface of 19-mm diameter or a square surface of 1451 mm^2. At the pole surface the magnetic flux density is greater than 0.47 T. A weakness of the NdFeB permanent magnet in comparison with the aluminum–nickel–cobalt alloy (Alnico) is the much lower Curie temperature; the maximum operational temperature of NdFeB is 180°C (Shang et al., 2005, 2008). This shortcoming is not overly detrimental in most laboratory testing conditions, but it may not be acceptable for practical applications.

The most widely used equipment for generating a magnetic field is a solenoid. The conventional alloy coils are usually connected in a series to have the electric resistance of 0.54 Ohm and a self-inductance of 80 millihenery. The power rates are usually limited to 10.6 kW (140 amp, 76 V). This type of magnetic field generator would not be suitable for practical flight operations without support from superconductivity technology, and this limiter must be removed by further development. The equipment in use in MHD channels for laboratory experiments usually has a pole gap of 10 mm and a pole diameter 25 mm, which can produce a magnetic field strength as high as 3.5 T. However, the field strength diminishes in proportional to the inverse square of the gap distance. Most of the measurements are collected across a hypersonic MHD channel with a width less than 150 mm. Under this arrangement, the nearly uniform transverse field along the centerline of the poles has a magnetic flux density of 1 T.

The profound impact of an externally applied magnetic field to a plasma field is well known. In recent computational simulations of a direct current discharge subjected to a transverse magnetic field, the magnetic field not only produced a substantial distortion of the discharge but also induces a side-shifting motion of the discharge column (Kimmel et al., 2006). In plasma actuator applications, similar behavior is also noted. Once a transverse magnetic field is applied across the plasma field, a movement of high-intensity attachment of charges on the anode appears. The attachment changes its position based on the polarity of the magnetic field. However, the most drastic phenomenon is the constriction of the discharge. Figure 9.13

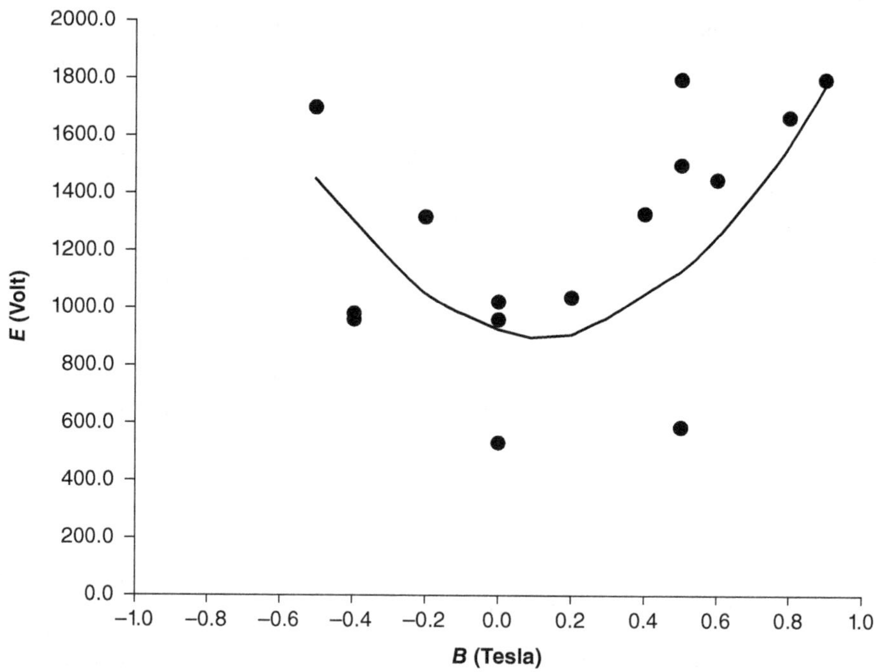

FIGURE 9.13 Modification to electric field potential by applied transverse magnetic field.

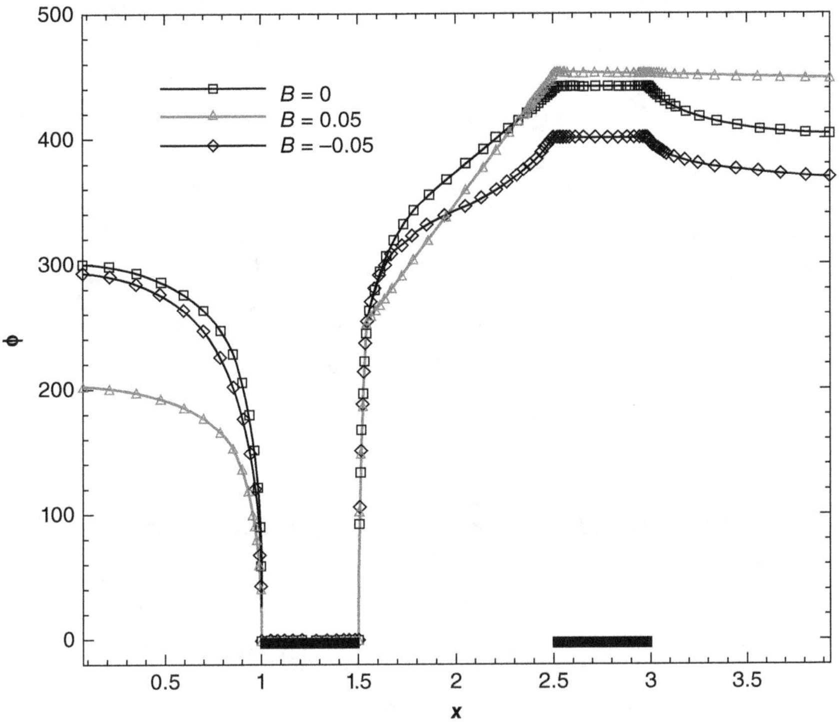

FIGURE 9.14 Electric potential over the electrodes with externally applied magnetic field.

describes the applied voltage for plasma generation at a constant circuit current versus the transverse magnetic field strength. The experimental data shows that the required voltage of the discharge increases as the magnitude of the magnetic field increases for both polarities, meanwhile the data-scattering band width also accentuated. Therefore, plasma generation in the presence of a magnetic field generally requires a higher electric field strength but with a high degree of uncertainty.

The computational surface electric potential distributions with/without externally applied transverse magnetic fields ($-0.05 < B < 0.05$) are given in Figure 9.14. When the externally applied magnetic field is absent, the required electric potential across the electrodes is 441.75 V to maintain a discharge current of 5.19 mA. When a negative polarity transverse magnetic field intensity is applied to expel electrons away from the electrodes and the insulator between them, the voltage required drops to a value of 401.0 V and the discharge current increases slightly to a value of 5.33 mA. When a magnetic field polarity is applied to suppress the charged particle movement near the surface, a higher voltage of 452 V is required to maintain a lower current in a DCD of 5.16 mA. Although simulations by the drift-diffusion model are applied beyond the original validated domain according to the similarity law by Townsend, the deviation represents only a small perturbation from its valid region. These numerical results are in general agreement with experiments of Kimmel et al.

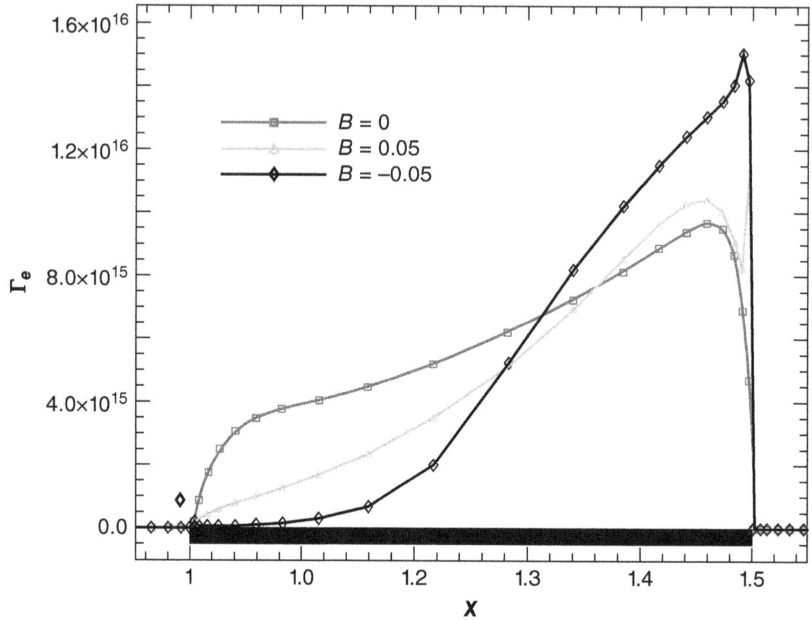

FIGURE 9.15 Electron number density over cathode with/without transverse magnetic field.

(2006). The effect of a transverse magnetic field to a DCD is important for flow control, but data and computational simulations are difficult to obtain with desired accuracy. A sustained research effort in this area is needed to offer understanding and insight for future advancement.

The electron number density flux, with and without an applied magnetic field of opposite polarities, on the cathode is given in Figure 9.15. The cathode is the black area shown just above the *x*-axis. The number density flux on the cathode has a particular physical significance because it reflects the secondary emission of the plasma-generating process. In the drift-diffusion model, a constant secondary emission coefficient of $\gamma = 0.1$ is often adopted in modeling this phenomenon (Raizer, 1991; Surzhikov et al., 2004). However, this coefficient is strongly dependent on the surface properties of an electrode (Von Engel and Steenbeck, 1932). The numerical results indicate that the weak transverse magnetic fields do not significantly alter the total magnitude of the number density fluxes over the cathode. However, according to computational electric field intensity distributions, the secondary emission has shifted the number density flux distributions over the cathode toward the edge closer to the anode.

A transverse magnetic field is applied across the electrode-embedded plate surface to assess the amplification and suppression of the electromagnetic–aerodynamic interaction. The current transverse to the magnetic field lines is given as $J_x = \sigma(E_x - \beta E_y)/(1 + \beta^2)$. By assuming that the medium conductivity has a weak dependence on the applied electric field, an effective Hall parameter has an approximated value of 1.3, at a magnetic field strength of 0.9 T and a current density of 200 mA. This value shall be comparable to a Hall parameter based on a collision frequency of 2.3×10^{10}/s in DCD. The definition of the Hall parameter $\beta = eB/(m_e v)$ yields a value of about 7 at $B_z = 0.9$ T. This is significantly higher than the value inferred from experimental measurements. Again, it should be noted that the experimental value is a bulk parameter, and the structure of the discharge varies significantly with a magnetic field. It is also questionable whether the conductivity remains constant in a rapidly changing electric field. The need of additional experimental effort for DCD in a magnetic field becomes obvious.

Images of the discharges for $B_z = 0$ and $B_z = \pm 0.9$ T in positive and negative directions with an electric current of 50 mA are shown in Figure 9.16. It is apparent that a transverse magnetic field indeed strongly influences the discharge in an expected manner. From experimental observations, the predominant effect of an externally applied magnetic field to a DCD is suppressing the visible discharge over the cathode layer. This observation is easily made by applying a strong and uniform transverse magnetic flux density of 0.9 T. Figure shows 9.16 that the DCD is sustained by an electrical field intensity of 1.2 kV. In the absence of an externally applied magnetic field, the discharge is the most visible over the cathode and concentrates above the insulator surface between electrodes. The plasma has also been conveyed downstream by the hypersonic stream beyond the embedded electrodes. The lowest panel of

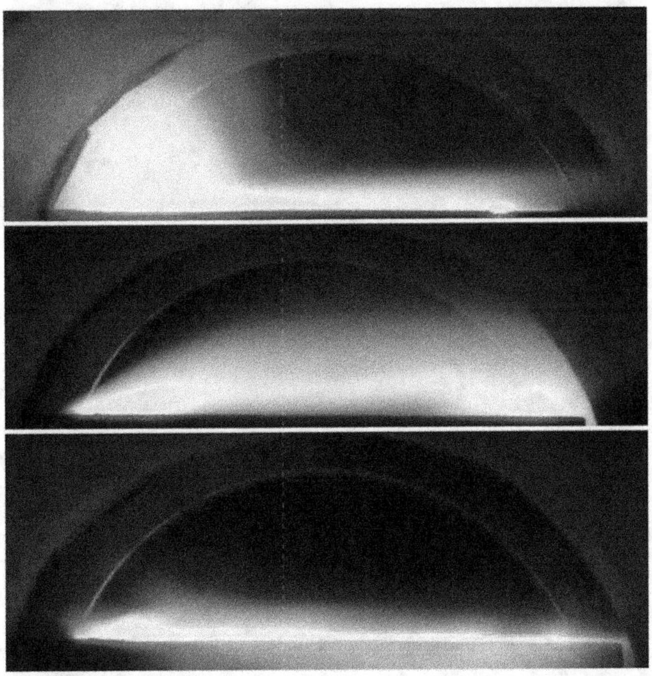

FIGURE 9.16 Effect of transverse magnetic field to the DCD, $B = 0$, $B = -0.9$ T, and $B = +0.9$ T.

Figure 9.16 shows that the Lorentz force vector ($J \times B < 0$) on the electrons is into the plate that suppresses the ability of the discharge to leave the surface of the plate. In fact, the discharge tends to spill over the plate surface and around the sides of the plate from the application of the magnetic force. Conversely, when a Lorentz force vector on the electrons directs away from the plate, it tends to move the glow off the plate surface. This is shown in the second panel of Figure 9.16. For this case, a dark region can be observed between the bright portion of the glow and plate surface. The discharge becomes more uniformly extended over the electrodes by an electron expelling Lorentz force ($J \times B > 0$). The visible plasma domain above the cathode is significantly suppressed by the applied magnetic field. These glow discharge characteristics persist and are similar over a wide range of the externally applied magnetic fields tested ($0.1 < |B| < 0.9$ T).

The accompanying computational simulations at identical experimental conditions are also performed using the drift-diffusion plasma model and results are shown in Figure 9.17. To alleviate the computational instability, simulations are limited to a lower magnetic intensity, -0.2 T $< B <$ 0.2 T. The polarities of the magnetic

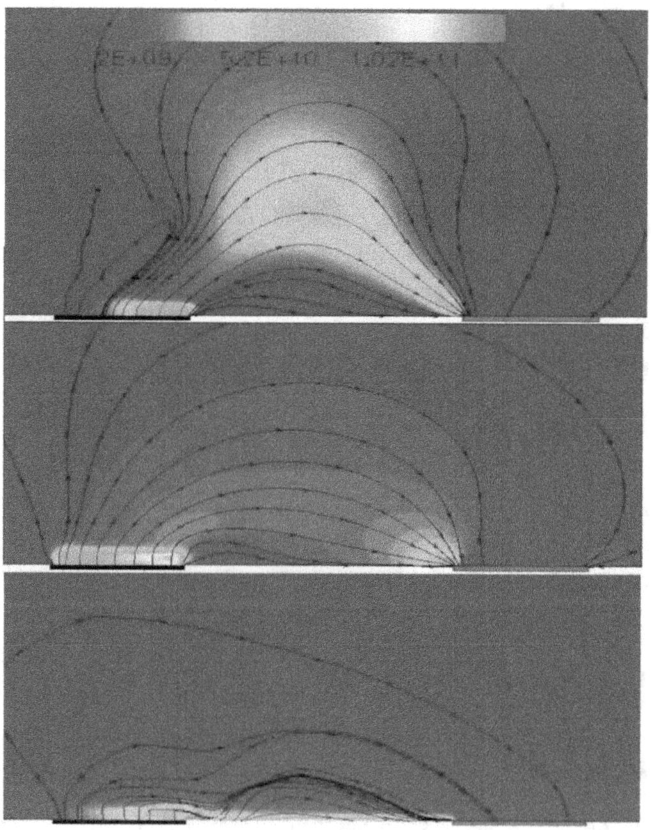

FIGURE 9.17 Simulated DCD in externally applied magnetic field -0.05 T $< B <$ 0.05 T.

field are reversed to generate two possible Lorentz force orientations. The force is directed away by letting ($J \times B > 0$) and toward the plate surface with ($J \times B < 0$). Presentation of computational results is arranged identically to that of experimental observations. The top insert is generated in the absence of a magnetic field, and results of $B = +0.05$ T and $B = -0.05$ T are inserted in sequence. The general behavior of numerical simulations is affirmed with experimental observations at a higher magnetic flux density. Detailed vectors of electron from numerical results are also displayed to reveal displaced electron particle trajectories by the applied transverse magnetic field. Under the stiff condition of the local high gradient region on the edges of the electrodes, the minimum grid point spatial spacing is reduced to 1.67×10^{-3} cm and the grid point density is also doubled. Numerical results obtained after the grid density refinement retain and slightly improved the definition of the essential features of the electromagnetic–aerodynamic interactions.

According to the testing conditions, the electron Larmor frequency at $B = 1.0$ T is 1.76×10^{11} radian/s and the electron-heavy particle collision frequency is estimated to be in a range from 3.9×10^{9}/s to 2.3×10^{10}/s. Under these conditions, the behavior of plasma in a hypersonic MHD channel is not collision dominated, and the maximum and effective Hall parameter's range is verified to have values ranging from 7.6 to 45.1.

Surface pressures were recorded for a variety of magnetic field strengths and electric currents for all combinations of electrode and magnet polarity. Current vectors in a collinear plane of the flat plate are directed either upstream or downstream, depending on the relative polarities of the magnetic field. The magnetic field vector in the vertical plane above the plate is transverse to the free stream velocity. The Lorentz force vector in this case is oriented either into or out of the plate surface, depending on the relative electric and magnetic field polarities. By implementing an electromagnetic force, the pressure induced by a DCD is produced by a modification of the boundary layer displacement thickness. When a Lorentz force is oriented into the plate, the pressure rise associated with Joule heating is suppressed by thinning the boundary layer displacement thickness. When a Lorentz force is directed away from the plate, the pressure rise is enhanced by deflecting the charged particles into the oncoming stream, the outward motion of ions provides additional coupling by momentum transfer collisions between ions and neutrals.

The evidence of amplified flow control effectiveness by applying a transverse magnetic field to DCD for high-speed flow control is presented by a series of side-by-side results from experiments in an MHD channel and accompanying computational results. The increased surface pressure by an externally applied magnetic field repelling charged particles away from the electrodes by the use of a Lorentz force is displayed in Figure 9.18. In this figure, computed results with and without the externally applied transverse magnetic field are accompanied by experimental data. The magnetic field intensity has a constant value of 0.2 T, and the result of the electromagnetic–aerodynamics interaction without the magnetic field, is included as the reference of comparison. In a DCD field, the electric field force dominates over that of the $J \times B$ acceleration in a nearly collisionless condition, and yet the effect of the Lorentz force is clearly demonstrated. The computed results

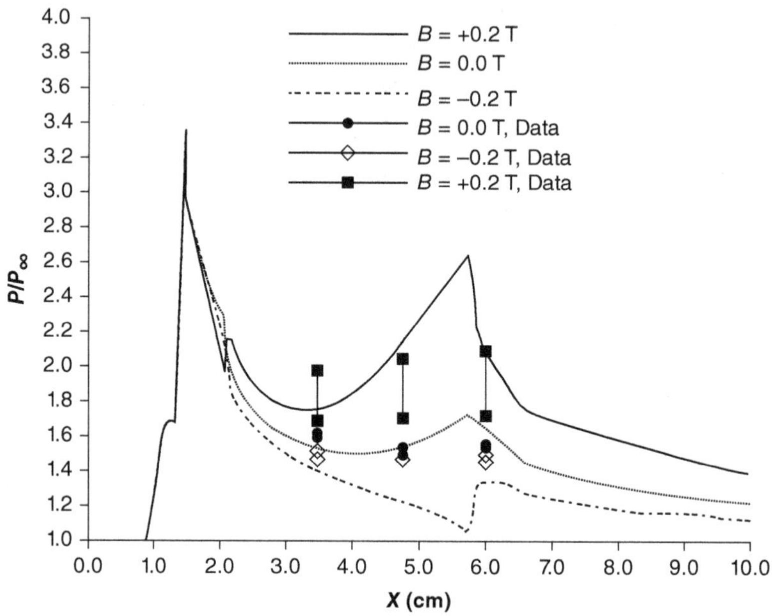

FIGURE 9.18 Amplification of induced surface pressure rise by applied magnetic field, $M = 5.15$, $\mathrm{Re} = 1.05 \times 10^5$.

generally overpredict pressure measurements, and the data also exhibit an unusually large scattering band due to the unstable DCD pattern in an applied magnetic field. As anticipated, the magnetic field generates a strong inviscid–viscous interaction that is further accentuated by the Lorentz force. The induced surface pressure is greater than the electro-aerodynamic interaction alone, at $B = 0.0$ T and under the identical electric field strength. The disparity between data and computational results using the drift-diffusion model is more pronounced for the Lorentz force that suppresses the outward motion of charged particles, $J \times B < 0$. Meanwhile, an increased effort is required to maintain computational stability in the presence of an applied magnetic field, because of the steep gradient region over the edges of the electrode as revealed by Figure 9.15. The numerical simulation at $B = -0.2$ T indicates a diminished surface pressure rise by suppressing the outward deflection of the charged particle streamlines. The numerical result tends to underpredict the measurements. Nevertheless, the numerical results using the drift-diffusion model capture the essential difference between applied magnetic fields of opposite polarities, and other phenomenological models are much less accurate in predicting the induced pressure with/without an applied magnetic field.

In summary, experimental data and computational results show that an externally applied transverse magnetic field can alter the charged particle trajectory between electrodes. By a suitable polarity of the applied magnetic field, the deflected charged species can further displace flow away from the electrode-embedded surface. The resultant local displacement thickness growth over the electrode generates stronger compression waves and leads to a more than 40% increase of local pressure rise over that by applying DCD alone. This behavior is equivalent to a virtual control surface with a greater angle of attack with respect to an oncoming stream. Admittedly, the quantification by experimental data and computational results contain uncertainties in accuracy, but the added mechanism for flow control is realizable for some special purpose applications. Amplifications to electromagnetic–aerodynamic interactions by a transverse magnetic field have been recognized for flow control and mitigation of stagnation point heat transfer on blunt bodies. Additional exploitation and path-finding research in this area are needed to achieve a better understanding for technology maturation.

9.4 VIRTUAL VARIABLE GEOMETRY COWL

The approach of utilizing electromagnetic–aerodynamic interactions for hypersonic flow control by an outward displacement of the flow path has been fully demonstrated near the leading edge of a control surface. This idea is equally applicable near the leading edge of a rectangular inlet as an MHD compressor. In a side-by-side experimental and computational investigation, a DCD is again adopted as a plasma actuator. The electrodes of the discharge are embedded in the sidewalls of a constant cross-sectional area inlet; this arrangement is nonintrusive when deactivated. Once the discharge is actuated, an increased slope of the displacement thickness through electromagnetic–aerodynamic interaction consistently generates oblique shocks acting like a side compression of a convergent inlet cowl. The compression is controlled by the plasma power input. The induced compression near the leading edge of the inlet emulates a variable geometry cowl.

Figure 9.19 presents the composite photograph of an activated DC discharge within a rectangular constant area inlet and a computationally simulated, activated DCD that duplicates experimental conditions (Shang, 2008). The inlet model is limited to a relatively small dimension of 10.16 × 3.81 × 3.11 cm ($x \times y \times z$). The four sidewalls of the model have sharp leading edges with an outward bevel angle of 20° at the entrance. The copper cathode and anode are embedded in opposite phenolic sidewalls parallel to the z coordinate. A direct current discharge is sustained by an electric potential of 800 V at an electrical current of 80 mA. All electrodes have identical dimensions of 3.175 × 0.64 × 0.311 cm (length × width × depth). The cathodes are placed at a distance of 0.79 cm downstream from the leading edge of the inlet, and the distance between the centerline of the cathode and anode is 2.22 cm. The electric current density on the anode has a value of 21.27 mA/cm^2 with the applied electric potential of 800 V. From this graph, the computational results duplicate charged species concentration distributions of the overall DCD pattern over the cathodes and anodes that correspond to the glow intensity of the DCD.

A geometrically simple rectangular inlet possesses intersecting shock wave structure within the hypersonic inlet. Four oblique shocks emanate from the mutually perpendicular leading edges of the rectangular inlet. The height of the inlet is slightly

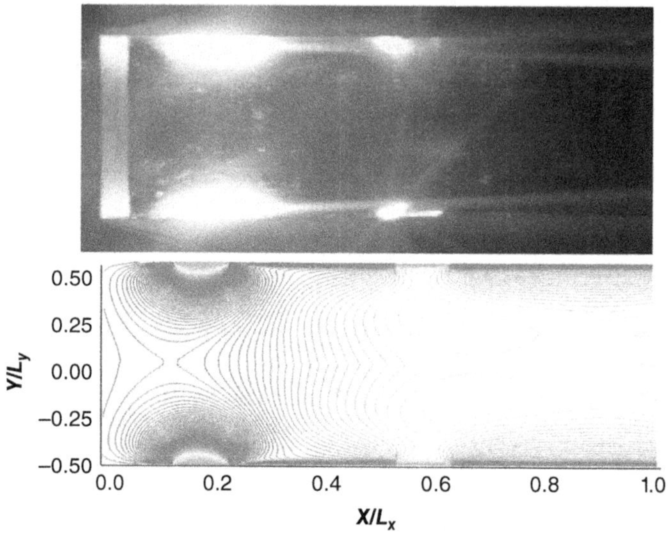

FIGURE 9.19 Side compression by DCD in a constant cross-sectional inlet, $M = 5.15$, Po = 370 torr, To = 300 k, EMF = 800 V, $I = 80$ mA.

greater than the width of the inlet (3.81 vs. 3.11 cm). Therefore, two shock waves that originate from the leading edge, parallel to the *z* coordinate, intercept upstream of the shocks parallel to the *y* coordinate. The pressure contours of the induced oblique shock waves, from the leading edges of the sidewalls, on the centerline of the *x*-*y* plane are depicted in Figure 9.20. Attention is drawn toward small perturbations generated by DBDs that are located downstream and near the entrance of the inlet. The compression waves generated by the DCD intersect and merge with shocks originating from the leading edge. This is revealed as a broadening region of compression in the pressure contours. This figure displays oblique shocks emanating from the leading edges and intersecting each other before leaving the exit plane. Coalescing and interacting compression waves are also visible upstream of the shock intersection points. Nearly identical wave structures exist in the *x*-*z* plane (also shown in Figure 9.20). Therefore, there are at least four oblique shocks that coexist in the cross-flow plane. In short, complex shock–shock and shock–boundary interactions are occurring within this simple inlet.

To highlight the concentrated discharge domain, only 40 constant-value density contours from $10^7/\text{cm}^3$ to a maximum of $10^{12}/\text{cm}^3$ are plotted. The globally neutral plasma with a charged particle concentration less than $10^8/\text{cm}^3$ is conveyed downstream and further diluted by diffusion and recombination. At a static pressure of 2.5 torr and an electric potential of 800 V, the highest charged particle concentration is located near the outer edge of the plasma sheath (around 0.5 mm). The DC discharge is operated at the unstable range and the data show perceptible fluctuations, but however, the numerical simulations remain steady. Nevertheless, the magnitude of charged particle densities agrees with detailed measurements at nearly identical

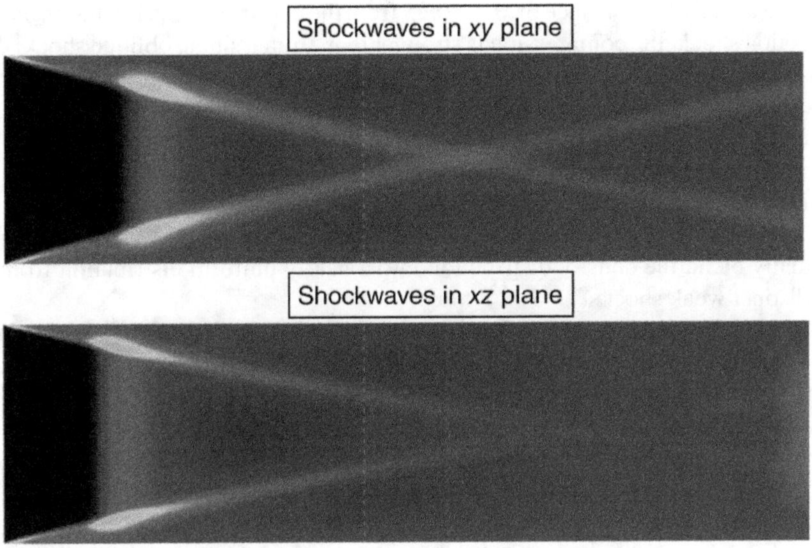

FIGURE 9.20 Perturbation generated by DCD to shock structures in a rectangular cross-sectional inlet, $M = 5.15$, Rey = 2.57×10^5, EMF = 800 V, $I = 80$ mA.

test conditions, and the overall structure of the DC discharge pattern bears close resemblance to the experimental observation.

In Figure 9.21, an intriguing flow field structure is revealed by the selected density contours at three streamwise locations. These numerical results were generated using a fine mesh system. Upstream of the intersection of four equal strength oblique shocks, a connecting shock wave must exist between two perpendicular families of shocks to satisfy the continuity condition (Shang, 1991, 2004). At the intersection of the oblique and the connecting shocks, two triplet points are formed. Therefore, a total of eight triplet points are contained in the shock formation of the cross-sectional

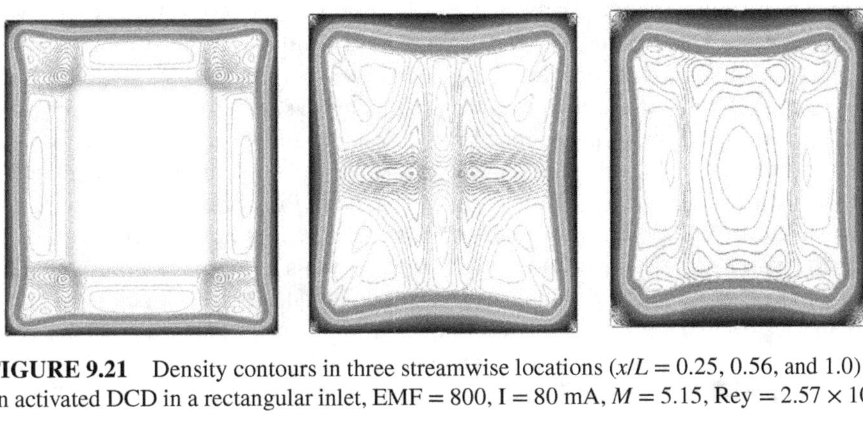

FIGURE 9.21 Density contours in three streamwise locations ($x/L = 0.25, 0.56$, and 1.0) of an activated DCD in a rectangular inlet, EMF = 800, I = 80 mA, $M = 5.15$, Rey = 2.57×10^5.

plane at a normalized distance of $x/L = 0.25$ from the entrance of the inlet. Beneath the connection shock, the compression is stronger than from a single oblique shock; thus, the viscous layer approaching the corner region is thinner than that on the sidewall. In the next two cross-sectional planes, at $x/L = 0.56$ and 1.0, the four oblique shocks are seen to intersect and diverge from each other toward the exit plane. Since the width of the inlet is smaller than the height, oblique shocks emanating from the leading edge parallel to the z coordinate intersect first and then diverge from each other downstream. At the exit plane, the viscous dissipation and wave attenuation eventually blend the density distribution into a nearly uniform distribution from the four distinct weak shocks.

The experimental data of a rather small inlet model is restricted to a Pitot pressure survey and optical observation. However, the performance of an inlet is measured by the static and stagnation pressure recovery. Therefore, computational simulation becomes pivotal to analyze the critical performance information that is not obtainable from experimental measurement. To achieve this goal, a substantial amount of effort was focused on validation of the physical phenomenon and calibration for different magnitudes between flow fields with and without an activated DC discharge. From both the experimental observations and computational simulations, the electromagnetic–aerodynamic interaction does not alter the basic flow field structure within the rectangular inlet. The four sharp and mutually perpendicular leading edges induce four oblique shocks over the side walls. Upstream of the first shock wave intersection, a total of eight triple points of the shock structure are formed. The effect of the side compression by the aerodynamic–electromagnetic interaction is made obvious by contracting the inlet cross-flow section in Figure 9.22. Two comparing pitot contours are generated under an identical flow condition with/without an activated DCD. All numerical results for the rectangular inlet are generated at an entering Mach number of 5.15 and the characteristic Reynolds number of 2.57×10^5, where the DCD is activated by an EMF of 800 V and a steady electric current of 80 mA.

Computational simulations show that downstream of the shock wave intersections, the flow field is characterized by reflected shock waves from the sidewalls and multiple shock-on-shock interactions. The Pitot pressure distributions along the centerline of the inlet with and without the activated DC discharge on the vertical walls are given in Figure 9.23. The oblique shocks, strengthened by the DC discharges over the vertical sidewalls, have steepened oblique shock angles from 16.9° to 18.0°. As a consequence, the shock interaction moves upstream and shows a greater compression ratio. In addition, a more pronounced pressure plateau appears in the streamwise pressure distribution. From viscous–inviscid interaction theory, this behavior reflects a steeper streamwise surface pressure gradient to support Pitot pressure measurements. Downstream of the shock wave intersections, the flow field generated by the DC discharge has a slightly lower Mach number and reveals a lower Pitot pressure than flow with a deactivated DC discharge. The computational simulations agree with data in both locations of the shock intersections and the relative magnitude of the Pitot pressures. The electromagnetic–aerodynamic interaction generates an 8.6% greater peak Pitot pressure at the shock intersection than its deactivated counterpart.

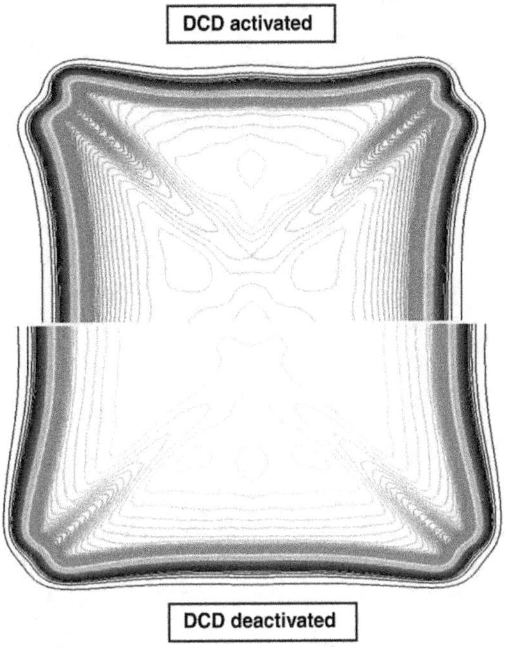

FIGURE 9.22 Shock structure at the inlet entrance with/without DCD side compression, EMF = 800 V, I = 80 mA.

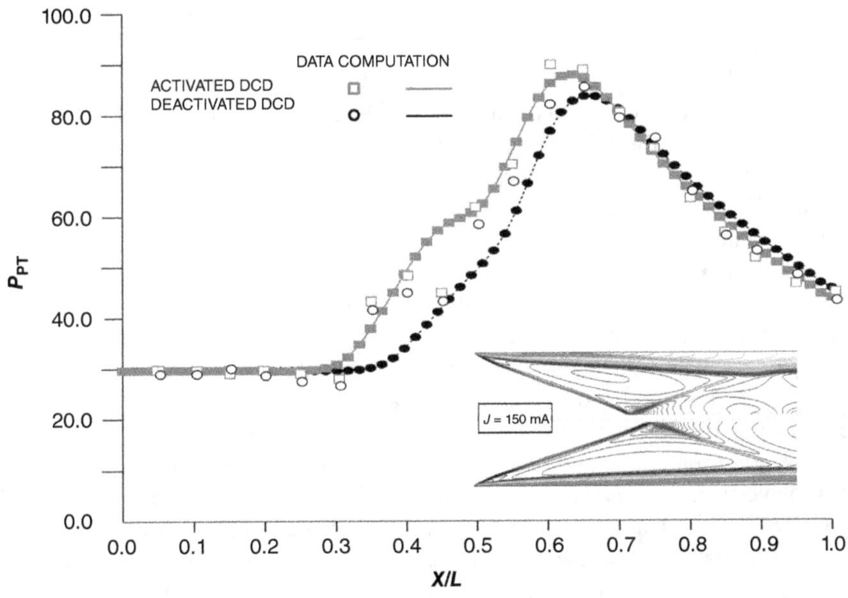

FIGURE 9.23 Comparison of Pitot pressure distributions with/without activated DCD, EMF = 800 V, I = 80 mA, M = 5.15, Rey = 2.57×10^5.

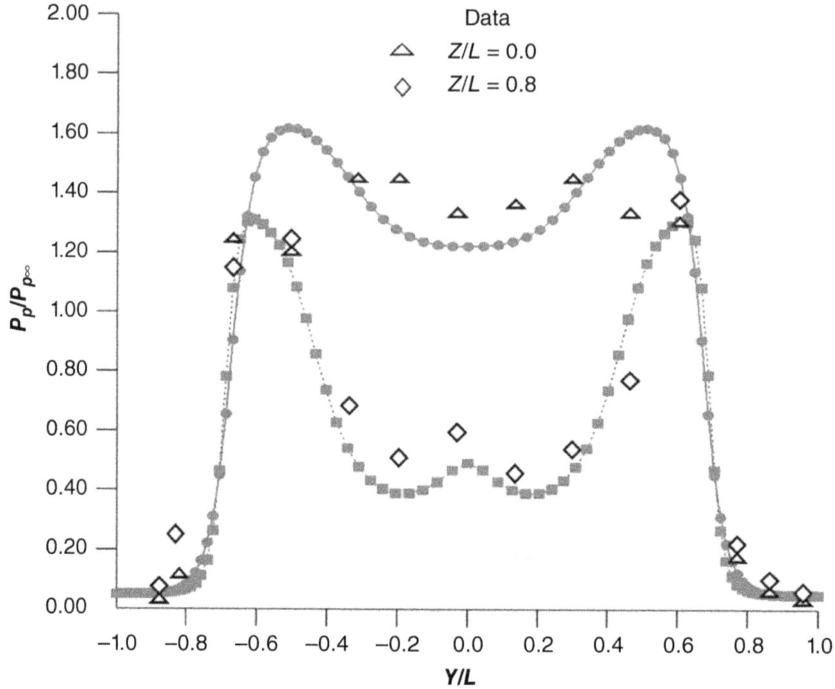

FIGURE 9.24 Validating Pitot pressure distributions at a location within the inlet close to the exit of $x/l = 0.95$, EMF = 800 V, $I = 80$ mA.

Extensive Pitot pressure surveys have been conducted for the electromagnetic–aerodynamic interaction within the inlet. The streamwise pressure distributions along the entire length of the inlet at different distances from the sidewall are recorded together with Pitot pressure profiles at the shock intersection and inlet exit. At two selected cross-flow planes, the behaviors are very similar; thus, only the pressure distributions near the exit of the inlet ($X/L = 0.95$) are depicted in Figure 9.24. Two distributions across the centerline ($Z/L = 0.0$) and near the horizontal wall ($Z/L = 0.80$) are compared with computational results. The overall agreement between data and the computed result is reasonable, particularly in view of the fact that the DC discharge is operated in an unstable mode and the Pitot data has a scattering band as high as 6%. In all, the simulated results capture a multiple shock-on-shock interaction. The maximum discrepancy between time-averaged measurement and computed simulation is determined to be less than 4.6%. The computed static pressure distribution across the vertical sidewalls and near the inlet exit exhibits similar behavior. Namely, when a DCD is activated, the intersection of steepened oblique shocks parallel to the electrode-embedded sidewall move upstream, in contrast to the unperturbed flow counterpart. The shock wave structure with the DCD, therefore, shifts forward along the inlet centerline in comparison with the unperturbed flow and with greater shock strength. The cross-flow field also reveals typical behavior of a viscous–inviscid interaction, in that the static pressure is elevated within the boundary layer. This similar behavior is also detected in the pressure distribution across the horizontal sidewalls.

In short, an additional compression induced by the DCD at the inlet exit is determined to be 11.6% by integrating the static pressure over the exit plane. This performance of a DCD at the entrance of a constant area rectangular inlet emulates a variable geometrical cowl with side compressions with an additional deflection angle of 1.2° at an entrance for a Mach number of 5.15. The power requirement for the MHD compression is nearly identical to that observed earlier for the virtual leading edge strake at a low value of 280 watts.

The effectiveness of MHD compression in a constant area rectangular inlet at different free stream Mach numbers has been investigated by four numerical simulations at Mach numbers of 5.0 and 6.0; the actuated DCD produces a greater compression than the deactivated counterpart. At a higher Mach number, energy input for plasma generation is kept at the same value as that for the lower Mach number flow. At the higher free stream velocity, 810.6 m/s versus 675.5 m/s, more electrode and Joule heating is convected away with the air stream and lower air temperatures are obtained. Therefore, the electromagnetic–aerodynamics interaction is diminished; as a consequence, the side compression by the DCD is not increased according to the cubic power of the Mach number as indicated by the Mach wave interaction theory at the leading edge. However, a higher compression can still be obtained by increasing the electric current density of the DCD. A critical issue to determine a scaling law for effective flow control via a virtual inlet cowl is unsolved and may be unsolvable by the nature of complex phenomena that demand much more accurate computational modeling and simulation capability.

To assess further the idea of additional compression through the electromagnetic–aerodynamic interaction, a DCD was also applied to a cylindrical inlet. Again limited by the small size of an MHD channel, the test setup must minimize the interference by measurement devices; thus, the validation is limited to a single Pitot probe survey experimental observation.

A photograph of an ignited DC discharge in the cylindrical inlet is presented in Figure 9.25. The discharge is sustained by an applied electric potential of 460 V and a current of 150 mA. Experimental data are collected at two different stagnation pressures (580 and 370 torr), which correspond to two different Reynolds numbers of 1.66×10^5 and 2.57×10^5. The DCD is viewed from the inlet exit and at an oblique angle from the axis. The dominant visual feature of the DCD is the glow over the electrodes and its reflection from the model sidewalls. Under this testing condition, the discharge current density on the anode is 6.78 mA/cm^2; this value is nearly three times lower than that of the rectangular inlet with a fixed size and discrete electrode arrangement (Shang et al., 2007; Shang, 2008). Thus, the DCD from electrode rings of a cylindrical inlet is essentially fluctuation free. This behavior persists even when the circuit current density on the anode reaches a value of 20.34 mA/cm^2 by increasing circuit current to 150 mA. Thus, the experimental measurements of this configuration have a significantly reduced data scatter than the previous results for a rectangular inlet.

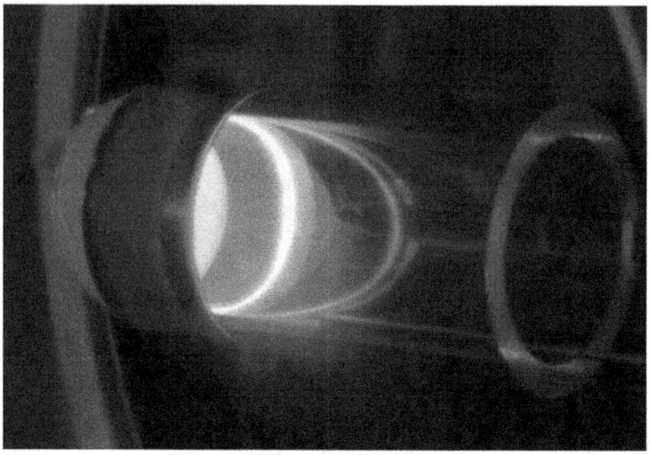

FIGURE 9.25 DCD in cylindrical inlet, EMF = 460 V, I = 150 mA, M = 5.15, $1.66 \times 10^5 < \mathrm{Re} < 2.57 \times 10^5$.

The density contours of the deactivated and the actuated DCD in the constant cross-sectional cylindrical inlet are presented in Figure 9.26. The copper cathode, anode, and phenolic separator rings have the length dimensions of 1.42, 1.56, and 0.63 cm, respectively. The cylindrical inlet has an inner diameter of 3.50 cm, a sharp beveled leading edge, and is designed to have a comparable cross-sectional area to that of the rectangular inlet of 11.8 cm^2. The entire flow field is dominated by a convergent and divergent conical shock, which is induced by the Mach wave interaction at the leading edge (Shang, 2008). The upstream conical shock converges to a focus and then reflects by a diverged conical shock toward the inlet exit. The comparative

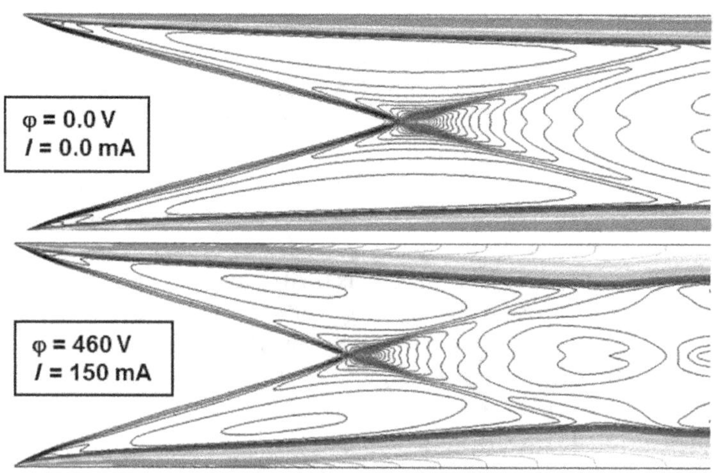

FIGURE 9.26 Density contours of cylindrical inlet with/without activated DCD.

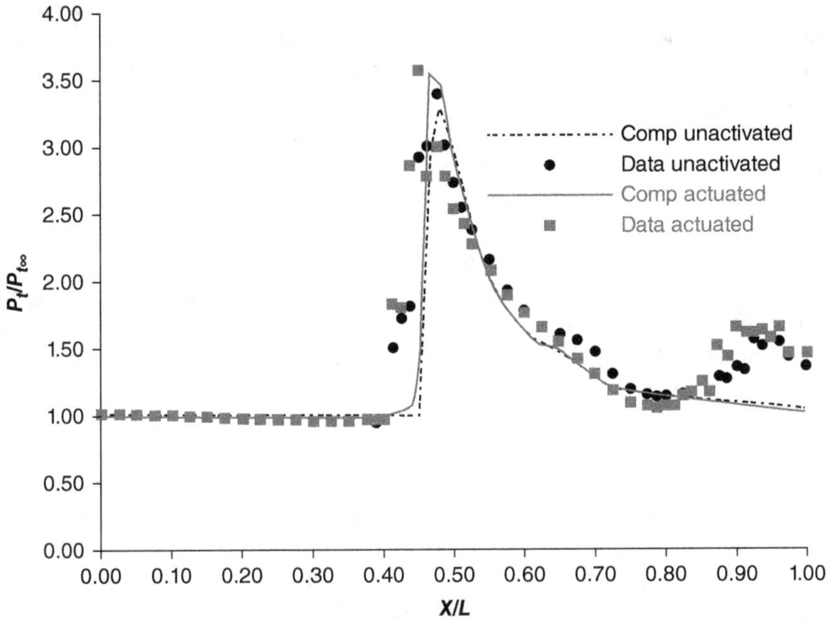

FIGURE 9.27 Axial Pitot pressure distributions with/without activated DCD, $M = 5.15$, $\mathrm{Re} = 1.66 \times 10^5$, EMF = 480, $I = 150$ mA.

study consists of two computed results; on the upper half-plane of the figure is the baseline case with the deactivated DCD and on the lower half-plane the solution is generated with an actuator powered by 69.0 watts, an applied electric field potential is 460 volts, and a current of 150 mA. When the DCD is activated, the strengthened and reflected conical shock impinges on the inner surface at a shorter distance from the inlet entrance and with a higher impingement pressure than that of the base case. At a lower Reynolds number condition, $\mathrm{Rey} = 1.66 \times 10^5$, the impinging shock generates an unexpected and additional shock-boundary–layer interaction near the inlet exit. The numerical simulation shows that the adverse pressure gradient triggers an incipient flow separation near the exit. This result is partially verifiable through the experimental observations by Pitot pressure measurements.

This specific phenomenon is verified in Figure 9.27 at a Reynolds number of 1.66×10^5 by comparison with streamwise Pitot pressure distributions along the axis of the model. The computed and measured results of the actuated DCD are designated by the solid line and filled square symbols, respectively, and the unperturbed condition by the dashed line and the filled circles. The computational simulations do not include the slightly blunt leading edge and thus have a weaker leading-edge oblique shock in comparison with the experimental data. As a consequence, the shock focus is consistently formed downstream of the experimental observation. More importantly, the divergent conical shock with deactivated DCD, after passing through the shock focus, does not impinge on the inlet side wall near the inlet exit. However, shocks studied by experimental observation impinge on the side wall and create an adverse pressure gradient near the inlet exit. This, in turn, has led to an incipient flow separation, and the flow becomes unsteady at the same Reynolds number. This behavior is different from the numerical simulations of the cylindrical inlet with a sharp leading edge for which the divergent shock exits the inlet nearly uninterrupted.

The important and precise comparison of pressure profiles near the cylindrical inlet exit plane is prevented by unsteady or an incipient separated flow behavior at the inlet exit. Two sets of data taken on different days exhibit a much larger difference. Nevertheless, the computational simulations of the cylindrical inlet indicate a definitive but smaller gain in DCD compressions than that of the rectangular configuration. All computational results are generated by a finite volume computer program with a drift-diffusion DCD model (Rumsey et al., 2007; Shang et al., 2007). From the computational investigation and verified experimental data for the virtual inlet cowl, the DCD in a rectangular inlet has produced an additional electromagnetic–aerodynamic compression of 11.7% with a plasma generation power of 64 watts compared to the unperturbed counterpart. The DCD in the cylindrical inlet produces less compression and only alters the compression by 6.7% under the identical free stream conditions of a Mach number of 5.15 and a Reynolds number of 2.57×10^5. The lower compression has two distinct contributors. First, the electrode electric current density is much lower for the cylindrical inlet. Second, the intrinsic difference in spatial dimension plays a role between three-dimensional rectangular and axisymmetric inlet cross sections. The Mangler scale of the $\sqrt{3}$ for an axisymmetric configuration actually reduces the outward boundary layer displacement thickness over a surface (Mangler, 1945). In summary, a DCD generates a lower electromagnetic–aerodynamic compression for the axisymmetric inlet. The compression is also supplied at a lower power input for plasma generation at 2.8 W/cm^2.

9.5 TRAILING EDGE OF AIRFOIL

The most effective flow control for altering lift by small perturbations of a subsonic flow over an airfoil is made by applying them to the trailing edge, because the Kutta condition determines the circulation around a lifting device (Von Von Mises, 1945). This condition stipulates a finite velocity leaving above the trailing edge that implies the vortex density vanishes locally, and there is no flow across the mean chord of a thin airfoil. In other words, the velocity leaves parallel to the trailing edge of the lifting surface. A DCD over the upper or suction side of an airfoil can modify its aerodynamic performance. Since the predominant effect of a DCD mainly evolves from Joule and electrode heating, this electromagnetic heating thickens the boundary layer displacement thickness, thus increasing skin friction and form drag, but it also alters orientation of the velocity leaving the airfoil. Although the physics of a DCD severely limits practical applications to a low-pressure environment, up to a few torr, the constraint can be removed by either pulsing the DCD or using an AC current DBD. Massines et al. (1998), Roth (2000), and more recently Enloe (2004) have carried out comprehensive studies of the electrodynamic structure of the DBD in an atmosphere condition. However, most DBD actuators focus on boundary-layer separation and laminar–turbulent transition such as work by Corke et al. (2010).

Despite the inherent limitation of a diffuse DCD to a low-pressure environment, some recent efforts have successfully simulated the electromagnetic–aerodynamic interaction with a DCD. The following discussion for flow control is along a similar vein, to exploit the fundamental mechanisms and flow structure alternations by the electromagnetic–aerodynamic interaction. The coupling between aerodynamic and electromagnetic fields at bifurcations lacks rigorous internal consistency. Thus, the present discussions are restricted to show how electromagnetic perturbations affect global structures through conventional aerodynamic parameters, such as the pressure, lift, and drag coefficients versus Mach and Reynolds numbers. Accurate computational simulations for flow control using a DBD or DCD at bifurcation points, such as the laminar–turbulent transition points, points of flow separation, and airfoil stall have to wait for future development.

The low magnetic field Reynolds magneto-aerodynamics equations retain essentially the basic features of the time-dependent, compressible Navier–Stokes equations. The Lorentz force and Joule heating appear as source terms in the momentum and energy conservation equations. The solving procedure is an implicit flux splitting, finite-volume method in conjunction with a limiter to approximate the Riemann problem. Thus, upwind-biasing is used for the convective and pressure terms, while central differencing is used for the shear stress and heat transfer terms. The drift-diffusion DCD model, equations (9.1), (9.2), and (9.3), are implemented to systematically analyze the effectiveness of a surface plasma actuator over a NACA-0012 airfoil. A pair of electrodes is placed near the trailing edge of the lifting surface in an attempt to alter the Joukowski–Kutta condition, thus the circulation of an airfoil. The electrodes have a combined width of 4.3% of the chord and are placed contiguously to each other. For DBD modeling, the relative position of the exposed and encapsulated electrode is interchangeable to create an induced streamwise wall jet or counterflowing velocity component above the electrodes.

The C type of grid topology is adopted with mesh points closely clustered near the airfoil surface. The nondimensional normal distance from the surface is scaled by the turbulent friction velocity u_τ and is uniformly less than unity over the entire airfoil. The outer computational boundary is placed at a distance of 40 chords away from the solid surface. Two grid systems, (321×105) and (321×121), were generated and the numerical resolution is measured by the total residue between the coarse and fine grid simulations for a zero-angle-of-attack simulation, $\alpha = 0.0$. The accepted error is less than a magnitude of 10^{-6} in normalized dependent variables. This criterion of numerical resolution applies to all computed results.

The characteristics of dynamic stall of an airfoil at low speed are sensitive to the Reynolds number. This behavior can provide additional insights for flow control via plasma actuators. Numerical simulations are conducted for a range of unperturbed free stream Mach numbers from 0.01 to 0.30 and at a range of Reynolds numbers from 1.2×10^4 to 1.2×10^6 based on the chord length of the airfoil. The simulated Mach

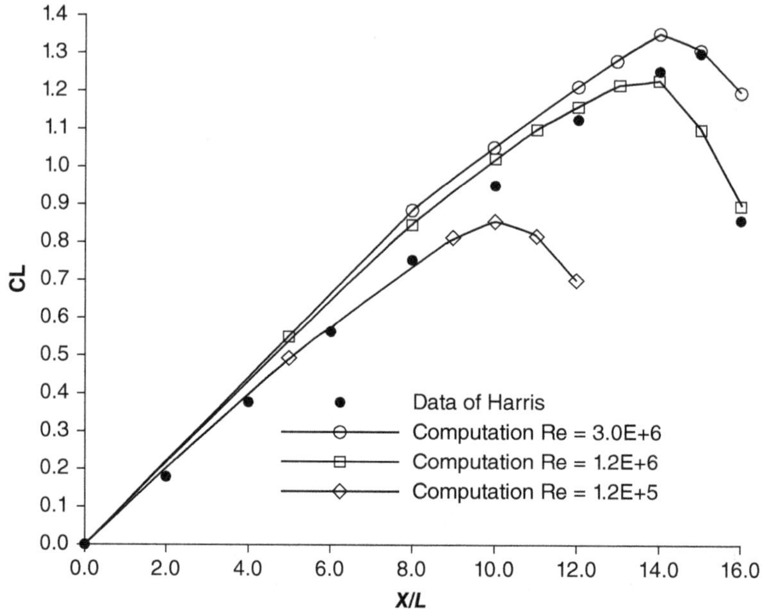

FIGURE 9.28 Comparison of lift coefficients with data.

and Reynolds numbers encompass nearly all documented experimental conditions (Harris, 1981). Figure 9.28 depicts comparisons of computed lift coefficients with data of Harris at a Reynolds number of 3.0×10^6 and a Mach number of 0.3. At the identical flow condition, the agreement between computational solutions and experimental observations is excellent over the complete range of angles of attack from zero incidence until stall. In the same figure, simulated results also reveal that the performance of the thin airfoil is strongly depended on Reynolds number. The stronger viscous–inviscid interaction at the lower Reynolds number begins to dominate flow field structures and results in an aerodynamic stall at lower Reynolds numbers. This observation suggests that there is an interesting scaling issue of applying a plasma actuator for flow control.

The density contours of the flow field around the NACA-0012 airfoil are depicted in Figure 9.29 at a Mach number of 0.2, a Reynolds number of 1.2×10^4, and an angle of attack of 2°. The flow approaching the trailing edge of the airfoil is profoundly affected by the modeled DCD and shows a large change in density over the electrodes by local heating. The energy input to the stream produces a perturbation beyond the electrode surface. The disturbance even propagates far upstream and downstream of the electrodes, indicating a strong viscous–inviscid interaction. The computational lift coefficient decreases by less than 2% and the drag coefficient increases 3.6% from the unperturbed flow. These observations are in general agreement with results generated at a lower Reynolds number of 2.5×10^3 and a larger electrode surface. The reduced influence of a DCD at a higher Reynolds numbers is reflected by a reduced cathode layer thickness. The modified flow field reveals that the viscous–inviscid

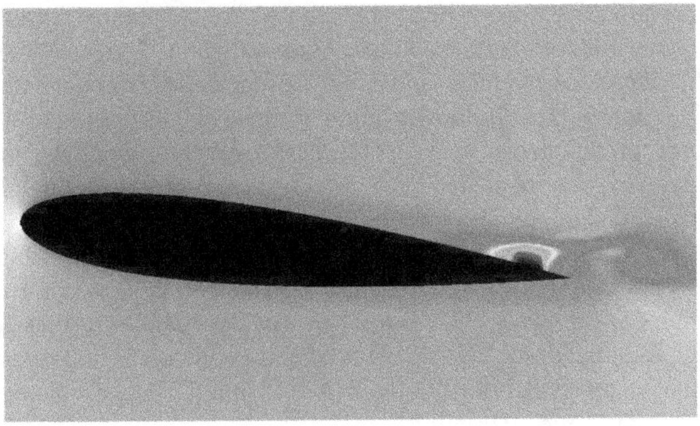

FIGURE 9.29 Density contour around NACA-0012 airfoil, $M = 0.3$, $Re = 1.2 \times 10^4$, $\alpha = 2$.

interaction is the fundamental mechanism for a plasma actuator and it is not overly dependent on the flight Mach number.

The pressure coefficient distribution, C_p, over the entire airfoil, with and without the DCD model, is given in Figure 9.30. These simulations are generated at a Mach number of 0.3, a Reynolds number of 1.2×10^4, and at a moderate angle of attack of 5°. Under a low Reynolds number condition, the modeled electromagnetic perturbation induces a significant change over the electrodes near the trailing edge,

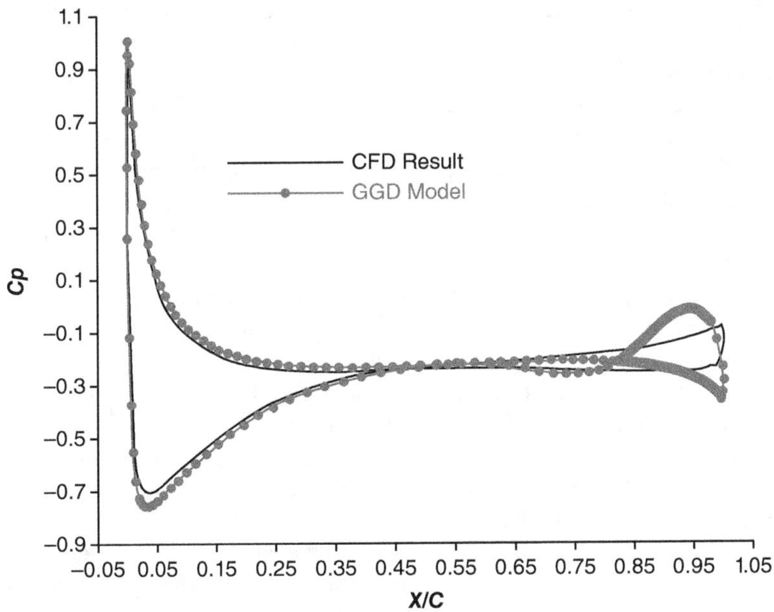

FIGURE 9.30 Comparison of C_p distributions with/without activated DCD, $M = 0.3$, Rey $= 1.2 \times 10^4$, $\alpha = 5$.

and the perturbation even propagates upstream to the stagnation point at the leading edge of the airfoil. As anticipated, the lift coefficient C_L reduces and the drag coefficient C_D increases by 1.2% and 1.0%, respectively. However, the important point for pondering is that the power input to the DCD is less than 64 watts but produces a perceptible change in the airfoil performance without any moving parts like present-day flow control devices. Admittedly, the present modeling of a DCD has significant limitations in physical fidelity, and yet it reveals several interesting issues for practical applications. Again, the scaling of Reynolds number is found to be most pronounced to indicate that the fundamental mechanism for flow control is built on electromagnetic perturbation. The control effectiveness can be further amplified by viscous–inviscid interactions. To achieve practical applications, the physical fidelity of the plasma actuator model must be improved to attain a decisive determination of the magneto-aerodynamic interactions.

9.6 HYDRODYNAMIC STABILITY AND SELF-OSCILLATION

Small perturbations in a boundary layer by a shallow bump can produce a profound change to the dynamic structure of a flow field (Gaster et al., 1994; Joslin and Grosch, 1995). A minuscule bump residing deep within the sublayer is easily mimicked by a thermal source to disturb velocity components. Similarly, a change in acoustic propagation speed within an open cavity can disrupt the pulsating pressure feedback loop to suppress or alleviate a self-sustained fluid oscillation (Rowley and Williams, 2006). The source of disturbance fits nicely within the capability of a DCD.

From the measurements of a velocity field perturbed by an oscillating bump within the boundary layer, the disturbance is entirely confined to the boundary layer, but the spanwise profile is substantially different moving downstream. This small disturbance to a boundary layer is studied to understand how such small perturbations can be a source for the laminar–turbulent transition. Using a direct numerical simulation, the experimental observation is duplicated to show that a pair of counterrotating streamwise vortices affects near-wall shear-layer structure. In theory, a thermal bump produced by a DCD through Joule heating will be more effective influencing flow stability over a wide Mach number range. Especially, since the stability of the shear layer is sensitive to surface heating. On the contrary, an unsteady localized arc filament plasma actuator has been shown to influence flow stability and offers the possibility for active control of a free jet at high speed and high Reynolds number. These two DCD applications for flow control are discussed next.

Yan and Gaitonde (2010) initiated a computation simulation for thermal bumps in steady and pulse modes. The studied configuration is a heating element placed in the center of a flat plate and the distance between the leading edge of a heat source and the leading edge of the plate is 7.63 mm, and the characteristic Reynolds number based on this running length is 3.721×10^5. From stability theory, at a free stream Mach number of 1.5 and a pulse frequency of 100 kHz, the wave length of the most unstable spanwise mode is 2 mm. The numerical results are generated on a maximum grid-point density of (854 × 287 × 81), using a high-resolution, upwind-biased third-order Roe scheme. Both steady and unsteady types of finite-span disturbances are used to generate streamwise vortices. The steady heating results in a decaying pattern downstream, whereas the pulsed thermal bump leads to a continuous vortex shredding and the magnitude of the streamwise vortical disturbance grows in the downstream direction. This pattern of disturbance propagation is consistent with linear stability analysis. The integrated disturbance energy also indicates that the energy growth of streamwise disturbances dominates in other directions. The numerical results reveal that the resulting three-dimensional flow distortion produces hairpin vortical formations that align with the streamwise direction. The computational results seem to indicate that the transition process bear analogy to the κ-type breakdown.

Numerical simulations are focused on the signature of vortical structures for turbulent flow as a perturbation moving downstream from the thermal disturbance. For a finite-span simulation, a series of counterrotating streamwise vortices is generated at the edges of the thermal bump. The initial vortical development is different from an infinite span simulation, but the vortex stretching by an induced, strong streamwise vorticity in the wall region is similar. In Figure 9.31, the hairpin-like vortices are presented in uneven stretched scales in the y and z coordinates and initiated across the span at a Reynolds number of $Re_x \approx 2.0 \times 10^6$. From an iso surface of the vorticity magnitude, the legs of the hairpin structures originating from the wall constitute a pair of counterrotating vortices that align with the streamwise location. The vortices are mainly composed of the x-component of the vortex, but the spanwise z-component is dominant in the boundary layer. These vortices also appear as heads of these hairpin structures in the graph. Their numerical results also demonstrate noticeable differences between finite and full span simulations. However, the appearance of the hairpin vortices confirms that the initial counterrotating streamwise vortices are not necessary for generating the hairpin vortices. It is important to note that the hairpin vortical structure is considered as one kind of the basic structures in turbulence and it is the mechanism for increasing momentum exchange in the near wall region. In summary, a thermal bump is easily generated by a DBD actuator and can be applied as an initiator for laminar–turbulent transition at a designated location.

A frequently overlooked DCD application is for suppression or alleviation of the pressure fluctuations in an open cavity. This type of oscillatory motion when amplified by a feedback loop can reach a resonating condition and cause catastrophic structural failures. This oscillatory phenomenon is not limited to an open cavity, but also occurs in combustor and nozzle exhausts. The self-sustained flow field oscillation is generated by an unstable free shear layer impinging on a solid wall and the pressure pulse is transmitted upstream by acoustic waves. Once the feedback loop is closed, the pulsations are amplified at discrete frequencies and reach a limiting cycle, which have been identified as different discrete or Rossitor modes. A localized arc filament plasma actuator with high-amplitude and wide-bandwidth has been applied to demonstrate that the peak tone, as well as the broadband pressure fluctuations in a strongly resonating cavity, can be significantly suppressed. At the same time, the resonance can be reestablished and significantly amplified in a weakly resonating cavity.

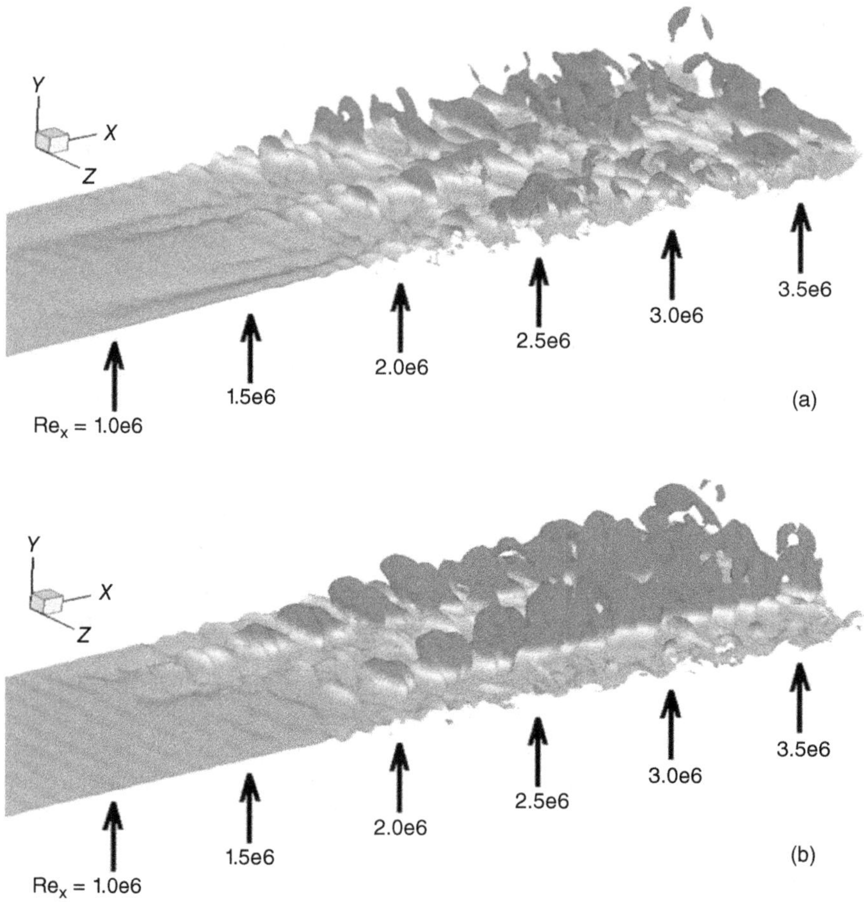

FIGURE 9.31 Hairpin vortex downstream of thermal bump. From Yan, H., and Gaitonde, D., *Phys. Fluid*, vol. 22, 2010. Used with permission from AIP.

Figure 9.32 displays the sound pressure level in decibels for several forcing frequencies by a localized arc filament plasma actuator. The electromagnetic interference is detected by a Kulite transducer. In the forcing frequency range tested, the maximum peak tone reduction of 10 to 20 dB is observed, as well as an averaged broadband noise reduction of almost 5 dB. The experimental results also reveal a unique resonant behavior at a commensurable, discrete frequency of the cavity oscillations (Yugu Yugulis et al., 2014).

This self-sustained phenomenon is known to not only exist in a cavity, but occurs in inlet buzzing, spike-tipped oscillation, and counterflow jet injection resonances. The pressure disturbances propagate upstream from the combustor through the subsonic portion of the boundary layer can disrupt a delicately balanced inviscid–viscous interaction generated by Mach wave interaction and compression ramps in a scramjet. For this reason, an isolator is a key component of a scramjet. An isolator consists

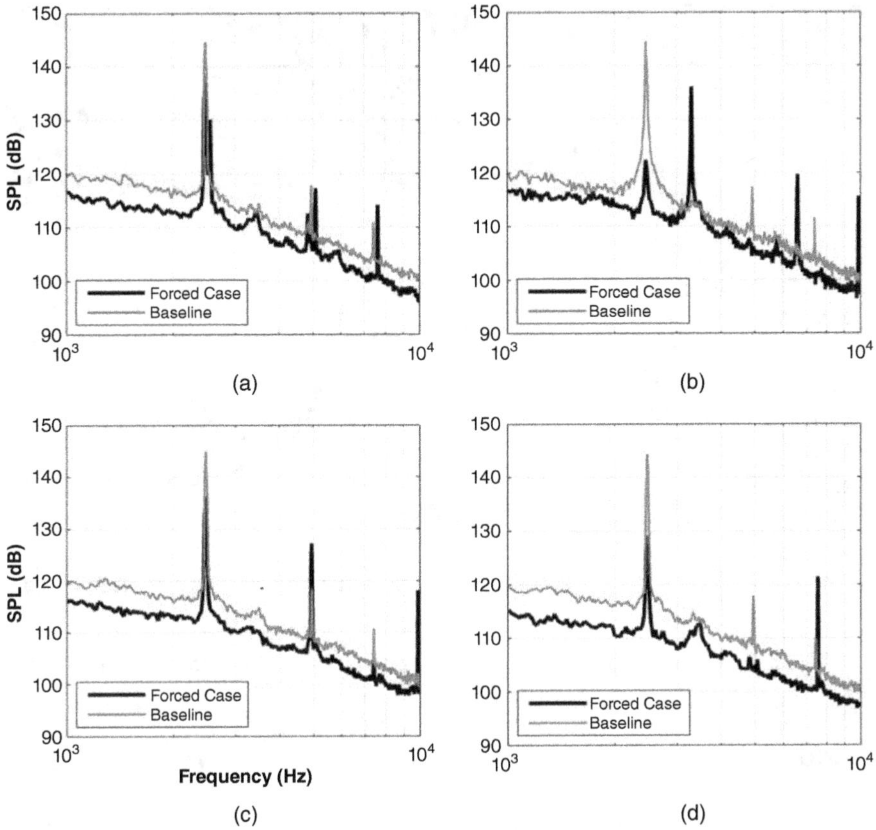

FIGURE 9.32 Suppressed sound pressure level at different Rossitor modes. From Yugulis, Y., Hansford, S., Gregory, J.W., and Samimy, M., *AIAA J.*, vol. 52, 2014. Used with permission from AIAA J.

of a constant-area supersonic diffuser serving to contain a complex shock system of decelerating oncoming flow from the inlet cowl from the entrance of the combustor. The major function of the isolator is to reduce the upstream back-feeding of the rising pressure within the combustion chamber. In gist, the isolator decouples the inlet flow field from the downstream combustion chamber. This decoupling is critical to prevent the inlet unstart at a moderate flight Mach number. At a Mach number greater than 8, this undesirable phenomenon will diminish. A good demonstration of the upstream pressure wave propagation in a scramjet isolator is given in Figure 9.33. In principle, this shock train may be interrupted by inserting a cavity in the isolator and using a pulsed DCD or arc filament. A demonstration from the work by Yugu Yugulis et al. shows that flow control with a DBD is a viable option; however, this endeavor presents itself as a formidable challenge for computational electromagnetic–aerodynamic interaction.

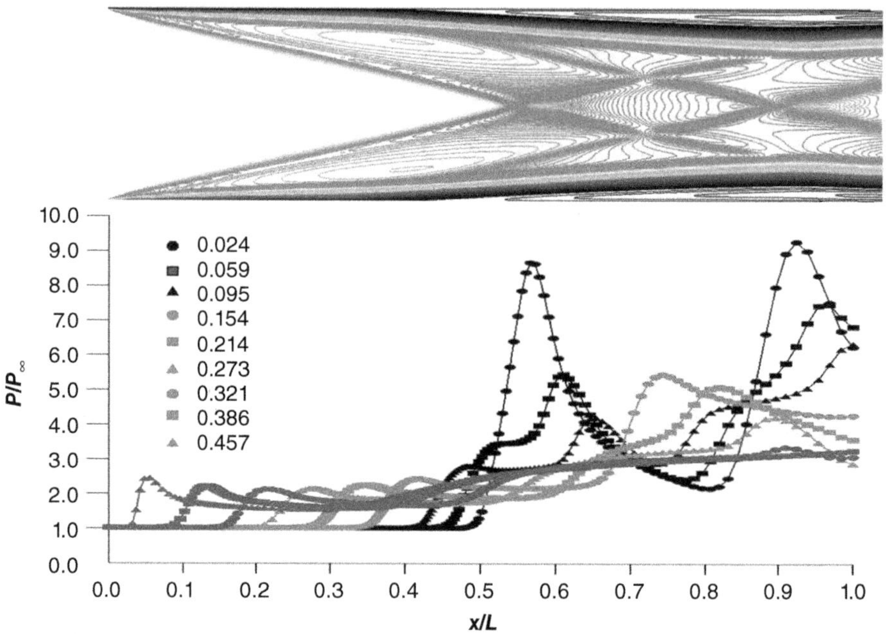

FIGURE 9.33 Upstream pressure propagation in an isolator, $M = 5.15$, Rey $= 1.13 \times 10^5$.

REFERENCES

Artana, G., D'Adamo, J., Léger, L., Moreau, E., and Touchard, G., Flow control with electrohydrodynamic actuators, *AIAA J.*, vol. 40, 2002, pp. 1773–1779.

Bityurin, V., Kilmov, A., Leonov, S., et al., Effect of heterogeneous discharge plasma on shock wave structure and propagation, *AIAA 99-4940*, 1999.

Bletzinger, P., Ganguly, B.N., Van Wie, D., Garscadden, A., Plasma in high-speed aerodynamics, *J. Phys. D: Appl. Phys.*, vol. 38, 2005, pp. R33–R57.

Boeuf, J.P., Lagmich, Y., Unfer, T., Callegari, Th., and Pitchford, L.C., Electrohydrodynamic force in dielectric barrier discharge plasma actuators, *J. Phys. D: Appl. Phys.*, vol. 40, 2007, pp. 652–663.

Borghi, C.A., Carraro, M.R., Cristofolini, A., et al., Magnetohydrodynamic interaction in the shock layer of a wedge in hypersonic flow, *IEEE Trans. Plasma Sci.*, vol. 34, 2006, pp. 2450–2463.

Corke, T.C., Enloe, C.L., and Wilkinson, S.P., Plasma actuators for flow control, *Annu. Rev. Fluid Mech.*, vol. 42, 2010, pp. 505–525.

Cristofolini, A., Borghi, C.A., Carraro, M.R., et al., Hypersonic MHD interaction on conical test body with a Hall electrical connection, *IEEE Trans. Plasma Sci.*, vol. 36, no. 2, 2008, pp. 530–541.

Enloe, C.L., McLaughlin T.E., Van Dyken, R.D., Kachner, K.D., Jumper E.J., and Corke, T.C., Mechanisms and responses of a single dielectric barrier plasma actuator: plasma morphology, *AIAA J.*, vol. 42, 2004, pp. 589–594.

Fraishtadt, V.L., Kuranov, A.L., and Sheikin, E.G., Use of MHD systems in hypersonic aircraft, *Tech. Phys.*, vol. 43., no. 11, 1998, pp. 1309.

Gaster, M. Grosch, C.E., and Jackson, T.L., The velocity field created by a shallow bump in a boundary layer, *Phys. Fluids*, vol. 6, 1994, pp. 3079.

Gulhan, A., Esser, B., Koch, U., et al., Experimental verification of heat-flux mitigation by electromagnetic fields in partially-ionized-Argon flows, *J. Spacecraft Rockets*, vol. 46, no. 2, 2009, pp. 274–283.

Gurijanov, E.P., Harsha, P.T., Ajax: new directions in hypersonic technology, *AIAA*, November 1996, pp. 1996–4609.

Harris, C.D., *Two-Dimensional Aerodynamic Characteristics of the NACA0012 Airfoil in the Langley 8-foot Transonic Pressure Tunnel*, NACA TM 81927, Hampton VA, 1981.

Hayes, W.D., and Probstein, R.F., *Hypersonic Flow Theory*, Academic Press, New York, 1959.

Hokinson, A.R., Hershkowitz, N., and Ashpis, D.E., Force measurements of single and double barrier DBD plasma actuators in quiescent air, *J. Phys. D: Appl. Phys.*, vol. 41, 2008, pp. 245209-1-9.

Joslin, R.D., and Grosch, C.E., Growth characteristics downstream of a shallow bump: computation and experiment, *Phys. Fluids*, vol. 7, 1995, pp. 3042.

Kimmel, R.L., Hayes, J.L., Menart, J.A., and Shang, J., Effect of magnetic fields on surface plasma discharges at Mach 5, *J. Spacecraft Rockets*, vol. 42, no. 6, 2006, pp. 1340–1346.

Lieberman, M.A., and Lichtenberg, A. J., *Principles of Plasma Discharges and Materials Processing*, John Wiley & Sons, Inc., New York, 1994.

Mangler, W., Boundary Layers on Bodies of Revolution in Symmetrical Flow, Ber. Aerodyn. Versuchsanst. Goettingen Germany, Rept. 45/A/17, 1945.

Massines, F., Rabehi, A., Decomps, P., Gadri, R.B., Segur, P., and Mayoux, C., Experimental and theoretical study of a glow discharge at atmospheric pressure controlled by dielectric barrier, *J. Appl. Phys.*, vol. 83, 1998, pp. 2950–2957.

Matsuda, A., Otsu, H., Kawamura, M., Konigorski, D., Takizawa, Y., and Abe, T., Model surface conductivity effect for the electromagnetic heat shield in re-entry flight, *Phys. Fluids*, vol. 20, no. 127103, 2008, pp. 127103, 1–10.

Raizer, Yu.P., and Surzhikov, S.T., Two-dimensional structure of the normal glow discharge and the role of diffusion in forming of cathode and anode current spots, *High Temp.*, vol. 26, no. 3, 1988.

Raizer, Yu.P., *Gas Discharge Physics*, Springer-Verlag, Berlin, Germany, 1991.

Resler, E.L., and Sears, W.R., The prospects for magneto-aerodynamics, *J. Aeronaut. Sci.*, vol. 25, 1958, pp. 235–245, and 258.

Rosa, R.J., *Magnetohydrodynamic Energy Conversion*, McGraw-Hill Book Company, New York, 1968, pp. 24.

Roth, J.R., Sherman, D.M., and Wilkinson, S.P., Electrohydrodynamic flow control with a glow-discharge surface plasma, *AIAA J.*, vol. 37, 2000, pp. 1166–1172.

Rowley, C.W., and Williams, D.R., Dynamics and control of high-Reynolds number over open cavities, *Annu. Rev. Fluid Mech.*, vol. 38, 2006, pp. 251–276.

Rumsey, C., Biedron, R., and Thomas, J., CFL3D: its history and some recent applications, *NASA TM-*, 112861, May 2007.

Shang, J.S., Chang, C., and Surzhikov, S.T., Simulating hypersonic magneto-fluid dynamic compression, *AIAA J.*, vol. 45, no. 11, 2007, pp. 2710–2720.

Shang, J.S., and Huang, P.G., Modeling AC dielectric barrier discharge, *J. Appl. Phys.*, vol. 107, no. 11, 2010, pp. 113302-1-7.

Shang, J.S., and Huang, P.G., Surface plasma actuators modeling for flow control, *Prog. Aerospace. Sci.*, vol. 67, 2014, pp. 29–50.

Shang, J.S., Electrostatic-aerodynamic compression in hypersonic cylindrical inlet, *Commun. Comput. Phys.*, vol.4, no. 4, 2008, pp. 838–859.

Shang, J.S., Roveda, F., and Huang, P.G., Electrodynamic force of dielectric barrier discharge, *J. Appl. Phys.*, vol. 109, no. 11, 2011, pp. 113301-1-8.

Shang J.S., and Surzhikov, S.T., Magnetoaerodynamic actuator for hypersonic flow control, *AIAA J.*, vol. 43, no. 8, August 2005, pp. 1633–1643.

Von Engel, A., and Steenbeck, M., *Elektrische Gasenladungen*, II, Springer, Berlin, Germany, 1932.

Von Mises, R., *Theory of Flight*, McGraw-Hill, New York, 1945.

Yan, H., and Gaitonde, D., Effect of thermally induced perturbation in supersonic boundary layers, *Phys. Fluids*, vol. 22, 2010, pp. 064101.

Yugu Yugulis, Y., Hansford, S., Gregory, J.W., and Samimy, M., Control of high subsonic cavity flow using plasma actuators, *AIAA J.*, vol. 52, 2014, pp. 1542–1554.

Ziemer, R.W., Experimental investigation in magneto-aerodynamics, *Am. Rocket Soc. J.*, vol. 29, 1959, pp. 642–647.

10

LORENTZ-FORCE ACTUATOR

INTRODUCTION

Fundamental mechanisms for plasma-based flow control actuators are built on the Joule heating $J \cdot E$ and Lorentz force $F = \rho_e(E + u \times B)$ effects. The components of the latter are usually further divided into a Coulomb force and a magnetic force that is exerted on charged particles by electromagnetic fields. In this regard, the Coulomb force is often referred to as the electrostatic force while the second term of the force is characterized as the magnetic force (Raizer, 1991). By definition of the conductive electric current, $J = \rho_e u$, an additional component of electric current will be induced by the gyrating charged particle motion in an externally applied magnetic field. The Lorentz force is thus expressible as $F = \rho_e E + J \times B$. This added physical dimension in aerodynamics is derived from the electromagnetic force/energy of an electrically conducting flow medium. In aerodynamic applications, the drastically altered transport properties of a fluid medium originate from excitation of internal degrees of freedom of the gas atoms/molecules via ionization. In practical applications, the ionization processes are limited to thermal and electron collisions. For both processes, a substantial amount of energy is required to excite atoms/molecules from a stable ground level to highly excite electronic states, and a part of this energy cascades into vibration excitation. Therefore, the effectiveness of flow control is closely tied to the energy required by ionizing processes for generating a sufficiently high charge number density. A significant electromagnetic–aerodynamics interaction must be built on

Computational Electromagnetic-Aerodynamics, First Edition. Joseph J.S. Shang.

self-sustained and amplifiable phenomena that are accentuated either by an externally applied magnetic field or by an ensuing aerodynamic interaction.

The first two topics in this chapter discuss the remote energy deposition via plasma injection or a microwave beam, and the resulting stagnation point heat transfer mitigation. It needs to be pointed out that the microwave energy deposition that is achieved by focusing a laser beam at a given upstream point and releasing energy by the inverse bremsstrahlung effect (Zel'dovich and Raizer, 2002). This is an electromagnetic interaction between the microwave and the gas, which eventually results in gas heating. However, this is not a Joule heating phenomenon as discussed in Chapter 9. In addition, this energy transfer process is entirely different from the energy deposition by a counterflowing jet as discussed by Shang (2002a,b). On the contrary, the mitigating stagnation point heat transfer is achieved by magnetically reducing gradients of velocity and temperature in the stagnation region of a blunt body via Lorentz forces.

For low-speed flow control when applied at the bifurcation points of a flow field in an atmospheric environment, all experimental evidence and numerical simulations have shown that a plasma actuator is most effective. By bifurcation it is meant that two distinct dynamic states of fluid motion coexist at a critical point according to a control parameter. Successful applications are noted at the separation point of flow over airfoils or near the dynamic stall of wings (Corke et al., 2010). Numerical solutions based on plasma models also indicate that the effectiveness of flow control is most strikingly obvious at the lower limit of Reynolds numbers. This observation is easily understood, because the electrostatic force is generated by charge separation in the plasma sheath and by charged particle collisions with neutrals over an encapsulated electrode. In a high Reynolds number environment, which is associated with a higher density or gas velocity, a thinning plasma sheath reduces the possible numbers of collision for momentum transfer between ion and neutral particles. The maximum magnitude of the force per unit volume caused by the Lorentz force can be large, but this force is sustained only during a very short period in a kHz AC duty cycle. The resultant time-averaged force per unit volume is significantly lower, around 10^3 N/m^3. This magnitude of force exists only at a perturbation level, but for certain circumstances it is sufficient to influence the hydrodynamic instability for the laminar–turbulent transition or boundary–layer separation.

An outstanding feature of plasma in an extremely strong externally applied magnetic field is that the magnetic lines are frozen in charged particle motion. This magnetic force restrains the small scale and random particle motions that occur in hydrodynamic instabilities and turbulent flows. A quantification of this phenomenon at lower magnetic flux density conditions was explored for incompressible flow nearly 40 decades ago. This finding is very attractive for any innovative processes to overcome a formidable challenge to improve aerodynamic performance, combustion, and hydrodynamic instability. To generate a very strong magnetic flux density, a high electric current must be maintained by a superconducting electric circuit. Unfortunately, to date very few efforts have been invested in this area of basic research.

10.1 REMOTE ENERGY DEPOSITION

The energy deposition for aerodynamic drag reduction by a counterflowing plasma jet issuing from the stagnation point has been advocated. However, there is uncertainty whether this phenomenon is really generated by the partially ionized gas. To clarify this issue, a series of side-by-side experimental and computational investigations was carried out (Shang, 2002a,b). The plasma is generated by a plasma cutting torch in a force measuring, hemispherical cylinder model. The plasma is generated by the second positive molecular nitrogen electronic transition $C^3\Pi_u \Leftrightarrow B^3\Pi_g$ and via the origin band (0,0) of ionized nitrogen N_2^+, $C^2\Pi_u^+ \Leftrightarrow X^2\Pi_g^+$. The plasma appears as a prominent feature in the spectrum range around 3400 and 4400 angstroms. An average vibration temperature is determined from a Boltzmann plot and blackbody modeling to be 4400 ± 400 K. The injecting plasma can be characterized to have a charge number density of $10^{12}/\text{cm}^3$ and an electronic temperature around 2.0×10^4 K. The near-field electric conductivity is estimated at 100 mho/m and an electric field intensity of 2.5 kV/cm. A detailed composite graph including a cutout mechanics drawing, schlieren image, and photograph of the counterflowing plasma torch is depicted in Figure 10.1. Indeed, the wave drag reduction has exceeded more than 50% for a hypersonic blunt body flow with a counterflow jet. However, the major contribution to the reduced drag is the result of shock wave degeneration through the viscous–inviscid interaction by the counterflowing jet and bow shock wave, rather than by the electromagnetic effect.

The control parameters of a counterflow injection are the jet mass flow rate, jet exit Mach number, and thermodynamic properties of the injectant (Finley, 1966; Shang

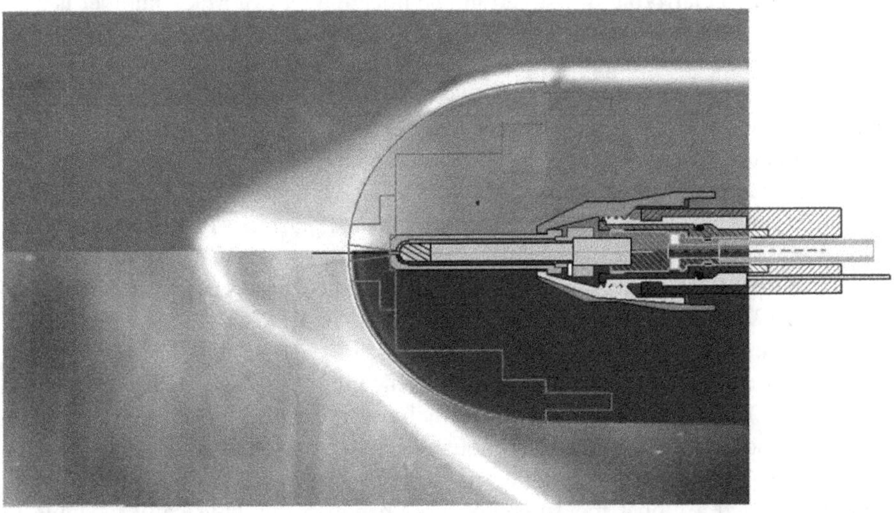

FIGURE 10.1 A composite cutoff layout of the counterflowing plasma injection jet, $M = 6.0$, $M_j = 3.26$.

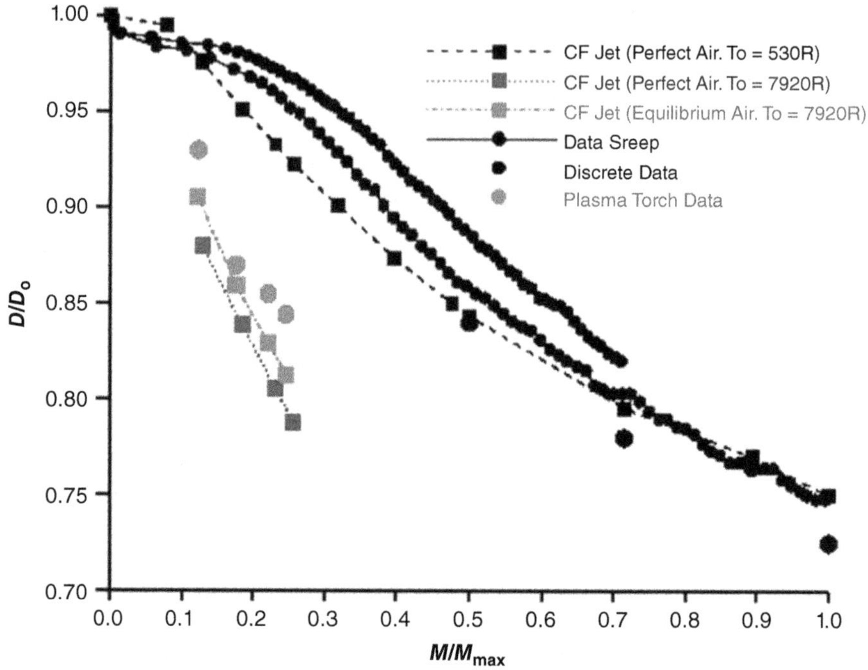

FIGURE 10.2 Compare drag reduction by counterflowing plasma and air injection.

et al., 2002). In a clarifying investigation, the injecting jet exit Mach number is 3.26 and exhausts into an oncoming free stream with a Mach number of 6.0. The injecting plasma jet is in a nonequilibrium thermodynamic state. An accurate description of the transient plasma is far from ascertained, but a reasonable measurement from the average global temperature is possible by emission spectra. Therefore, only the thermal effect of the plasma injection can be isolated and analyzed by solving the mass-averaged Navier–Stokes equations with a known equilibrium air composition. The equilibrium chemical composition of the injected plasma is determined in terms of local mass fractions near the Mach disk in the shock layer at a temperature of 4400 K and pressure of 10.5 kPa. The calculated mass flow rates yield a range of values of 0.13×10^{-3}, 0.19×10^{-3}, 0.23×10^{-3}, and 0.56×10^{-3} kg/s, which correspond to the injecting stagnation pressure ratios of 1.4, 2.0, 2.5, and 2.8, respectively. The reduced drag data as a function of the injecting mass flow rate are given in Figure 10.2. The experimental data also includes measurements by different data collection processes, either by a continuous sweep or by discrete point measurements separately. The discrepancies in values represent the time lag from time-dependent phenomenon and a global measurement error band. In this physically meaningful format, the contribution by the drag reduction can be easily compared. First, the plasma injection at a higher stagnation temperature of 4400 K generates a greater drag reduction than the room-temperature counterpart. The calculated drag force by equilibrium air composition follows closely the measurements of the counterflowing plasma jet to indicate that a local thermodynamic equilibrium assumption is reasonable. The drag reduction by the room-temperature air injection is lower, but a major portion of the difference is created by expansion from a higher stagnation temperature. In addition, the total energy for the plasma medium is distributed among different internal degrees of excitation; thus, the effect by different stagnation temperatures is diminished. Under this condition, the difference in drag reduction between calculated results and experimental measurements is confined within the band of data uncertainty. Second and finally, at identical mass injection rates, the contribution by the plasma medium for reduced drag is within the range of data uncertainty.

Nevertheless, it is still logical to seek aerodynamic performance enhancement through electromagnetic–aerodynamic interactions by taking advantage of a weakly ionized gas as an electrical conducting medium. The altered transport properties, together with the intrinsic interaction with an electromagnetic field, introduce extra mechanisms for momentum and energy transfer to a global flow field. The idea of drag reduction or flow control using remote energy deposition is very attractive. Energy deposition techniques have a wide range of varieties, from plasma arcs, laser pulses, microwaves, electron beams, glow discharges, and plasmoids (Shang et al., 2005). The intended applications encompass an equally wide array from drag reduction, lift and moment enhancement, energetic mixing, to shock structure modification. The mechanism of remote energy deposition for flow control has since become more sophisticated; the energy is introduced into the flow field through either a laser beam as done by Adelgren, Yan, Elliott, Knight, Beutner, and Zheltovodov (Adelgren et al., 2005) or using broad-band microwaves as done by Kolesnichenko (2004) are especially noteworthy.

In most remote energy deposition applications, a high-speed vehicle is immersed in the high-temperature wake of an intense energy source. The flow medium is ionized upstream of a bow shock, contrary to all conventional hypersonic flow environments. If the wake is supersonic and possesses a higher acoustic speed, then the dispersive characteristic of wave propagation in plasma yields an advantage for wave drag reduction. In addition, the energy deposition is self-adjusted, which means that the energy deposition rate is controlled by the local properties of the flow medium. This behavior is especially noticeable when the energy deposition is achieved by electron or laser beams (Raizer, 1966; Morgan, 1975). For laser energy deposition, the initial energy is released to electrons by multiphoton ionization, and then electrons absorb the laser radiation as a result of electron inverse Bremsstrahlung collisions. The suddenly released energy at a predetermined upstream location to a hypersonic vehicle triggers a blast wave that modifies the thermodynamic properties of the oncoming stream. The released thermal energy leads to a lower local Mach number and complex ensuing shock-on-shock interactions. The pulsating chemical kinetics and electromagnetic–aerodynamic interactions involve complex phenomena of an electromagnetic wave propagation in plasma and discharge in space. For viable applications, these phenomena still require a better basic understanding.

A focused laser beam energized by a sufficient power source can achieve a high radiative flux density that leads to an electric discharge. The pressure and temperature of the surrounding gas absorb the released laser energy at the optical focal point and behave like a blast wave from a point energy source. The sequence of events has been studied by Raizer (1966) and Morgan (1975) and it can be described as the photoionization initiated by seeding electrons, followed by the cascading process. The electromagnetic-wave–induced detonation shock by absorption and reflection of laser energy propagates along the beam path, along the axis from the source of the emitting beam to the focal point. In turn, the blast wave compresses and heats the surrounding gas medium; meanwhile, the ionized gas transitions to a lower excited state. The local flow field receiving an extra point source of energy suddenly must adjust accordingly. The vortex ring formation will appear by the curvature of the blast wave (Adelgren et al., 2005).

The time sequence of instantaneous schlieren photographs is depicted by Figure 10.3 for laser energy deposition from a 25.4-mm-diameter spherical-nose cylinder to a distance of 1 diameter upstream (d = 25.4 mm). The experimental observations are collected from a wind tunnel that operates at a Mach number of

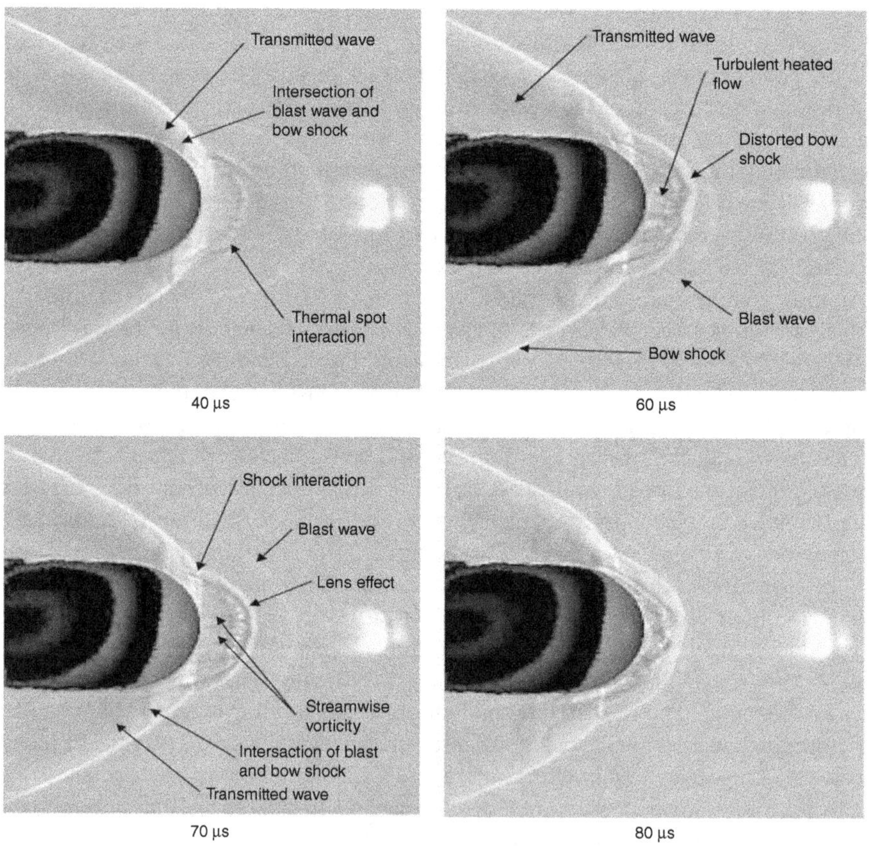

FIGURE 10.3 Time sequence of schlieren images of laser remote energy deposition, t = 60–70 μs. From Adelgren, R.G., Yan, H., Elliott, G.S., Knight, D.D., Beutner, T.J., and Zheltovodov, A.A., *AIAA J.*, vol. 43, 2005

3.45. The energy of the electromagnetic wave is measured at 283 mJ/pulse, and the volume of the energy deposition focal point is estimated at 3 ± 1 mm^3. The remotely deposited laser energy is maintained at a frequency of 10 Hz. Between pulses the flow travels 64 m. The remarkable sequences of schlieren images by Adelgren et al. (2005) during an elapsed time of 60–70 μs after the pulse are given in Figure 10.3. The complex wave structure associated with remote energy deposition by a pulsed electromagnetic wave also stands out for understanding. Specifically, the thermal spot-generated streamwise vorticity begins to break down into smaller scales of eddies. The distorted bow shock over the sphere begins to collapse backward and the faster-moving blast wave coalesces with the bow shock. The unsteady wave motion traveling in the heated wake creates a rapidly changing flow field and leads to significant change in aerodynamic forces and heat transfer.

A microwave discharge is a mechanism that can enhance combustion stability by acting as a catalyst and imparting additional energy to overcome the activity energy barrier of chemical reactions. One general feature unifies self-ignition and simulated microwave discharges that propagate toward the radiating microwave source. This phenomenon is revealed by the waveguide technique in which a transmitting, high-power microwave causes breakdown to leading to plasma creation and propagation in a wave guide. The most effective microwave application is derived from the resonant interaction. At low pressures, when the angular frequency of the microwave, ω, exceeds the electron elastic collision frequency, ν_c, the resonance with the plasma field will lead to a vacuum torch. At high pressure, when $\omega \ll \nu_e$ the resonance is suppressed by dissipation and ceases. In general, the microwave structure consists of curved and branched channels. The continuous elongation of the channel in space is displayed in Figure 10.4 by Kolesnichenko (Kolesnichenko, 2004). A discharge channel in an electromagnetic field with an electric intensity significantly lower than the breakdown value is often referred to as a streamer, and the mechanism of propagation is also known as a streamer.

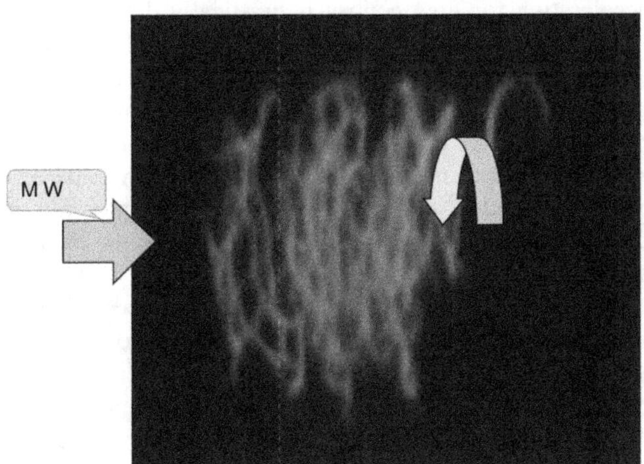

FIGURE 10.4 Streamer structure of microwave. From Kolesnichenko, Y. private communication.

At a constant electromagnetic field intensity, pulse duration, and gas pressure environment, the discharge structure varies from pulse to pulse. This behavior reflects the stochastic nature in the formation of the discharge structure, which is specific to a system formed by the streamer mechanism. The random streamer formation also exists in the dielectric barrier discharge (DBD) that has been discussed in Chapters 2 and 8. A most astonishing element of these structures is the existence of the regular structure of a constant sinusoid wavelength aligning with the polarity of the applied electric field. The discharge propagation velocities may be presented as $u \sim E^{\mu}/p^{\nu}$, similar to the reduced electric field intensity. Additional research in microwave discharge for rapid ignition and enhancing combustion stability is warranted.

10.2 STAGNATION POINT HEAT TRANSFER MITIGATION

For most hypersonic flights, the gas mixture bounded by a bow shockwave and vehicle surface exists in highly excited internal energy modes. As the temperature of the air exceeds 2000 K, a fraction of the dissociated molecules will start to shed their electrons. The ionized air mixture is characterized by a finite value of electric conductivity, which can have values greater than 100 mho/m, depending on the flight speed and altitude. The interaction between the charged particle motion and the electromagnetic field generates a Lorentz force and Joule heating as additional mechanisms influencing aerodynamic performance of flight vehicles. Resler and Sears (1958) examined a dimensionless parameter that governs the relative magnitude of the Lorentz and conventional aerodynamic force. They concluded that the Lorentz force becomes dominant at a high altitude and high speed. This environment is naturally encountered in hypersonic flight.

One of the possible uses of the Lorentz force is to accelerate or decelerate gas continuously at subsonic or supersonic speeds within inlets without choking the flows, even in constant cross-sectional channels. This implementation of Lorentz acceleration/deceleration for flow control is drastically different from altering the displacement thickness via Joule heating like that presented in the previous chapter. The coupling of velocity and temperature of an electrically conducting flow medium through electromagnetic fields makes manipulation of the flow field possible (Meyer, 1958). The flow orientation of a charge-carrying medium can also be altered by the inherent property of Hall currents and the curvature of charged particle trajectories. Hall effects can be applied as the driving force to divert the direction of ionized gas for combustion enhancement or to reduce the length of the inlet compression or distortion.

According to the classic MHD formulation, Lorentz acceleration always has a component, $\sigma(E + u \times B) \times B$, that will decelerate flow velocity (Resler and Sears, 1958). Across a nonuniform velocity profile of a shear layer, the decelerating force changes magnitude with distance from a solid surface. In general, the net result is a reduction in the velocity and temperature gradients of a shear layer. Meanwhile, the normal component of Maxwell's stress tensor also increases the standoff distance of the bow shock. The flattening of both velocity and temperature profiles diminishes

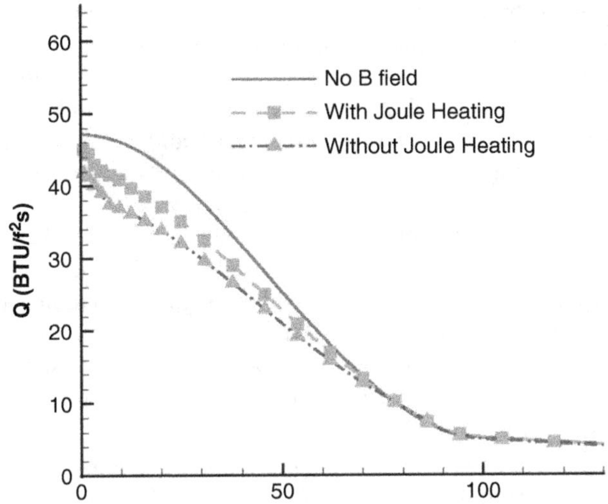

FIGURE 10.5 Idealized stagnation region heat transfer mitigation by a magnetic field.

both skin friction and heat transfer to the vehicle surface. Under some conditions, the eddy current induced by the turbulent fluid particles crossing lines of magnetic field tends to damp the turbulent oscillations. Even at a moderate value of Hartmann number, the wall shear of a turbulent flow will be modified.

An idealized upper theoretical limit for stagnation point heat transfer mitigation is demonstrated by Gaitonde and Poggie (2003) by solving the MHD equations with a value of electric conductivity determined by a square power-law dependence to local temperature (using a reference value of $\sigma_{\text{ref}} = 100$ mho/m). The externally applied magnetic field mimics a solenoid coil with pole aligned to the body's axis of symmetry. At the pole, the maximum flux density attains a value of 0.12 T; this value is realistically achievable by a conventional solenoid coil. The simulated flow condition is described by a free stream Mach number of 16.34 and the Reynolds number based on the diameter of the cylinder is 2.952×10^5. The computational results are compared with the classical solution by Fay and Riddell (1958) and are presented in Figure 10.5. The solution generated in the absence of a magnetic field is given by a solid red line that is essentially the classical aerodynamic solution. Two other numerical results with and without Joule heating are presented on discrete grid points along the blunt forebody surface. All numerical simulations with an applied magnetic field show lower heat transfer rates at the stagnation point than a value of 49 BTU/ft^2-s given by the classical aerodynamic solution. The interacting solution with an electromagnetic field, including Joule heating, exhibits a clear reduction of heat transfer rate over the entire stagnation region in comparison with the classical aerodynamic result.

The boundary–layer approximation at the stagnation region adopted by Fay and Riddell is valid for compressible hypersonic flows in disassociated air. The simplified approximations are restricted only for a gas medium of binary mixture so the diffusion coefficient is practically independent of composition. Since one of the fundamental assumptions is that the boundary layer is much thinner than the longitudinal radius of curvature of the body, the optimal coordinate system for the boundary–layer equation is Cartesian. Therefore, the theoretical results are valid for a wide range of flow conditions and configurations. The demonstrative calculation by Gaitonde actually indicates a reduction in stagnation point heat transfer by 6% or more, including the Joule heating contribution. The computational results are degraded by the well-known carbuncle numerical artifact at the axis of symmetry that has been eliminated since then. In a following computational simulation by Poggie and Gaitonde (Poggie et al., 2002), the reduction in stagnation point heat transfer is reaffirmed and has revealed that a variable electric conductivity for ionized gas tends to diminish the heat transfer rate reduction. Nevertheless, the Lorentz acceleration is demonstrated to mitigate heat transfer in the stagnation region of a blunt body.

Practical applications using an externally applied magnetic field for stagnation point heat transfer mitigation are much more complex than an idealized computational simulation. First, the location of the solenoid coil or magnets with respect to the charged particle motion plays a pivotal role in the magnitude of the Lorentz force. The magnetic flux density decays according to an inverse square of the distance from the magnetic pole. Second, and most critically, the scalar relationship between electric current and electric field intensity by the generalized Ohm's law is no longer valid. The produced electrical conductivity by both electron impact and thermal ionization is also rather low in all circumstances. The electric conductivity in the shock layer has a value lower than 100 mho/m and varies rapidly from the immediate downstream location of an enveloping bow shock to the body surface. Therefore, the most reliable physical data must be acquired by nonintrusive experimental techniques and by estimation.

Gulhan et al. (2009) have provided some concrete experimental data by conducting measurements in an arc heated facility. The flow medium is ionized argon generated by a maximum power input of 1.4 mW, at a stagnation pressure, and specific enthalpy of 37.5 kPa and 2.0 mJ/kg, respectively. The free stream velocity of 2350 m/s is measured by diode laser absorption spectroscopy, and the partially ionized argon has a range of electron number densities from $2.50 \times 10^{11}/\text{cm}^3$ to $2.75 \times 10^{11}/\text{cm}^3$ as measured by a Langmuir probe and microwave interferometer. Over a quartz covered hemispheric cylinder and a flat frontal surface cylinder, the externally applied magnetic field revealed a reduction of surface temperature as high as 16% and 44%, respectively, by an applied magnetic flux density at their testing conditions.

In another combination of experimental and computational investigations by Matsuda et al. (2008), a weakly ionized supersonic plasma flow with a Hall parameter ($\beta \sim 300$) has also been performed over hemispherical cylindrical models with both a conductive and insulated surface in an arc heated tunnel. Their major findings substantiate the results by Ziemer (1959) of the outward displacement of the bow shock standoff distance by an externally applied magnetic field. In addition, their data also show a dependence of the model surface conductivity. The observed shock layer thickness is greater for an insulated surface model than that of a conductive surface model. A quantification of a reduced heat transfer in the stagnation region is obscured by the electromagnetic wave interference to their spectral measurement techniques. In all, the quantification for mitigation of the stagnation point heat transfer is extremely difficult, even by the most sophisticated measurement techniques. This formidable challenge is extravagated by interferences created by an applied magnetic field to measuring instrumentations.

10.3 FEATURES OF DIELECTRIC BARRIER DISCHARGE

Perceptible flow control effectiveness using surface plasma perturbations has been demonstrated by direct current discharges (DCDs) and corona discharges from low density subsonic to hypersonic flow regimes. The most promising approach for flow control at atmosphere pressure is demonstrated by Corke et al. (2010), who adopt the DBD for surface plasma generation. This particular gas-ionizing process via electron impact is also known as the silent discharge and has been applied to ozone production since the middle nineteenth century. In applications, a DBD operates at atmospheric pressure with a gap distance between electrodes on the order of 1 mm to a few centimeters (Moreau, 2007). At a relatively large combined value of pressure and gap distance ($p \times d$), the BDB operates in a streamer mode. An alternating voltage is required to support the random transient streamer formations in the electrode gap and quenching by localized charge build-up on a dielectric surface. In essence, the DBD discharges are operated in the micro-discharge mode where the streamer discharge's lifetime is governed by the gas medium, electric field intensity, and capacitance of the dielectric barrier. An important feature of plasma generation by a DBD is that the low-pressure restriction imposed by DCDs (Roth et al., 2000) is removed.

The nonequilibrium chemical kinetics associated with the DBD is well known and is thoroughly studied by treating 143 reactions among 30 reacting species (Elisson et al., 1991). The complex chemical composition of a weakly ionized gas is fully recognized, but the most profound feature for aerodynamic applications is the transport property of electrical conductivity. In spite of the long history of DBD applications, the critically important electromagnetic field parameter for electromagnetic–aerodynamic interactions is still largely uncertain, especially the quantification of momentum transfer between charged and neutral particles. Nevertheless, the electromagnetic–aerodynamic interaction through inelastic collisions between ion and neutral particles produces a wall-jet-like air stream. The jet stream is also referred to as the electric wind of a DBD and is located around the plasma sheath. The maximum speed of the wall jet can be as high as few meters per second. This energetic wall jet has been shown to be very effective for flow control. There is a very limited amount of basic knowledge on the interaction of the charged and neutral particles via collisional momentum exchanges in an electromagnetic field. But the source of the driving force is completely understood from charge separation over electrodes that have played a dominant role for DBD applications.

In DBD operation, Townsend's mechanism controls the secondary emission and multiple primary avalanches and ultimately maintains the discharge. Impressive pioneering computational simulation efforts by Massines et al. (1998) have gained helpful insights. In all their approaches, they assume that the electrons and ions under the atmospheric pressure condition are in equilibrium with the electric field. All previous investigators understood the key elements of a discharge; namely, the temporal–spatial variation of the discharge physics must be analyzed in multidimensional space. Their early efforts also investigated the basic phenomenon during a single AC cycle after the initial transient has subsided. The two-dimensional analyses were conducted for a parallel electrode discharge that is unsuitable for flow control applications. For flow control applications, the side-by-side electrode arrangement must be adopted to fully benefit from the wall-jet, so-called electric wind, effect for flow field modifications.

In an AC cycle, when the externally applied electric field intensity exceeds the electric breakdown voltage threshold, the plasma will ignite. Ignition depends on the gas medium, electrode gap distance, and the electric permittivity of the dielectric. In dry air, the breakdown voltage of a typical DBD for flow control has a value slightly more than 2.0 kV. After the breakdown, conductive current appears between electrodes by electron motion. The electric field intensity across the electrode and the dielectric drops from the EMF of the external source but retains nearly a constant value because of the self-limiting characteristics of a DBD that prevent the discharge from transitioning to an arc. This unique feature has been discussed previously by the combined effects of the reduced electric field intensity by charge accumulation over the dielectric surface and a voltage drop in the closed electric circuit. The discharge ceases when the applied voltage falls less than the breakdown value, and the voltage-current relationship returns to an applied AC field. The phenomenon repeats itself as soon as the absolute magnitude of the externally applied voltage is greater than the breakdown value in both positive and negative polarities. This voltage–current relationship is routinely recorded by experimental observations (Stanfield et al., 2009) and computational simulations with physics-based plasma models (Shang et al., 2014). The cyclic voltage and current relationship is displayed in Figure 10.6.

Figure 10.7 depicts a detailed composition of the electrical current components during a complete AC cycle. The experimental data are observed over more than 500 AC cycles to ensure repeatable behavior. The solid line denotes the constant displacement electrical current within the AC circuit operating over a spectrum upward from 5.0 kHz. According to the generalized Ampere's circuit law, the magnitude of this component of electric current is linearly dependent on the rate of change of the electric voltage displacement with respect to time. Thus, the voltage–current relationship traces a straight line with a constant slope. When the measured total electric current departs from the straight line, this is an indication that breakdown occurs. The magnitude of the conductive current during discharge is the difference between the reference conditions and the displacement current. However, at the breakdown voltage, the magnitude of the conductive current is extremely small as has been shown by the results of computational simulations. Therefore, current is not an accurate criterion to determine the exact breakdown voltage but presents irrefutable evidence of the DBD. The calculated conductive electrical current is in good agreement with

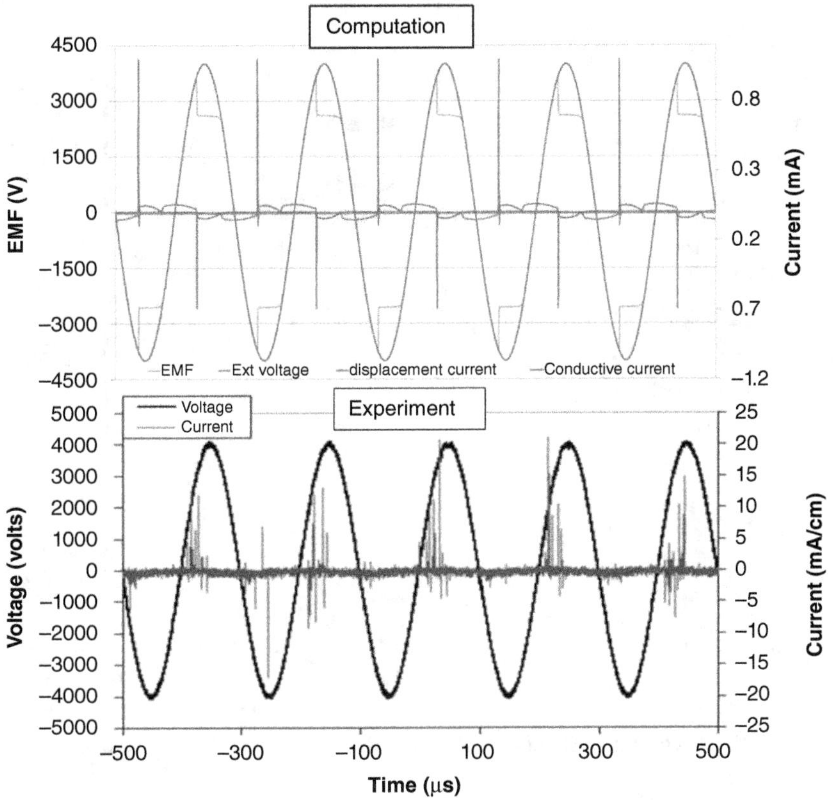

FIGURE 10.6 Voltage–current curve of a DBD, EMF = 5.0 kV, ω = 5 kHz.

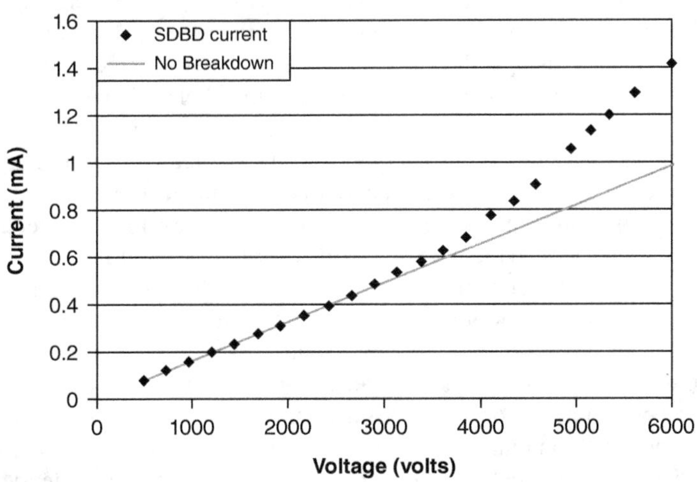

FIGURE 10.7 Electrical voltage and current data for DBD.

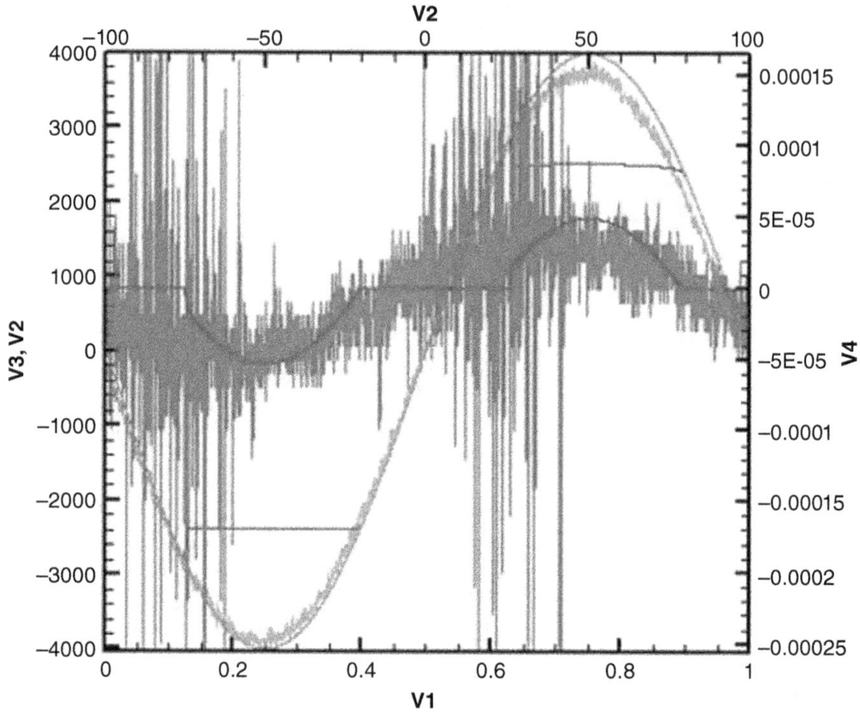

FIGURE 10.8 Effect of EMF on DBD, EMF = 4.0 kV, $f = 5$ kHz.

the data. The integrated conductive electrical current through plasma is less than a few milliamperes and is comparable to the magnitude of a DCD.

The DBD must operate with an AC cycle, generally with a voltage up to 20 kV and within frequencies in the microwave spectrum; therefore, all computational simulations are conducted using a time-dependent procedure (Shang, 2005; Shang et al., 2014; Moreau, 2007; Hoskinson et al., 2008). A typical electric voltage and current relationship during an AC cycle is depicted in Figure 10.8. Because of the charge accumulation on the dielectric surface and the resistance of the discharge gas in an electric circuit, the electric field potential across the electrodes is reduced. The reduced electric potential across the electrodes leads to a self-limiting electric potential across the electrodes, which prevents a discharge transition into an arc (Raizer, 1991). In this figure, the fundamental self-limiting characteristic of the DBD is convincingly captured by the computational simulation at an externally applied electrical potential of 4.0 kV across overlapping electrodes. More importantly, computational results also demonstrate that physical fidelity can be achieved by the drift-diffusion theory and electron impact ionization using quantum chemical-physics modeling. In spite of the fact that computational simulations are incapable of duplicating multiple micro-discharges, the calculated conductive current component denoted by the blue line replicates experimental observations in both magnitude and duration during

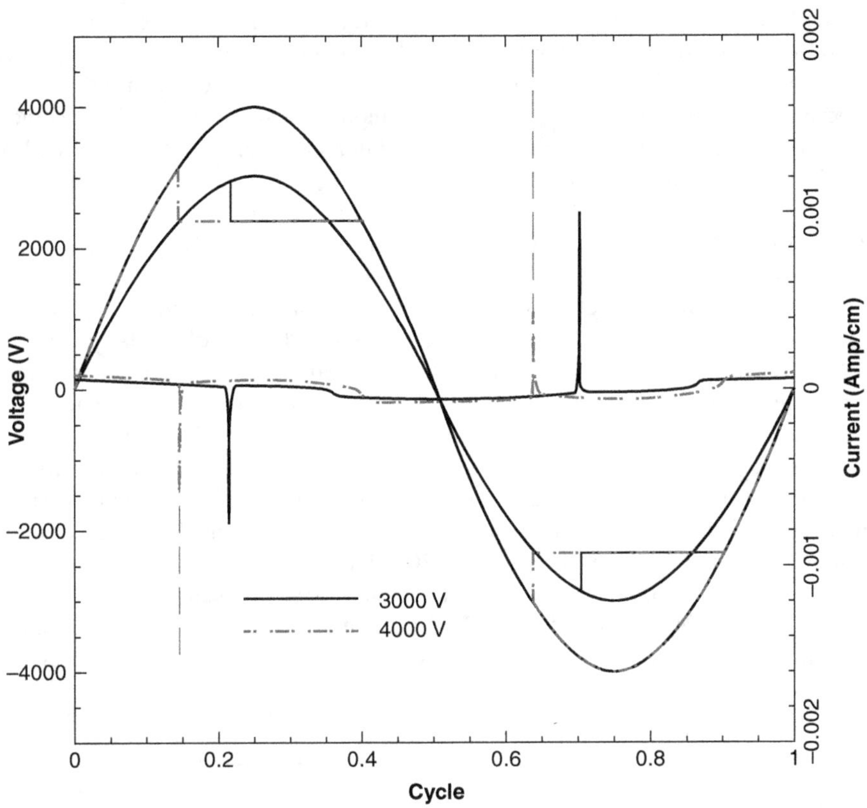

FIGURE 10.9 Breakdown voltage versus externally applied electric field intensity.

discharge periods. In essence, the macroscopic and global behaviors of DBD are captured by the physics-based modeling.

From additional computational simulations and validation by experimental observations, the period of discharge within a DBD cycle is strongly influenced by the externally applied EMF, gap distance of electrodes, and electric permittivity of the dielectric film. At the constant breakdown voltage of a gas mixture, such as air, and across a fixed electrode gap distance, the voltage–current relationship predicted by the drift-diffusion model is depicted by Figure 10.9. The fundamental self-limiting characteristic of the DBD is actually duplicated by numerical simulations at the externally applied electrical potentials of 3.0 and 4.0 kV across overlapped electrodes. At quiescent atmospheric conditions, the discharge breakdown takes place at a voltage of 2.2 kV, before the externally applied electrical field reaches its peak values for both polarities (Huang et al., 2011; Shang et al., 2011). Then a lower and nearly constant electric potential is maintained by the conductive current flowing through the discharge and surface charge accumulation over the electrodes within the discharge domain. In the positive polarity phase, the discharge occurs when the positive-going potential exceeds the breakdown voltage and continues until the decreasing external field falls beneath the breakdown threshold. The identical behavior is also observed for the negative polarity phase. During the initial breakdown process, a sudden drop of the electric potential with respect to time induces a surge of the displacement current in the discharge. In experimental measurements, this surge is indicated by the existence of multiple micro-discharges and also leads to a train of pulsations in the electrical circuit. Similarly, the sudden drop of the electrical potential has also generated nonphysical oscillations in most numerical results known as the Gibbs phenomenon. From all computational results, including the previous simulation in Figure 10.8, the conductive electrical current, although it is minuscule, is seen to exist only during the discharge and vanishes when the applied field voltage is beneath the breakdown value. This numerical behavior has revealed the fact that an elevating, large AC voltage may not increase the period of discharge without upper bound. The reason for a greater integrated periodic electrostatic force at a higher AC voltage is derived from extended discharge duration and a greater charged particle number density.

Additional substantiations of the important dependence of DBD effectiveness for flow control at an electrode gap distance of 1.0 mm are presented by another high-resolution simulation shown in Figure 10.10 (Shang et al., 2014). At the quiescent atmospheric condition, the discharge breakdown of dry air takes place at a voltage of 2.5 kV, at EMFs of 5.0 and 3.0 kV, respectively, before the externally applied electrical potential reaches its peak values for both polarities. Then a lower and

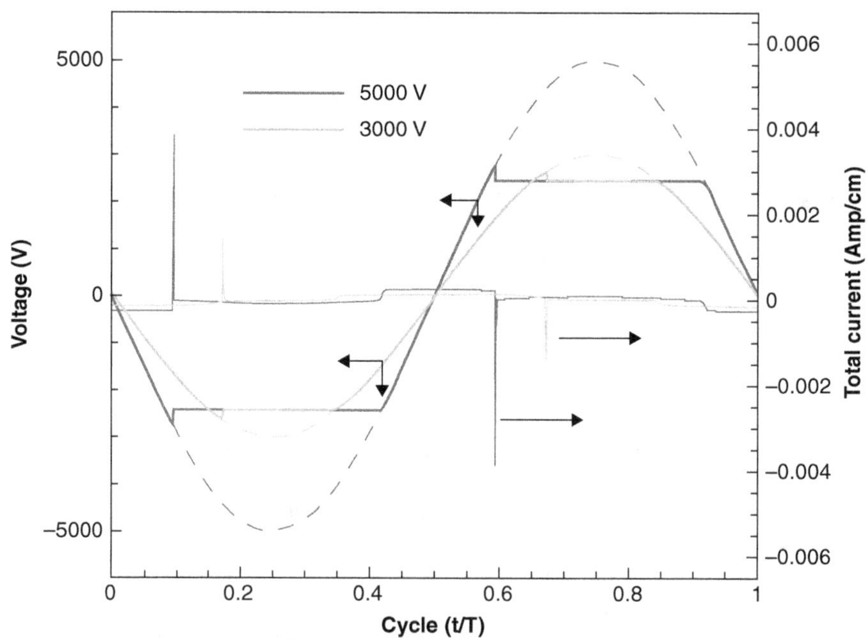

FIGURE 10.10 Effect of EMF on DBD, EMF = 3.0 and 5.0 kV, $f = 5$ kHz.

constant electric potential of 2.5 kV is maintained by the conductive current and the surface charge accumulation on the electrodes. The numerical result duplicates the pattern of experimental surface potential measurements in Figure 10.10. In the positive polarity phase, discharge occurs when the positive-going potential exceeds the breakdown voltage and continues until the negative-going external field falls beneath the breakdown threshold. The identical behavior is also observed for the negative polarity phase but at a reverse varying voltage sequence. By virtue of high-resolution numerical algorithms, the Gibbs phenomenon in the computational simulations is completely eliminated to demonstrate clarity at electric breakdown as shown in both Figures 10.9 and 10.10. The surges of electric conductive current at the breakdown thresholds appear as spikes. The Gibbs phenomenon, however, is accepted numerical behavior in most computational simulations (Golubovskii et al., 2002; Boeuf et al., 2007; Likhanskii et al., 2008; Solov'ev et al., 2008). From the high-resolution computational results, the increased effectiveness of DBDs for flow control is reaffirmed at a greater applied AC voltage. Since the breakdown voltage is maintained at a constant value at a fixed electrode gap distance, the greater externally applied voltage prolongs the discharge period. The electric breakdown takes place sooner and ceases later, leading to a longer sustained periodic electrostatic force from charge separation.

The sizes of ion clouds at the peak positive voltage are compared for EMFs of 3, 5, and 8 kV in Figure 10.11. The discharge domain retains similar shapes, but their sizes are noticeably affected by the driving voltage. The maximum positively charged ion number density increases from 1.6×10^{11}, 5.8×10^{11}, to 1.2×10^{12} cm^{-3} with increasing amplitudes of the applied voltage. The magnitude of the electron density near the electrodes also increases proportionally with a rising applied voltage, but at a lower value than that of the ions because of a greater mobility of electrons. Faster propagation speeds of electrons move them farther away from electrodes. The combination of higher charge number density and longer discharge duration leads to a greater periodic electrostatic force to agree with experimental observation. The maximum instantaneous magnitude of the force generated by 8 kV is more than double that at 5 kV. The value is as high as 2.5 N/cm^3 (Shang et al., 2011). From graphic results, the high-intensity discharge domain is closely clustered to the corner between the exposed metallic electrode and the encapsulated dielectric. The exposed electrode has a height of 1.0 mm, and the highly concentrated, separated free charges are embedded deeply in the plasma sheath. The computational simulations also show that the charged ion density drops exponentially from its peak value near the corner between the exposed electrode and dielectric electrode toward the ambient. The periodic Coulomb force only exists immediately adjacent to the plasma sheath of the electrodes.

Simulations have also been performed to investigate the dependence of discharge structure and force field by the driving voltage. The computational results show that a major effect of a higher applied voltage leads to a higher charged particle number density and longer discharge duration. The discharge period increases from 66 to 78% of the entire AC cycle with elevating values of applied voltage from 5 to 8 kV. On the contrary, the electrical potential within the discharge is unaffected by the range of voltages studied. This behavior of breakdown voltage is consistent

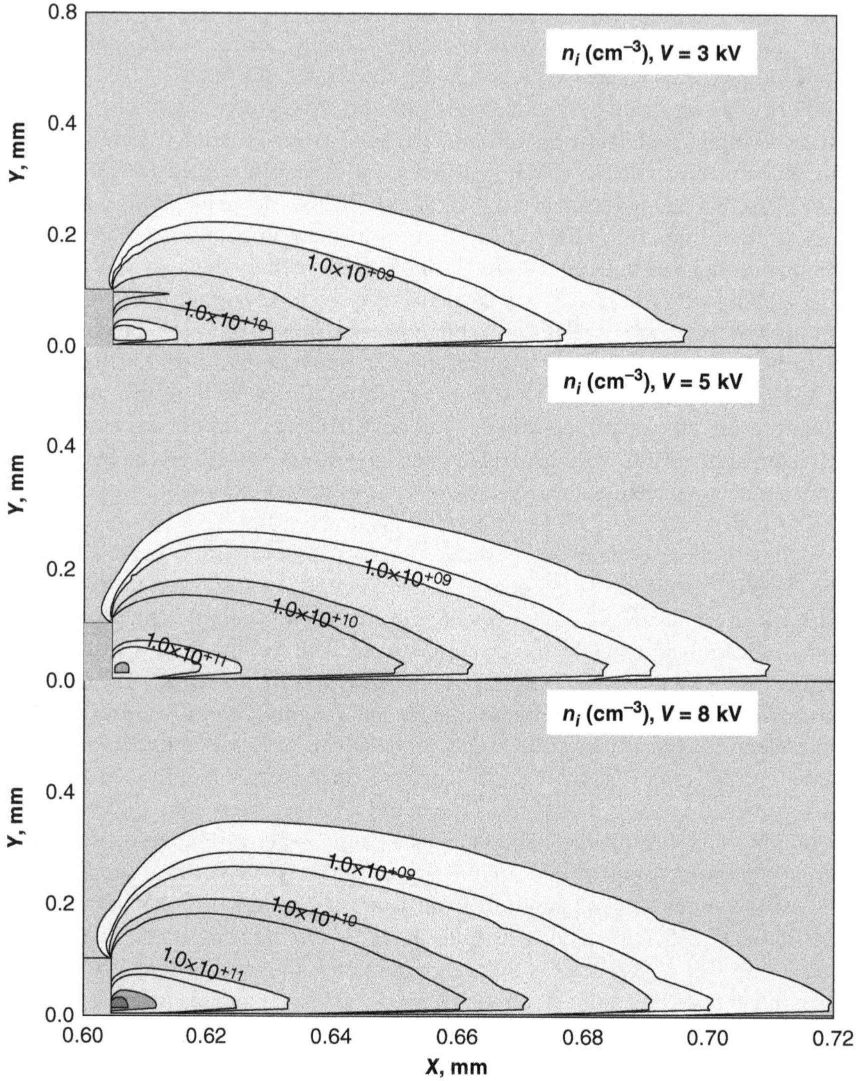

FIGURE 10.11 Ion number density at EMFs of 3.0, 5.0, and 8.0 kV.

with results discussed just, showing that the discharge voltage depends strongly on the dielectric permittivity. The surge peak displacement current is also shown to be strongly dependent on the driving voltage and can be as high as a value of 4 mA/cm with an applied EMF of 8 kV.

Figure 10.12 summarizes the relationship between the breakdown voltage and the gap distance between the exposed electrode and the dielectric barrier surface. Three simulations are produced under an identical externally applied AC electrical potential of 3.0 kV. The gap distances are placed from an overlapping position to

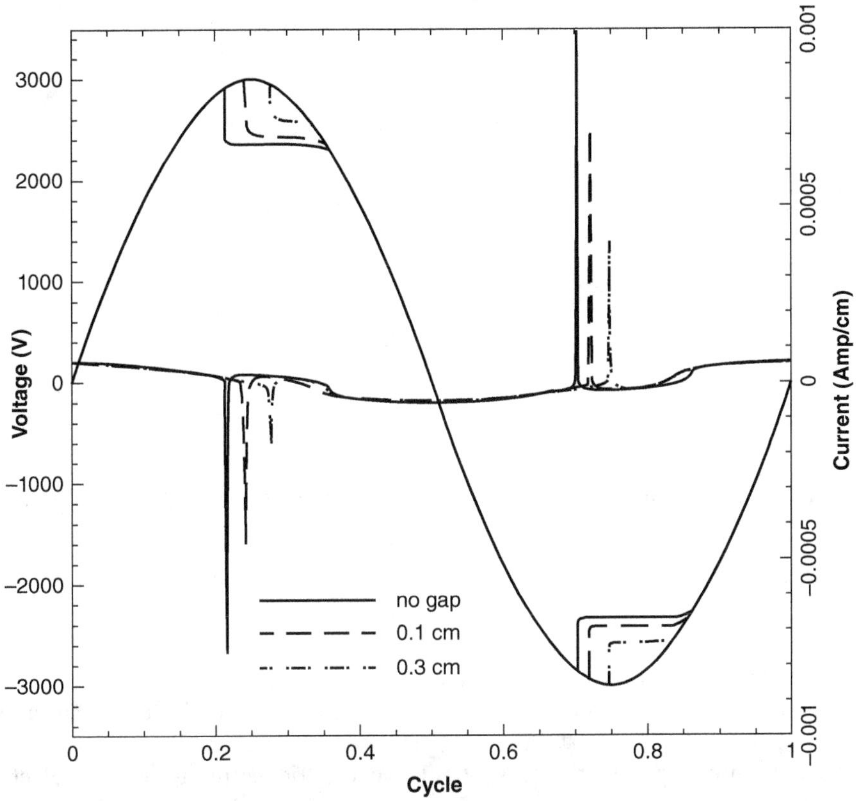

FIGURE 10.12 Breakdown voltage versus electrode gap distance.

distances from 1 to 3 mm. The surge of the displacement currents at breakdown is clearly displayed at the ignition of the discharge for all three different electrode positions. The breakdown voltages of different gap distances increase as the distance is increased. The lowest breakdown voltage is 2.28 kV when the gap distance vanishes, and the highest voltage is around 2.60 kV when the gap distance is raised to 3.0 mm. These observations are identical for both the positive and negative polarities of the AC cycle. An important observation from these numerical results can be made that honors Paschen's law, namely for a fixed product of ambient pressure and gap distance ($p \times d$), the breakdown voltage is fixed. For all the above simulations, the pressure remains at a constant value of one atmosphere. Therefore, the important parameter, the quotient E/P for energy gain between collisions for ionization, is exclusively dependent on the simulated electric field intensities.

The Coulomb force between charged particles is inversely proportional to the electric permittivity of a medium and proportional to the electric field intensity in the discharge. The periodic force is synchronized with the AC polarization. Therefore, the effect of electric permittivity can dominate over electric field interaction with the

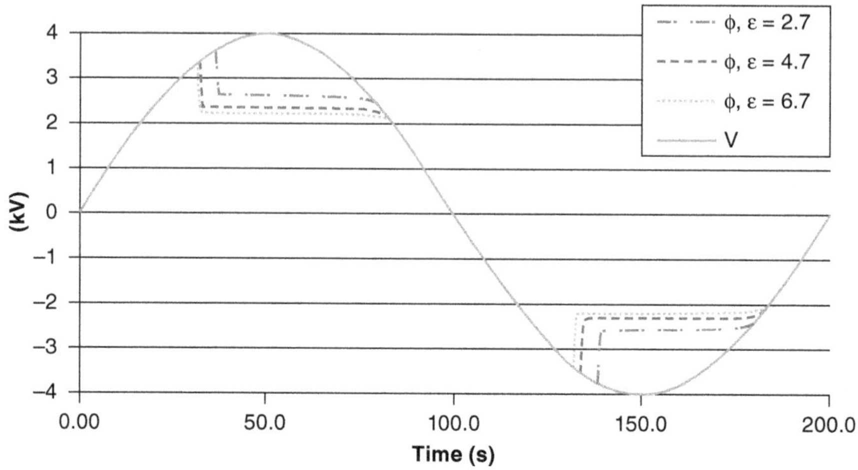

FIGURE 10.13 Dependence of breakdown voltage of the electric permittivity of dielectrics.

surface discharge. From experimental observations, the value of the relative dielectric permittivity of dielectric materials exerts profound influence on the characteristic of the discharge. However, experimental data have not produced a clear quantification of this surface material property associated with discharge dynamics of a DBD. For this reason, the dependence of discharge properties on the dielectric material is studied by modeling. Calculations were performed at an externally applied voltage of 4 kV for three different values of relative electric permittivity for dielectric surfaces, $\varepsilon = 2.7$ (polystyrene), $\varepsilon = 4.7$ (glass), and $\varepsilon = 6.7$ (rubber). Figure 10.13 depicts the dependence of the breakdown voltages on different dielectric materials at an externally applied voltage of 4.0 kV. Increasing the dielectric permittivity leads to a lower breakdown voltage threshold and a greater magnitude of conductive current. The breakdown voltage drops from a value of 3.6 kV at $\varepsilon = 2.7$–3.3 kV at $\varepsilon = 4.7$, but does not significantly reduce between the relative permittivities of 4.7 and 6.7. When the value of electrical permittivity changes from 2.7 to 6.7, the electrical field intensity E in the cathode layer increases its range from 350.0 to 409.0 V/cm. The corresponding conductive electric current also elevates to maximum values of 4.4–6.0 mA accordingly. In essence, the discharge breakdown occurred at a lower voltage for a higher value of relative permittivity and leads to an earlier ignition and a delayed extinguishment; thus, the discharge duration is prolonged.

The effect of electric permittivity of the encapsulated electrode to the size of the discharge is presented in Figure 10.14. The discharge domain shows that increasing the dielectric permittivity leads to a lower breakdown voltage threshold and a reduced discharge domain. The shortened electric current passage between electrodes increases electrical field intensities when the electrical permittivity increases from 2.7 to 6.7. The corresponding conductive electric current also elevates to a higher maximum value mentioned previously. The positively charged ion densities are elevated slightly over the maximum value of $10^{11}/\text{cm}^3$ to reach a value of $1.24 \times 10^{12}/\text{cm}^3$

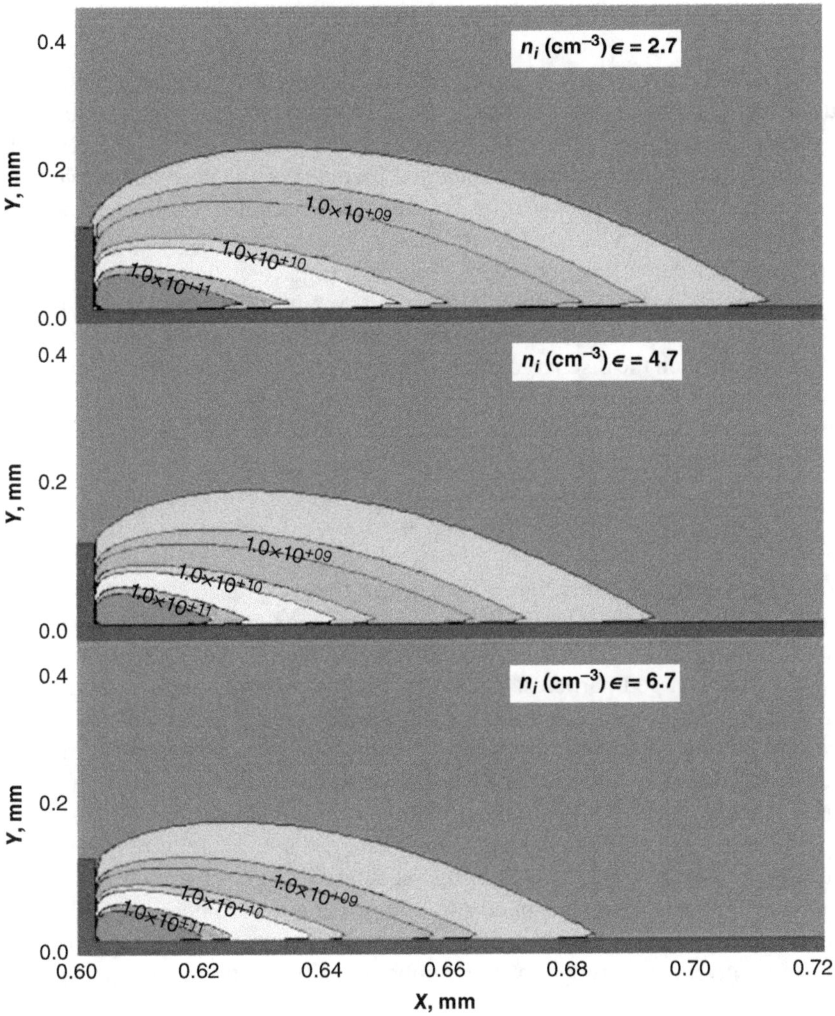

FIGURE 10.14 Effect of permittivity on DBD, $2.7 < \varepsilon < 6.7$, EMF = 4.0 kV, $\omega = 5.0$ kHz.

with greater values of electric permittivity of the dielectric surfaces. The high ion concentration region locates on the dielectric surface immediately adjacent to the intersection of the electrodes. In essence, the discharge breakdown occurs at a lower voltage for a higher value of the relative permittivity of the dielectric surface and reveals an earlier ignition and a delayed extinguishment. These combined effects extend the discharge duration and experiments support the computational findings. However, higher electric permittivity also produces a more compact discharge domain as shown in Figure 10.14. Unfortunately, there is no direct confirmation of this effect from experimental observations.

10.4 PERIODIC ELECTROSTATIC FORCE

The interactions of moving charged particles in an electromagnetic field always induce Lorentz forces. In case of free space charge separation, the electrostatic force is present. The total force in a fully ionized gas consists of a shear stress tensor including the magnetic pressures of charged particles or the normal Maxwell stress components, and body force arising from the gravitation field. In the case being discussed here, this gravitational force is negligible. In partially ionized plasma, the charged number densities produced by electron impact or thermal excitations are so low to appear as trace amounts in comparison with the neutral particle number density. The discharge separation domain is also extremely compact over the solid surfaces upon which the discharge resides and it is confined within the plasma sheath. According to the governing formulation of electromagnetic–aerodynamics, only the Lorentz force on a charged particle in an electromagnetic field is significant and is given as (Boeuf et al., 2007) (Shang et al., 2010)

$$F = q(E + u \times B) \tag{10.1}$$

The electromagnetic force has a component that is aligned collinearly with the external applied electric intensity qE and the another component oriented in the direction perpendicular to both the applied magnetic flux density B and instantaneous velocity of the charged particles. Thus, the resultant charged particle motion is not rectilinear but follows a rather complex spiral trajectory. In the absence of an externally applied magnetic field, a periodic electrostatic force becomes a simpler and dominant dynamic event of a DBD. Over and between discharging electrodes, charge separation takes place. The basic mechanism generating the electric wind is momentum transfer by collisions between positively charged ions and neutral particles. The periodic electrostatic force derives from free space charge separation during cyclic dynamic events caused by an applied AC voltage (Boeuf and Pitchford, 2005; Huang et al., 2011; Shang et al., 2014). The dominant dynamic force at a given point in time can be given as a product of the number density of separated charges and the electrical field intensity vector.

$$F \approx e(n_+ - n_- - n_e)E \tag{10.2}$$

According to equation (10.2), the force vector must follow the combined sign of electric field vector and a net difference of positive ions and negative ions and electrons. A study to show how the force vector is modified by the separated free-space charges, namely the net balance of positively and negatively charged number densities, has been conducted. If one neglects the net balance of charged densities, the electric potential field is described by the instantaneous AC voltage applied across the electrodes. The electric potential distribution across the electrodes is simply a solution to the Laplace equation that is exclusively dictated by the externally applied AC field, as shown by panels (a-1) and (b-1) in Figure 10.15. In direct contrast to

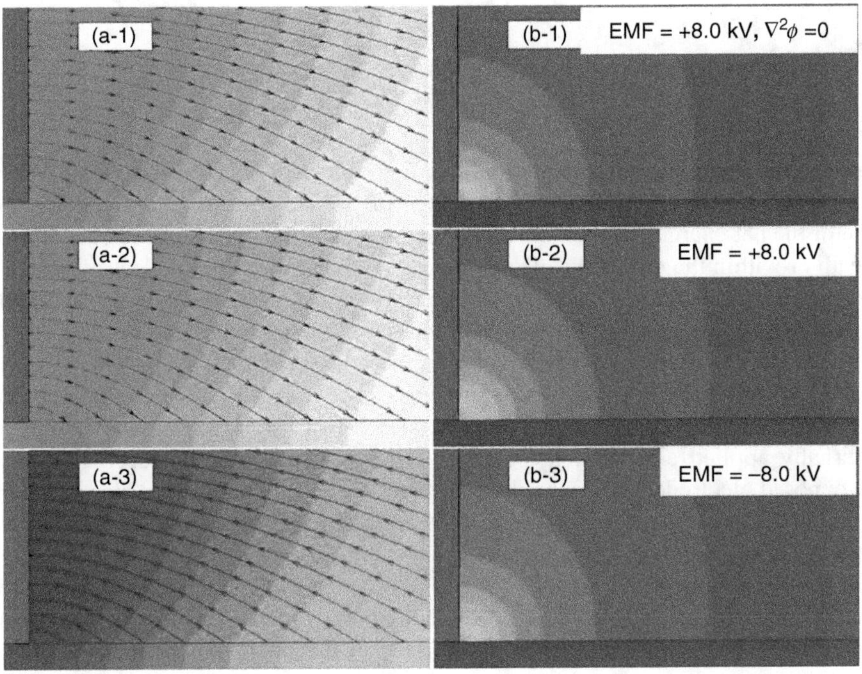

FIGURE 10.15 Perturbation to electric field strength by charge separation in a DBD.

panels (a-1) and (b-1), contours of the magnitudes of the electric fields obtained by solving the plasma dynamics Poisson equation with sums of separated charged particle number densities are displayed in panels (a-2), (b-2) and (a-3), (b-3). The computational results display the same dimensionless quantities when the AC cycle at an applied EMF of 5 kHz are in their maximum and minimum values of ±8.0 kV. As it can be seen from this figure, the separated charge number densities are too small to significantly alter solutions of the electric field intensity. The instantaneous electric force field is therefore mostly acting according to the externally applied electric potential and potential alters direction when the polarity on the exposed electrode changes sign.

For globally neutral plasma and in the absence of induced magnetic field intensity, the electrical field intensity equals the negative gradient of the electrical potential $\vec{E} = -\nabla\phi$. In a three-dimensional formulation, the periodic electrostatic force has three components:

$$F_x = e(n_e + n_- - n_+)\frac{\partial\phi}{\partial x} \tag{10.3a}$$

$$F_y = e(n_e + n_- - n_+)\frac{\partial\phi}{\partial y} \tag{10.3b}$$

and

$$F_z = e(n_e + n_- - n_+)\frac{\partial \phi}{\partial z} \tag{10.3c}$$

From these equations, the orientation of a periodic electrostatic force in a DBD is determined by the product of the derivative of the electric field potential and the sign of the net charged particle density. Charge separation occurs only in the plasma sheath or within the gap of the electrodes and a greater mobility of electrons leave a greater number density of ions in this region. From the ionization kinetics, equation (8.28), the number density of positively charged ions is much greater than that of the negatively charged counterparts. So the net charge particle density in equation (10.3a) through (10.3c) in the sheath region is mostly positive. The orientation of the periodic electrodynamic force is thus dictated by the sign of the gradient of the externally applied electric field potential. The electrical potential decreases from the exposed electrode toward the dielectric surface when the exposed electrode acts as an anode. Therefore, the gradient of the electric potential or the electric field intensity is negative. The resultant force from equation (10.3a) has a positive sign; thus, the x-component electrostatic force is directed away from the exposed electrode toward the dielectric covered electrode. In other words, the x-component of the force is directed toward the dielectric surface in the positively biased phase. For the positively biased phase, the separated ions propagate from the anode toward the dielectric encapsulated cathode. The moving ions collide and transfer momentum to stationary neutral particles to generate the so-called electric wind.

The process and motion of positively charged ions reverse its direction during the negatively biased AC phase. The force vectors for the positive and negative polarity phases of the AC cycle are depicted in the upper and lower panels of Figure 10.16 at the peak values of applied voltages, ±5.0 kV. By convention of a flow control actuator, the electrostatic force component in this dynamic event is the so-called push-and-pull type in reference to the edge of the exposed electrode on the dielectric covered electrode side (Enloe et al., 2004). It shall also be noted that during the negative AC polarity phase, the electrostatic force is exerted toward the vertical face of the exposed metallic electrode. The momentum transfer from the positively charged ions to the neutral particles is rather ineffective to reverse the motion of neutral particles from the prior phase of the cycle. Therefore, the commonly observed electric wind is mostly flowing away from the exposed electrode over the dielectric encapsulated electrode.

Meanwhile the y-component force is always pointing downward toward the electrodes, and the z-component is parallel to the intersection of the electrodes. The span effective of a DBD will be noticeable if end walls are installed or very short widths of electrodes are used. The results of computational simulations based on the formulation of equations (10.3a) through (10.3c) are in agreement with experimental observations, but there is a small difference in magnitude. The computational results produce only the periodic electrostatic force of the DBD, which is the driving mechanism of the electric wind, but the experimental data are the measured resultant force through the interaction of the charged and neutral particles. The total force of the

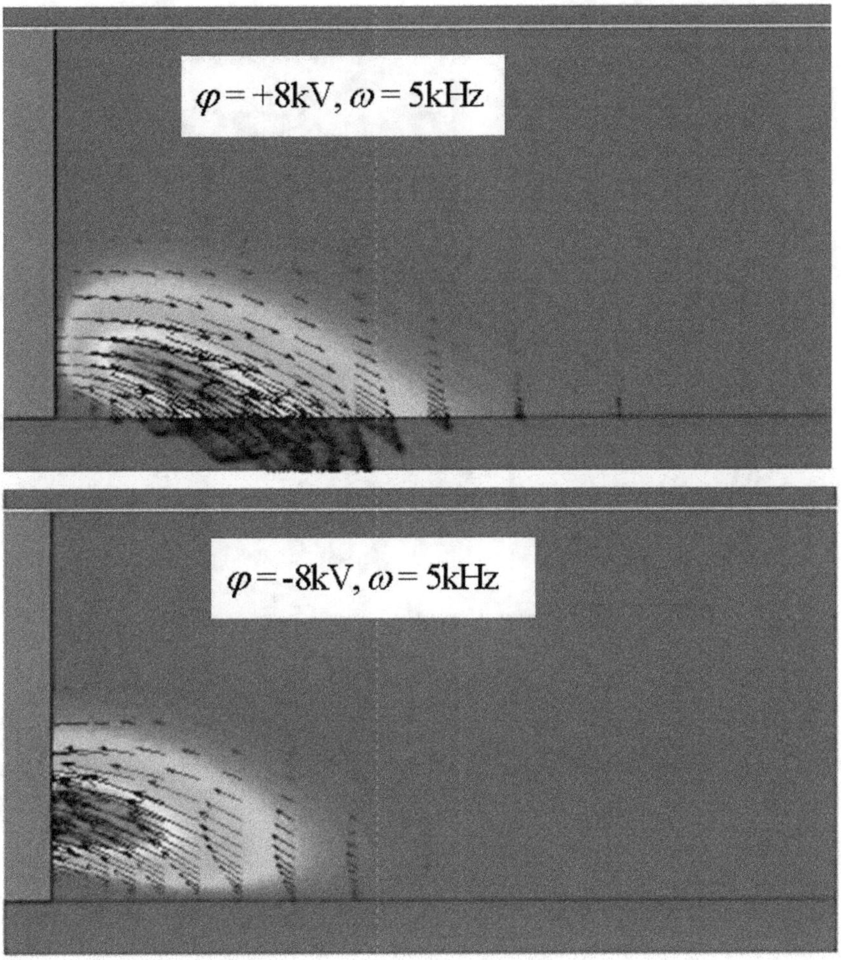

FIGURE 10.16 Vector force field of DBD at peak vales of EMF.

electric wind generated by a DBD can be obtained only by integrating the periodic electrostatic force into the computational aerodynamic–electromagnetic equations, equations (9.1) through (9.3).

The instantaneous x-components of the periodic electrostatic force over the entire AC cycle are sampled at different locations in Figures 10.17 and 10.18. The depicted force distributions correspond to the instantaneous values and with the maximum magnitude of force of 2.5 N/cm^3. This value is comparable with nearly all computational results in the open literature (Shang et al., 2011, 2014). In each figure, the sampling locations are indicated by the symbols of dotes in each respective discharge domain. In general, the magnitude of the x-component force component drops rapidly with increasing distance above the encapsulated electrode. This phenomenon is easily

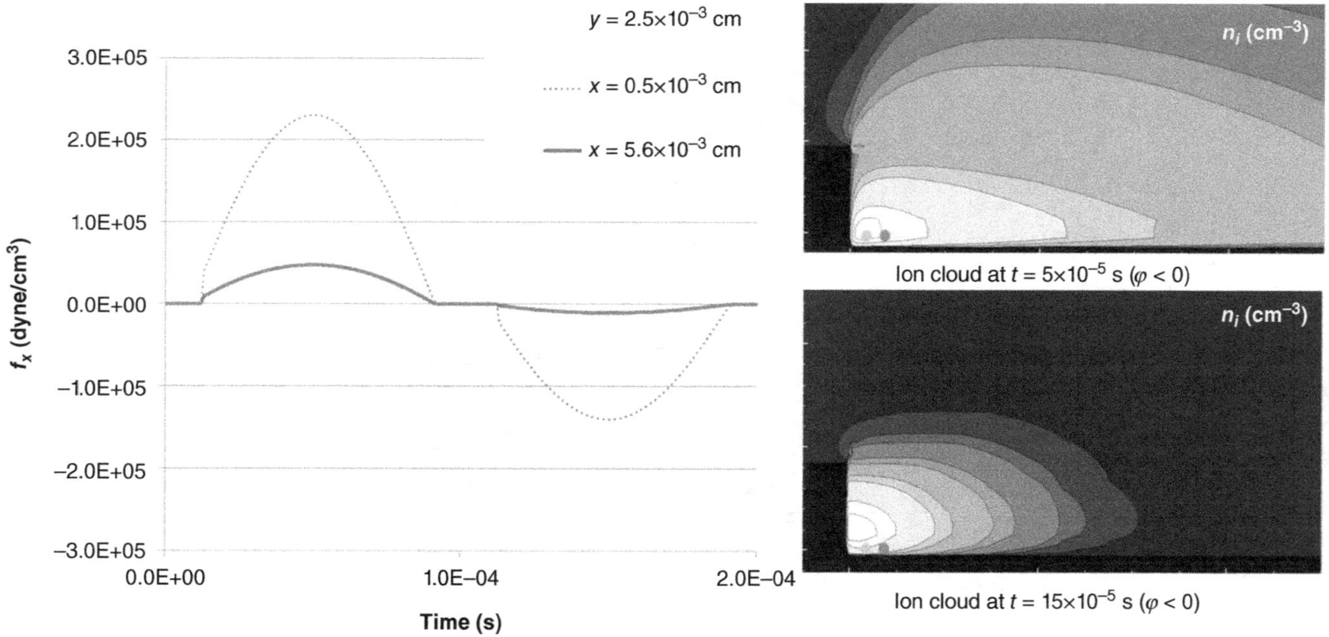

FIGURE 10.17 x-component force sampled at different x locations, EMF = 8.0 kV, f = 5.0 kHz, $\varepsilon = 2.7$.

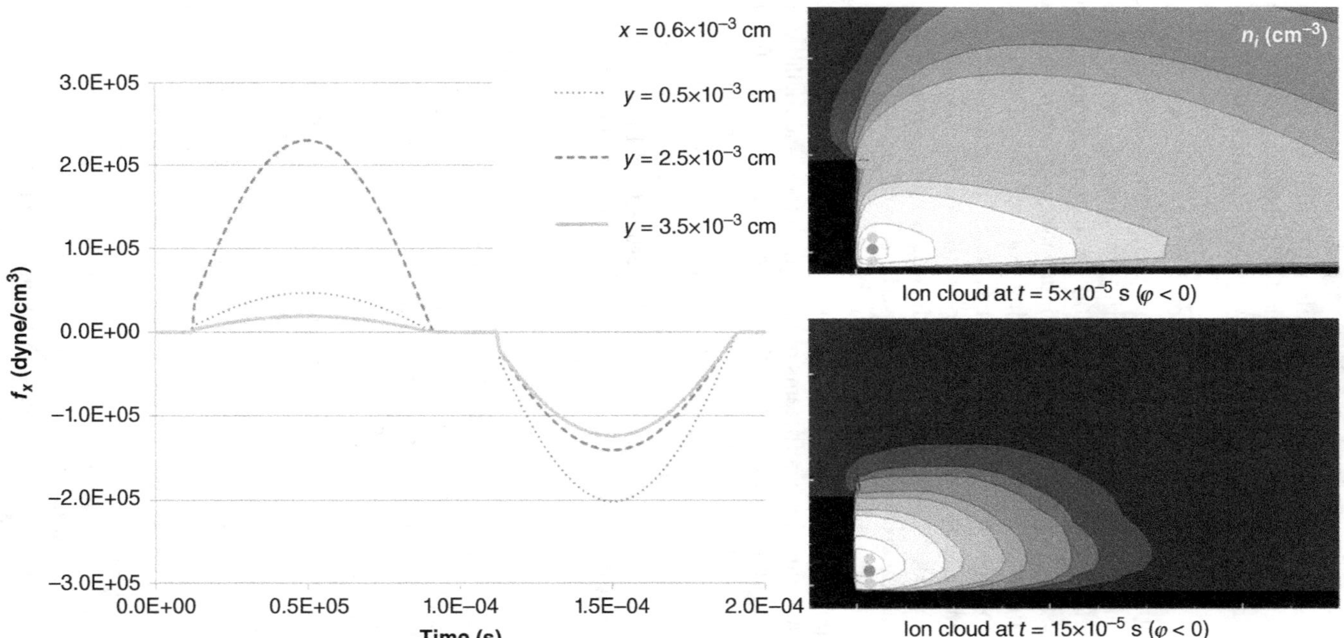

FIGURE 10.18 x-component force sampled at different y locations, EMF = 8.0 kV, f = 5.0 kHz, $\varepsilon = 2.7$.

understood as the location is farther away from the junction of the electrodes and this point is increasingly moving beyond the very thin plasma sheath. Similarly, the force component is also decreasing perceptively when the sample location is farther removed from the intersection of the electrodes. This behavior is obvious, because the free space charge separation can exist only in the immediately adjacent location of the electrodes.

In Figure 10.17, the magnitude of the x-component of the periodic electrostatic force at two locations $x = 0.050$ and 0.056 from the corner of the exposed electrode, but at the identical elevation above the dielectric, are sampled. The data are recorded continuously for the complete AC cycle of 200 μs. Two instantaneous ion number density contours are also appended to this figure taken at the peak values of the EMF when the times elapsed from the starting point of the AC cycle are 50 and 150 μs. The x-component force attains its maximum instantaneous magnitude of 2.46 N/cm^3 when the exposed electrode receives the peak value of the externally applied voltage of 8.0 kV (Shang et al., 2011). It is also observed that the x-component of force is persistently greater when the exposed electrode acts as the anode. When the AC polarity is reversed, the exposed electrode acts as a cathode, and the maximum local force component is reduced to 1.47 N/cm^3. The difference is generated by the different secondary emission coefficients over the dielectric and metallic surfaces.

A similar presentation, but depicting the x-component force at different heights over the dielectric surface, is given in Figure 10.18. The instantaneous force is calculated at the same x location of 0.056 mm from the edge of the exposed electrode for three different heights above the encapsulated surface of 0.025, 0.035, to 0.065 mm. The general behavior is similar to the results from Figure 10.17. Namely, the instantaneous force dropped rapidly as the distance is increased from the electrodes junction during the entire AC cycle—the maximum force is indicated at the position of $x = 0.056$ mm, $y = 0.025$ mm, and $z = 0.0$ mm when the exposed electrode receives the peak EMF value. There is a large drop in the force as the height above the electrodes increases from 0.025 to 0.035 mm, but then there is an increase in the force from 0.035 to 0.050 mm. The trend in the instantaneous force distribution is not consistent according to the different heights of its location to show the dynamic structure of the discharge pattern and the influence of the geometric constraint by the corner configuration.

The most important observation from the computational simulations is that the net force parallel to the dielectric surface during an AC cycle is the net sum of components in opposite orientations. For this reason, the integrated x-component of force developed by a DBD over a complete AC cycle that can be used for flow control is substantially reduced by the negating counteraction. The vastly different time scales, on the order of tens of microseconds for the instantaneous electromagnetic force and on the order of fractions of second for the inertia response, lead to the well-known directional biased electric wind of the DBD. This momentum transfer collision between charged and neutral particles has been consistently observed. From the knowledge of a total collision cross section of ions with neutral particles of 10^{16} cm^2, the collisions with neutral particles are the principal momentum transfer mechanism for the so-called electric wind. Therefore, the most effective frequency for momentum transfer between charged and neutral particles is sonic or on the order of tens of milliseconds. The global force of the DBD on this time scale can be evaluated by integrating the total electrostatic force over an entire AC cycle and the complete discharge domain. In numerical analysis, the rather arbitrary discharge domain is defined by a volume in which a charged number density is within 1% of the maximum value,

$$F = \frac{1}{V\tau} \iint \vec{F}(t, x, y) dt dx dy dz. \tag{10.4}$$

In this equation, the period of an AC cycle and the volume of a three-dimensional discharge domain are denoted as τ (200 μs) and V, respectively. The temporal–spatial averaged electrodynamic force thus represents the total possible net resultant force during a complete AC cycle.

Two-dimensional numerical results of the time-space averaged force of a DBD over the range from 3 to 8 kV with a simple harmonic sinusoidal waveform are depicted in Figure 10.19. The relatively small net x-component force during the discharge has an averaged value from 2.0×10^4 to 2.4×10^3 N/cm^3. The y-component force indicates a persisted downward direction with a maximum magnitude of 3.0×10^3 N/cm^3 at an externally applied voltage of 8 kV. The magnitude of the force components is small and almost linearly proportional to the value of the externally applied voltage. However, a DBD can generate a 16.7 times greater averaged electrodynamic force by using a rectified AC cycle with the identical applied voltage magnitude, in contrast to a

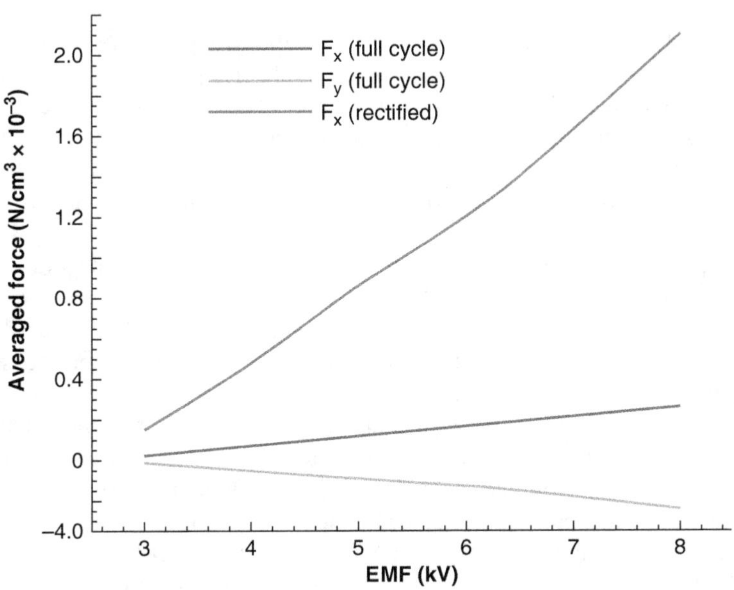

FIGURE 10.19 Temporal–spatial averaged force of DBD, 3.0 < EMF < 8.0 k EMF, $\omega = 5.0$ kHz, $\varepsilon = 2.7$.

sinusoidal waveform. This result is easily demonstrated by eliminating the negatively biased phase of the AC cycle and using a rectified voltage to generate the DBD. A substantial enhancement is realizable only by the rectified AC cycle with only a positively biased phase and by completely depleting the accumulated surface charge on the dielectric surface after each discharge (Opaits et al., 2008). An ingenious process to neutralize the positively charged ions over the dielectric surface after each discharge can be challenged, but this huge gain in electrodynamics force for flow control should not be ignored.

Another interesting issue is the effect of negatively charged ions on the periodic electrostatic force. To address this question, one needs to examine the generation and depletion processes of negatively charged ions. The theory of nonequilibrium chemical kinetics indicates that the key mechanisms are the electron attachment, detachment, and ion–ion recombination. From most detailed analyses, the number density of the negatively charged ions in a DBD is much lower in comparison with its positively charged counterpart: it generally appears as a fraction of a few percent (Bogdanov et al., 2008; Shang et al., 2014). Again, for all computational simulations with an EMF ranging from 3.0 to 8.0 kV, the number densities of the negatively charged ions are nearly two orders of magnitude lower than there positively charged counterpart; the typical values are around $(n_+)_{max} \sim O(10^{12}\,\mathrm{cm}^{-3})$ and $(n_-)_{max} \sim O(10^{10}\,\mathrm{cm}^{-3})$, respectively. These values agree well with the most elaborate calculations using chemical kinetic models (Bogdanov et al., 2008). The quantified difference in charged particle number densities has been further confirmed by the electrical current–voltage curve during an AC cycle of 5 kHz and driven by an EMF of 8 kV. The computational results from the two- and three-component ionization models are indiscernible.

Figure 10.20 displays the comparison between the two (panels a) and three (panels b) species ionization models at the instance the electric potential over the exposed electrode is positive and reaches its maximum value. The data are depicted in a composite presentation for the electric field intensity vector E, the net balance of different number density of charged particles $(n_+ - n_e)$, and $(n_+ - n_- - n_e)$, as well as the instantaneous electrostatic force vector at an applied EMF of 8.0 kV. First of all, the discharge patterns are unaltered; all the electric field intensity vectors point along the electric field lines, shown in (a-1) and (b-1). The instantaneous and maximum value of $(n_+ - n_e)$ over the dielectric barrier attains a value of $3.96 \times 10^{12}\,\mathrm{cm}^{-3}$. On the contrary, the $(n_+ - n_e - n_-)$ obtains a value of $3.97 \times 10^{12}\,\mathrm{cm}^{-3}$, which reflects a relatively insignificant negatively charged ion number density. The resultant force, according to equation (10.4), is uniformly directed from the exposed electrode to the dielectric barrier and has an instantaneous maximum magnitude of 7.097×10^5 dyne/cm^3. The numerical result is nearly identical to the result of 7.066×10^5 dyne/cm^3 from the two charged species modeling. This result is greater than the maximum force of 2.765×10^5 dyne/cm^3 generated by the EMF at 4 kV and of 4.468×10^5 dyne/cm^3 by the EMF at 6.0 kV (Huang et al., 2011; Shang et al., 2014). However, it needs to be remembered that the maximum electrodynamic force exists only within the thin plasma sheath and for a very short period during the discharge.

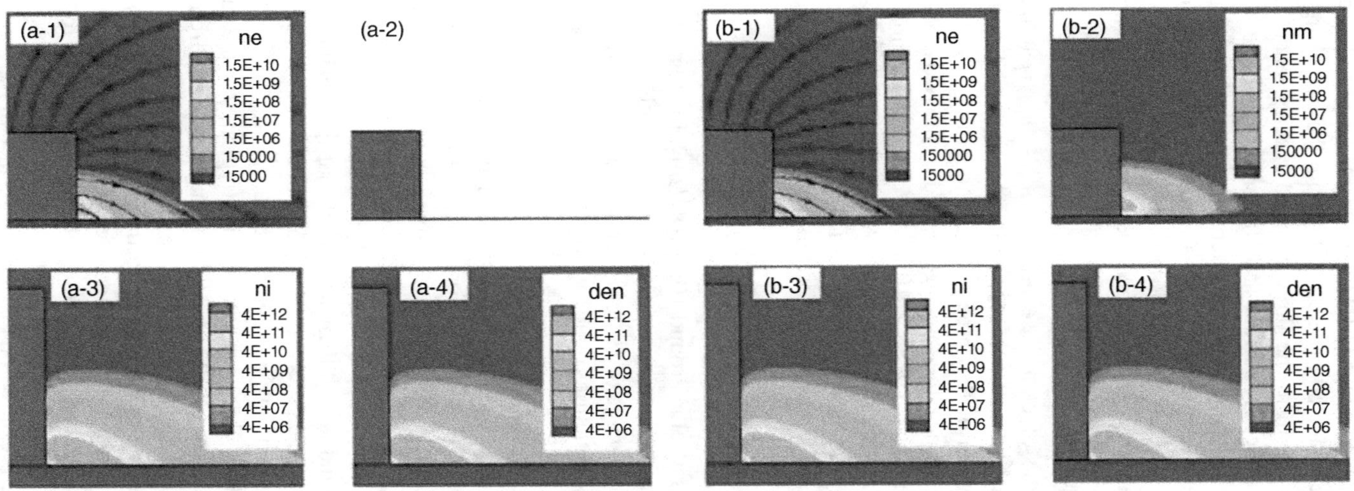

FIGURE 10.20 Comparison of two- and three-component ionization models; n_e, n_-, and n_+, at $\varphi = +8.0$ kV, $f = 5.0$ kHz.

Figure 10.21 displays the electric potential and charged particle number density distributions for the instantaneous DBD field when the exposed electrode is acting as the cathode. The panels labeled (a) are results from the two charged species model and those labeled (b) are from the three charged species model. All positive ions are repulsed from the dielectric surface and clustered around the lower corner of the exposed electrode. At the same instant, a large number of faster moving electrons are propagating further over the dielectric barrier away from the corner region of the electrodes. The sign of the electric field intensity is now negative, pointed from the dielectric barrier toward the exposed electrode. According to computational results, the plasma sheath is extremely thin (<0.002 mm) and its thickness decreases with a lower value of the electron secondary emission coefficient.

The maximum value of charge separation ($n_+ - n_e$) near the intersection of the electrodes is 4.4018×10^{12} cm^{-3}, and the corresponding net number density of charged separation of the three-species model ($n_+ - n_- - n_e$) is 4.4059×10^{12} cm^{-3}. As a consequence, the forces generated by the two models are very close, 1.0012×10^3 dyne/cm^3 and 0.9988×10^3 dyne/cm^3 for the three and two species models, respectively. At this instance of the AC cycle, the peak instantaneous electrostatic force is greater than when the electrical potential on the exposed electrode is positively biased, but the maximum force is exerted on the vertical surface of the exposed electrode.

From all chemical kinetics studies, negatively charged ions are usually two or three orders of magnitude lower than positively charged ions. The positively charged number density n_+ has a value around 10^{12}–10^{13}/cm^3. In most computational simulations using all known chemical kinetic models, the negatively charged ions number density n_- is capped by a value of 10^9/cm^3. This pattern is also directly substantiated by the computational simulations done by Bogdanov et al. (2008) and Solov'ev et al. (2008).

The time-averaged periodic electrostatic forces by the two- and three-charged species ionization models over an AC cycle and the entire discharge domain are depicted in Figure 10.22. The overall force structure is indistinguishable from the two ionization models due to the fact that negative ions are not significant as compared with positive ions. In short, by considering the presence of the negatively charged ions with a chemical-physics kinetic model, the averaged force that produces the well-known electric wind is less than without it, but it is insignificant. However, it must be emphasized that the precise quantification is not the state of the art for DBD simulations and still uncertainty must be accepted.

Form the vector plots; there are two distinct regions for the direction of time-averaged forces separated by a null-value line cutting diagonally from the corner of the exposed electrode to the dielectric surface. The region below the line of demarcation is the positive force creating a mean pushed motion in the downstream direction, which is a wall-jet-like force from the exposed electrode toward the dielectric covered electrode. It becomes the time-space integrated driving force for the well-known electric wind of a DBD. The region above the line of separation has a negative force and its maximum force is exerted on the vertical surface of the exposed electrode and diminishes rapidly toward the outer edge of the exposed electrode. This force exerted

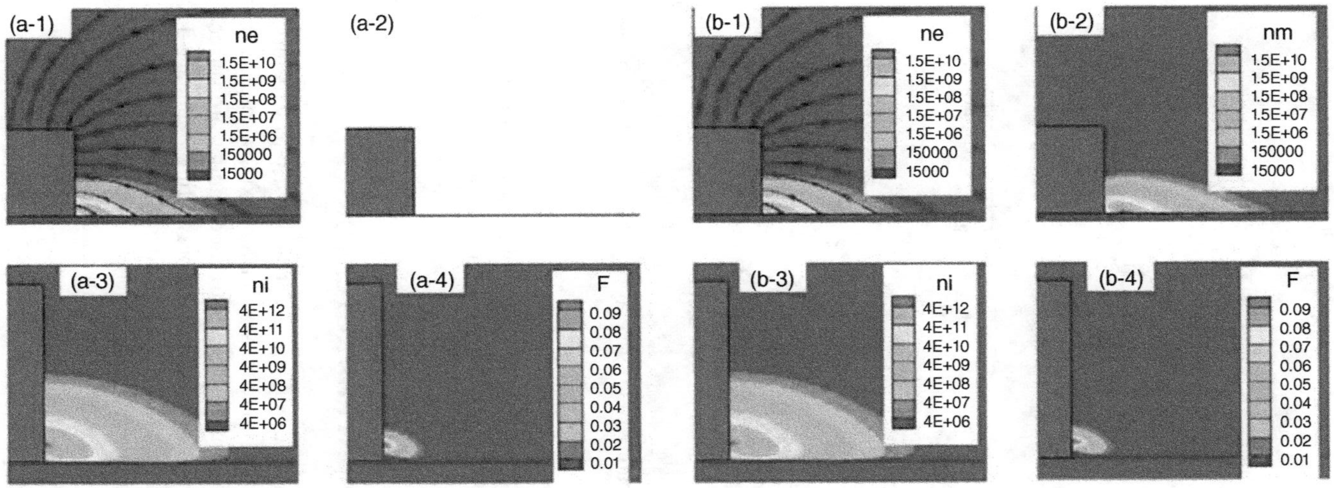

FIGURE 10.21 Comparison of two- and three-component ionization models; n_e, n_-, and n_+, at $\varphi = -8.0$ kV, $f = 5.0$ kHz.

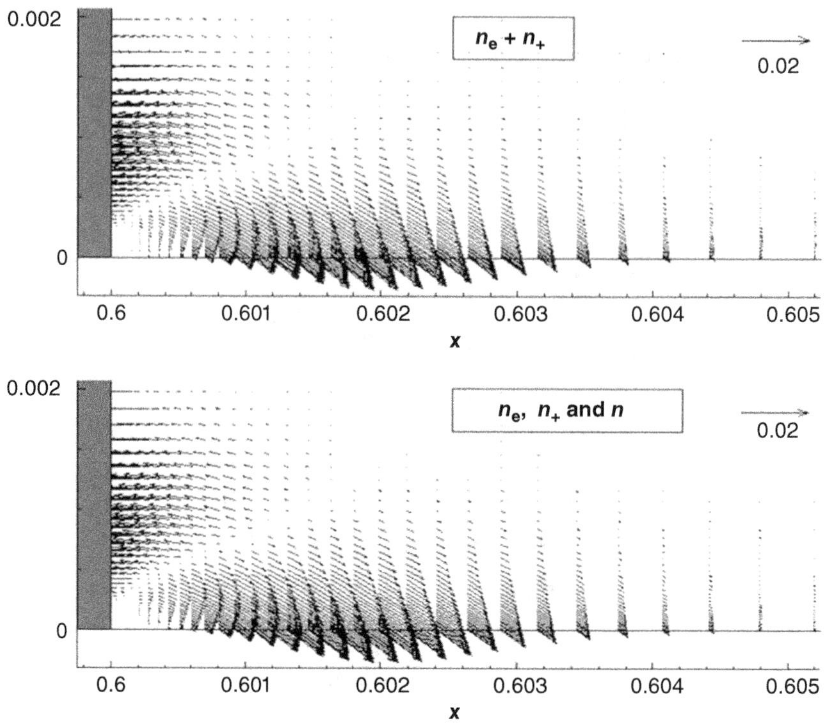

FIGURE 10.22 Comparing time-space averaged force of DBD by two- and three-species ionization models.

on the vertical electrode surface may generate a recirculating fluid motion locally, but this flow structure requires a coupled solution of the computational electromagnetic-aerodynamics.

10.5 DBD FLOW CONTROL ACTUATOR

The fluid boundary layer exhibits a great sensitivity to structural change by disturbances introduced into the inner most sublayer next to a solid surface. For this reason, frequently applied boundary–layer control techniques are surface injection, slotted wing, and wall suction. The processes of preventing boundary–layer separation are achieved by supplying kinetic energy to fluid particles that are retarded by the viscous shear. Then the accelerating fluid can overcome adverse surface pressure gradients and remain attached to the surface. In all cases, additional energy or momentum is imparted to fluid particles near the solid surface. The electric wind of a surface DBD offers an attractive alternative to mechanical means.

The temporal–spatial averaged particle motion induced by a single asymmetric DBD has been investigated thoroughly by Corke et al. (2007). A wall-jet-like

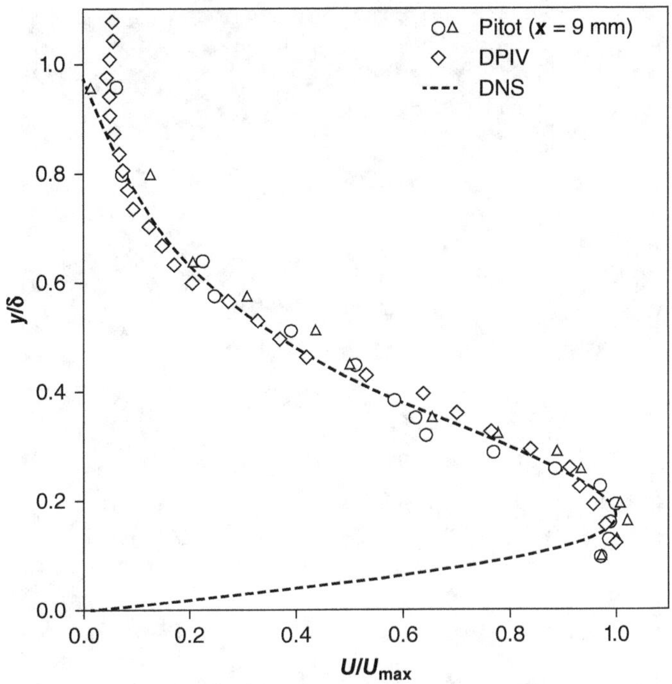

FIGURE 10.23 Wall jet induced by DBD in still air. From Corke, T.C., Post, M.L., and Orlov, D.M., *Progress in Aerospace Science*, vol. 43, 2007. Used with permission from Elsevier.

velocity profile parallel to a solid surface is compiled from a collection of data obtained using particle image velocimetry. Two sets of data points using a pitot probe are also included as a check on measurement accuracy. A solution by a direct numerical simulation (DNS) is also included in Figure 10.23. The experimental data reveal that the magnitude of the maximum velocity of a single asymmetric electrode configuration has a strong dependence on the EMF of the sinusoidal waveform and the size of the exposed and encapsulated electrode areas. This behavior is consistent with the ionization by electron impact in which the charge accumulation and secondary emission lead to massive cascading.

A striking experimental demonstration of a DBD as a flow control device is the suppression of the von Karman street downstream of a circular cylinder (Thomas et al., 2008). By the direct contrast of two particle–image–velocimetry images for flow at a Reynolds number of 3.3×10^4, four activated DBD actuators keep flow attached on the lee side of a cylinder. The DBD actuators are placed at 90, 135, 225, and 270° measured clockwise from the forward stagnation point of the cylinder. After the activation of the DBD actuator, the alternating vortexes shedding from the cylinder are completely suppressed by the induced velocity on downstream side of the cylinder. The flow in the wake region is rejoined on the centerline of the symmetric flow field without a separation region.

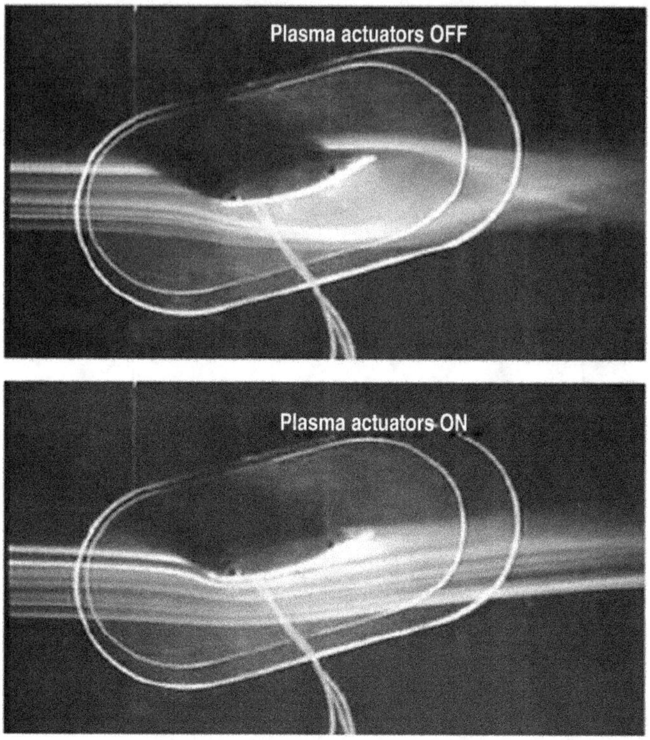

FIGURE 10.24 Leading-edge separation control by DBD actuator. From Post, M. PhD dissertation, University of Notre Dame, 2004. Used with permission.

The flow visualization recorded by Post (2004) and presented in Figure 10.24 illustrates the effectiveness of the DBD actuator in controlling leading-edge separation on an airfoil. These images are generated at a relatively low Reynolds number of 1.58×10^5. The upper insert of the photograph image displays smoke traces of a flow field formation without an activated plasma actuator. The separation domain over the entire windward side of the airfoil is clearly visible and is accentuated further by an extensive wake region. When the DBD actuator is applied, the boundary layer over the airfoil remains attached from the leading edge and extends toward the trailing edge. The laminar structure of the flow from the unperturbed upstream region is maintained to the far wake region.

DBD actuators have also been applied for controlling airfoil leading-edge separation at high angles of attack. Numerical and experimental investigations have been conducted for NACA 0009, 0012, 0015, 0021, 66_3-018, and HS3412 airfoils. Experimental studies have been done for Reynolds numbers based on the chord length covering a range from 1.7×10^4 to 4.6×10^5 and encompassing free-stream speeds from 10 to 30 m/s. The effectiveness of DBD flow control is confirmed by experimental observations for the NACA 0015 airfoil. From the smoke streakline patterns, the leading-edge stall leads to a full leading-edge separation and the separated flow region covers the entire suction surface of the airfoil (Corke et al., 2007). The DBD actuator extends the on-set stall angle of this airfoil from approximately 14 to 16° until post stall. This experimental observation demonstrates that the wall jet of an activated DBD postpones an airfoil's stall at high angles of attack. A modeled numerical simulation for a NACA 0021 airfoil at a free stream speed of 35 m/s and a chord Reynolds number of 8.0×10^5 also shows that leading-edge stall has been delayed from the angle of attack of 16° to beyond 23°.

Flow control using a DBD is equally effective to control flow separation in a dynamic environment. DBD actuators have successfully been applied to control dynamic stall of oscillating airfoils at a Strouhal number of unity, $N_{st} = c\omega/u$. The integrated complete-oscillatory-cycle lift by a leading-edge DBD actuator achieves up to 10% improvement for a NACA 0015 airfoil with a 12.7-cm chord and span of 25.4 cm. Flow control is achieved by controlling the dynamic stall vortex originating from the leading edge. Experimental evidence is also collected by using a DBD for flow separation control over a turbine blade in the low pressure stage at a low Reynolds numbers based on the chord length of the blade.

One of the very few DBD applications for high-speed flow control is demonstrated by Nishihara et al. (2011). A perturbation to the supersonic–hypersonic shock structure has been introduced by a pulsed microwave discharge, laser optical breakdown, and DCD. All these perturbation mechanisms to the flow field are pure thermal effects. A different source of perturbation by Nishhara's effort is by a diffuse nanosecond pulse DBD actuator generated on a cylinder model immersed in a Mach 5 free stream. The discharge is maintained by an AC frequency of 100 kHz and the energy coupled to the actuator is 7.3–7.8 mJ/pulse. The plasma temperature as determined from the nitrogen emission spectra indicates a temperature of 340 ± 30 K, which is the typical value for a DBD in the atmosphere. The standoff displacement and structure of the bow shock are recorded by phase-locked schlieren images. Compression waves associated with each individual nanosecond discharge pulse are observed, and the interaction of the induced wave and a steady bow shock in front of the cylindrical model increases the standoff distance by up to 25%. In view of a great disparity in conductive and displacement electric currents of DBDs and by Ampere's circuit law, the nansecond periodic pulse generates a significant normal component of the Maxwell stress, $B_x^2/2\mu$. The net result is similar to the externally applied magnetic field like the observation made by Ziemer (1959).

From most experimental observations, a unique property of the DBD as a flow control actuator emerges in that electromagnetic effects are most effective in controlling perturbations at aerodynamic bifurcations and under conditions in which the dimension of the plasma sheath is comparable to the surrounding shear layer (Shang et al., 2014). These conditions for effective flow control using plasma-based mechanisms are nearly identical to DCDs, but DBDs are free from the restriction of low-density or low-pressure operation. It should be recognized that the laminar–turbulent transition and the random fluctuating cascading eddies of turbulent flows fit nicely into these conditions.

10.6 LAMINAR–TURBULENT TRANSITION

Mitchner and Kruger (1973), in their study of Hartmann flow, have noted that the magnetic field effects on the wall shear stress of turbulent flows are substantially different in comparison with that of laminar flows. Under some conditions, the surface shear stress in turbulent flows is actually reduced at a moderated value of Hartmann number. An explanation is offered as the magnetic field tends to damp turbulent motion to decrease the surface shear stress. This flow behavior is also pointed out that the phenomenon is the result of a restraining force on the turbulent motion via fluctuating fluid particles from crossing the lines of the magnetic field. In recent technology developments for aerodynamic drag reduction, riblets are realized as one of the viable techniques for viscous drag reduction. Riblets are simply a series of parallel fine groves on a surface that is just another mechanical device for reducing surface shear. The flow control is achieved by applying micro/macrovortex generators to alter characteristics of the laminar–turbulent transition and turbulent fluctuations (Walsh, 1983; Bechert et al., 1997). Although these two mechanisms for turbulent flow control are entirely different from each other, the consequences are identical and the results are obtained by controlling random small-scale vortical structures of a flow field.

When environmental disturbances are small, the laminar–turbulent transition process is governed by the spatial growth of unstable eigenmodes. The development of spanwise periodic disturbances is a consequence of a secondary instability of a base flow of Tollmein–Schlichting waves. Streamwise magnetic field lines may restrain the rapid growth of two-dimensional disturbances to large amplitudes and trigger a breakdown to random three-dimensional turbulent motion. Once the transition to turbulence is complete, the turbulent boundary layer is characterized by packets of hairpin vortices that induce a low-speed region with regular spanwise spacing. The detached small eddies consist of hairpin vortices that separate from the solid surface and are responsible for the outer turbulent layer structure (Garcia-Mayoral et al., 2011; Jimenez, 2013). The basic idea for suppressing laminar–turbulent transition and turbulent shear stress reduction is achieved by the force of magnetic field lines to restrain or suppress high-frequency and random fluctuation of the minuscule eddies.

In investigation of the hydrodynamic instability, the innovation by Herbert using the parabolized stability equation (PSE) to take into account the influence of surface curvature, fine-scale surface features, and boundary–layer growth by an amplification of disturbances is helpful (Herbert, 1997). The basic idea is based on negligible dependence of the convective disturbance on the downstream condition of the flow field. The disturbance growth mechanisms include both algebraic and exponential components through primary and higher instabilities. This approach is in general agreement with Taylor's hypothesis on the characterization of streamwise vortical dynamics, for which the vortices or turbulence can be assumed to be homogenous in the streamwise direction, which evolves statistically in time by a convective velocity. The results of the PSE have been shown to agree with DNS for incompressible and compressible transitional flows even over a swept wing at a modest computational expense (Moin et al., 1998; Arnal and Casaalis, 2000). The most convincing results by PSE for predicting laminar–turbulent transition is provided by Arnal. Excellent agreement is displayed between the results of the DNS and the PSE for simulating cross-flow instability for different RMS values of the streamwise disturbance velocity component for different modes of instability. These findings provide impetus for innovations by applying electromagnetic forces for controlling laminar–turbulent transitions and turbulent shear stress reduction.

Numerical simulations have been conducted by modeling the plasma actuator using the generic, as well as the experimental derived, body force distributions. The fidelity to physics is uncertain, because a fixed relation between the wavelength of the instability and the strength of forcing no longer holds. Only partial success in suppressing the unstable Tollmien–Schlichting waves is realized, but the simulations provide further insight for practical application of plasma actuators for laminar–turbulent transitions. This research area is critical for future basic research for plasma-based actuators (Albrecht et al., 2006; Kotsonis et al., 2013).

An impressive achievement for suppressing laminar–turbulent transitions is depicted in Figure 10.25. The computational simulation has been accomplished by solving a high-resolution, implicit, large-eddy numerical algorithm by Gaitonde (2010). His numerical result simulate a NACA 0015 wing section at an angle of attack of 18°, free-stream Mach number of 0.1, and a nominal Reynolds number of 4.5×10^4 on a $(308 \times 75 \times 145)$ O-topology mesh with a branch cut from the trailing edge. The numerical procedure couples unsteady electromagnetic forces with a multifluid model. The plasma-based actuator is modeled with a phenomenological model. A major flow field formation appears from the existence of relatively large convection coherent vortices outside the wall-shear region produced by the impulsive force. A set of vortices is also generated from the surface of the wing. The detailed vortical structures of a wide range of sizes are captured by the numerical solution. It is observed that flow separation occurs near mid-chord. The effect of the actuator near the leading edge generates a laminar attached flow by the impulse initial transition. At the same time, turbulence originating in the shear layer downstream of the trailing edge propagates upstream after several DBD duty cycles. In an asymptotic state, the flow transitions to turbulent near mid-chord but remains attached in the mean motion. By exciting instability in the separated shear layer, fine scale features arise through transition to turbulence that are absent in the steady structure. Future work is required to generalize the analysis from a monochromatic microwave signal to wide-band and complex wave forms. However, the reader must bear in mind that the numerical results are generated for the purpose to seeking insights for the most complicated computational electro-aerodynamics phenomenon.

Finally, a fundamental challenge exists in coupling a fast, electromagnetic force with a much slower fluid motion dominated by inertia. Many computational simulations for DBDs as a flow control actuator have been carried out by very sophisticated numerical procedures via DNS or LES (large eddy simulation) in the open literature, but not equally matched in physical fidelity in electromagnetic modeling. These findings unavoidably pose a serious challenge for validation and to convert this information into basic knowledge and must be improved for future progress.

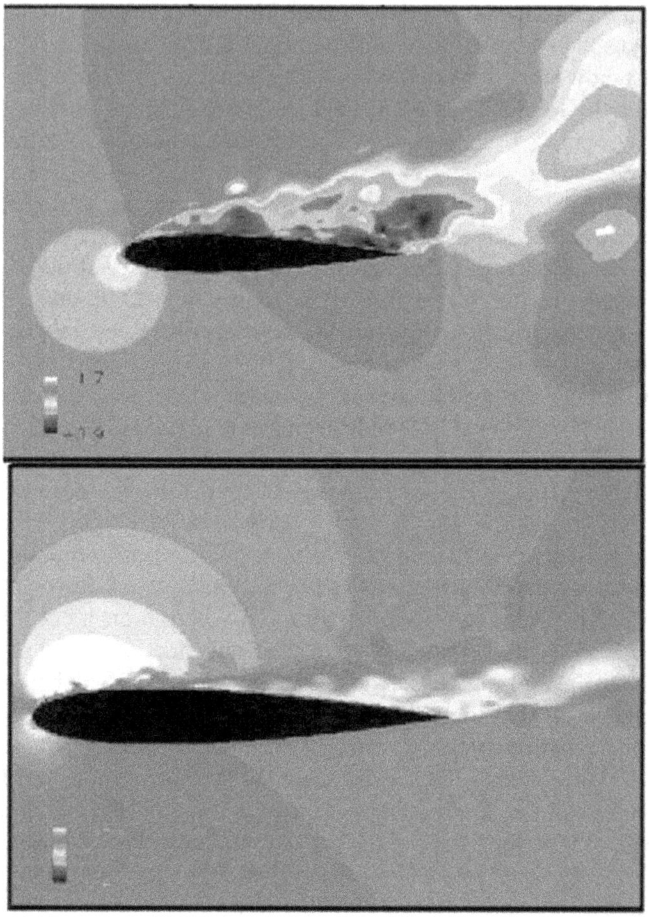

FIGURE 10.25 Simulations of a DBD for laminar–turbulent transition. From Gaitonde, D.V., *Intern. J. Comp. Fluid Dyn.*, vol. 24, 2010. Used with permission.

10.7 ION THRUSTERS FOR SPACE EXPLORATION

The ion engine is one of the few known bona fide practicing technologies using electromagnetic forces to accelerate ions for propulsion. Ion engines routinely are used for station keeping for commercial and military communication satellites in geosynchronous orbits, such as ESA Artemis and AEHF-1 (Brewor, 1970; Gallimore et al., 1994). However, electromagnetic propulsion systems are limited by the thrust per unit area of the engine's exhaust and the space charge, limits as described by Child's law. From the third law of Newton dynamics, a spacecraft's final velocity is linearly proportional to the exhaust velocity vector but depends logarithmically on the exhausting mass. The ion thruster is characterized by an input power up to tens of kilowatts with an exhaust velocity around 20–50 km/s (Jahn, 1968). The resultant thrust is in the range from 20 to 500 mN with a propulsive efficiency up to 80%. However, the most impressive performance criterion is its extremely high specific impulse, $I = u_e/g$, where the gravitational acceleration is assigned the value at sea level. In fact, the specific impulse of an ion thruster has a range from 1000 to 10,000 s using propellants of xenon, bismuth, liquid cesium, and argon.

There are two types of ion thrusters either by the Coulomb force to accelerate ions either in the same direction as an applied electric field or by the Lorentz force. The first experimental test of an ion engine is attributed to Goddard in 1916–1917 (Gallimore et al., 1994). The major components of any ion engine consist of an ion source, accelerator grids, and neutralizer. The ionization is achieved by electron impact into cesium, argon, krypton, and most commonly xenon because of these atoms' relatively low ionization potentials. A contemporary implementation of an ionization chamber employs a hollow cathode inserted into the ionizing chamber that provides the secondary electron emission. High-energy electrons from the hollow cathode collide with neutrals in the ionization chamber and produce positively charged ions. The positively charged ions are extracted from the ionization chamber and are accelerated downstream of the engine by a system of grids. The electrostatic potential difference on the grids achieves high exhaust velocities of the ions and with minimum beam impingements on grids. The grids serve two major functions: first, keeping neutral particles in the ionization chamber while allowing ions to be extracted, and second, accelerating the extracted ions to high velocities. The perforation of the grids focuses the ion stream into an array of smaller beams with minimum impingement on the solid portion of the girds. The last device of an ion thruster is the neutralizer that prevents charge accumulation on the vehicle. It is basically another hollow cathode that emits electrons into the positively charged ion beam to prevent the ion engine from obtaining a large negative charge that would prevent additional ions from leaving the ionization chamber.

The typical operating conditions in an ionization chamber of ion thruster usually have the maximum electron number density around $10^{13}/cm^3$ ($10^{19}/m^3$), and the electron temperature is on the order of 3 eV or greater. Thus, the Debye length in the high number density region of an ionization chamber is just a few microns and the plasma frequency is estimated to be about $10^{11}/s$. The plasma in the discharge chamber exists in a similar state of most plasma actuators for flow control. However, the gas mixture is always in a rarefied condition and measured by the Knudsen number (Hayes et al., 1959);

$$K_n = \frac{\lambda}{L}, \quad \lambda = 1/\sqrt{2}n\sigma \tag{10.5}$$

In this equation, λ is the mean free path between collisions, L denotes a characteristic length of the chamber, n and σ are the number density of gas particles and the collision cross-sectional area for an ionization process, respectively. At a xenon neutral particle number density of $10^{19}/m^3$ and a collision cross section of 5×10^{20} m^2, respectively, the mean free path yields a value about 1.4 m within a typical ionization chamber. In general, this value is several times greater than most diameters of a typical ionization chamber; therefore, a computational simulation is beyond the continuum regime. With the ionization mean free path longer than the diameter of the ionization chamber, some means of electron confinement is required. This is done by setting up a magnetic field long the walls of the ionization chamber. This magnetic field reflects electrons back into the discharge chamber, giving them a much longer mean free path.

Most current methods of computational electromagnetic-aerodynamics within an ionization chamber adopt a particle-in-cell (PIC) approach in combination with a Monte Carlo collision (MCC) technique. In a full PIC–MCC simulation, all particles, including the neutral and charged species, are tracked by the macroparticle assumption. The basic governing equations consist of the classic Newton–Lorentz equations for charged particle motion and the electric field distribution by Poisson equation;

$$
\begin{aligned}
m_i \frac{d^2 x_i}{dt^2} &= q\left(E + \frac{dx_i}{dt} \times B\right) \\
\nabla^2 \varphi &= \frac{e}{\varepsilon}(n_- - n_+)
\end{aligned}
\tag{10.6}
$$

In a full PIC–MCC computational simulation, all charged particles are tracked by the macroparticle assumption. Each macroparticle represents billions to trillions of real particles of the rarefied plasma; thus, the total number of the tracked particles can easily exceed 10^8. The required huge computational resource must satisfy the peculiar and stringent computational stability criteria of PIC–MCC algorithms in time and space. The numerical procedure suffers from statistical noise of a large number of sampled macroparticles. On the contrary, physical fidelity of modeling imposes severe resolution requirements because of the Debye length shielding in space and temporal restrictions because of the plasma frequency. In all, the computational simulations by the Lagrangian formulation are essential in analyzing the plasma dynamics in the rarefied gas regime, and the computational resource requirement is extremely demanding.

There are also numerous techniques that have been implemented to reduce the computational resource limitations, including the hybrid approach by formulating some types of particles by the Eulerian formulation. Another technique is by artificially decreasing the mass of the heavier particles, increasing the electrical permittivity, or by scaling the characteristic length to reduce the size of the computational domain. In cases where the electric field distribution can be input into the computation from experimental measurements or by estimation, the restrictions on the governing equations are reduced by decoupling the tracking of the particles from the calculation of the electric field. All these numerical techniques have demonstrated the ability to cut the computational resources significantly and are still a vibrant research area (Taccogna et al., 2005).

One type of electrostatic ion thruster is depicted in Figure 10.26. The plasma is generated by ionizing the propellant by electron impact, and a beam of positively charged ions is separated from the electrons at the downstream end of the ionizing chamber, a collisionless stream of ions is then accelerated by an electrostatic field between two permeable grids. After an array of ion beams is stabilized, electrons

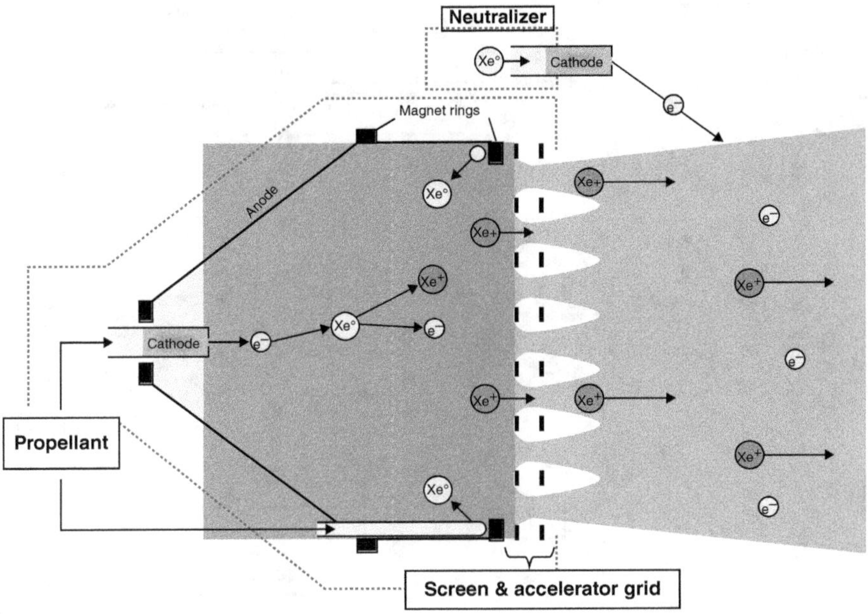

FIGURE 10.26 Schematic of electrostatic ion thruster.

are injected into the beam, downstream of the grids, with a neutralizing cathode to produce a high speed exhaust of zero net electrical charge. From a simple one-dimensional analysis, the maximum ion current density can be sustained across the electrode gap d by an applied electric field potential φ and is expressible as

$$J = (4\varepsilon/9)(2e/m_i)^{1/2}(\varphi^{3/2}/d^2) \tag{10.7a}$$

From the thrust density relationship, the exhaust velocity is dependent on the square root of the product of ion charge-to-mass ratio and the applied electric field potential,

$$u_e = [(2q/m_i)\varphi]^{1/2} \tag{10.7b}$$

which indicates that an extremely high exhaust speed exceeding 10 km/s is achievable.

The electromagnetic thruster, on the contrary, relies on the interaction of conducting propellant charged motion with a magnetic field to provide the acceleration by the Lorentz force including the component of $J \times B$. The applied electric field E and the magnetic flux density B are perpendicular to each other and to the ionized propellant velocity u, as shown in Figure 10.27. The streamwise body force accelerates the propellant along the thruster and transfers streamwise momentum to neutral particles by collisions and by a microscopic polarized field. As a consequence, the thruster densities are greater than that of the electrostatic ion thruster, but at a cost

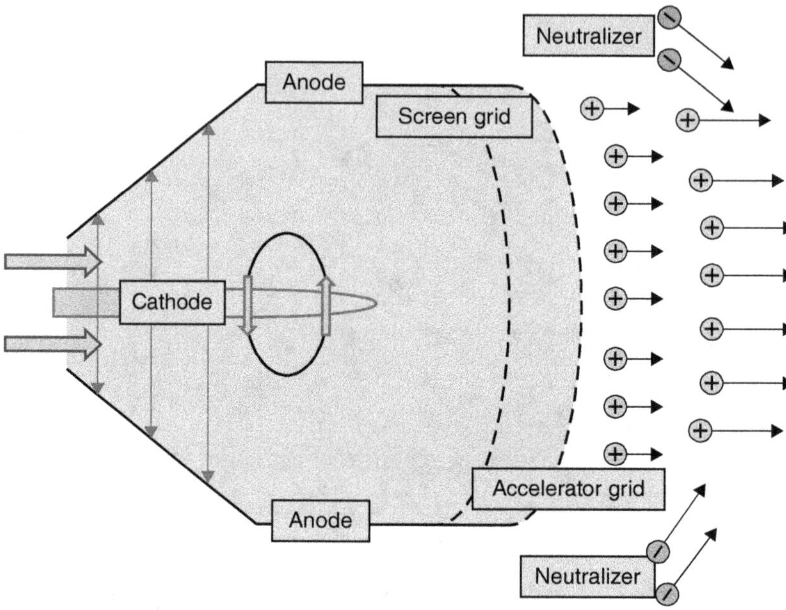

FIGURE 10.27 Schematic for electromagnetic ion thruster.

of much more complex electromagnetic–aerodynamic interactions. It is important to note that the ionized propellant in an electromagnetic thruster is basically macroscopically neutral; therefore, it is not constrained by the space charge limitations as in the electrostatic accelerators.

There are also wider variations of electromagnetic thruster configurations such as the magneto-plasma-dynamic thruster. In this arrangement, the electrodes are arranged by a central cathode and an annular anode. The propellant is ionized by an azimuthal electric arc in the coaxial gap between the electrodes, and the high arc current acts in conjunction with the azimuthal magnetic field to accelerate the hot plasma downstream. At a low plasma density, but high-intensity magnetic flux density, the electric current will divert from the alignment of the applied electric field by the Larmor gyration and acquires a component in the direction of $E \times B$—the Hall Effect. An electromagnetic engine operated in this mode is known as a Hall thruster. The cycloidal motion of current-carrying electrons locks the electrons into a nearly collisionless cross stream drift, and in turn, this leaves the positively charged ions free to accelerate downstream. In this sense, the Hall thrusters are hybrid electrostatic–electromagnetic accelerators with neutralized space charge by the drifting electrons.

The ion and electromagnetic thrusters exemplify complex electromagnetic–aerodynamic interactions; it is very effective but is severely limited in the resultant force magnitude, mostly due to space charge limitations. Therefore, it mostly is relegated to special purpose applications.

10.8 PLASMA MICRO JET

The simplest application of a plasma device via dielectric barrier or corona discharge is the micro jet. This device spans the widest engineering application range from the thermoacoustic instability suppression in a swirl-stabilized, lean-premixed combustion to catalyzing complex behaviors in biological systems (Fridman et al., 2008; Sands et al., 2008; Jansky et al., 2010), or the specialized purpose of reducing thermal NOx production. Lean-premixed combustion emerges to be the most promising approach for future gas turbine applications. From experimental observations, the combustion instability arises at discrete frequencies with large amplitude pressure oscillations. An inner recirculation zone of the swirling flow field can play a dominant factor by coupling the acoustics and heat release that leads to enhance combustion stability. The counterrotating circulation generated by a group of plasma micro jets has shown to slow down the primary swirl and to effectively suppress the combustion instability. In fact, the instability has transitioned to a higher-frequency instability mode. The microjet can also alter the flame shape for nonpremixed flames by decreasing the flame length and the amplitude of the buoyancy-induced flickering.

The history of applying a DBD for sterilization and pasteurization is well known; however, the wound healing and tissue regeneration applications are beginning to attract attention from the medical community (Fridman et al., 2008). Specifically the nonthermal oxygenic nitrogen oxide, NO, reduces the inflammatory reaction in purulent wounds. The plasma can also heal proinflammatory diseases of soft tissue and has been demonstrated for purulent wound. The micro jet has also been applied to sterilization of living tissue and blood coagulation by cauterization through nonthermal plasma stimulation of a natural mechanism. A most noteworthy application is treating cardiovascular obstacles by disposing of the plaque in a noninflammatory manner. The micro jet actually opens a new and exciting area for plasma medicine.

The dielectric micro jet can be constructed by a glass or isolator capillary of a few millimeters in diameter with copper ring electrodes surrounding the capillary, as shown by Figure 10.28. The anode ring is placed near the exit of the capillary and is separated from the cathode ring by a distance less than 1 cm. The electrodes are often

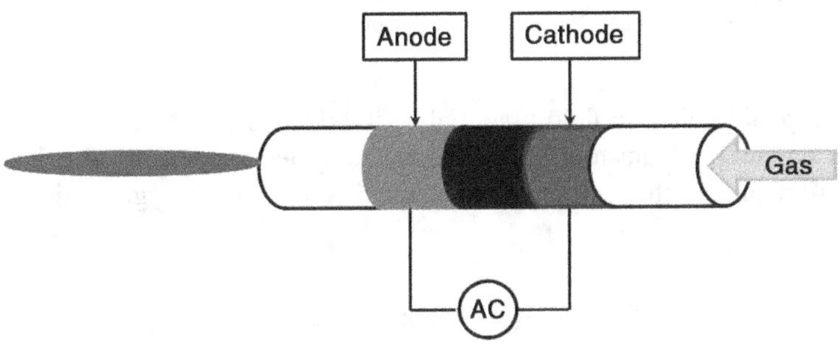

FIGURE 10.28 Schematic of a plasma micro jet.

isolated by nylon nuts to prevent arcing externally to the capillary. When energized, this DBD is excited between the electrode gap and the charged particles accelerate and exhaust from the capillary. The micro jet is terminated through a virtual ground plane because there is no physical electrode beyond the capillary. The electric pulse can also be supplied by a positive-going unipolar voltage in the 10 kV range with a microwave frequency. The discharge condition reveals that the plasma micro jet initiates independent of the discharge inside the capillary and is self-sustained. Under this arrangement, the plasma micro jet has properties consistent with a cathode-directed streamer discharged in free space. The most important feature of the plasma jet is the presence of a weakly ionized channel in the wake of the advancing ionization front. The propagation speed exhibits a speed in a range up to 10^7 cm/s and with an electron temperature around 1 eV or 2950 K in a nonequilibrium thermodynamic state.

From the physical evidence by experimental observation, the plasma micro jet constitutes a fluid dynamic motion of exhausting jet with a weakly ionized gas front. The energized jet contains only a trace amount of charged particles that are accelerated by an externally applied electric field within the capillary; thus, the micro jet can be modeled by the low magnetic Reynolds electromagnetic–aerodynamic formulation, equations (7.7) through (7.9) with further simplified approximations. Since the ionizing energy is rather limited to a trace amount of ionized species, the energy of the internal modes of excitation is negligible from the energy conservation equation perspective. In essence, the continuity equation is replaced by the drift-diffusion theory to account for the charged species of the three-fluid model; see equation (8.28). The governing equations for the micro jet are

$$\begin{aligned}
&\frac{\partial n_e}{\partial t} - \nabla \cdot (n_e \mu_e \vec{E} + d_e \nabla n_e) = \alpha(|\vec{E}|)|\vec{\Gamma}_e| - \beta n_+ n_e \\
&\frac{\partial n_+}{\partial t} + \nabla \cdot (n_+ \mu_+ \vec{E} - d_+ \nabla n_+) = \alpha(|\vec{E}|)|\vec{\Gamma}_e| - \beta n_+ n_e \\
&\frac{\partial n_n}{\partial t} - \nabla \cdot (d_n \nabla n_n) = 0
\end{aligned} \tag{10.8}$$

$$\frac{\partial \rho \vec{u}}{\partial t} + \nabla \cdot (\rho \vec{u}\vec{u} + p\vec{\vec{I}} - \vec{\vec{\tau}}) = \rho_e \vec{E} \tag{10.9}$$

$$\frac{\partial \rho e}{\partial t} + \nabla \cdot \left[\rho e \vec{u} - \kappa \nabla T + \sum \rho_i u_i h_i + \vec{u} \cdot p\vec{\vec{I}} + \vec{u} \cdot \vec{\vec{\tau}}\right] = \vec{E} \cdot \vec{J} \tag{10.10}$$

The compatible electric field potential of the discharging field is determined by Poisson's equation of plasma dynamics, and the electric density is calculated by the definition of the charge number density flux to close the governing equation system:

$$\begin{aligned}
&\nabla^2 \varphi = \frac{e}{\varepsilon}(n_e - n_+); \qquad E = -\nabla \varphi \\
&J = e(\Gamma_+ - \Gamma_e)
\end{aligned} \tag{10.11}$$

From a close examination, the identical equations are used for calculating the periodic electrostatic force in the DBD; that is the force exerted on neutral particles by charged particles in an externally applied electric field. The initial values and boundary conditions for the electric field are straightforward, as previously discussed for DBD simulations, and the controlling aerodynamic phenomenon are the nonslip conditions for the velocity components and the vanishing pressure gradient condition for density next to a solid surface. A piecewise discrete surface temperature within the capillary can also be imposed as necessary. Finally, the far field boundary condition is described as the unperturbed asymptotic state.

The only complication that may arise is from the vastly different time scales in the electromagnetic forces and the slow flow inertia. This concern can be eliminated by performing a time integrated force input to the global aerodynamic phenomenon. There is no known computational simulation existing in the literature. A numerical simulation of micro jets will definitely open a new avenue for research into enhancing combustion stability, plasma-assisted ignition, and wound pathologies.

REFERENCES

Adelgren, R.G., Yan, H., Elliott, G.S., Knight, D.D., Beutner, T.J., and Zheltovodov, A.A., Control of Edney IV interaction by pulsed laser energy deposition, *AIAA J.*, vol. 43, February 2005, pp. 256–269.

Albrecht, T., Grundmann, R., Mutschke, G., and Gerth, G., On the stability of the boundary layer subject to a wall-parallel Lorentz force, *Phys. Fluids*, vol. 18, no. 92006, 2006, pp. 098103–098107.

Arnal, D., and Casaalis, G., Laminar–turbulent transition prediction in three-dimensional flows, *Prog. Aerosp. Sci.*, vol. 36, 2000, pp. 173–191.

Bechert, D.W., Bruse, M., Hage, W., and Van der Hoeven, J.G.T., Experiment on drag-reducing surfaces and their optimization with an adjustable geometry, *J. Fluid Mech.*, vol. 338, 1997, pp. 59–87.

Boeuf, J.P., Lagmich, Y., Unfer, T., Callegari, Th., and Pitchford, L.C., Electrohydrodynamic force in dielectric barrier discharge plasma actuators, *J. Phys. D: Appl. Phys.*, vol. 40, 2007, pp. 652–663.

Boeuf, J.P., and Pitchford, L.C., Electrohydrodynamic force and aerodynamic flow acceleration in surface dielectric barrier discharge, *J. Appl. Phys.*, vol. 97, 2005, pp. 103307-1-10.

Bogdanov, E.A., Kudryavtsev, A.A., Kuranov, A.L., Kozlov, L.E., and Tkchenko, T.L., 2D Simulation of DBD plasma actuator in air, *AIAA* 2008-1377, 2008.

Brewor, G.R., *Ion Propulsion Technology and Application*, Cordon and Breach Science Publishers, 1970.

Corke, T.C., Enloe, C.L., and Wilkinson, S.P., Plasma actuators for flow control, *Annu. Rev. Fluid Mech.*, vol. 42, 2010, pp. 505–525.

Corke, T.C., Post, M.L., Orlov, D.M., SDBD plasma enhanced aerodynamics: concepts, optimization and applications, *Prog. Aerosp. Sci.*, vol. 43, 2007, pp. 193–217.

Elisson, B., and Kogelschatz, U., Nonequilibrium volume plasma chemical processing, *IEEE Trans. Plasma Sci.*, vol. 19, 1991, pp. 1063–1077.

Enloe, C.L., McLaughlin T.E., Van Dyken, R.D., Kachner, K.D., Jumper E.J., and Corke, T.C., Mechanisms and responses of a single dielectric barrier plasma actuator: plasma morphology, *AIAA J.*, vol. 42, 2004, pp. 589–594.

Fay, J.A., and Riddell, F.R., Theory of stagnation point heat transfer in dissociated air, *J. Aero. Sci.*, vol. 25, 1958, pp. 73–85, 121.

Finley, P.J., The flow of jet from a body opposing a supersonic free stream, *J. Fluid Mech.*, vol. 26, 1966, pp. 337–368.

Fridman, G., Fridman, G., Gastol, A., Shekhter, A.B., Nasilets, V.N., and Fridman A., Applied plasma medicine, *Plasma Proc. Polymars*, vol. 5, 2008, pp. 503–533.

Gaitonde, D.V., and Poggie, J., An implicit technique for 3-D turbulent magneto-aerodynamics, *AIAA J.*, vol. 41, no. 11, 2003, pp. 2179–2293.

Gaitonde, D.V., Three-dimensional plasma-based flow control simulations with high-fidelity coupled first-principles approaches, *Int. J. Comp. Fluid Dyn.*, vol. 24, 2010, pp. 259–279.

Gallimore, A., Meyer, R., Kelley, A., and Jahn, R., Anode power deposition in applied-field segmented anode MPD thruster, *J. Propul. Power*, vol. 10, 1994, pp. 262–268.

Garcia-Mayoral, R., and Jimenez, J., Hydrodynamic stability and breakdown of the viscous regimes over riblets, *J. Fluid Mech.*, vol. 678, 2011, pp. 317–347.

Golubovskii, Yu.B., Maiorov, V.A., Behnke, J., and Behnke, J.F., Influence of interaction of interaction between charged particles and dielectric surface over a homogeneous barrier discharge in nitrogen, *J. Phys. D: Appl. Phys.*, vol. 35, 2002, pp. 751–761.

Gulhan, A., Esser, B., Koch, U., Siebe, F., and Riehmer, J., Experimental verification of heat-flux mitigation by electromagnetic fields in partially-ionized-argon flows, *J. Spacecraft Rockets*, vol. 46, no. 2, 2009, pp. 274–282.

Hayes, W.D., and Probstein, R.F., *Hypersonic Flow Theory*, Academic Press, New York, 1959.

Herbert, T. Parabolized stability equations, *Ann. Rev. Fluid Mech.*, vol. 29, 1997, pp. 245–283.

Hoskinson, A.R., Hershkowitz, N., and Ashpis, D.E., Force measurements of single and double barrier DBD plasma actuators in quiescent air, *J. Phys. D: Appl. Phys.*, vol. 41, 2008, pp. 245209–1–9.

Huang, P.G., Shang, J.S., and Stanfield, S.A., Periodic electrodynamic field of dielectric barrier discharge, *AIAA J.*, vol. 49, no. 1, January 2011, pp. 119–127.

Jahn R.G., *Physics of Electric Propulsion*, McGraw-Hill, New York, 1968.

Jansky, J., Tholin, F., Bonaventura, Z., and Bourdon, A., Solution of the discharge propagation in a capillary tube in air at atmospheric pressure, *J. Phys. D: Appl. Phys.*, vol. 43, 2010, pp. 395210–1–13.

Jimenez, J., Near-wall turbulence, *Phys. Fluids*, vol. 25, 2013, pp. 101302–1–18.

Kolesnichenko, Y., Basics in beams MW energy deposition for flow/flight control, AIAA 2004-0669, Reno, NV, January 2004.

Kotsonis, M., Giepman, R., Hulshoff, S., and Veldhuis, L., Numerical study of the control of Tollmein–Schliciting waves using plasma actuators, *AIAA J.*, vol. 51, 2013, pp. 2353–2364.

Likhanskii, A.V., Shneider, M.N., Macheret, S.O., and Miles, R.B., Modeling of dielectric barrier discharge plasma actuator in air, *J. Appl. Phys.*, vol. 103, 2008, pp. 055505-1-13.

Massines, F., Rabehi, A., Decomps, P., Gadri, R.B., Segur, P., and Mayoux, C., Experimental and theoretical study of a glow discharge at atmospheric pressure controlled by dielectric barrier, *J. Appl. Phys.*, vol. 83, 1998, pp. 2950–2957.

Matsuda, A., Otsu, H., Kawamura, M., Konigorski, D., Takizawa, Y., and Abe, T., Model surface conductivity effect for the electromagnetic heat shield in re-entry flight, *Phys. Fluids*, vol. 20, no. 127103, 2008, pp. 127103, 1–10.

Meyer, R.C., On reducing aerodynamic heat-transfer rates by magnetohydrodynamic technique, *J. Aero. Sci*, vol. 6, 1958, pp. 561–572.

Mitchner, M., and Kruger, C.H., *Partially Ionized Gases*, John Wiley & Sons, New York, 1973.

Moin, P., and Mahesh, K., Direct numerical simulation; a tool in turbulent research, *Ann. Rev. Fluid Mech.*, vol. 30, 1998, pp. 539–578.

Moreau, E., Airflow control by non-thermal plasma actuator, *J. Phys. D: Appl. Phys.*, vol. 40, 2007, pp. 605–636.

Morgan, G.C., Laser-induced breakdown of gases, *Rep. Prog. Phys.*, vol. 38, 1975, pp. 621–665.

Nishihara, M., Takashima, K., Rich, J.W., and Adamovich, I.V., Mach 5 bow shock control by a nanosecond pulse surface dielectric barrier discharge, *Phys. Fluids*, vol. 23, 2011, pp. 066-101-1-11.

Opaits, D., Shneider, M., Miles, R., Likhanskii, A., and Macheret, S., Surface charge in dielectric barrier discharge plasma actuators, *Phys. Plasma*, vol. 15, pp. 073505, 2008.

Poggie, J., and Gaitonde, D.V., Magnetic control of flow past a blunt body: numerical validation and exploration, *Phys. Fluids*, vol. 14, no. 5, 2002, pp. 1720–1731.

Post, M., *Plasma Actuators for Separation Control on Stationary and Oscillating Wings*, PhD dissertation, University of Notre Dame, 2004.

Raizer, Yu.P., Breakdown and heating gases under the influence of a Laser beam, *Sov. Phys.-Usp.*, vol. 8, 1966, pp. 650–673

Raizer, Yu.P., *Gas Discharge Physics*, Springer-Verlag, Berlin,Germany, 1991.

Resler, E.L., and Sears, W.R., The prospects for magneto-aerodynamics, *J. Aeronaut. Sci.*, vol. 25, 1958, pp. 235–245, 258.

Roth J.R., Sherman, D.M., and Wilkinson, S.P., Electrohydrodynamic flow control with a glow-discharge surface plasma, *AIAA J.*, vol. 37, 2000, pp. 1166--1172.

Sands, B.L., Ganguly, B.N., and Tachibana, K., A streamer-like atmospheric pressure plasma jet, *Appl. Phys. Lett.*, vol. 52, 2008, pp. 151503.

Shang, J.S., Hayes, J., and Menart, J., Hypersonic flow over a blunt body with plasma injection, *J. Spacecraft Rockets*, vol. 39, no. 3, May–June 2002, pp. 367–375.

Shang, J.S., Plasma injection for hypersonic blunt body drag reduction, *AIAA J.*, vol. 40, no. 6, 2002a, pp. 1178–1186.

Shang, J.S., Shared knowledge in computational fluid dynamics, *Electromag .Magnetobiol.-Aerodyn., Prog. Aerosp. Sci.*, vol. 38, no. 6/7, 2002b, pp. 449–467.

Shang J.S., and Surzhikov, S.T., Magnetoaerodynamic actuator for hypersonic flow control, *AIAA J.*, vol. 43, no. 8, August 2005, pp. 1633–1643.

Shang, J.S, Surzhikov, S.T., Kimmel, R., Gaitonde, D., Menart, J., and Hayes, J., Mechanisms of plasma actuators for hypersonic flow control, *Prog. Aerosp. Sci.*, vol. 41, 2005, pp. 642–668.

Shang, J.S., Solving schemes for computational magneto-fluid-dynamics, *J. Sci. Comput.*, vol. 25, no. 1, October 2005, pp. 289–306.

Shang, J.S., and Huang, P.G., Modeling AC dielectric barrier discharge, *J. Appl. Phys.*, vol. 107, no. 11, 2010, pp. 113302-1-7.

Shang, J.S., Roveda, F., and Huang, P.G., Electrodynamic force of dielectric barrier discharge, *J. Appl. Phys.*, vol. 109, no. 11, 2011, pp. 113301-1-8.

Shang, J.S., Andrienko, D.A., Huang, P.G., and Surzhikov, S.T., A computational approach for hypersonic nonequilibrium radiation utilizing space partition algorithm and Gauss quadrature, *J. Comp. Phys.*, vol. 266, 2014, pp. 1–21.

Shang, J.S., and Huang, P.G., Surface plasma actuators modeling for Flow control, *Prog. Aerosp. Sci.*, vol. 67, 2014, pp. 29–50.

Solov'ev, V., Konchakov, A.M., Krivtsov, V.M., Aleksandrov, N.L., Numerical simulation of a surface barrier discharge in air, *Low-Temperature Plasma*, vol. 34, no. 7, 2008, pp. 594–608.

Stanfield, S.A., Menart, J., Dejoseph, C., Kimmel, R.L., and Hayes, J.R., Rotational and vibrational temperature distributions for a dielectric barrier discharge in air, *J. AIAA*, vol. 47, 2009, pp. 1107–1115.

Taccogna, F., Longo, S., Capitelli, M., and Schneider, R., Self-similarity in Hall plasma discharge: applications to particle models, *Phys. Plasmas*, vol. 12, 2005, pp. 053502.

Thomas, F., Kozlov, A., and Corke, T., Plasma actuators for bluff body flow control, *AIAA J.*, vol. 46, 2008, pp. 1921–1931.

Walsh, M., Riblets as a viscous drag reduction technique, *AIAA J.*, vol. 21, 1983, pp. 485–486.

Zel'dovich, Ya.B., and Raizer, Yu.P., *Physics of Shock Waves and High-Temperature Hydrodynamic Phenomena*, Dover Publication, Mineola, 2002.

Ziemer, R.W., Experimental investigation in magneto-aerodynamics, *Am. Rocket Soc. J.*, vol. 29, 1959, pp. 642–647.

INDEX

Computational Electromagnetic-Aerodynamics, First Edition. Joseph J.S. Shang.

IEEE Press Series on

RF and Microwave Technology

Series Editor: **George Kizer**, *Telecommunications Consultant, Dallas, Texas, USA*

Wireless technology has become a worldwide game-changer. It has redefined how we—and machines—interact in ways unimaginable a few years ago. The IEEE Press Series on RF and Microwave Technology spans the latest development in this area. While integrating established practices with emerging developments, the series encompasses devices, systems and enabling technology over the wide span of activity. This series will be of interest to students, practicing engineers and anyone interested in how this world-changing technology is evolving.

Computational Electromagnetic-Aerodynamics · Joseph J.S. Shang

www.ingramcontent.com/pod-product-compliance
Lightning Source LLC
LaVergne TN
LVHW011903200926
848484LV00002B/519